D0216223

STRUCTURAL SYSTEMS DESIGN

STRUCTURAL SYSTEMS DESIGN

ROBERT A. COLEMAN, P.E.

Enwright Associates, Inc.

PRENTICE-HALL, INC., *Englewood Cliffs, N.J. 07632*

Library of Congress Cataloging in Publication Data

Coleman, Robert A.
Structural systems design.

Includes bibliographical references and index.
1. Structural design. I. Title.
TA658.C56 1983 624.1'771 82–11259
ISBN 0–13–853978–2

Editorial production/supervision by Ellen Denning
Interior design by Ellen Denning
Page layout by James M. Wall
Manufacturing buyer: Anthony Caruso

Printed in the United States of America

10 9 8 7 6 5 4 3 2 1

ISBN 0-13-853978-2

PRENTICE-HALL INTERNATIONAL, INC., *London*
PRENTICE-HALL OF AUSTRALIA PTY. LIMITED, *Sydney*
EDITORA PRENTICE-HALL DO BRASIL, Ltda *Rio de Janeiro*
PRENTICE-HALL CANADA INC., *Toronto*
PRENTICE-HALL OF INDIA PRIVATE LIMITED, *New Delhi*
PRENTICE-HALL OF JAPAN, INC., *Tokyo*
PRENTICE-HALL OF SOUTHEAST ASIA PTE. LTD., *Singapore*
WHITEHALL BOOKS LIMITED, WELLINGTON, *New Zealand*

To Molly, Catherine, and Charlotte

CONTENTS

4 ELEMENTS 89

5 SUBSYSTEMS 162

6 SYSTEMS 228

7 CREATION OF SYSTEMS 285

8 CONCLUSION 310

PREFACE

The structural engineer plays a major role in a problem-plagued construction industry, an industry that is one of the poorest-performing industries in the country. Productivity is low, wages high, and costs alarming. The cost of new construction has escalated to the point that serious questions are being raised about the availability of future capital to finance construction. To make matters worse, increasing society concerns about energy and the environment are likely to add further to this escalating cost.

Society is, at the same time, becoming concerned about the quality of the built environment. "Limited growth" and "no growth" policies are increasingly being mentioned as serious planning criteria. It is a fairly easy transition from this planning perspective to one concerned with the construction quality of what *is* built. Craftsmen are no longer a readily available resource.

A consumer-conscious public has added a third problem to the already plagued construction industry-legal damage suits. Those who design buildings and those who construct them are increasing targets for liability claims. Although many of these claims are not legitimate, some are unfortunately the result of negligence on the part of the designer or builder.

Architects and engineers faced with this background of problems are seeking new ways to improve the process of design and construction. *Innovative solutions* are being sought to achieve function at lower cost, without sacrificing quality. However, we are literally flooded with information regarding construction products, design methods, framing systems, and recommended construction practices! Without an organized process for handling this large volume of information, often we must resort to using familiar and "tried and tested" designs, which in many cases reflect the technology of 20 years ago. This spectrum of "over-choice" is further complicated by our analytical training, which has not equipped us to deal *conceptually* with alternatives.

Our problem with innovation is that it has been an elusive quality and one rarely captured in our educational process. We have devoted most of our education and research to developing our analytical skills, and very little to developing their opposite, synthesis (or inductive) skills. "Advancements in engineering science have contributed heavily to the development of sophisticated methods of analysis and to optimization methods for *well structured problems.* However, the conceptual selection of an engineering system is by and large still acceded to subjective decisions. Although creativity has been recognized as the ability to conceive and generate design alternatives, there is no established procedure for generating innovative and highly successful alternatives."[1] The traditional educational approach has concentrated on design of elements (or components) with most of the design criteria given (e.g., loads, material, joint conditions, etc.)—the well-structured problem. This has tended to emphasize memorization of design formulas, an approach Charles Kettering called "stenographic engineering." Conceptual ability requires both a "feeling" for behavior and approximate analysis/design skills.

Structural Systems Design addresses the need for conceptual design skills. It is organized to serve as a text for senior and graduate-level classes on structural systems and as a professional reference book.

A *structural systems course* may be taught using this book alone, or in conjunction with the laboratory manual. The creative process develops through individual experience with "qualitative" models, lecture discussions, and problem exercises. The "systems design model" defines a systematic process for defining the design problem (requirements and loads), establishing alternative concepts, and arriving at preliminary design solutions. The course follows the outline of the model, progressing from early sessions on requirements through the entire process of structural system assembly. The lecture sessions describe the history and development of various system components, concepts of behavior, and approximate analysis/design methods. Example problems illustrate this development throughout the text, and a final design project provides the opportunity for teams of students to create conceptual designs and make presentations to a design "jury."

The laboratory model series is part of the total creative process developed in the course. Students construct the models with basic materials and conduct experiments designed to illustrate principles of structural behavior. The model series begins with element models—transitions into subsystem models—and concludes with total system models. In this way, students "experience" bending, stress, deformation, and failure. Analytical/abstract concepts are brought to life by visual, sometimes dramatic observations of actual behavior under load.

For the *practicing professional,* this book presents an organized approach to understanding and working with structural systems to produce preliminary designs. The explanation of structural behavior builds through discussion, visual techniques such as the "load path" concept, and the illustrations. Approximate analysis/design methods serve as the basic mathematical tools for dealing conceptually with alternative framing systems, and the Chapter appendix material contains concise references on loads, materials, requirements, and system selection guidelines. The book establishes the concept that structural systems may be developed based on a systematic generation of alternatives, comparison of alternatives, and selection based on value engineering concepts.

The product of this selection is a structural system with preliminary member sizes. The key resources for design reference are:

- Defined process for systems design
- Systems data base (for preliminary screening of alternatives)
- Latest information on wind, earthquake, and snow loads
- Approximate analysis/design methods (including new methods for columns, beam-columns, and frames)
- Reference material in appendices (including material properties, member properties, and fire protection)
- Behavior discussion of various systems

One major objective is to provide a ready reference source to assist the engineer during preliminary design sessions, hopefully minimizing the frantic search for guidance as well as retreats into detailed design formulas.

The structural engineer who can creatively organize the vast resources of time, space, money, and materials can be a leader in the construction industry. It is the author's hope that this book can assist in that effort.

The following material is reproduced with permission from American National Standard *Minimum Design Loads for Buildings and Other Structures,* ANSI A58.1–1982, copyright 1982 by the American National Standards Institute. Tables 3–1 through 3–9 and tables 3–12 through 3–14; Equations 3–1 through 3–5 and equations 3–15 through 3–25; Figures 3–14, 3–16, 3–18, 3–20, 3–50, 3–51, and 3–73; Article 3–15; and Appendix 3–C. Copies of this standard may be purchased from the American National Standards Institute, 1430 Broadway, New York, N.Y. 10018.

This book represents the combined efforts of several key individuals, as well as classes of students and practicing engineers and architects at Enwright Associates, Inc., and Clemson University. Special acknowledgment and sincere appreciation to Joan Reid for the typing, Pat King for administrative assistance, Bill Fowler for the art work, and Molly and Catherine Coleman for assistance with the illustrations and model materials.

ROBERT A. COLEMAN

[1] Tung Au, "Heuristic Approach to System Design," *Journal of Engineering Education,* Vol. 59, No. 7, March 1969, p. 861.

STRUCTURAL SYSTEMS DESIGN

1

INTRODUCTION

1-1 BACKGROUND

Long before engineers had the power of computers and sophisticated mathematics at their disposal, structures were built that reflected a basic understanding of structural behavior. The arches of the Roman Aqueduct, the cathedrals of seventeenth-century Europe, and the buildings of ancient Greece were marvels of engineering. Their form and proportion could hardly be improved upon, even today. Some still stand—dramatic, visual evidence of the intuitive knowledge of their designers.

This intuitive knowledge, a basic tool of early engineers, was developed by working with models, by trial and error, and by empirical methods. It gave to these engineers an ability to create structural forms and to understand their behavior. Today, we use words like *synthesis* and *design* to describe this ability to create form. In contrast to analysis, which is a deductive technique, design is inductive or creative in nature.

Unfortunately, emphasis on refinement of analytical tools—from mathematics to computers—has overshadowed development of our design ability. To quote Pier Luigi Nervi:

> It would be absurd to deny the usefulness of that body of theorems, mathematical developments, and formulas known by the rather inaccurate name of "Theory of Structures." But we must also recognize and state unequivocally that these theoretical results are a vague and approximate image of physical reality. We come nearer to this reality only by adding the results of experiments to the mathematical results, by observing the actual phenomena, by establishing a conceptual basis for these phenomena, and above all by understanding intuitively the static behavior of our works.
>
> The fundamental assumption of the theory of structures is that structural materials are isotropic and perfectly elastic. But the most commonly used building materials, like masonry and concrete, are far from being isotropic and elastic. . . . No soil is perfectly stable nor settles uniformly as time goes by. All building materials, and particularly masonry and concrete, flow viscously.
>
> The preeminence given to mathematics in our schools of engineering, the purely analytical basis of the theory of elasticity, and its intrinsic difficulties persuade the young student that there is limitless potency in theoretical calculations, and give him blind faith in their results. Under these conditions neither students nor teachers try to understand and to feel intuitively the physical real-

> ity of a structure, how it moves under a load, and how the various elements of a statically indeterminate system react among themselves. . . . The formative stage of a design during which its main characteristics are defined and its qualities and faults are determined once and for all cannot make use of structural theory and must resort to intuition and schematic simplifications.
>
> The essential part of the design of a building consists in conceiving and proportioning its structural system; in evaluating intuitively any dangerous thermal conditions and support settlements, in choosing materials and construction methods best adapted to the final purpose of the work and to its environment; and, finally, in seeking economy. When all these essential problems have been solved and the structure is thus completely defined, then and only then can we and should we apply the formulas of the mathematical theory of elasticity to specify with greater accuracy its resisting elements.[1]

Even our analytical training has not been "systems" oriented—concentrating mainly on analysis and design of components (elements). With load assumptions, material selection, and mathematical model usually given, engineers learn to apply mathematical theory to only a limited portion of a complete structure. We are thus faced not only with a need to develop the intuitive ability envisioned by Nervi but also with a need to fill in some "gaps" in the analytical process.

1–2 OBJECTIVES

The major goal of this book is to teach and serve as a reference for *structural design:* to develop a systematic process for integrating, distilling, and refining structural knowledge until a structural form emerges that represents an optimum solution to a usually many-faceted problem. The use of the word "problem" is appropriate—engineering design is seeking solutions to "problems." Problems must, of course, be initially defined, so the process must start with *problem definition.*

Problem solving (design) proceeds from *problem definition to generation of alternatives, comparison of alternatives, system selection, mathematical modeling,* and then to *final analysis* and *member proportioning.* The place of *analysis* and *member proportioning* in this process suggests that the design process really *ends* at the *beginning* of our initial structural knowledge and training. This, of course, was Nervi's point.

The structural designer must develop certain skills and understanding in several areas in order to proceed with problem solving. These include the following:

- Structural requirements
- Loads
- Structural materials and structural systems and an intuitive feeling for their behavior
- Approximate analysis/design skills
- Data base of structural system characteristics
- Methodology for generating and comparing alternatives
- Mathematical modeling
- Analytical training
- Member proportioning

These subjects (except for final analysis and member proportioning) are covered in this book as specific objectives within the overall framework of the *systems design approach* described in the following section.

1–3 THE SYSTEMS APPROACH

Design Model

The use of the term *systems approach* indicates that there is a defined process for proceeding from problem definition right through the various stages of problem solving. The statement is true—there is a defined process—but it is not as straightforward as the description might imply. Structural design requires a search for ideas and the molding of these ideas into structural form. This creative process continually tests ideas against project requirements, with even the requirements themselves subject to change depending on the system selected. Herein lies the apparent "mystery" of structural design. Dead load, for example, is not known until the system is selected—but the system cannot be selected unless the dead load is known! Certain assumptions must be made and later tested for validity as the design progresses.

The structural systems design model (Fig. 1–1) can be visualized as a series of screens (progressively finer in size) which refine information on loads and structural assemblies until a final structural system is selected and designed. The screens represent requirements and various stages of evaluation and comparison. The model contains the following parts: structural requirements, loads, elements, subsystems, systems, selection guidelines, approximate analysis/design, value engineering comparison, final analysis, and member proportioning. These parts interact with each other and generally follow a common path leading from problem definition to problem solving, with several stops and retracing of steps along the way as information is refined and tested against project requirements. The three physical parts (elements, subsystems, and systems) constitute a hierarchy of structural assembly which we will use later in this book for study of structural systems. An understanding of *behavior,* including mathemat-

[1] Pier L. Nervi, *Structures* (New York: Copyright by F. W. Dodge Corp., 1956), pp. 13–14.

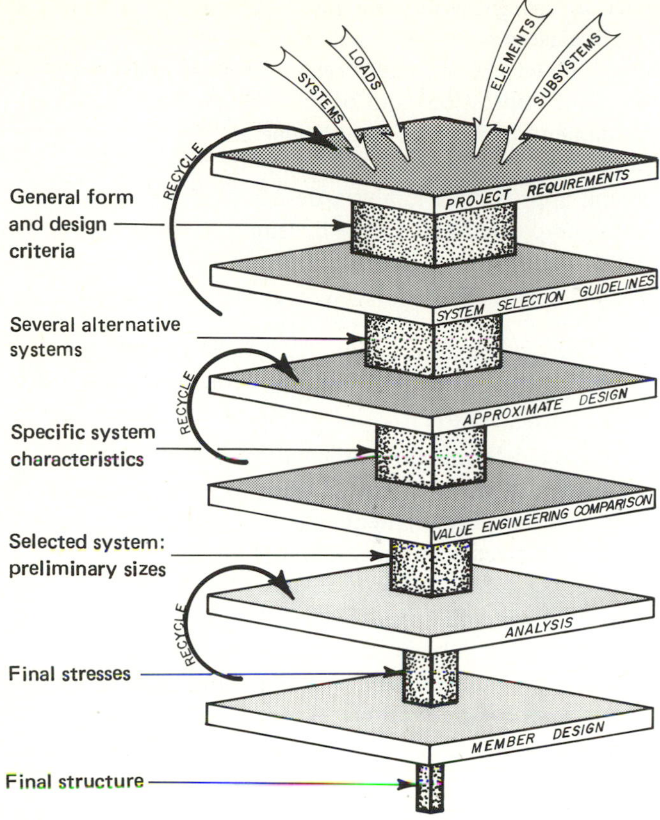

Figure 1–1. Systems Design Model

ical modeling skills, is a necessary companion on this journey from *idea* to *conceptual form.* Behavior is the background "theme" to the design process.

Organization and Approach of the Book

Structures are assemblies of components, connected in such a way as to form a stable system capable of resisting applied loads (Fig. 1–2). The components may be studied individually (as elements), or as subassemblies (subsystems) or as a whole (complete systems). Our approach will be to study structural behavior in this exact sequence, building on our knowledge of element behavior to understand subsystems and building on our knowledge of subsystem behavior to understand systems. Within the framework provided by the systems design model, structural systems will thus be studied as an *assembly of parts.* This approach to the study of behavior will also provide us later with a convenient framework for selecting structural systems. A data base will be constructed using functional elements and subsystems as the basic building modules.

A typical chapter contains a discussion of the subject, approximate analysis/design techniques, example problems, problems, an appendix of reference information, and a list of references.

Physical models and other visual techniques, such as the load path diagram, appear throughout the book as *visual tools,* intended to develop "feeling" for behavior and provide a basis for formulating mathematical models. The following sections describe the general background to these techniques.

Figure 1–2. Structural Assembly Components

1–4 METHODS

Approximate Analysis/Design

Conceptual design ability is related to a *feeling* for the physical behavior of structural systems and for the *mathematical boundaries* of the particular problem. Mathematical boundaries are limits of loads and corresponding member forces. If upper and lower limits to the mathematical possibilities can be established, reasonable approximations can generally be made somewhere in between. These approximations will allow a rapid determination of possible structural alternatives as well as a preliminary comparison of selected ones. Still later in the design process, these same approximations will allow "order-of-magnitude" checking of final analysis/design data, computer results, and so on.

Approximate analysis/design methods involve certain simplifying assumptions regarding loads, the mathematical model of the structural system, and member proportioning formulas. Ideally, they should involve only very basic simple formulas which can be easily applied without reference to detailed design texts and codes. Engineers in conceptual design sessions (group or individual) need a *feeling* for the *stress* and *deformation* levels that might exist in potential members, in order to match appropriate materials and make evaluations. Axial stress (tension or compression) and bending stress calculations are fundamental parts of the engineer's background, expressed in the familiar P/A and MC/I equations. By basing approximate analysis/design on these basic equations, simple methods can be developed and readily incorporated and retained in our *structural memory.* Formulas involving only P/A and MC/I terms are established in this book.

As an example of this approach to approximate analysis/design, the steel continuous beam in Fig. 1–3 can be designed by (1) assuming inflection points as shown to create a determinate mathematical model, (2) converting the series of concentrated loads to an equivalent uniform load for easy moment calculation, and (3) calculating the required section modulus using an "average" allowable bending stress of 18,000 psi (124 MPa) and using a minimum depth based on limiting deflection criteria. Approximate methods can also be supplemented with charts and tabulated data, and these are included in the text in appropriate sections.

Figure 1–3. Approximate Design of Continuous Steel Beam

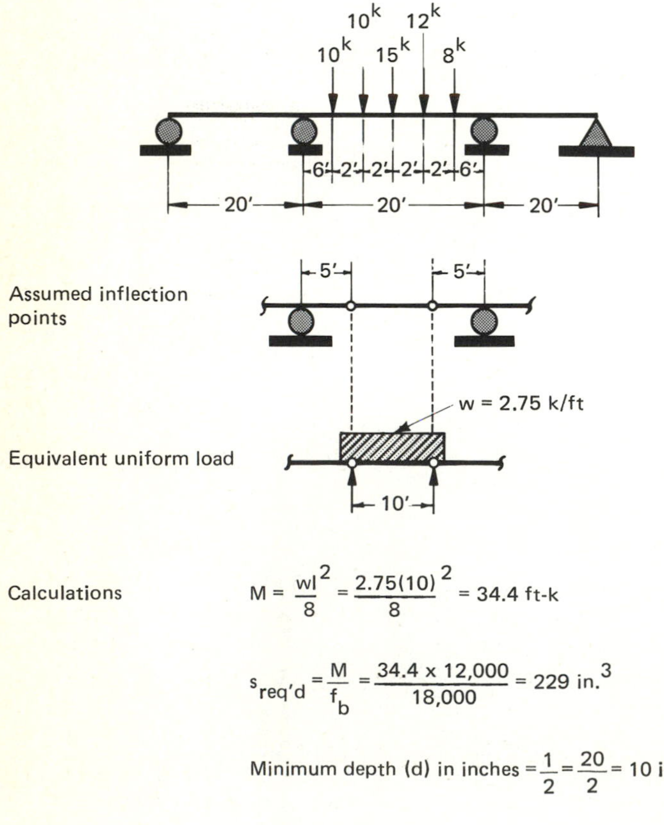

Mathematical Models

Mathematical models are idealized mathematical representations of the physical prototype (Fig. 1–4). The model allows the engineer to predict analytically the response of structural members to load. All of our structural theory utilizes various structural models to simulate actual structures. Mathematical models usually require certain simplifying assumptions in order to formulate the mathematical expressions. For example, Fig. 1–5 shows the actual construction of a beam–column connection and the corresponding mathematical model. Does the fact that ideal hinged conditions do not really exist affect the performance of the beam? of the column? What assumptions can be made that are reasonable for design practice? It is important to understand the answers to these questions and to be able to correlate the mathematical model with the behavior of the prototype.

The requirements for a mathematical model include system geometry, member properties, applied loads, and connection assumptions (hinged or fixed). "Free-body diagrams" are a form of mathematical model. The mathematical model (similar to the free-body diagram) represents a system in equilibrium with the applied load.

Sometimes it is not clear just what the mathematical model should be, as in the case of a wall penetrated with openings and loaded with gravity loads and wind loads. Fortunately, almost any model will do (as long as it is in equilibrium and sufficient material ductility exists) be-

Figure 1–4. Mathematical Model of Beam

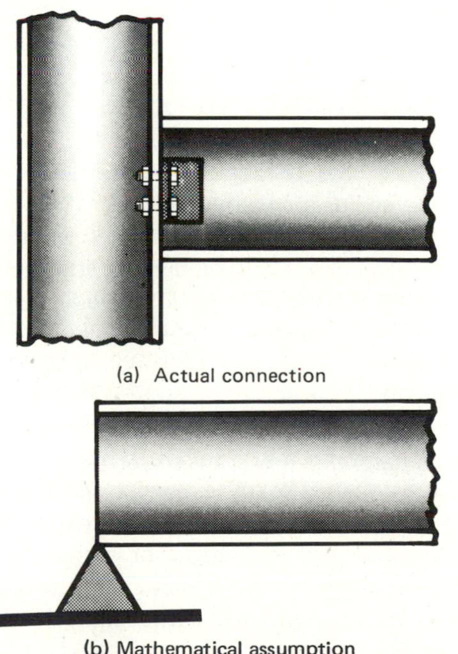

Figure 1–5. Hinged Connection: Actual Construction and Mathematical Model

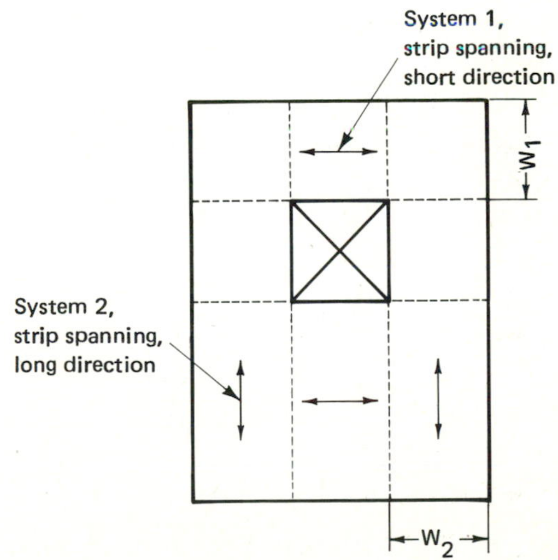

Figure 1–6. Selection of Mathematical Model

cause of a principle of structural mechanics called the *lower bound theorem* (Ref. 4). The lower bound theorem states:

> *If an equilibrium distribution of stress can be found that balances the applied load and is everywhere less than or equal to the yield stress, the structure will not fail. At most, it will just have reached the plastic limit load.*

Generally, it is advisable to select the stiffest system available in order to achieve the most economical design. For example, system 1 in Fig. 1–6 is far more economical to use than system 2.

Each of the systems studied in the text will be examined finally from the mathematical standpoint, so that we can "close the gap" between our analytical training and actual structural behavior.

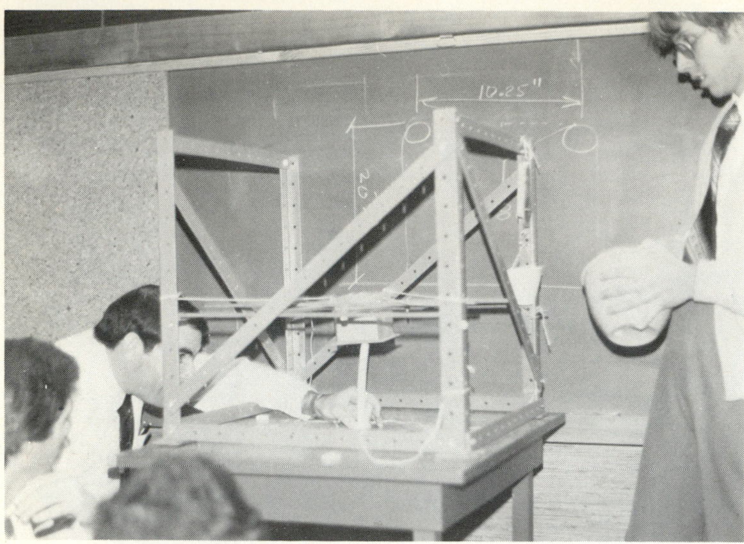

Figure 1–7. Structural Model Test

Physical Models

Physical models used for teaching of behavior (rather than for exact experimental analysis) are termed "qualitative" models. The ideal qualitative model should be simple to construct and should deform noticeably under load (to demonstrate behavior clearly).

Styrofoam is a material that is easy to fabricate into models. It has a low modulus of elasticity [approximately 1300 psi (8.96 MPa)], which makes loading and observation of deflections easy. For example:

Prototype:

$$\text{Modulus of elasticity } E_p = 3{,}000{,}000 \text{ psi } (20.7 \text{ GPa})$$

$$\text{Load } W_p = 100 \text{ psf } (4788 \text{ N/m}^2)$$

Styrofoam model:

$$\text{Modulus of elasticity } E_m = 1300 \text{ psi } (8.96 \text{ MPa})$$

$$\text{Load } W_m = \frac{1300}{3{,}000{,}000}(100)$$

$$= 0.043 \text{ psf } (2.05 \text{ N/m}^2)$$

The load W_m of 0.043 psf (2.05 N/m^2) is therefore relatively easy to apply and often is made substantially higher so that deflection behavior can be observed. If, for example, the load is 2.0 psf (the approximate weight of selected steel washers), the deflection in the model will be

$$\Delta_m = \frac{2.0}{0.043}\Delta_p = 46.5\,\Delta_p$$

Care must be taken to use models only in the range of stresses that are compatible with the prototype material. Otherwise, the behavior may not be the same. The model may fail in buckling while the prototype fails in compression, or vice versa.

Styrofoam and other model materials are used in the laboratory manual[2] that was developed to supplement this book for college courses incorporating parallel laboratory and lecture sessions. The manual contains a series of model experiments that track the systems development in this book. A *model support rig* is used to aid in the assembly and load testing of the various models. The model experiments may either be used as classroom demonstration aids or may be constructed by students (in laboratory sessions). Construction by students is highly recommended as the preferred method. The resulting "hands-on" experience is invaluable (Fig. 1–7).

Load Path Diagrams

Load path diagrams are drawings or sketches that show the deformed shape of the structure under load as well as the *load path* the forces take to the ground. For example, the load path diagram for a frame is shown in Fig. 1–8. The ability to construct load path diagrams usually is acquired through experience with physical models. The rules for constructing load path diagrams are:

A. Draw *two-dimensional* structures using two dimensions (length and depth) for all members.

   1. Show all applied loads using ——→ for concentrated loads, ▯▯▯▯▯▯▯ for uniform loads and ↻ for moments.

[2] Robert A. Coleman, *Structural Systems Design, Laboratory Manual* (Greenville, S.C.: Enwright Associates, Inc., 1982).

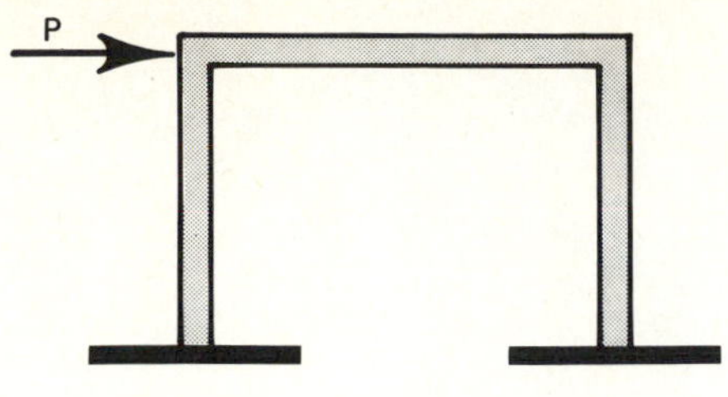

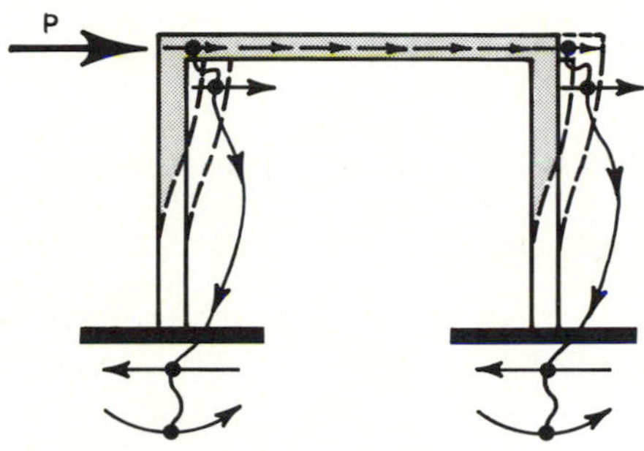

Figure 1-8. Load Path Diagram for Frame

2. Draw the deformed shape with dashed lines to illustrate bending behavior.
3. Show the axial force load path with a series of arrows (⟶ ⟶ ⟶).
4. Show shear forces as ⟶ and bending moments as) .
5. Connect shear forces resulting from axial loads (or vice versa) with a curved line and connect shear forces (transmitted along members) to other shear forces and reactions with a similar line.
6. Shear is always accompanied by moment—at rigid joints and at interior points of maximum stress. Show moment notations at interior points of maximum stress. At rigid (moment-resisting) joints show single) in the direction of initial unbalanced joint moment (the moment that will cause the joint to rotate). Show moments and shears as they *act on the joint. Note:* One method for determining proper directions for shear and moment on individual members is to isolate each joint as shown in Fig. 1–9. Show the initial unbalanced joint moment and applied shears; then show the resisting forces.

Figure 1-9. Joint Forces

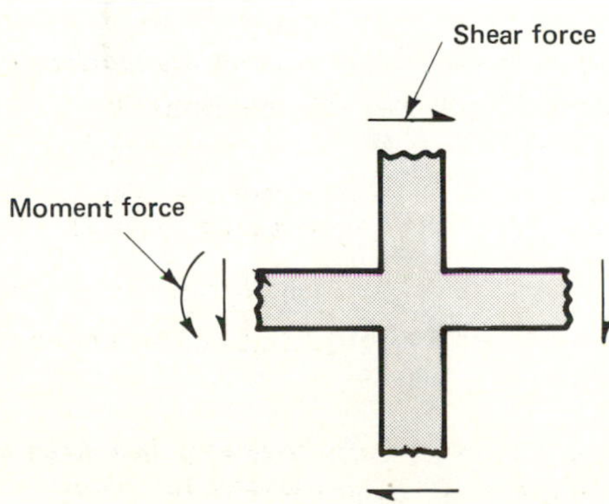

B. Draw *three-dimensional* structures by dividing the structure into planes and drawing load path diagrams for each plane. Supplement these diagrams with isometric drawings of the structure where possible.
   1. For vertical planes, follow the rules outlined above for two-dimensional structures.
   2. For horizontal planes, the following modifications are necessary:
      a. Show applied concentrated loads with ● .
      b. Show uniform loads by shading the area over which the load is applied.
      c. Show in-plane applied loads with ⟶ .
      d. Show axial force load path with ⟶ .
      e. Show in-plane shear forces as ⟶ and in-plane moments as) .
      f. Show shear forces perpendicular to the plane as ⊗ and moments as ⟶⟶ .
      g. Follow steps outlined in part A for establishing the load path.

C. All applied loads must be connected to a continuous load path and must terminate in the form of ground forces (reactions). Show these reactions and equate them to the applied loads as a check.

D. Forces that have no completed load path usually indicate instability.

Figure 1–10 shows several examples of load path diagrams. This load path technique is used as a visual tool in this book and to record observations made of the model tests in the laboratory manual.

1-5 THE STARTING POINT

With the *approach* and *methods* established, we can now start the development of the design process. As stated earlier, this process must start with a *definition of the problem*—or what we shall call *structural requirements.* The next chapter covers this subject and begins a series of chapters that discuss the various parts of the systems design model.

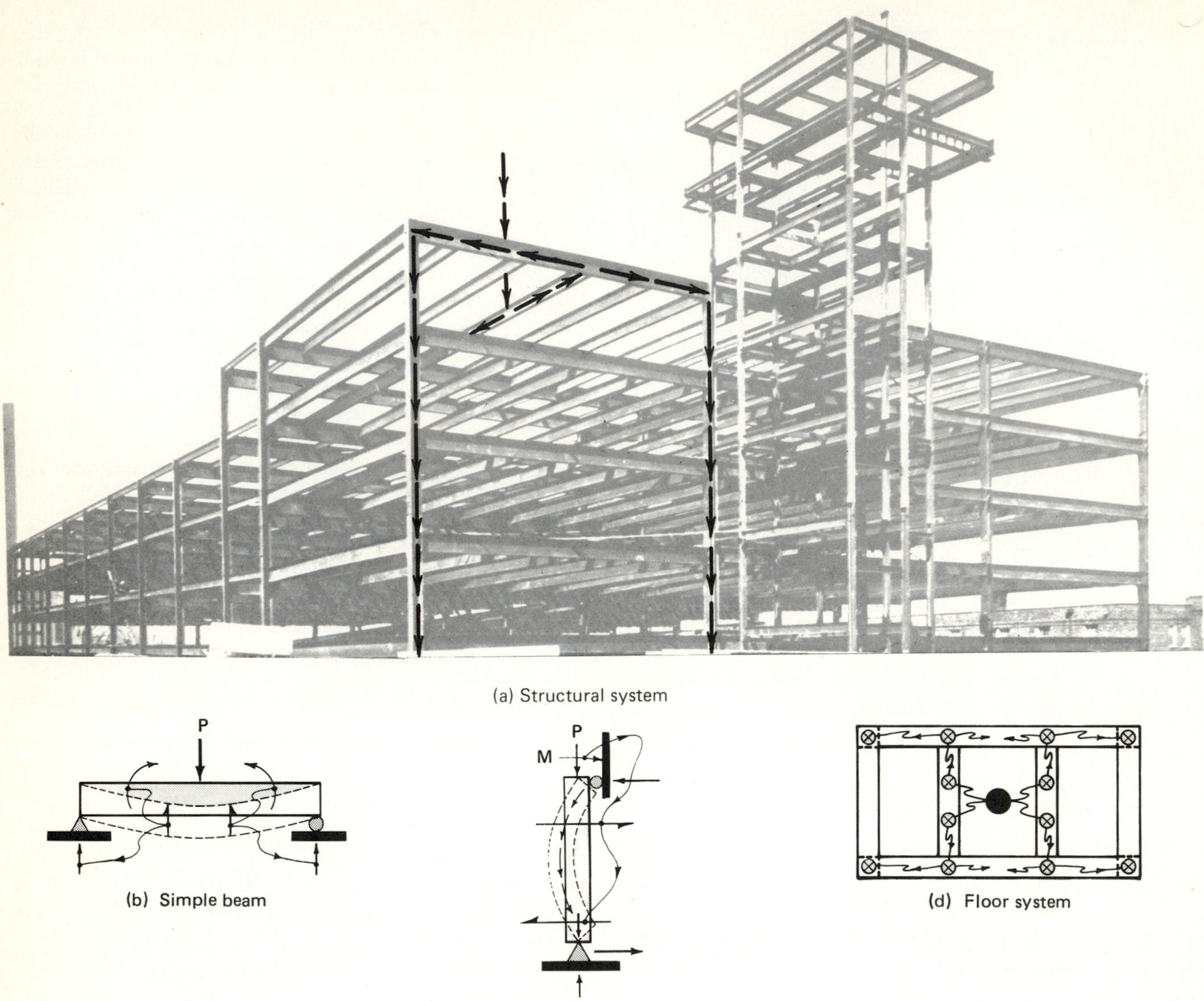

(a) Structural system

(b) Simple beam

(c) Column

(d) Floor system

Figure 1–10. Load Path Diagram Examples

REFERENCES

1. CARPENTER, SAMUEL T., *Structural Mechanics.* New York: John Wiley & Sons, Inc., 1960.
2. CHARLTON, T. M., *Model Analysis of Plane Structures.* Oxford: Pergamon Press Ltd., 1966.
3. COWAN, H. J., J. S. GERO, G. D. DING, and R. W. MUNCEY, *Models in Architecture.* Essex, England: Elsevier Publishing Company, Ltd., 1968.
4. GREENBURG, H. J., and W. PRAGER, "Limit Design of Beams and Frames," *Transactions of the ASCE,* Vol. 117 (1952), pp. 447–484.
5. HINSON, BARRY, *Basic Structural Behaviour via Models.* New York: John Wiley & Sons, Inc., 1972.
6. HOHAUSER, SANFORD, *Architectural and Interior Models.* New York: Van Nostrand Reinhold Company, 1970.
7. NERVI, P., *Structures.* New York: F. W. Dodge Corp., 1956.
8. PAHL, PETER JAN, and KETO SOOSAAK, *Structural Models for Architectural and Engineering Education.* Cambridge, Mass.: The MIT Press, 1963.
9. SALVADORI, MARIO, and ROBERT HELLER, *Structure in Architecture.* Englewood Cliffs, N.J.: Prentice-Hall, Inc., 1975.
10. SALVADORI, MARIO, and MATTHYS LEVY, *Structural Design in Architecture.* Englewood Cliffs, N.J.: Prentice-Hall, Inc., 1967.
11. WHITE, RICHARD N., PETER GERGELY, and ROBERT G. SEXSMITH, *Structural Engineering,* Vol. 3: *Behavior of Members and Systems.* New York: John Wiley & Sons, Inc., 1974.

STRUCTURAL REQUIREMENTS

2–1 INTRODUCTION

The earliest structures made were constructed to provide shelter for people and their families. The structures provided relatively weatherproof enclosures and were capable of supporting their own weight, as well as a certain amount of imposed load from snow and wind. Later, people began constructing structures to serve purposes other than shelter: foot bridges across streams, small dikes to impound water, and temples for worship. The development of structural form expanded to cover a myriad of purposes: the monumental purpose typified by structures such as the great pyramids of Egypt, the religious purpose typified by the cathedrals of Europe, the transportation purpose typified by the early bridges across the great rivers of Europe. People's shelter began taking on other characteristics that made it more durable, longer lasting, and more esthetically pleasing. Structures emerged to house factories and manufacturing operations, offices for business, transportation terminals, and museums.

The development of structures has been a response to the continuing evolution of social and cultural aspirations. The choice of a structural system today starts with a determination of the *requirements* (or purposes) for which the structure is to be built. This purpose is often expressed in a set of goals that are established by the owner or developer of the structure. These goals incorporate concepts such as the function of the structure, form and appearance, economy, time to construct, and so on. Depending on the structure, other goals may include the performance of the structure under varying conditions, such as dynamic behavior under machine loading, deflection under live load, and fire resistance. As a broad classification, structural requirements may be broken down into four categories:

1. Functional requirements
2. Esthetic requirements
3. Serviceability requirements
4. Construction requirements

Functional requirements relate to the fundamental purpose for which the structure exists. Esthetic requirements relate to esteem value. Serviceability relates to the expected performance of the structure, and construction requirements relate primarily to time, cost, and quality considerations.

This chapter describes these requirements as a very

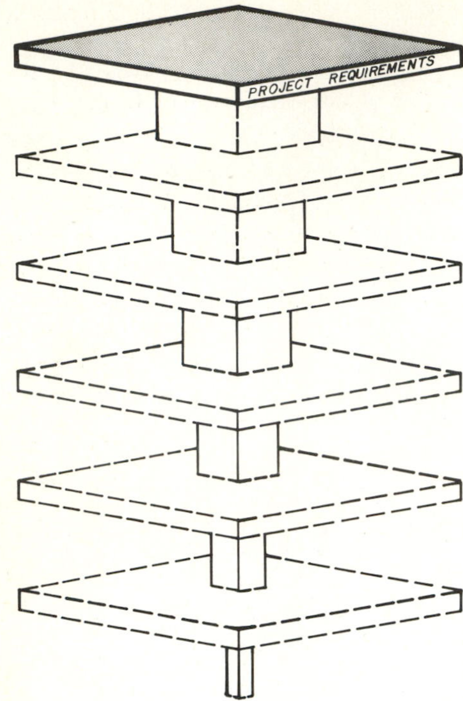

Figure 2–1. Systems Design Model

important first step in structural systems design (Fig. 2–1). Indeed, it has been demonstrated that requirements established during the early phases of predesign and design by the owner, regulatory agencies, and the designer contribute more to construction costs than any other single element (see Fig. 2–2). For example, the requirement by an owner that no columns exist in the work space can have significant cost implications for a project. Still other requirements may not be considered during design and may later prove to be devastating: for example, the omission of a flexibility check for ponding of rainwater on a roof structure and the resulting collapse of the roof during a rainstorm.

Figure 2–2. Impact of Decisions and Requirements on Construction Cost

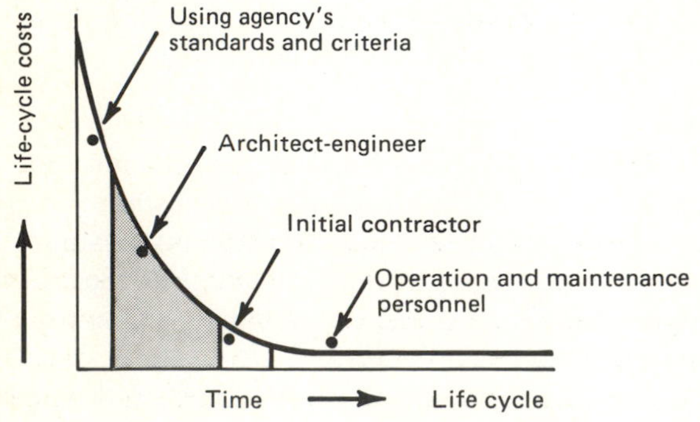

2–2 FUNCTIONAL REQUIREMENTS

The basic functional purpose of structures has remained pretty much the same since the construction of the early crude shelters. Expressed as a twofold mission, the basic purpose of structures is to *enclose and define space* and *carry loads.*

Spatial Function

The form a structure takes is directly influenced by the *define space* part of the basic purpose. Several examples serve to illustrate this point: If a warehouse storage system requires 20 ft (6.09 m) of clear height, this requirement defines the height of the structure. If a river crossing is 300 ft (91.4 m) wide, this requirement defines the length of the bridge. If an efficient office layout dictates a rectangular building, the shape of the structure is thus established.

The definition and arrangement of space is a far more involved process than these simple illustrations indicate. The architectural challenge is to create *building forms* that satisfy the functional requirements, yet are esthetically and economically successful. The structural challenge is to create *structural forms* that accommodate the functional requirements while supporting the imposed loads as efficiently as possible.

Load-Carrying Function

Form Follows Force. After satisfying the spatial requirements, a structural system then must *carry loads* (the second part of the basic function). The loads that a structure carries are the result of its own dead weight, superimposed loads that it must carry due to environmental effects and the loads generated by the particular function of the structure. The loads may involve water pressure on the side of a dam, soil pressure against a retaining structure, or office equipment in an office building. The structural form selected to carry the loads may be limited by the spatial requirements, or there may be freedom to describe a geometry closely following the most efficient force or *load path.*

Load path (by definition) is the path that loads (or stresses) take in traveling from their point of application on a structure to the ground. Much like an electric current that flows until it is "grounded," loads must travel until they meet ultimate resistance in the ground. The most efficient load paths are those that activate the unique strengths of the structural material: tension in steel, compression in concrete, and so on. Bending (by contrast) is a relatively inefficient way to resist load. This concept is illustrated by Fig. 2–3, which compares the cost/strength ratio for steel in tension and flexure.

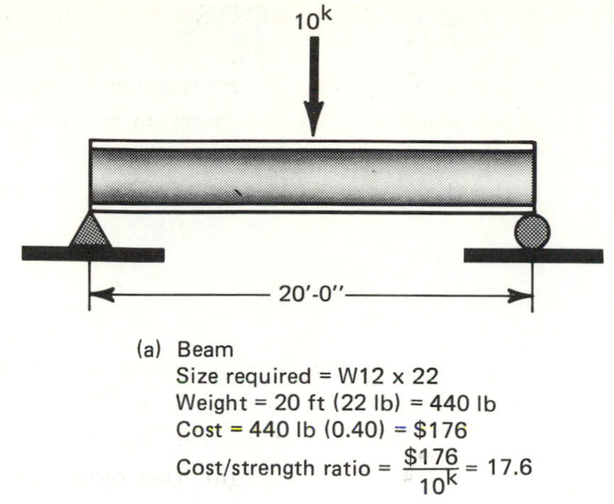

(a) Beam
Size required = W12 x 22
Weight = 20 ft (22 lb) = 440 lb
Cost = 440 lb (0.40) = \$176
Cost/strength ratio = $\frac{\$176}{10^k} = 17.6$

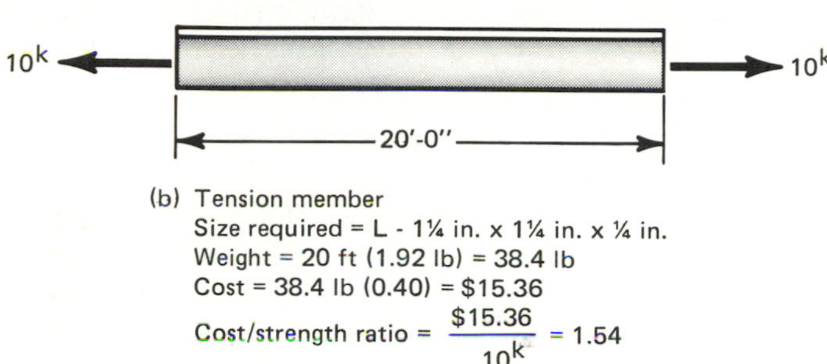

(b) Tension member
Size required = L - 1¼ in. x 1¼ in. x ¼ in.
Weight = 20 ft (1.92 lb) = 38.4 lb
Cost = 38.4 lb (0.40) = \$15.36
Cost/strength ratio = $\frac{\$15.36}{10^k} = 1.54$

Figure 2–3. Comparisons of Cost/Strength Ratio for Different Steel Shapes

The structures of Nervi (Fig. 2–4) vividly capture the concept of load path in their design. In fact, it was Nervi who pioneered this philosophy of structural design back in the 1950s. He drew many of his ideas from the efficient structural forms found in nature—the leaf, for example.

Figure 2–4. Gatti Wool Plant in Rome (From Robert E. Fischer, ed., *Architectural Engineering—New Structures*, McGraw-Hill Book Company, New York, 1964)

Strength. The general geometric forms as well as the more specific dimensional characteristics are strongly influenced by the requirement to carry loads without failure and at a reasonable safety factor. Failure may either be a material failure (crushing, tearing, bearing capacity failure of soil) or may be a significant change (often sudden) in the geometry of the structure. Safety against material failure requires *strength.* Strength is a product of material properties and material arrangement. For example, a relatively weak material such as wood, 2600 psi (17.9 MPa) working stress, when arranged as shown in Fig. 2–5, can be appreciably stronger under bending loads than steel, 18,000 psi (124 MPa) working stress.

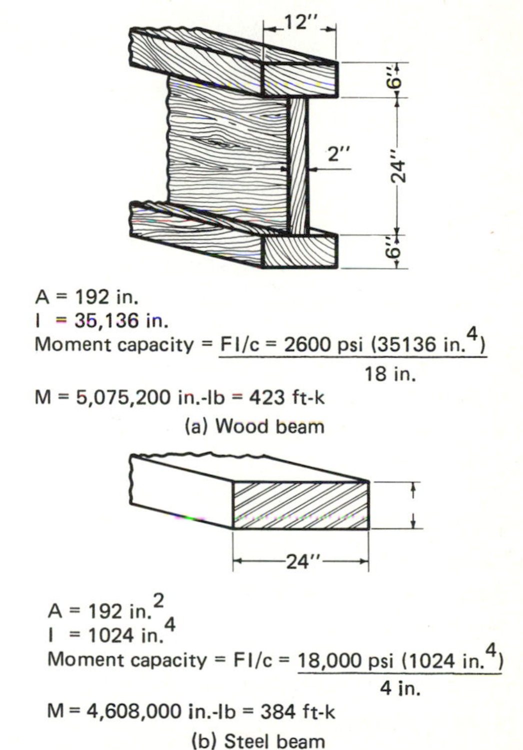

A = 192 in.
I = 35,136 in.
Moment capacity = $FI/c = \frac{2600 \text{ psi } (35136 \text{ in.}^4)}{18 \text{ in.}}$
M = 5,075,200 in.-lb = 423 ft-k
(a) Wood beam

A = 192 in.^{2}
I = 1024 in.^{4}
Moment capacity = $FI/c = \frac{18{,}000 \text{ psi } (1024 \text{ in.}^4)}{4 \text{ in.}}$
M = 4,608,000 in.-lb = 384 ft-k
(b) Steel beam

Figure 2–5. Effect of Material Arrangement on Strength

Adequate strength may be defined in several ways depending on the design code and design methods being used. Normally, strength is defined by specifying one or more of the following *limits:*

- Plastic limit
- Elastic limit
- Fracture limit
- Fatigue limit

Plastic limits are usually associated with design methods that utilize the ultimate strength of the material: plastic design in steel, ultimate strength and limit design in concrete, and so on. *Elastic limits* are associated with design methods that are based on the material reaching

(a) Column buckling

(b) Frame buckling

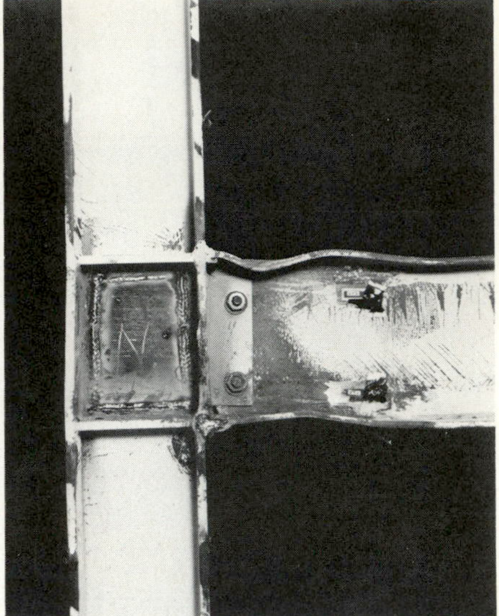

(c) Flange buckling

Figure 2–6. Types of Buckling Behavior (Courtesy of Dr. Le-Wu Lu, Fritz Engineering Laboratory, Lehigh University)

its yield point or some specified approximation: allowable stress design in steel, working stress design in concrete, allowable stress design in masonry, and so on. *Fracture limits* normally incorporate provisions that minimize the possibility of brittle fracture of the material: underreinforcement of concrete to assure ductile failure, welding procedures that protect against fracture of metals, and so on. *Fatigue limits* are specified when structural elements will be subjected to repeated applications of the live load (sometimes accompanied by reversals of stress) during their life: bridge members, structural members supporting machinery, and so on.

Stability. Safety against sudden geometric change, called *instability,* depends on the material arrangement and support conditions. Basic stability is ensured when a structure is in "static equilibrium" under the applied loads—that is, the three equations of statics are satisfied. The problem of stability arises, however, when a compressive stress is applied to a slender structural element. The element, which is stressed initially in uniform compression, must remain perfectly straight if it is to be in equilibrium and have only pure compressive stress. Any tendency of the shape (due to imperfect geometry, load not applied exactly concentrically, etc.) to depart from the line of action of the compressive stress induces bending in the element, and the bending enhances the tendency to move farther from the initial position. At a certain stress level, the member undergoes sudden large lateral displacement, termed *buckling.* When this happens (and it can be predicted mathematically), a column bends out of plane, a beam flange "crinkles," or a frame sways. These types of buckling behavior are illustrated in Fig. 2–6. *Buckling* is related to dimensions (thickness, length, etc.) and to material properties such as modulus of elasticity and creep.

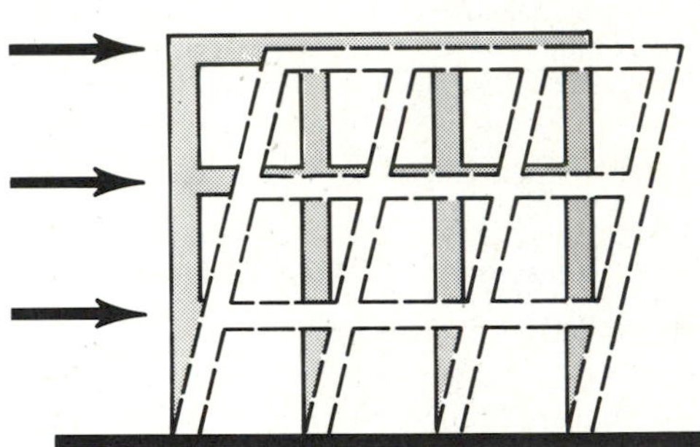

Figure 2–7. Lateral Load Stability

Stability must also be provided for the effect of *lateral load* on the structure (Fig. 2–7). This stability is normally achieved by vertical bracing systems, shear

Figure 2–8. Progressive Collapse

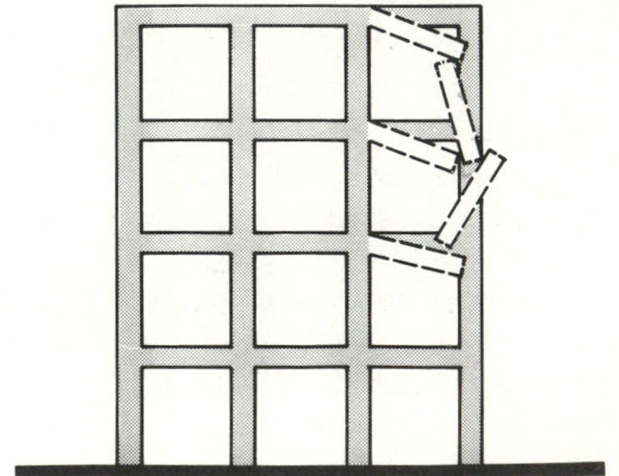

walls, moment resisting frames, or combinations of these systems. Taller structures require innovative approaches to achieving lateral stability, as the cost of providing this stability is a significant portion of the total cost.

Recent events have emphasized the critical importance of still another type of instability called *progressive collapse.* Progressive collapse is a chain reaction of failures following damage to only a relatively small portion of a structure (Fig. 2–8). Figure 2–9 shows the disastrous consequences of this type of instability.

Figure 2–9. Ronan Point Collapse (Courtesy of London Express News and Feature Services)

2–3 ESTHETIC REQUIREMENTS

The earlier discussion described the influence of functional requirements on the form that a structural system takes. Development of the structural form is also directly influenced by the architectural concern for esthetic appeal. The structural form may be expressed as part of the architectural design or it may be completely hidden from view.

Figure 2–10. Pontiac Stadium (Courtesy of Owens-Corning Fiberglas Corporation)

The appearance of a structural system is normally important when it is exposed. In fact, expression of the structural system has been used effectively as an architectural design theme.

> "Finally, in considering the aesthetics of a building, one must carefully distinguish between those buildings in which the structure is relatively unimportant and, hence, not uniquely determined and those in which the structure is essential in determining the appearance of the building. There are buildings in which, in the limit, the appearance is almost exclusively determined by structural considerations. A one-family house can be built with a wood structure, a steel or a concrete structure, while a suspension bridge is uniquely determined by its structural requirements. All buildings in between these two extremes have a structural component that is bound to influence, in varying degrees, their appearance and they influence the aesthetic response of their users.[1]

Pontiac Stadium (Fig. 2–10) is an example of one of the most exciting structural forms (air-supported roof) created in recent years. Its appearance is almost exclusively determined by the structural requirements.

2–4 SERVICEABILITY REQUIREMENTS

Serviceability (performance) requirements are more detailed design criteria that specify the standards a struc-

[1] Mario G. Salvadori, *The Message of the Structure* (New York: American Society of Civil Engineers, 1977). (Paper Presented at the ASCE Spring Convention and Exhibit. Dallas, Tex., April 25–29, 1977.)

tural system must meet in fulfilling its basic purpose. These include considerations such as fire protection and durability. These requirements are normally considered "secondary requirements" (the value engineering definition) because they usually do not contribute to the basic function of the structure. The following sections describe many of the more common serviceability requirements encountered in structural systems design.

Deflection

Vertical Deflection. The downward displacement of members as gravity load is applied is the most common form of deflection. The extent of deflection is related to structural *stiffness,* commonly expressed in terms of the length of the member, the cross-sectional area, and the moment of inertia. Attached or adjacent nonstructural

Table 2–1 Deflection Limitations[a]

Reason for limiting deflection	*Examples*	*Deflection limitation*	*Portion of total deflection on which the deflection limitation is based*
	Sensory Acceptability		
Visual	Droopy cantilevers and sag	By personal preference	Total deflection
Tactile	Vibrations of floors and sag can be felt	$L/360$	Full live load
	Lateral building vibrations	No recommendation	Gust portion of wind
Auditory	Vibrations producing audible noise	Not permitted	
	Serviceability of Structure		
Surfaces that should drain water	Roofs, outdoor decks	$L/240$	Total deflection
Floors that should remain plane	Gymnasia and bowling alleys	$L/360$ + camber or $L/600$	Incremental deflections after floor is installed
Members supporting sensitive equipment	Printing presses and certain building mechanical equipment	Manufacturer's recommendations	Incremental deflections after equipment is leveled
Members supporting industrial cranes	Light to medium crane girders	$L/600$	Crane loads
	Heavy cranes	$L/1000$	Crane loads
	Effect on Nonstructural Elements		
Walls	Masonry and plaster (including lintels)	$L/600$ or 0.30 in. (7.6 mm) max. or $\phi = 0.00167$ rad	Incremental deflections after walls are constructed
	Metal movable partitions and other temporary partitions	$L/240$ or 1 in. (25.4 mm) max.	Incremental deflections after walls are constructed
	Lateral building movement	0.15 in. (3.8 mm) offset per story 0.002 × (height)	Five minutes sustained wind load
	Vertical thermal movement	$L/300$ or 0.60 in. (15.2 mm)	Full temperature differential
Ceilings	Plaster	$L/360$	Incremental deflection after ceiling is built
	Unit ceilings such as acoustic tile	$L/180$	

[a] ACI Committee 435, *Deflections of Concrete Structures,* SP-43 (Detroit, Mich.: American Concrete Institute, 1974).

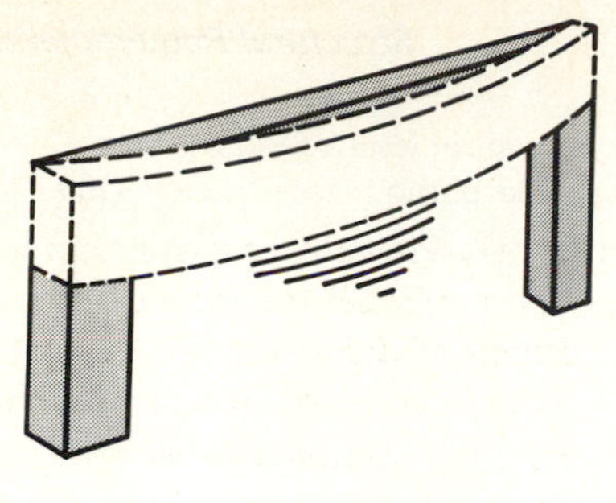

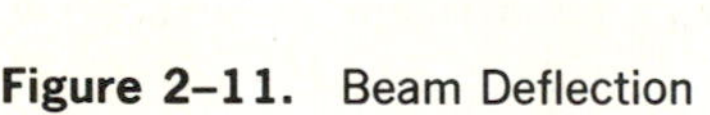

Figure 2–11. Beam Deflection

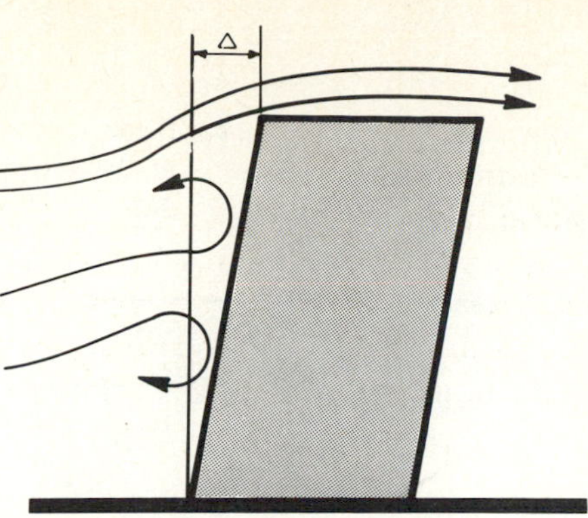

Figure 2–12. Wind Drift

Figure 2–13. Ponding of Water

Table 2–2 Deflection-Index Guide

Type of building	Type of construction: Walls	Type of construction: Floors	Exposure	Code wind require-ments	Deflection index
Office building	Curtain wall	Steel, protective ceiling	Minimum	Moderate	0.0025
Hotel building	Masonry	Concrete	Maximum	Extreme	0.0025
Office building	Masonry	Steel, concrete encased	Average	Moderate	0.003
Residence	Masonry	Concrete	Minimum	Extreme	0.0025
Office building	Curtain wall	Steel, sprayed fire protection	Maximum	Minimum	0.0015

elements (partitions, windowpanes, etc.) can be damaged by excessive deflection. In addition to purely physical effects, deflection behavior can have important visual and psychological consequences (Fig. 2–11). There is perhaps nothing that inspires less confidence in a structure than the sight of a "sagging" beam over your head! Table 2–1 contains deflection limits recommended by ACI Committee 435 and others and is a good guide for establishing deflection limits.

Wind Drift. This form of deflection is associated with lateral wind loads (Fig. 2–12). As with vertical deflection, drift affects both nonstructural elements and human comfort. The amount of *sway* due to wind can be approximately calculated based on equivalent static loads applied to the structure. Drift is not strictly a static problem, however, as the dynamic behavior (frequency and amplitude with which the deflection occurs) can have important consequences. In very tall buildings, the total dynamic behavior must be considered, not just the maximum drift. Table 2–2 lists limits on wind drift for those structures for which such static limits are sufficient.

Ponding. Ponding is a form of deflection behavior associated with roof members and rainwater (Fig. 2–13). It is the trapping of rain due to progressively increasing deflection—called *ponding instability*—as well as clogged roof drains, settlement of column footings, or normal deflection. The addition of water load above the usual roof design live loads can have disastrous results (see Fig. 2–14).

Ponding can be prevented or minimized by providing a *positive roof slope* so that water runs off. The

Figure 2–14. Roof Collapse Due to Ponding

slope must be established considering the effects of deflection. The following examples illustrate the procedure for determining required roof slope:

EXAMPLE PROBLEM 2-1: Roof Slope Requirements

Determine the required roof slope for the framing system shown in Fig. 2-15. Assume that the masonry wall completely restricts the deflection of the eave girder.

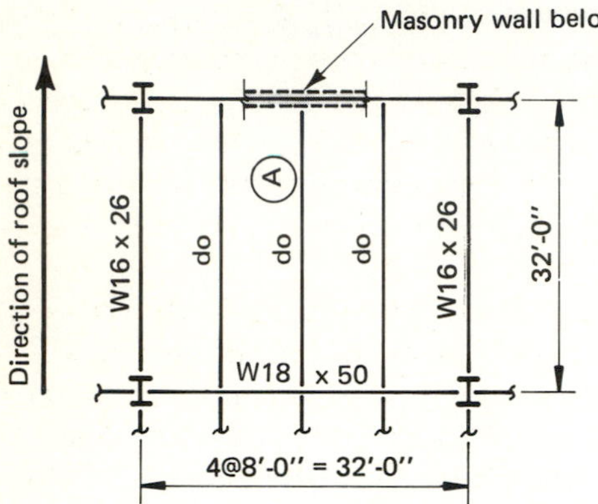

Figure 2-15. Framing System: Example Problem 2-1

Solution

1. The first step is to calculate the midspan deflection (Δ) for the W18 $\times$ 50 girder and end slope (θ) for purlin *A*.

$$P = 0.040 \text{ ksf} \times 8 \text{ ft} \times 32 \text{ ft} = 10.24 \text{ kips } (45{,}547 \text{ N})$$

$$R = \tfrac{3}{2} \times 10.24 \text{ kips} = 15.36 \text{ kips } (68{,}321 \text{ N})$$

where P = purlin reaction to W18
R = end reaction of W18

$$M = 15.36 \text{ kips} \times 16 \text{ ft} - 10.24 \text{ kips} \times 8 \text{ ft} = 162 \text{ ft-kips } (219{,}672 \text{ Nm})$$

$$f = \frac{M}{S} = \frac{162 \text{ ft/kips} \times 12 \text{ in./ft}}{89 \text{ in.}^3} = 21.9 \text{ ksi } (151 \text{ MPa})$$

where M = bending moment
f = bending stress

Δ = coefficient $\times$ span length (*AISC Manual,* p. 2–130)[2]
$= 0.95 \times 0.0398 \times 32 \text{ ft} = 1.21$ in. (30.7 mm)

$$\text{End } \theta \text{ (W16} \times 26) = \frac{wl^3}{24EI} = \frac{0.32 \text{ kip/ft} \times (32 \text{ ft})^3 \times 144 \text{ in.}^2/\text{ft}^2}{(29 \times 10^3 \text{ ksi}) \times 300 \text{ in.}^4 \times 24} = 0.0073 \text{ rad} = 0°25'$$

2. Select roof pitch to provide at least a dead level condition at the lower end of the purlin. Draw sketch (Fig. 2–16).

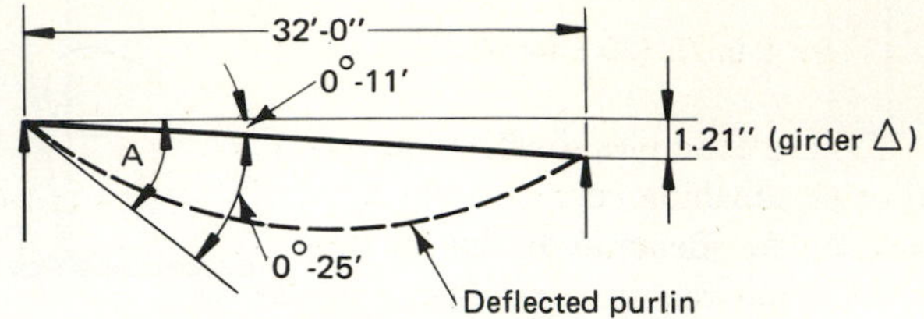

Figure 2-16. Deflection: Example Problem 2-1

$$\text{Angle } A = 0°25' + 0°11' = 0°36'$$

Use $\frac{1}{8}$ in./ft (10.4 mm/m) pitch ($A = 0°35'49''$).

EXAMPLE PROBLEM 2-2: Roof Slope Requirements

Determine the required roof slope for the framing system shown in Fig. 2-17. Assume that the masonry wall completely restricts the deflection of the W16 $\times$ 26 eave purlin.

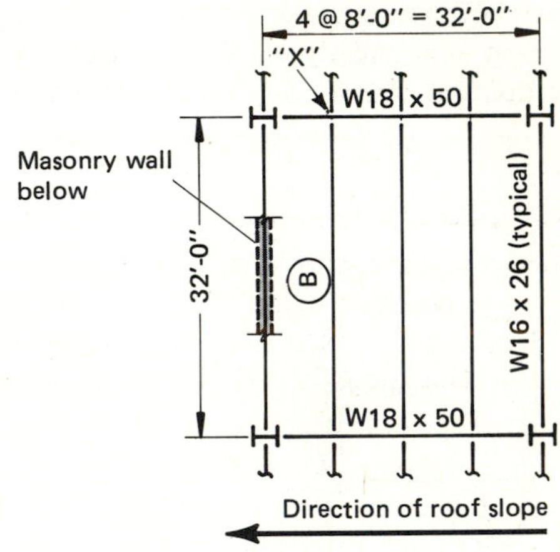

Figure 2-17. Example Problem 2-2

Solution

1. Calculate midspan deflection for purlin *B* and deflection at point *X* on the W18 $\times$ 50 girder.

$$\Delta_B = \frac{5wl^4}{384EI} = \frac{5(0.040 \text{ ksf} \times 8 \text{ ft})(32 \text{ ft})^4(1728 \text{ in.}^3/\text{ft}^3)}{384(29 \times 10^3 \text{ ksi})(300 \text{ in.}^4)} = 0.87 \text{ in. } (22.09 \text{ mm})$$

$\Delta_X = 0.88$ in. (22.3 mm) by separate calculation (not shown)

2. Neglect deflection of roof deck and select roof pitch so that midpoint of purlin *B* is at least the same elevation as the eave purlin. Note that the eave purlin is assumed to have zero deflection.

[2] *Manual of Steel Construction,* 8th ed. (Chicago: American Institute of Steel Construction, Inc., 1980), pp. 2–130.

$$\text{Total } \Delta = 0.88 \text{ in.} + 0.87 \text{ in.} = 1.75 \text{ in. (44.4 mm)}$$

$$\text{Min. pitch} = \frac{1.75 \text{ in.}}{8 \text{ ft}} = 0.22 \text{ in./ft}$$

Use $\frac{1}{4}$ in./ft (20.8 mm/m).

These examples illustrate an approach to the problem of establishing correct roof pitch. Although the two examples are identical except for the orientation of the framing, one requires almost two times as much roof pitch as the other for proper drainage. Actual problems involving interior drains, warped roofs, cantilevered and continuous framing systems, and so on, will complicate the calculations. However, the basic concept of computing deflections and slopes in order to determine required roof pitch should be applicable to all situations.

A *secondary escape system* can also be provided in case roof drains become clogged. Scuppers (which are penetration openings through parapet walls), low parapets, and open pipes are examples of such secondary escape systems (Fig. 2–18). On flat roofs and on roofs where some accumulation of rainwater is possible, *ponding instability* must be checked and the members designed for the maximum impounded water depth. References 1, 10, and 14 contain procedures for checking ponding instability.

Figure 2–18. Secondary Escape System for Roof Drainage

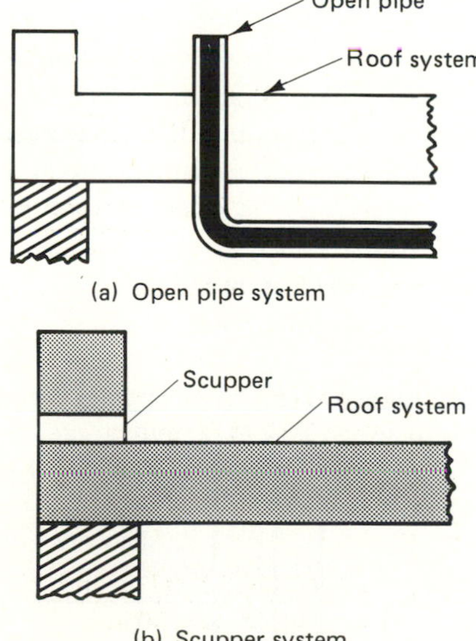

Vibration

Vibration is the response of a structure to a dynamic (rapidly applied) load (Fig. 2–19). The response is in the form of repeated cycles of movement (oscillations about the original position). The dynamic load can be due to machines, such as reciprocal compressors, weaving machines, and so on; people can also become the dynamic load—walking on lightweight floor systems and similar structures. Military troops "break step" while marching across bridges to prohibit establishing a rhythmic load that could set up vibrations in the bridge. Large partition free areas of lightweight floor systems have frequently been susceptible to vibration problems. The response of the structure is dependent on the forcing frequency (of the load), the natural frequency of the structural system, and damping.

Figure 2–19. Vibration

Vibration can have undesirable effects on people and equipment. Manufacturing equipment with sensitive calibration is particularly susceptible to support vibration. Human perception and comfort during vibration has been measured as a basis for establishing vibration limits (Fig. 2–20). References 13 and 8 contain procedures for considering vibration response during design.

Figure 2–20. Human Perception of Vibration (Modified Reiher–Meister Scale)

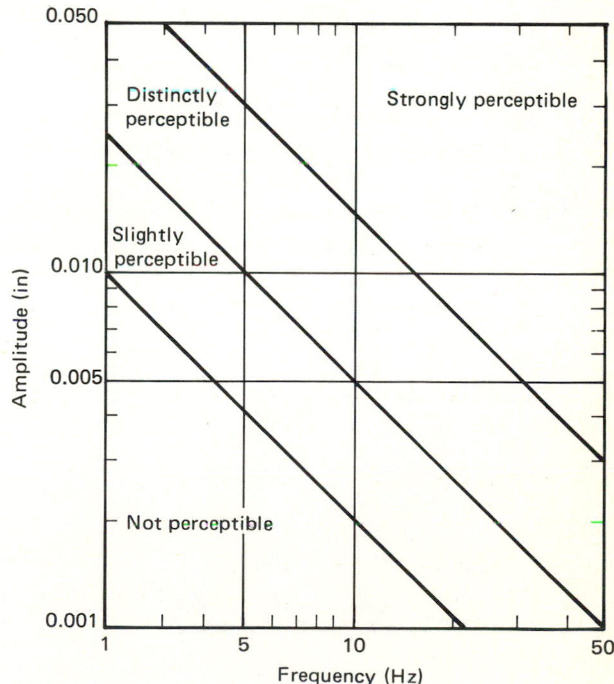

Sound Transmission

Structural systems transmit *sound* due to the compression and expansion of the material and due to vibration (Fig. 2–21). The amplification due to vibration can be great if the natural frequency of the structure is too close to

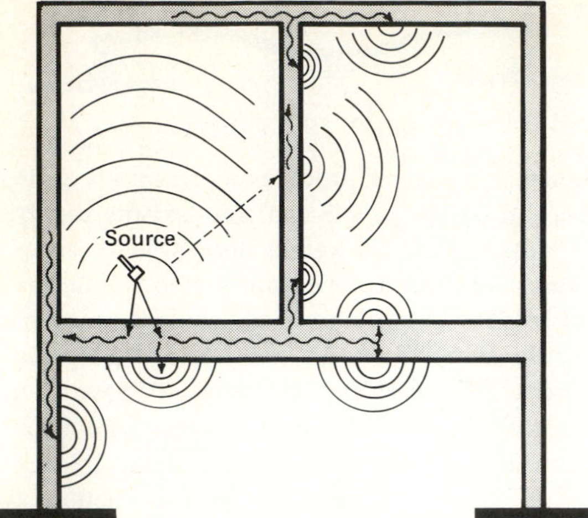

Figure 2–21. Sound Transmission

the frequency of the sound. There is a church balcony railing in a South Carolina church that vibrates violently every time a certain note is played on the organ! Sound transmission is usually minimized by barriers, porous materials, vibration damping, vibration isolation, or by quieting the source.

Sound transmission requirements are normally stated in terms of a required *Sound Transmission Class* (STC) rating for the wall, floor, or roof system. Sound transmission varies with the *frequency* of sound and the *weight* (mass) and *stiffness* of the construction. Transmission loss (TL) tests, standardized by ASTM E 90, determine the actual attenuation for specific types of construction. The STC rating system compares the TL test curve for a construction element with a "standard contour" which reflects known human subjective response to the TL performance. The sound transmission loss at all frequencies, from 125 to 4000 Hz, is important (in varying degrees), so the standard contour is fitted to the entire TL test curve of the actual construction, and the relative vertical position of the contours determines the Sound Transmission Class for the element and compares its effectiveness with that of different construction elements. See Fig. 2–22, for example. Table 2–3 contains a listing of STC requirements for various building occupancy types.

Figure 2–22. Sound Transmission Curves

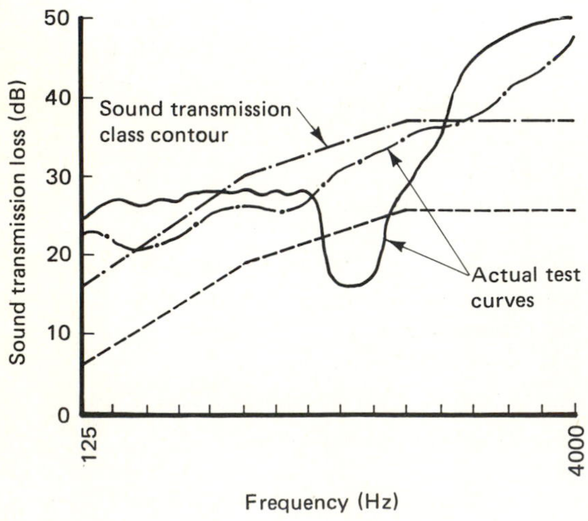

Watertightness

Watertightness can be an important consideration for roofs and exterior walls of buildings, water and waste treatment structures, and so on (Fig. 2–23). Exclusion of water depends not only on the properties of the material itself but also on the extent to which joints and connections can be made watertight.

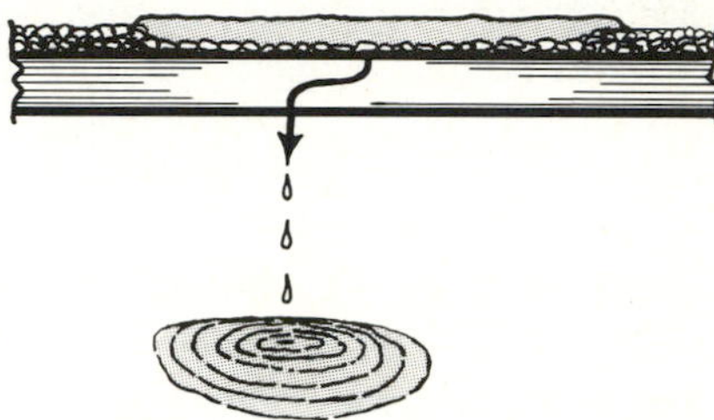

Figure 2–23. Watertightness

Materials such as concrete and masonry are permeable due to the presence of voids and cracks in their internal structure. Water also penetrates interfaces between mortar and masonry units where bonding between the materials is not adequate to prevent seepage. Special mix designs and detailing (to control cracking) improve the watertightness of concrete, and double-wythe masonry walls provide a cavity to intercept moisture, directing it toward "weep holes" (Fig. 2–24) at lower points in the walls. Waterproofing, often added to wall surfaces to prevent water penetration, includes asphalt and felt membranes, special sealers, and paint coatings.

Figure 2–24. Weep Holes

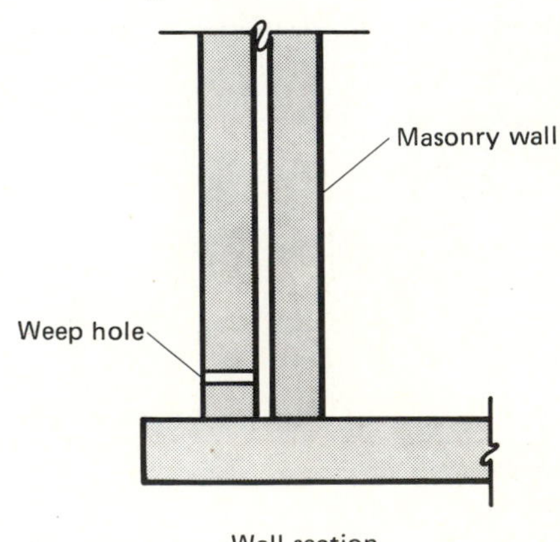

Table 2–3 Sound Transmission Requirements[a]

Type of occupancy	*Room considered*	*Adjacent area*	*Sound isolation requirement*
Normal office areas, normal privacy requirements	Office	Adjacent offices	STC 37
		General office areas	STC 37
		Corridor or lobby	STC 37
		Washrooms and toilet areas	STC 42
		Exterior of building	STC 37
		Kitchen and dining areas	STC 42
		Manufacturing areas and mechanical equipment rooms	STC 47
Any normal occupancy, using conference rooms for group meetings or discussions	Conference rooms	Other conference rooms	STC 42
		Adjacent offices	STC 42
		General office areas	STC 42
		Corridor or lobby	STC 42
		Washrooms and toilet areas	STC 47
		Exterior of building	STC 37
		Kitchen and dining areas	STC 47
		Manufacturing or other noisy interior areas	STC 47
Normal business offices, drafting areas, banking floors, etc.	Large general office areas	Corridors or lobby	STC 32
		Exterior of building	STC 32
		Data processing areas	STC 37
		Manufacturing areas and mechanical equipment areas	STC 42
		Kitchen and dining areas	STC 37
Office in manufacturing, laboratory, or test areas	Shop and laboratory offices	Adjacent offices	STC 37
		Manufacturing, laboratory, or test areas	STC 42
		Washrooms and toilet areas	STC 37
		Corridor or lobby	STC 32
		Exterior of building	STC 32
Motels and urban hotels	Bedrooms	Adjacent bedrooms, separate occupancy	STC 47
		Bathrooms, separate occupancy	STC 47
		Living rooms, separate occupancy	STC 47
		Dining areas	STC 47
		Corridor, lobby, or public spaces	STC 47
		Mechanical equipment rooms	STC 52

[a] Charles G. Ramsey and Harold R. Sleeper, *Architectural Graphic Standards,* 6th ed. (New York: John Wiley & Sons, Inc., 1970), pp. 504–505.

Table 2–3 (*Continued*)

Type of occupancy	*Room considered*	*Adjacent area*	*Sound isolation requirement*
		Exterior of building	
		Normal street or highway noise	STC 42
		Heavy highway traffic	STC 47
		Airport noise	STC 47
Apartments, multiple-dwelling building	Bedrooms	Adjacent bedrooms, separate occupancy	STC 47
		Bathrooms, separate occupancy	STC 47
		Bathrooms, same occupancy	STC 37
		Living rooms, separate occupancy	STC 47
		Living rooms, same occupancy	STC 42
		Kitchen areas, separate occupancy	STC 47
		Kitchen areas, same occupancy	STC 42
		Mechanical equipment rooms	STC 52
		Corridors, lobby, public spaces	STC 47
		Exterior of building	STC 42
Normal school buildings without extraordinary or unusual activities or requirements	Classrooms	Adjacent classrooms	STC 37
		Laboratories	STC 42
		Corridor or public areas	STC 37
		Kitchen and dining areas	STC 42
		Shops	STC 47
		Recreational areas	STC 47
		Music rooms	STC 47
		Mechanical equipment rooms	STC 52
		Toilet areas	STC 42
		Exterior of building	STC 37
Any occupancy where serious performances are given; requirements may be relaxed for elementary schools or noncritical types of occupancy	Theaters, concert halls, lecture halls	Adjacent similar areas	STC 52
		Corridors and public areas	STC 47
		Recreational areas	STC 52
		Mechanical equipment spaces	STC 52
		Classrooms	STC 47
		Laboratories	STC 47
		Shops	STC 52
		Toilet areas	STC 52
		Exterior of building	STC 52

Waterproofing of joints and connections requires special details such as those shown in Fig. 2–25. These include flashing (thin metal sheeting installed to bridge joints), waterstops, and roof expansion joint materials. Sealing of roof systems at the perimeter and at penetration points (pipes, vents, etc.) has historically been a problem.

Perhaps the most plaguing problem in the construction industry is the leaking roof! The roof system is a

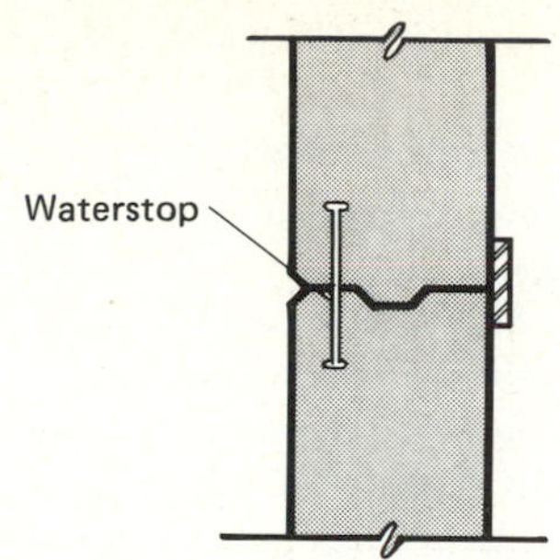

Figure 2–25. Waterproofing at Joints

composite of several materials (felt, tar or asphalt, gravel, insulation, structural deck, flashing, etc.) and is, perhaps, one system where our technology has not recognized the practical realities of field construction. This multilayer system must be installed with dry materials, at prescribed temperatures, and with unusual care. The lesson to be learned is that reliance on top-quality materials and installation for several interrelated components of a weatherproofing system is unrealistic. We need to develop simpler, more foolproof materials.

Volume Change

Volume-change behavior is a factor in most structural systems (Fig. 2–26). Volume changes occur due to temperature expansion and contraction, creep and shrinkage, moisture expansion, and strain due to applied load. Expansion and contraction is critical in extreme temperature applications, such as furnaces and boilers. Temperature and other volume-change influences may have an influence on the choice of structural materials, but more often provisions must be made for these effects in the form of expansion/contraction joints and control joints.

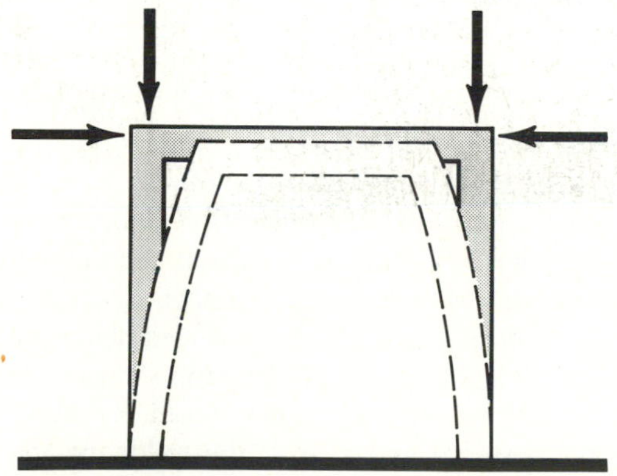

Figure 2–26. Volume Change

Structural Joints. Expansion/contraction joints in the *frames* of buildings, bridges, and other structures are located at intervals that minimize the buildup of restrained volume change stresses and at points of discrete geometry change (Fig. 2–27). If the forces generated by temperature rise are significantly greater than shrinkage and creep forces, a true "expansion joint" is needed. However, in *concrete* structures, true expansion joints are seldom required. Instead, joints that permit contraction of the structure are needed to relieve the strains caused by decreases in temperature and restrained creep and shrinkage, which are additive. Such joints are properly called contraction or control joints but are commonly referred to as expansion joints.

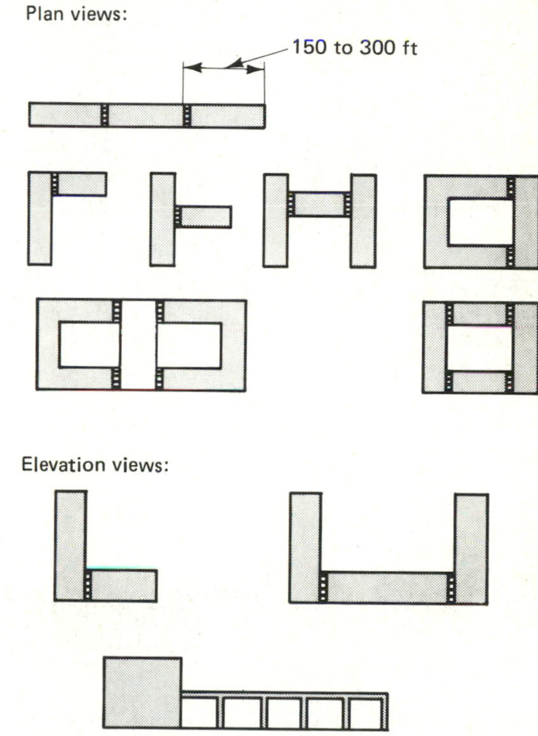

Figure 2–27. Expansion and Contraction Joint Locations

There is a wide divergence of opinion concerning the spacing of expansion joints. Typical practice in steel and concrete structures is to locate expansion joints at distances between 150 and 200 ft (45.7 to 60.9 m). However, buildings exceeding these limits have performed well without expansion joints. Recommended joint spacings for buildings are generally based on experience and may not consider several important items. Among these items are the types of connections used, the column stiffnesses in simple span structures, the relative stiffness between beams and columns in framed structures, and the weather exposure conditions. Nonheated structures such as parking garages are subjected to greater temperature changes than occupied structures, so shorter distances between expansion joints are warranted.

The need for thermal expansion joints in buildings may be determined either by an empirical method or by an analytical method. An *empirical* method, recommended by the Federal Construction Council,[3] based on a study completed in 1974, is described in the following paragraphs:

1. For buildings having a beam-and-column or slab-and-column structural frame, the maximum length of the building without expansion joints may be determined in accordance with Fig. 2–28 on the basis of the design temperature change (Δt) in the locality of construction.

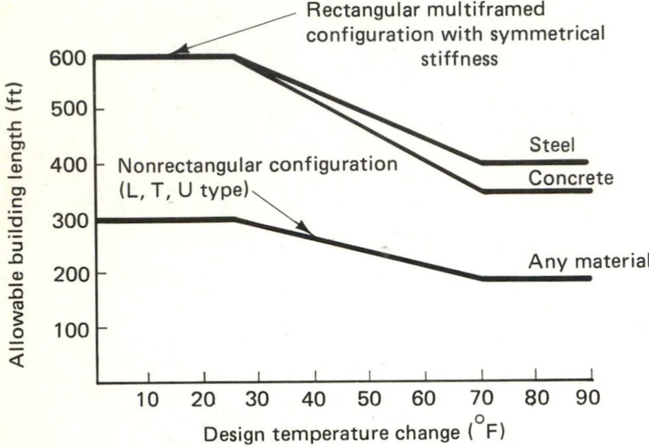

(a) If the building will be heated only and will have hinged-column bases, use the allowable length as specified.

(b) If the building will be air conditioned as well as heated, increase the allowable length by 15% (provided the environmental control system will run continuously).

(c) If the building will be unheated, decrease the allowable length by 33%.

(d) If the building will have fixed-column bases, decrease the allowable length by 15%.

(e) If the building will have substantially greater stiffness against lateral displacement at one end of the plan dimension, decrease the allowable length by 25%.

Figure 2–28. Joint Spacing. Maximum allowable building length without use of expansion joints for various design temperature changes. Applicable to buildings of beam and column construction, hinged at the base, and with heated interiors. (Extracted from *Expansion Joints in Buildings,* 1974, by permission of the National Academy of Sciences, Washington, D.C.)

2. The design temperature change (Δt) is computed in accordance with the formula

$$\Delta t = (T_w - T_m) \quad \text{or} \quad (T_m - T_c) \qquad (2\text{–}1)$$

whichever is greater, where:

T_m = the *mean* temperature during the normal construction season in the locality of the building. The normal construction season for a locality is defined as that contiguous period in a year during which the *minimum* daily temperature equals or exceeds 32°F. [For example, the normal construction season for Anchorage, Alaska, is $5\frac{1}{2}$ months (April 24–October 8) and for Birmingham, Alabama, is year-round (January–December).]

T_w = the temperature exceeded, on the average, only 1% of the time during the summer months of June through September in the locality of the building. (In a normal summer there would be approximately 30 hours at or above this design value.)

T_c = the temperature equaled or exceeded, on the average, 99% of the time during the winter months of December, January, and February in the locality of the building. (In a normal winter there would be approximately 22 hours at or below this design value.)

3. Figure 2–29 presents a graphical plot of Δt values for the United States.

4. For buildings supported by continuous, exterior unreinforced masonry, expansion joints should be placed at intervals not exceeding 200 ft (60.9 m). In addition, intermediate subjoints should be positioned and spaced in accordance with the recommendations of the Brick Institute of America (BIA) and the National Concrete Masonry Association (NCMA).

This procedure assumes that structures will be built when the minimum daily temperatures are above 32°F (0°C). The design temperature change is then based on differences from the mean (T_m) rather than on the difference between the extreme high and low temperature (T_w and T_c).

Other Joints. In addition to joints in the structural frame, joints must be provided in *wall* and *floor systems* to minimize the buildup of stresses that could cause cracking. Two types of joints are utilized for this purpose: *expansion joints,* which permit relative expansion movement, and *control joints,* where weakened planes influence shrinkage cracks to occur at the joint. These joints are located at intervals of length dictated by the characteristics of the material and also at the following locations:

1. At all abrupt changes in wall height or slab width
2. At all changes in wall or slab thickness
3. Above joints in foundations and floors
4. Below joints in floors and roofs that bear on walls

[3] Federal Construction Council, *Expansion Joints in Buildings,* Technical Report No. 65 (Washington, D.C.: National Academy of Sciences, 1974).

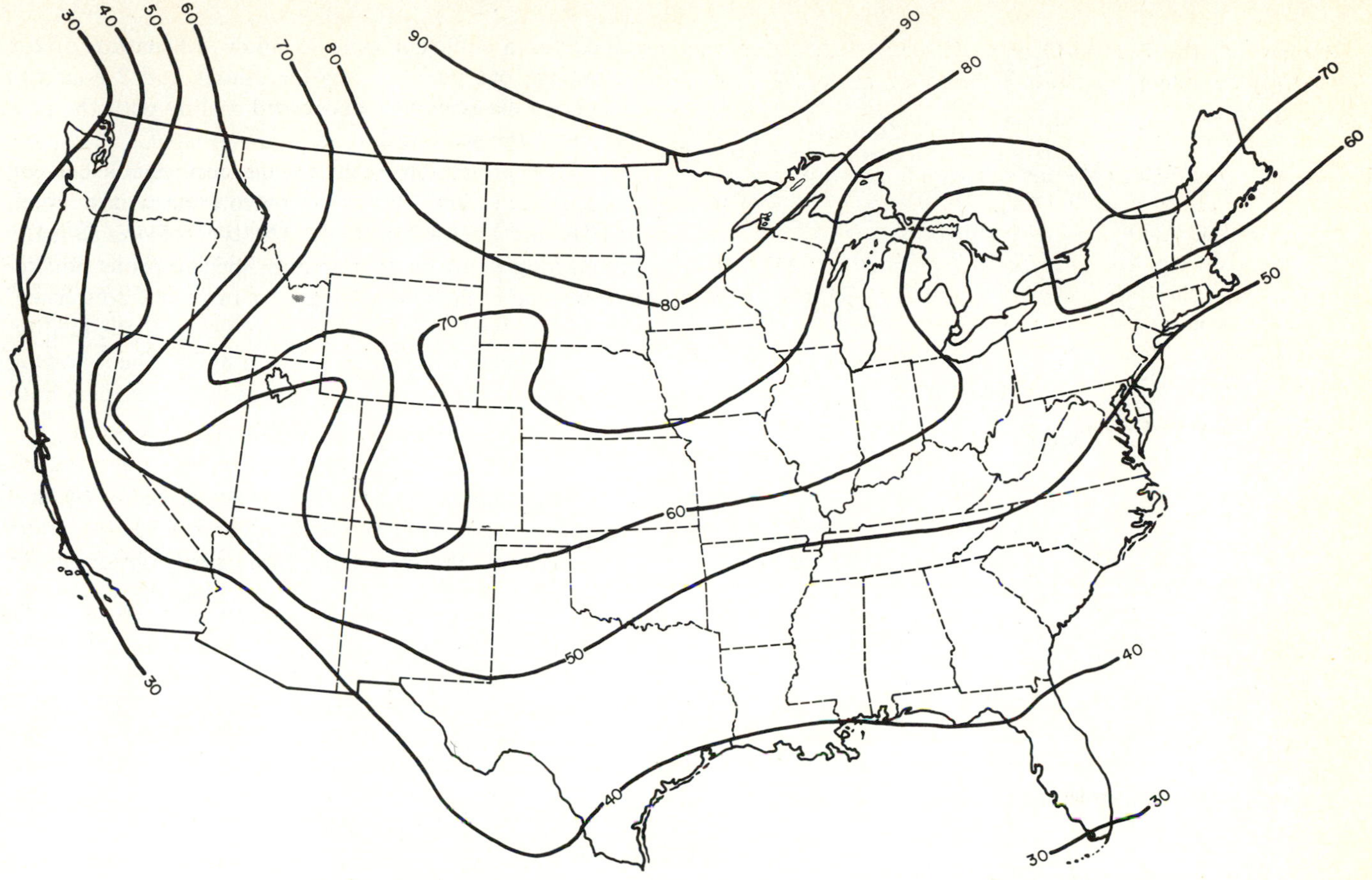

Figure 2–29. Maximum Seasonal Climate Temperature Change, °F (From *PCI Manual for Structural Design of Architectural Precast Concrete,* Prestressed Concrete Institute, Chicago, 1977, p. 5–60)

5. Near bonded wall intersections or corners, at a distance not exceeding one-half the regular joint interval
6. At one or both sides of door and window openings in walls, unless other crack control measures are used such as joint reinforcement or bond beams

Concrete masonry walls undergo drying shrinkage which requires control joints spaced at intervals between 40 and 60 ft (12.1 to 18.3 m) depending on the extent of joint reinforcement and wall panel size. Expansion joints are normally required only in conjunction with overall building expansion joints.

Brick masonry walls experience little shrinkage but rather swell considerably due to absorption of moisture. Expansion (rather than control) joints must be provided at intervals ranging from 30 to 100 ft (9.1 to 30.5 m), depending on the accumulated expansion that can be contained within the normal 1-in. (25.4-mm) joint width. The Brick Institute of America (BIA)[4] recommends the following formula for calculating accumulated expansion:

$$w = [0.0002 + 0.000004(T_{\max} - T_{\min})]L \quad (2\text{–}2)$$

where L = length of wall, in.
$T_{\max}$ = maximum mean wall temperature, °F
$T_{\min}$ = minimum mean wall temperature, °F
w = total expansion of wall, in.

Contraction due to *concrete shrinkage* requires frequent joints in concrete systems. In fact, of all structural materials concrete requires the most careful attention to avoid cracking. Concrete shrinks as it hardens, it creeps (plastic flow) under load, and it is often bonded to other structural materials with different expansion and contraction characteristics (coefficients of expansion). *Reinforced concrete walls* normally require control joints at 20- to 25-ft (6.1- to 7.6-m) intervals and expansion/contraction joints at 80- to 100-ft (24.4- to 30.5-m) intervals. *Concrete slabs on grade* require control joints (in each direction) at spacings of 15 to 25 ft (4.6 to 7.6 m), as well as isolation joints at walls and around columns (Fig. 2–30).

[4] "Differential Movement," Technical Notes on Brick and Tile Construction, No. 18A (Washington, D.C.: Brick Institute of America, 1963).

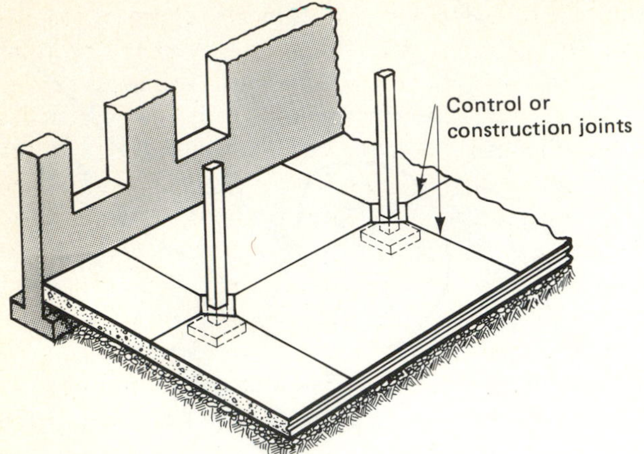

Figure 2–30. Concrete Slab Joints

Strain. Strain due to applied load can create secondary stresses in structural systems—axial shortening of truss members, for example. A unique feature of tall buildings is the importance of axial shortening of the columns—particularly the differential strain between columns leading to bending and shear stresses in the floor system (Fig. 2–31).

Figure 2–31. Column Shortening Effects

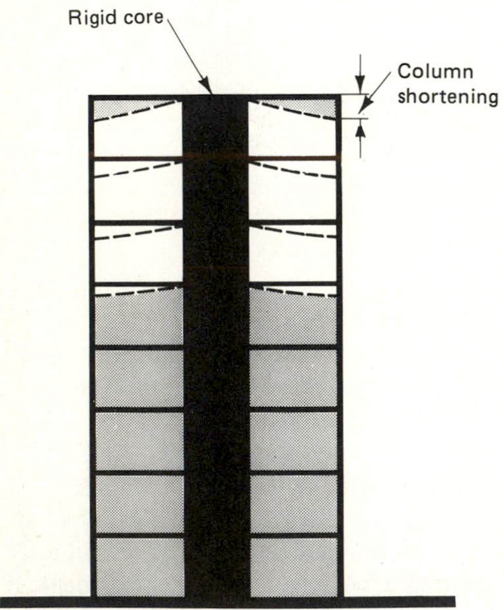

Fire Protection

Many structural systems require some form of protection from *fire.* Fire protection is generally required for two reasons: protection of *people* and protection of the *structure.* These requirements overlap to some extent, but protection for people is normally concerned with proper exits and prevention of flame and smoke spread, while protection for the structure is concerned with sustaining property value. Certainly, structural integrity must be maintained for a sufficient time to allow evacuation of the occupants, but requirements for 1-hour and 2-hour fire protection are obviously concerned mainly with the protection of the structure.

Fire protection requirements are established by building codes and by fire insurance underwriters (such as FIA and Factory Mutual). The type of *occupancy* is the starting point for determining specific protection requirements as far as building structures are concerned. Occupancy type defines the primary usage of the facility, such as educational, residential, and so on. The following paragraphs describe the basic approach to providing both people and structural fire protection.

Protection of People. People are protected by having both an exit from the building and an access route to that exit, jointly referred to as a *means of egress.* Means of egress requirements normally include the following:

- Maximum travel distance to an exit
- Minimum number of exits
- Capacity of the means of egress
- Minimum fire ratings for corridor walls, doors, and stairway walls which protect the exit and access route

An example of these requirements (Standard Building Code) is shown in Tables 2–4 and 2–5.

Protection of the Structure. The structure is protected by limitations placed on *allowable building heights* and *floor areas,* in conjunction with limitations regarding the *type of construction.* The type of construction is a

Table 2–4 Exit Requirements[a]

	Maximum distance of travel of an exit (ft)	
Occupancy	*Unsprinklered*[b]	*Sprinklered*
Group A: Assembly	150	200
Group B: Business	150	200
Group E: Educational	150	200
Group H: Hazardous	NP	75
Group F: Factory–industrial	150	200
Group I: Institutional	NP	200
Group M: Mercantile	150	200
Group R: Residential	150	200
Group S: Storage	150	200
Open parking structures	200	200

Minimum number of exits	*Occupancy load*
2	50– 500
3	501–1000
4	More than 1000

[a] *Standard Building Code* (Birmingham, Ala.: Southern Building Code Congress International, Inc., 1979).

[b] NP, not permitted.

Table 2–5 Means of Egress Capacity Requirements[a]

Occupancy	*Persons allowed per unit (22 in.) of exit width** — *Level travel (corridors, doors, ramps, etc.)*	*Stairs*
Group R: Residential	100	75
Group B: Business	100	60
Group E: Educational	100	75
Group I: Institutional (restrained)	30	22
Group I: Institutional (unrestrained)	45	35
Group M: Mercantile	100	60
Group A: Assembly	100	75
Group S: Storage	100	60
Group F: Factory–industrial	100	60
Group H: Hazardous	50	30

* Minimum number of persons per floor area shall be calculated in accordance with the following:

Use	*Area per occupant (sq ft)*
Assembly without fixed seats	
Concentrated	7 net
Standing space	3 net
Unconcentrated	15 net
Assembly with fixed seats	b
Bowling alleys, allow five persons for each alley, including 15 ft of runway and for additional areas	7 net
Business areas	100 gross
Small restaurants (without fixed seats)	15 net
Small restaurants (with fixed seats)	b
Court rooms, other than fixed-seating areas	40 net
Educational	
Classroom areas	20 net
Shops and other vocational areas	50 net
Industrial areas	100 gross
Institutional	
Sleeping areas	120 gross
Inpatient treatment and ancillary areas	240 gross
Outpatient area	100 gross
Library	
Reading rooms	50 net
Stack area	100 gross
Mercantile	
Basement and grade-floor areas	30 gross
Areas on other floors	60 gross
Storage, shipping area	100 gross
Parking garage	200 gross
Residential	200 gross
Storage area, mechanical	300 gross

[a] *Standard Building Code* (Birmingham, Ala.: Southern Building Code Congress International, Inc., 1979).

[b] The occupant load for an assembly area having fixed seats installed shall be determined by the number of fixed seats.

rating classification based on the degree of protection provided (Type I incorporates the most protection, for example). Table 2–6 is an example of heights and areas permitted for each occupancy type and type of construction by the Standard Building Code. Five interrelated factors are shown: occupancy classification, type of construction, building height, maximum area per floor, and use/nonuse of sprinklers. Fire protection requirements are the most stringent for Type I construction, with requirements diminishing for each higher numbered type until Type IV, which has the least stringent requirements (see Table 2–7). Normally, building occupancy classification and tentative height and area are known, so we can work from right to left in the table to find the highest numbered type of construction that permits the height and area that have been tentatively chosen. The definition for the type selected must also fit the actual material and systems being proposed (i.e., Type III is for heavy timber systems, etc.).

If initial trials with Table 2–6 yield fire protection requirements that are judged too severe, heights and areas of the structure may be adjusted to lessen the requirements. Alternatively, building materials and systems may be changed to comply with the more stringent requirements. The hour rating finally established requires the structural member to withstand 650°F heat for the indicated duration without significant structural deterioration. The "Standard Fire Test" (ASTM E119) is used to evaluate proposed protection systems (Fig. 2–32). The following example illustrates the procedure for determining fire protection requirements.

Figure 2–32. Fire Test (Courtesy of Underwriters' Laboratories, Inc.)

Table 2–6 Allowable Heights and Areas[a,b]

Occupancy	Height	Type of Construction									
		Type I		Type II		Type III		Type IV			
								1-hour		Unprotected	
		p Unspk	Spk	p Unspk	Spk	p Unspk	Spk	p Unspk	Spk	p Unspk	Spk
Business	UH	UA	UA								
	12	UA	UA								
	80 ft	UA	UA	UA	UA						
	5	UA	UA	UA	UA	25.5	51.0	25.5	51.0		34.0
	4	UA	UA	UA	UA	25.5	51.0	25.5	51.0		34.0
	3	UA	UA	UA	UA	25.5	51.0	25.5	51.0		34.0
	2	UA	UA	UA	UA	25.5	51.0	25.5	51.0	17.0	34.0
Multistory	1	UA	UA	UA	UA	25.5	51.0	25.5	51.0	17.0	34.0
One story only		UA	UA	UA	UA	25.5	76.5	25.5	76.5	17.0	51.0
Educational	UH	UA	UA								
	80 ft	UA	UA	UA	UA						
	2	UA	UA	UA	UA	18.0	36.0	18.0	36.0		
Multistory	1	UA	UA	UA	UA	18.0	36.0	18.0	36.0	NP	NP
One story only		UA	UA	UA	UA	18.0	54.0	18.0	54.0	12.0	36.0
Hazardous	4	NP	11.5								
	3	NP	11.5	NP	8.3						
	2	NP	11.5	NP	8.3	NP	7.5				
Multistory	1	NP	11.5	NP	8.3	NP	7.5	NP	NP	NP	NP
One story only		NP	11.5	NP	8.3	NP	7.5	NP	5.0	NP	5.0
Factory–industrial	UH	UA	UA								
	80 ft	UA	UA	UA	UA						
	6	UA	UA	UA	UA		63.0				
	5	UA	UA	UA	UA		63.0				
	4	UA	UA	UA	UA		63.0		63.0		42.0
	3	UA	UA	UA	UA	31.5	63.0		63.0		42.0
	2	UA	UA	UA	UA	31.5	63.0	31.5	63.0	21.0	42.0
Multistory	1	UA	UA	UA	UA	31.5	63.0	31.5	63.0	21.0	42.0
One story only		UA	UA	UA	UA	31.5	94.5	31.5	94.5	21.0	63.0
Mercantile	UH	15.0	UA								
	80 ft	15.0	UA	15.0	UA						
	5	15.0	UA	15.0	UA	13.5	27.0	13.5	27.0		18.0
	4	15.0	UA	15.0	UA	13.5	27.0	13.5	27.0		18.0
	3	15.0	UA	15.0	UA	13.5	27.0	13.5	27.0		18.0
	2	15.0	UA	15.0	UA	13.5	27.0	13.5	27.0	9.0	18.0
Multistory	1	15.0	UA	15.0	UA	13.5	27.0	13.5	27.0	9.0	18.0
One story only		15.0	UA	15.0	UA	13.5	40.5	13.5	40.5	9.0	27.0
Residential	UH	UA	UA								
	12	UA	UA								
	80 ft	UA	UA	UA	UA						

[a] *Standard Building Code* (Birmingham, Ala.: Southern Building Code Congress International, Inc., 1979).

[b] Height for types of construction is limited to the number of stories shown, or height in feet. Allowable areas are shown in thousands of square feet. Maximum heights: Type I, no limit; Type II, 80 ft; Type III, 65 ft; Type IV, 65 ft (1 hour), 55 ft (unprotected). UA, no limit of floor area; UH, no height limit; NP, not permitted.

Table 2–6 (*Continued*)

		Type of Construction									
								Type IV			
		Type I		*Type II*		*Type III*		*1-hour*		*Unprotected*	
Occupancy	*Height*	*p Unspk*	*Spk*	*p Unspk*	*Spk*	*p Unspk*	*Spk*	*p Unspk*	*Spk*	*p Unspk*	*Spk*
	5	UA	UA	UA	UA			18.0	36.0		24.0
	4	UA	UA	UA	UA			18.0	36.0		24.0
	3	UA	UA	UA	UA	18.0	36.0	18.0	36.0		24.0
	2	UA	UA	UA	UA	18.0	36.0	18.0	36.0	12.0	24.0
Multistory	1	UA	UA	UA	UA	18.0	36.0	18.0	36.0	12.0	24.0
One story only		UA	UA	UA	UA	18.0	54.0	18.0	54.0	12.0	36.0
Storage	UH	UA	UA								
	6	UA	UA	30.0	60.0		48.0				
	5	UA	UA	30.0	60.0		48.0				
	4	UA	UA	30.0	60.0		48.0		48.0		32.0
	3	UA	UA	30.0	60.0		48.0		48.0		32.0
	2	UA	UA	30.0	60.0	24.0	48.0	24.0	48.0	16.0	32.0
Multistory	1	UA	UA	30.0	60.0	24.0	48.0	24.0	48.0	16.0	32.0
One story only		UA	UA	30.0	90.0	24.0	72.0	24.0	72.0	16.0	48.0

Table 2–7 Required Fire Ratings[a,b]

				Type IV		*Type V*		*Type VI*	
Structural element	*Type I*	*Type II*	*Type III*	*1-hour protected*	*Unprotected*	*1-hour protected*	*Unprotected*	*1-hour protected*	*Unprotected*
Party and fire walls	4	4	4	4	4	4	4	4	4
Interior bearing walls									
Supporting more than one floor, columns, or other bearing walls	4	3	2	1	NC	1	0	1	0
Supporting one floor only	3	2	1	1	NC	1	0	1	0
Supporting a roof only	3	2	1	1	NC	1	0	1	0
Columns									
Supporting more than one floor or other columns	4	3	H	1	NC	1	0	1	0
Supporting one floor only	3	2	H	1	NC	1	0	1	0
Supporting a roof only	3	2	H	1	NC	1	0	1	0

[a] *Standard Building Code* (Birmingham, Ala.: Southern Building Code Congress International, Inc., 1979).

[b] NC, noncombustible; NL, no limits; H, heavy timber sizes.

Table 2–7 (*Continued*)

Structural element	Type I	Type II	Type III	Type IV 1-hour protected	Type IV Unprotected	Type V 1-hour protected	Type V Unprotected	Type VI 1-hour protected	Type VI Unprotected
Beams, girders, trusses, and arches									
Supporting more than one floor or columns	4	3	H	1	NC	1	0	1	0
Supporting one floor only	3	2	H	1	NC	1	0	1	0
Supporting a roof only	$1\frac{1}{2}$	1	H	1	NC	1	0	1	0
Floor construction	3	2	H	1	NC	1	0	1	0
Roof construction	$1\frac{1}{2}$	1	H	1	NC	1	0	1	0
Exterior bearing walls: horizontal separation (distance from common property line or assumed property line)[c] (ft)									
0–3	4(0%)	3(0%)	3(0%)	2(0%)	1(0%)	3(0%)	3(0%)	1(0%)	1(0%)
Over 3–10	4(10%)	3(10%)	2(10%)	1(10%)	1(10%)	2(10%)	2(10%)	1(20%)	0(20%)
Over 10–20	4(20%)	3(20%)	2(20%)	1(20%)	NC(20%)	2(20%)	2(20%)	1(40%)	0(40%)
Over 20–30	4(40%)	3(40%)	1(40%)	1(40%)	NC(40%)	1(40%)	1(40%)	1(60%)	0(60%)
Over 30	4(NL)	3(NL)	1(NL)	1(NL)	NC(NL)	1(NL)	1(NL)	1(NL)	0(NL)
Exterior nonbearing walls: horizontal separation (distance from common property line or assumed property line)[c] (ft)									
0–3	3(0%)	3(0%)	3(0%)	2(0%)	1(0%)	3(0%)	3(0%)	1(0%)	1(0%)
Over 3–10	2(10%)	2(10%)	2(10%)	1(10%)	1(10%)	2(10%)	2(10%)	1(20%)	0(20%)
Over 10–20	2(20%)	2(20%)	2(20%)	1(20%)	NC(20%)	2(20%)	2(20%)	1(40%)	0(40%)
Over 20–30	1(40%)	1(40%)	1(40%)	NC(40%)	NC(40%)	1(40%)	1(40%)	0(60%)	0(60%)
Over 30	NC(NL)	NC(NL)	NC(NL)	NC(NL)	NC(NL)	NC(NL)	NC(NL)	0(NL)	0(NL)

[c] %, percent of wall opening permitted.

EXAMPLE PROBLEM 2–3: Fire Protection Requirements

Determine the fire resistance ratings required (Standard Building Code) for the columns, beams, floors, and roof of a single-story manufacturing plant containing 30,000 sq ft (2787 m^2) of area.

Solution Enter Table 2–6 opposite the Factory–Industrial building classification. For a one-story structure, minimum protection of Type IV (1-hour) is required for unsprinklered conditions (up to 31,500 sq ft) and Type VI (unprotected) is required for sprinklered conditions (up to 30,000 sq ft).

Assume that sprinklers will not be used and enter Table 2–7. Type IV construction requires the following ratings:

Columns: *1 hour*
Beams: *1 hour*
Floors: *1 hour*
Roofs: *1 hour*

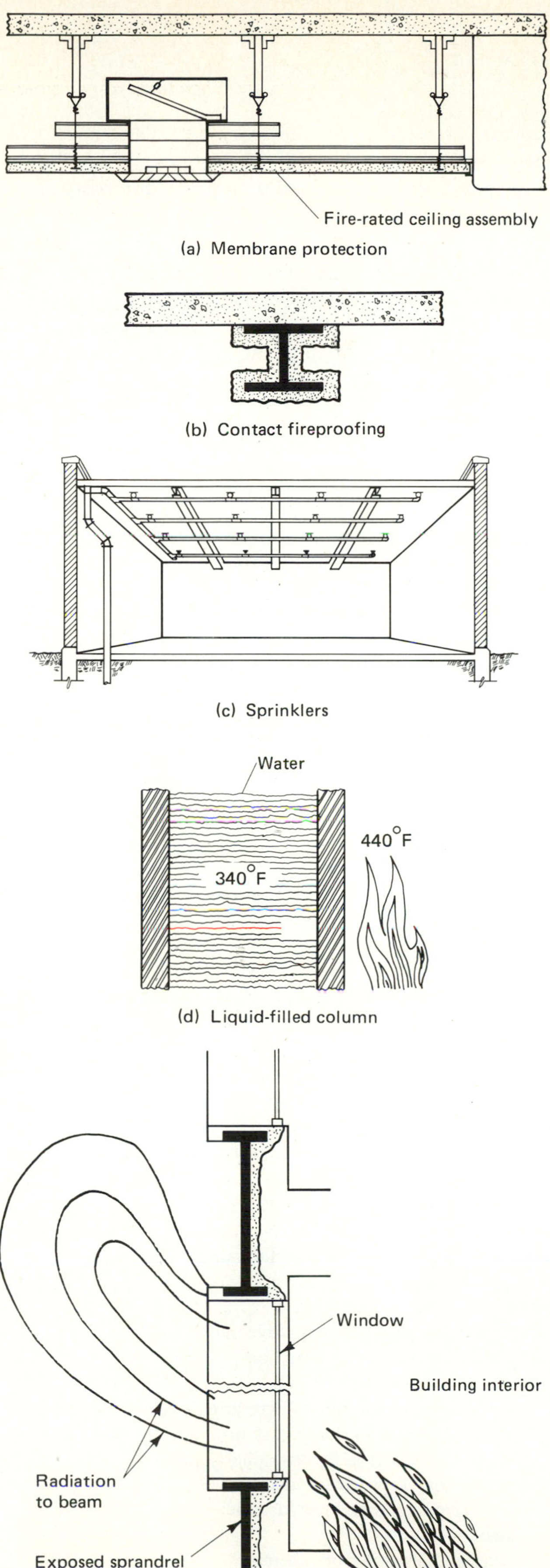

(a) Membrane protection

(b) Contact fireproofing

(c) Sprinklers

(d) Liquid-filled column

(e) Flame shielding

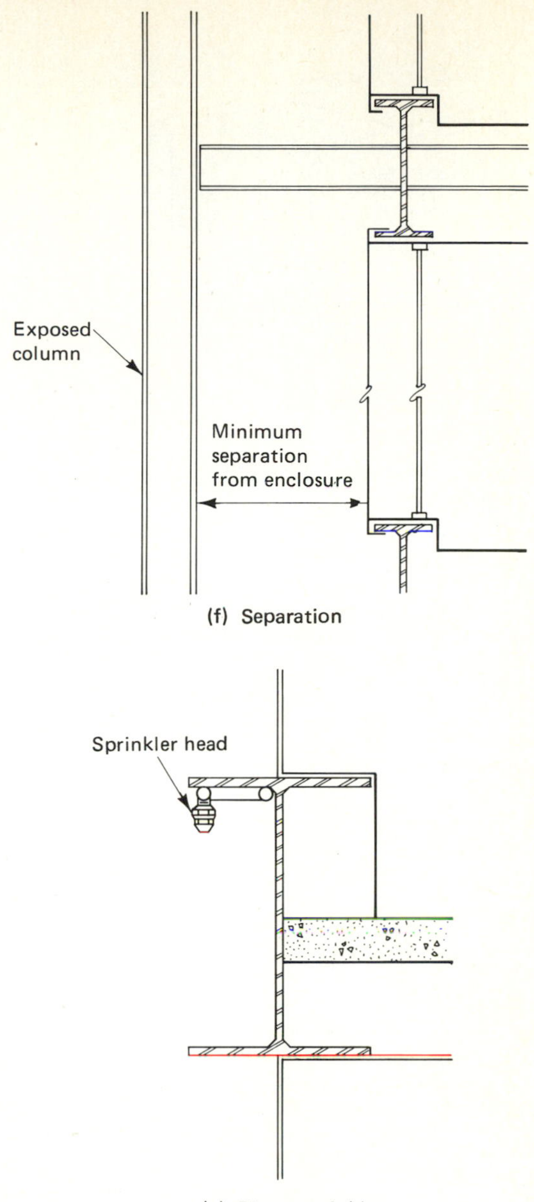

(f) Separation

(g) Direct sprinkler

Figure 2–33. Fire Protection Methods

Numerous methods are available for fire protection, and many have been tested and certified by Underwriters' Laboratories, Inc., which is a testing agency specifically established to evaluate such things as quality of fire protection. The various fire protection methods can be grouped into six categories. These are illustrated in Fig. 2–33 and Appendix 5C.

Durability

The exposure of the structure to weather, marine water, insects, and so on, must be considered and the proper materials and protection provided to assure *durability* (Fig. 2–34). Durability is the "long-lasting" quality of certain materials that minimizes the effects of corrosion, wear, rot, decomposition, and so on. The durability of materials is a function of chemical structure, fabricating

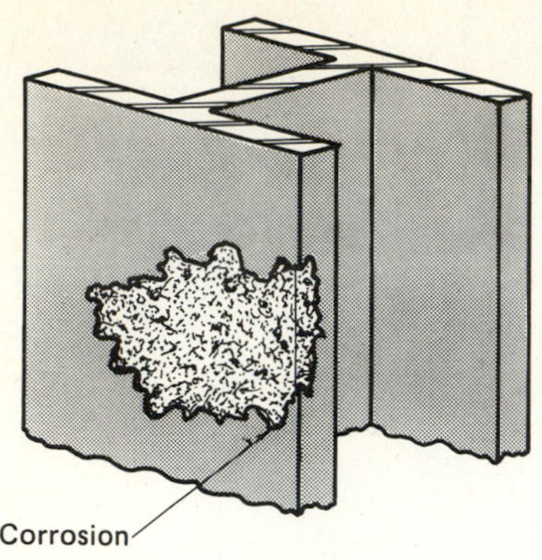

Figure 2–34. Durability Affected by Corrosion

process, protection, and environment. Various alloying methods, heat control during fabrication, and coating systems (paint, etc.) can be used to improve durability. Table 2–8 describes various types of deterioration and their prevention. Deterioration of some materials may exclude their use in certain environments where replacement would be extremely expensive (or impossible). For example, the initial life span of fabric roof systems was 7 years. Even at relatively low initial cost, the cost of replacement at 7 years prohibited many engineers and owners from using this system.

Table 2–8 Types of Deterioration of Structural Materials and Their Prevention[a]

Material	*Type of deterioration*	*Prevention or remedy*
Wood	Dry rot, fungus	Chemical coating or impregnation
		Ventilate structure
		Avoid wetting or contact with ground
	Insect, rodent attack	Avoid contact with ground
		Initial coating or impregnation with repellent
		Periodic treatment with repellent
	Wear	Paint or other surface finish or applied finish elements
		Overdesign for some loss
Concrete	Freezing, thawing cycles	Avoid crack development as much as possible
		Use air-entraining cement or additive
		Paint or plastic coating to avoid moisture penetration
	Corrosive gases, liquids	Use corrosion-resistant cement
		Coat or artificially harden surface
	Wear	Coat, cover, or artificially harden surface
		Overdesign for some loss
		Use metal nosings and edgings
Metals	Oxidation	Paint, galvanize, etc., to protect surface
		Use nonprogressive oxidizing alloys
		Use specified minimum thickness for safety against dimensional loss
	Corrosive gases, liquids	As for oxidation; plastic coatings effectively used
	Electrolysis	Avoid contact of dissimilar metals, especially where water is present
	Wear	Coat, cover, or harden surface
Plastic	Decomposition, chemical change	Avoid contact with noncompatible materials, especially adhesives, joint-sealing compounds, cleansing agents; avoid conditions of exposure not suitable to material (e.g., temperature, sun, corrosive atmosphere)
	Thermal extremes	Select material with tested adequacy for extremes of heat or cold anticipated
	Wear	Use surface hardening or coating

[a] Edwin H. Gaylord, Jr., and Charles H. Gaylord, *Structural Engineering Handbook* (New York: McGraw-Hill Book Company, 1968).

Insulation

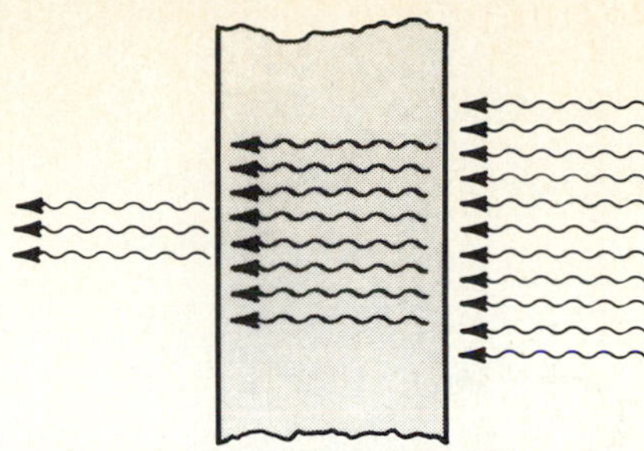

Figure 2–35. Heat Transmission through Structural Elements

The *insulation* value of structural systems and components is an ever-increasing requirement as the emphasis on energy conservation increases (Fig. 2–35). The heat resistance value (R), measured in Btu per hour per square foot per degree difference in temperature (between warmer and cooler sides) for a stated thickness of material is the industry standard for measuring the relative insulation value of building materials. The reciprocal of the R value is the U value, which measures total Btu transmitted per unit. The R value is convenient to use for adding the combined resistances of two or more layers of material (U values cannot be added directly, similar to electrical conductances). Required U values typically range from 0.13 to 0.50 for walls and 0.04 to 0.10 for roofs. Requirements may be dictated by the applicable building code or by energy-conscious design criteria.

Minimum requirements for thermal design of the exterior envelope of buildings are specified by ANSI/ASHRAE Standard 90A-80, which has been adopted by many building codes. The Standard allows two different approaches: (1) a *prescriptive* method, utilizing equations and charts for establishing maximum thermal transmittance values (U_0); and (2) a *systems analysis* method, which uses computerized building simulation programs to model the energy performance. The second method specifies no specific insulation requirements for individual building components but requires that building energy consumption be equal to or less than that of a building designed by the prescriptive method.

The prescriptive method is particularly useful for preliminary design purposes. It establishes maximum average U values (termed U_0) for the exterior envelope based on either heating requirements or cooling requirements (whichever controls). Averages for wall surfaces include wall, window, and door areas; averages for roof surfaces include roof and skylight areas. Heating requirements (insulation requirements for the heating season) are based on the "degree-day method," in which the difference between the mean daily temperature and 65°F is used as a measure of fuel consumption. Cooling requirements (insulation requirements for the cooling season) are based on either degree-days or geographical location, expressed in terms of degrees north latitude.

Since the cooling requirement calculations involve several charts and formulas, it is more convenient to base preliminary estimates of thermal requirements just on the heating criteria. U_0 values can be obtained directly using the annual degree-day curves of Fig. 2–36 and the

Figure 2–36. Degree-Day Chart (Data from *ASHRAE Systems Handbook and Product Directory,* American Society of Heating, Refrigerating and Air Conditioning Engineers, Inc., New York, 1980)

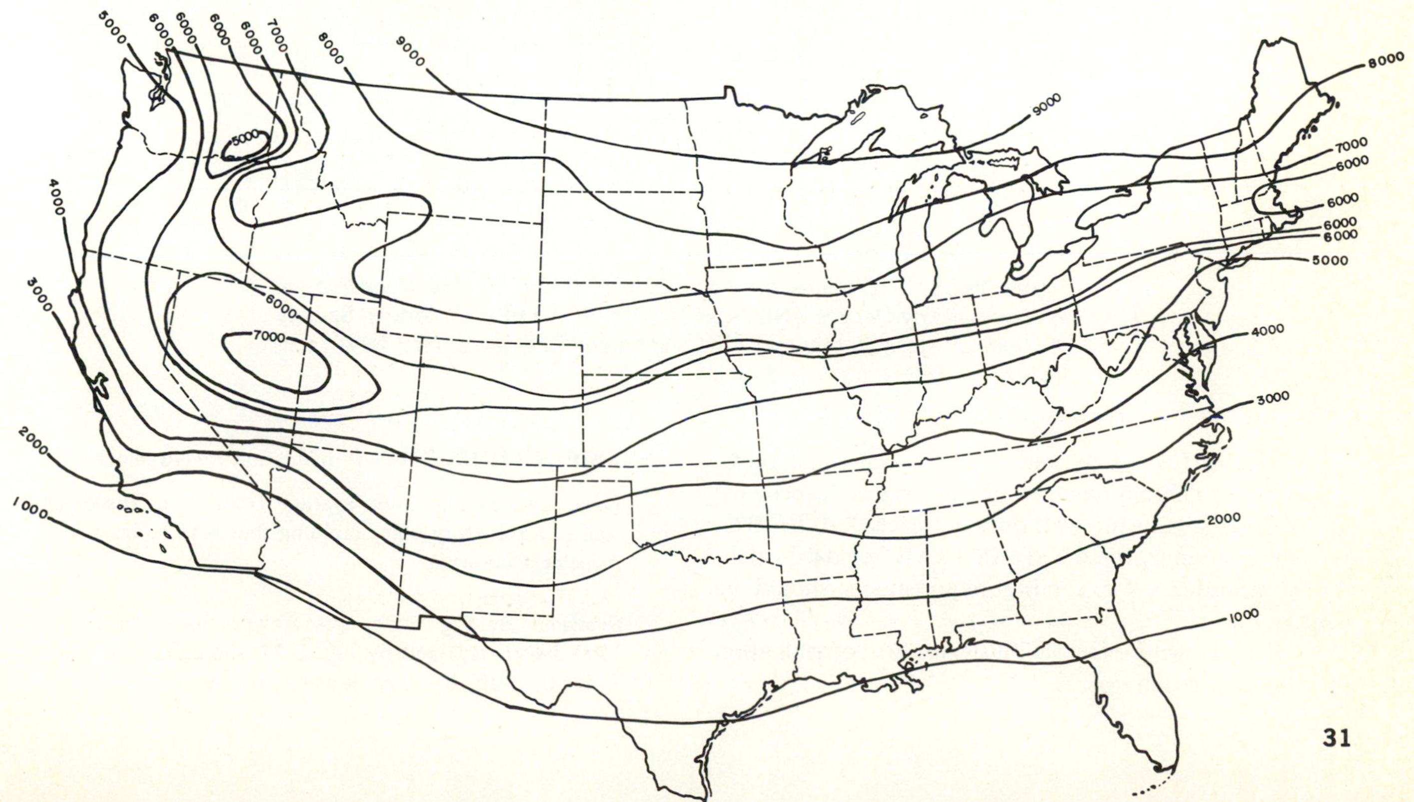

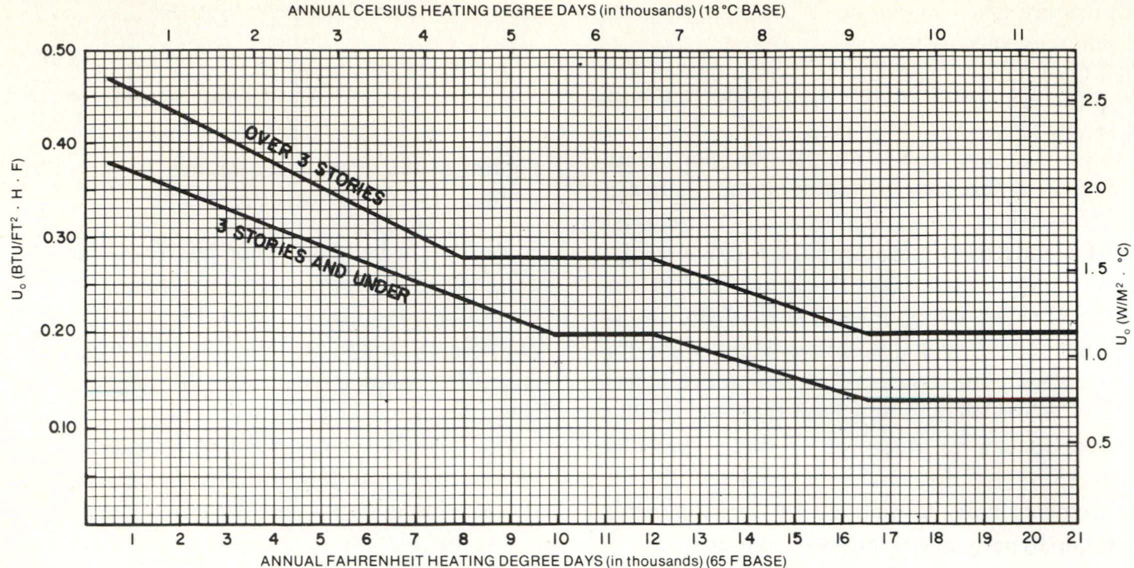

Figure 2–37. U_0 Values for Walls (From *Energy Conservation in New Building Design,* ANSI/ASHRAE/IES 90A-1980, American Society of Heating, Refrigerating and Air Conditioning Engineers, Inc., New York, 1980)

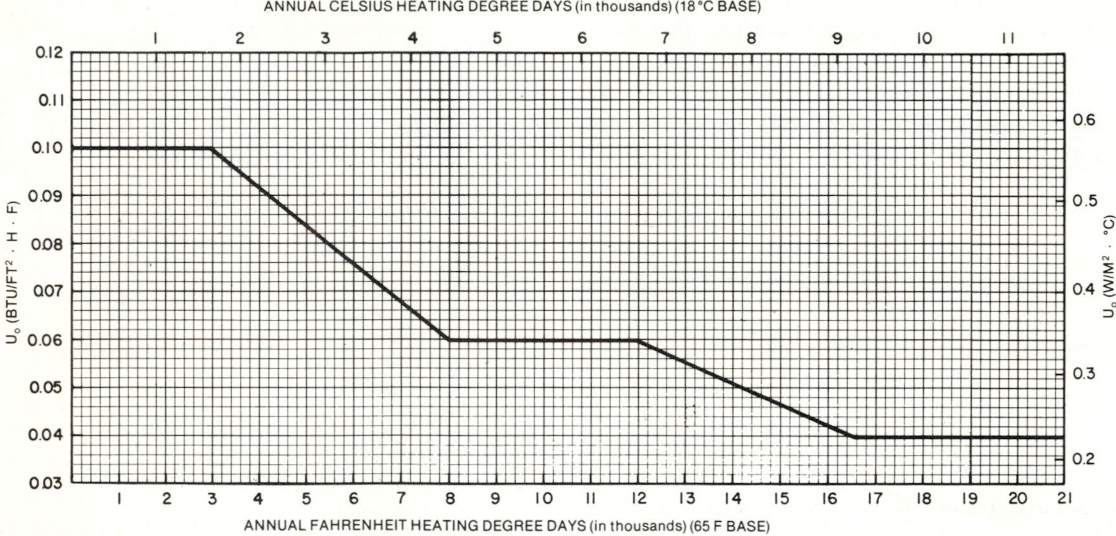

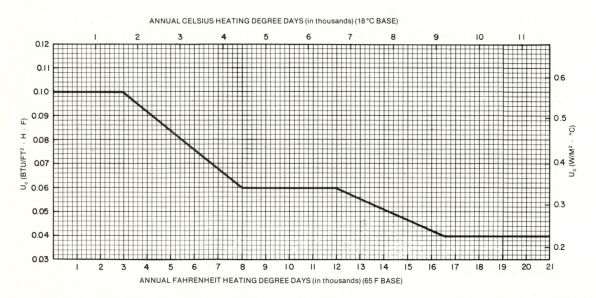

Figure 2–38. U_0 Values for Roof/Ceilings (From *Energy Conservation in New Building Design,* ANSI/ASHRAE/IES 90A-1980, American Society of Heating, Refrigerating and Air Conditioning Engineers, Inc., New York, 1980)

U_0 curves of Figs. 2–37 and 2–38 for walls and roofs, respectively. This approach is valid as long as the percentage of glass area in the wall does not exceed 20 to 30%. Beyond this range, the ANSI/ASHRAE Standard calculation procedures for cooling requirements should be used.

The following example illustrates the determination of insulation requirements:

EXAMPLE PROBLEM 2–4: Insulation Requirements

Determine the maximum average U value for the exterior wall of a one-story office building that is to be built in Los Angeles, California.

Solution By Fig. 2–36, Los Angeles has approximately 1300 degree-days and by Fig. 2–37, the maximum average U for the walls based on heating criteria is 0.365.

Formulas are given in the ANSI/ASHRAE Standard for combining various wall and roof system components to yield the required average U values for the total assembly. Actual selection of wall subsystems is covered in Chapter 5 and reference material is provided in Appendix 5B.

2-5 CONSTRUCTION REQUIREMENTS

Construction requirements describe the financial and constructability constraints imposed on structural system selection. These requirements essentially include *time, cost,* and *quality* objectives and constitute important criteria for design.

Economy

Achieving function at *optimum cost* is always an objective of structural design (Fig. 2–39). Optimum cost, however, often involves many considerations, several of them quite outside the field of structural design. Cost certainly includes the cost of the structural system: both the *initial cost* (including necessary fire protection, weather protection, cost of openings, etc.) and the *life-cycle cost.* Life-cycle cost (by definition) includes the initial cost plus the future costs of maintenance, repair, replacement, fuel costs, and investment payment. It is entirely possible for an alternative structural system to have a lower initial cost but a higher life-cycle cost when compared to a given system.

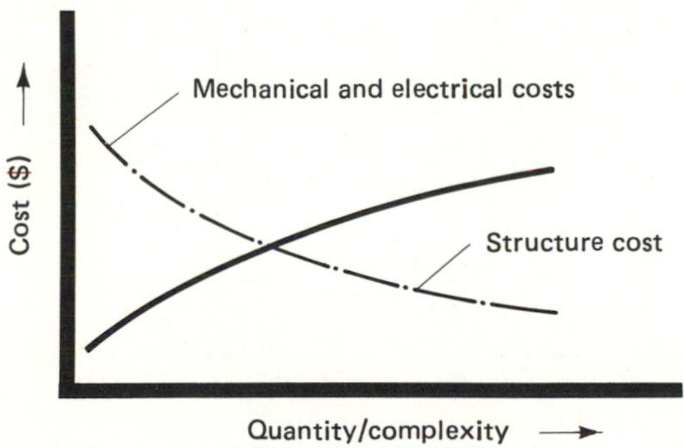

Figure 2–39. Cost Considerations

Cost also includes the cost effect on other systems. Such factors as the depth of the structural system, the insulating value of wall elements, encasement, and so on, will affect the costs of the electrical, mechanical, and architectural systems. The true cost of the structural system must reflect the differential cost effect on these other systems.

Time

The importance of *time* in the choice of structural materials and systems is often extremely important (Fig. 2–40).

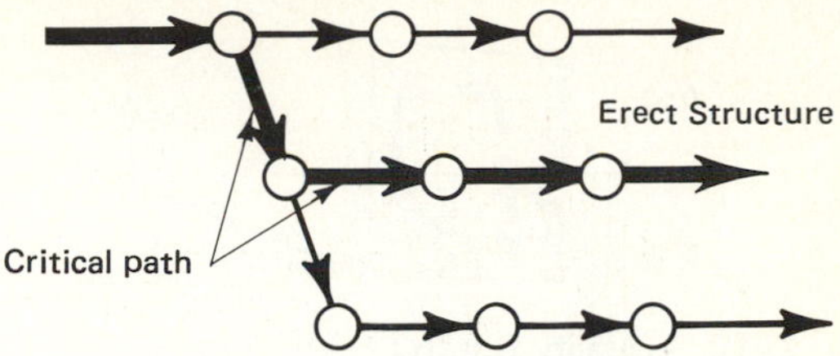

Figure 2–40. Critical Time Requirements for Structural System

Often the structural system is on the "critical path" of many projects—meaning that each month of structural time translates into an equal amount of project time. The owner's interest payments (on the construction loan) are related directly to time, as is facility startup and beginning of product sales. Thus the availability of selected structural materials, delivery times, and erection speed can all have important consequences.

Flexibility

It may be important that the structural system provide the *flexibility* to allow future rearrangement of interior partitions and equipment. This may dictate large column-free areas together with consideration of a uniform partition load in design of all floor members. It also may be important that a future addition can be added easily (Fig. 2–41). This may eliminate the possibility of using a load-bearing wall at places where future expansion is planned. It may also mean that bracing systems (for lateral stability) should be located with future expansion in mind.

Figure 2–41. Flexibility Considerations

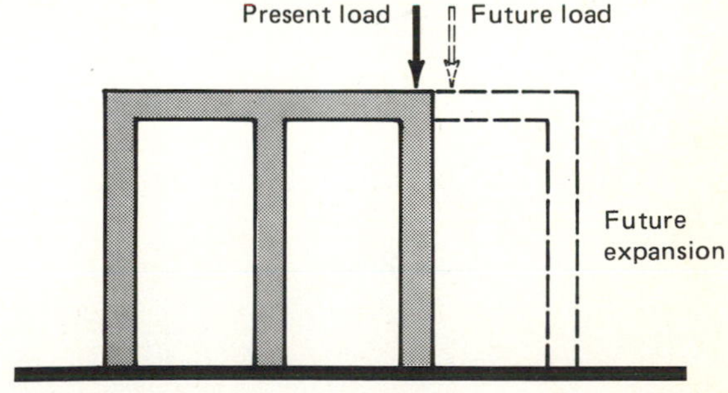

Quality Control

The skill of the labor force has a significant impact on the quality of the structural system. The structural system selected should be consistent with the expected standards of construction *quality.* Structural forms that require clockwork precision and sophistication have no place in a construction environment void of skilled craftsmen. A reinforced concrete beam in the hands of a contractor who has never used concrete before can be a disaster

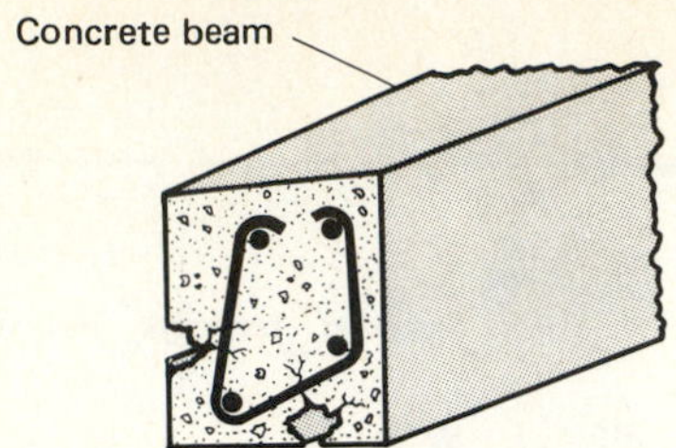

Figure 2–42. Quality Control Considerations

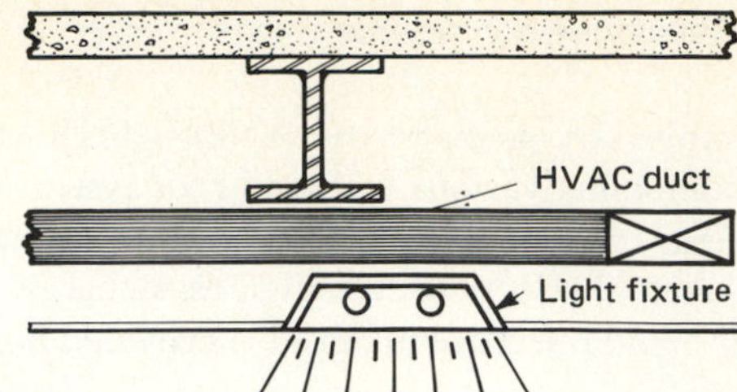

Figure 2–43. Relation of Structural System to Other Systems

(Fig. 2–42)! Erection tolerances, degree of difficulty, and familiarity with the materials can affect the degree of quality actually obtained in the field.

The problems with roof systems were mentioned earlier. It is worth repeating that complex, close tolerances, and quality-sensitive structural systems can be engineered to precision—but will fail if they cannot be built with the labor available.

Relation to Other Systems

The structural system often affects the cost of other systems, as well as the constructability of these systems. Clearances for HVAC duct, lights, piping, and so on, are directly influenced by the depth of the structural system (Fig. 2–43). Penetrations through floor and roof systems, walls, and even foundations are sometimes required

Table 2–9 Mechanical Space Requirements[a] (Operating Weights in Parentheses in Kips)

Refrigeration Machine Room

	Machine room area (sq ft)				*Heat-removal equipment dimensions (W × L × H)(ft)*	
Cooling load (tons)	*Direct expansion system*	*Reciprocating chiller*	*Centrifugal chiller*	*Absorption chiller*	*Cooling tower*	*Air-cooled condenser*
Up to 50	160	350	— (4.3)	—	7 × 6 × 7 (3.6)	7 × 16 × 7 (2.3)
50–100	160	400	480 (8.8)	420 (13.7)	12 × 8 × 8 (10.0)	8 × 16 × 6 (3.5)
100–250	—	530	620 (17.9)	640 (26.7)	19 × 13 × 9 (26.7)	—
250–500	—	—	960 (35.8)	1,100 (53.4)	26 × 14 × 13 (30.0)	—
500–750	—	—	1,160 (39.4)	1,400 (89.2)	26 × 21 × 13 (43.0)	—
750–1,000	—	—	1,500 (73.3)	1,500 (98.3)	26 × 27 × 13 (60.0)	—
1,000–1,500	—	—	1,640 (101.0)	1,680 (137.0)	26 × 42 × 13 (86.0)	—

Boiler Room

	One boiler		*Two boilers*	
Heating load (100 mbh)	*Room area (sq ft)*	*Boiler weight (lb)*	*Room area (sq ft)*	*Weight per boiler (lb)*
Up to 5	130	1,680	200	2,050
5–10	170	2,740	240	3,360
10–15	200	4,340	260	4,580
15–20	230	4,930	290	5,480
20–30	260	7,140	320	8,680
30–40	290	8,680	380	9,860
40–50	370	13,060	420	11,960

[a] Charles G. Ramsey and Harold R. Sleeper, *Architectural Graphic Standards* (New York: John Wiley & Sons, Inc., 1970), pp. 622–623.

Table 2–9 (*Continued*)

Fan Room

Supply air (1000 cfm)	Packaged unit: Room area (sq ft)	Packaged unit: Unit weight (lb)	Built-up system: Room area (sq ft) Single duct	Built-up system: Room area (sq ft) Double duct	Built-up system: Equipment weight (lb)
Up to 10	210	3,400	290	310	2,800
10–20	320	5,500	350	380	4,400
20–30	430	9,100	470	510	6,300
30–50	—	—	710	780	10,700
50–75	—	—	980	1,050	15,200
75–100	—	—	1,290	1,370	22,000
100–150	—	—	1,510	1,600	30,000

to accommodate mechanical and electrical systems. Encasement of pipes and electrical conduit within floor systems is still another common requirement. Finally, the structural system must provide space and support for equipment specified by the mechanical and electrical engineers.

The relationship of the structural system to the mechanical and electrical systems quite often establishes requirements for both space and clearances that must be considered in comparing alternate systems and in final design. The best approach for determining these requirements is *teamwork* among the structural engineer, the architect, and the mechanical engineer. Tables 2–9 and 2–10 give preliminary guidelines for these requirements.

Table 2–10[a] Mechanical Clearance Requirements

Clearance in Mechanical Equipment Rooms

Type of building and total gross floor area (sq ft)	*Percentage floor area required for mech. space*	*Min. clear height—air handling equipment (ft-in.)*	*Min. clear height—heating plant (ft-in.)*	*Min. clear height—refrigeration plant (ft-in.)*
Commercial buildings				
10,000	10	8–0	8–0	8–0
50,000	8	9–0	10–0	9–0
100,000	6	10–0	12–0	14–0
Institutional buildings				
10,000	12	8–0	10–0	8–0
50,000	10	10–0	12–0	12–0
100,000	8	10–0	14–0	14–0

Clearances in Ceilings for Low-Velocity Ductwork

Approximate area served per shaft or local equip. room (sq ft)	*With return air ductwork (in.): With lighting below ceiling*	*With return air ductwork (in.): Recessed lighting fluorescent*	*No Return Air Ductwork (in.): With lighting below ceiling*	*No Return Air Ductwork (in.): Recessed lighting fluorescent*	*Deduct if system serves interior areas only (in.)*
2,500	20–30	24–34	14–18	18–22	4–6
5,000	30–36	34–40	18–22	22–26	4–7
10,000	36–44	40–48	22–26	26–30	6–8

[a] Charles G. Ramsey and Harold R. Sleeper, *Architectural Graphic Standards* (New York: John Wiley & Sons, Inc., 1970), pp. 622–623.

2–6 SUMMARY

This chapter described the basic function of structural systems: the *enclosure and definition of space* and *support of loads.* Numerous other esthetic, serviceability, and construction requirements are quite often important and were also described.

It is essential that we establish only those requirements that are necessary. Many structures have been designed with unrealistic or unnecessary requirements—only because the engineer did not take the time to investigate fully the situation. This leads to costly designs, a luxury the construction industry can ill afford.

The next chapter continues the development of the *systems design model,* with a description of *structural loads.*

PROBLEMS

2–1. A proposed office building is to be built in the local community. The building will be rectangular in plan, two stories in height, for a total height of 24 ft (7.3 m), and will contain 10,000 sq ft (929 m^2) per floor. The mechanical room will be located on the roof and will contain air-handling units and refrigeration compressors. Establish the following information:

1. The fire rating classification and the specific protection requirements for the major structural elements. Use the local building code.
2. The deflection limits for the floor and roof system.
3. The limit for wind drift.
4. The extent to which noise and vibration should be considered in the design.
5. The need for a building expansion joint.
6. The estimated time for delivery and erection of structural steel, assuming that steel requirements are 5 lb per building square foot (239 N/m^2). The estimated time for installing a concrete system, assuming 0.50 cu ft (0.014 m^3) of concrete per building square foot. Consult local contractors and estimating systems.
7. The relative difficulty with quality control that exists in the local community. Compare steel systems and concrete systems in terms of quality of construction in the local area.
8. Thermal insulation requirements for the walls and roof/ceiling system.

Describe your conclusions (or assumptions) and why you believe they are proper.

2–2. An industrial plant will be built in the local community. The building will be divided into three sections: (1) 8000 sq ft (743 m^2) of office space; (2) 50,000 sq ft (4645 m^2) of manufacturing; and (3) 50,000 sq ft (4645 m^2) of warehouse. The building will be sprinklered. Building heights are: office 20 ft (6.1 m), manufacturing 30 ft (9.1 m), and warehouse 50 ft (15.2 m). The manufacturing process will produce high humidity and toxic fumes. Extremely sensitive equipment will control the manufacturing process. Determine the following information:

1. The fire rating classification and the specific protection requirements for the major structural elements. Use the local building code.
2. The extent to which durability should be considered in the material selection.
3. The extent to which flexibility should be considered in the design.
4. The need for building expansion joints.
5. The importance of building vibration.
6. Thermal insulation requirements for the walls and the roof/ceiling system.

Describe your conclusions (or assumptions) and why you believe they are proper.

REFERENCES

1. "Almost-Exact Formulas for Ponding of Simply-Supported Slab on Simply-Supported Girders," Paper Presented at AISC National Engineering Conference, St. Louis, Mo., 1975.
2. BERANEK, LEO L., *Noise and Vibration Control.* New York: McGraw-Hill Book Company, 1971.
3. CHANG, F.-K., "Human Response to Motions in Tall Buildings," *Journal of the Structural Division, ASCE,* Vol. 99 (June 1973), pp. 1259–1272.
4. CHAPMAN, ROBERT E., et al., *Economics of Protection against Progressive Collapse.* Washington, D.C.: National Bureau of Standards, 1974.
5. CHEN, P. W., and L. E. ROBERTSON, "Human Perception Thresholds of Horizontal Motion," *Journal of the Structural Division, ASCE,* Vol. 98 (August 1972), pp. 1680–1695.
6. COHN, BERT M., *Design of Fire-Resistive Assemblies with Steel Joists,* Technical Digest No. 4. Arlington, Va.: Steel Joist Institute.
7. *Fire-Resistant Steel Frame Construction,* 2nd ed. Washington, D.C.: American Iron and Steel Institute, 1974.

8. GALAMBOS, THEODORE V., *Vibration of Steel-Joist-Concrete Slab Floors,* Technical Digest No. 5. Arlington, Va.: Steel Joist Institute.
9. HANSEN, R. J., J. W. REED, and E. H. VANMARCKE, "Human Response to Wind-Induced Motion of Buildings," *Journal of the Structural Division, ASCE,* Vol. 99 (July 1973), pp. 1589–1605.
10. HEINZERLING, JOHN E., *Structural Design of Steel Joist Roofs to Resist Ponding Loads,* Technical Digest No. 3. Arlington, Va.: Steel Joist Institute, May 1971.
11. KHAN, F. R., and R. A. PARMELEE, "Service Criteria for Tall Buildings for Wind Loading," *Proceedings, Third International Conference on Wind Effects on Buildings and Structures,* Tokyo, September 1971, pp. 401–407.
12. KOSSOVER, D., and R. A. PARMELEE, *An Analysis of Human Response to Steady-State Vibrations.* Evanston, Ill.: Department of Civil Engineering, Northwestern University, August 1971.
13. MARINO, FRANK J., "Ponding of Two-Way Roof Systems," *AISC Engineering Journal,* July 1966.
14. MCGUIRE, W., and E. V. LEYENDECKER, *Analysis of Nonreinforced Masonry Building Response to Abnormal Loading and Resistance to Progressive Collapse.* Washington, D.C.: National Bureau of Standards, November 1974.
15. MURRAY, THOMAS M., "Design to Prevent Floor Vibrations," paper presented at AISC National Engineering Conference, St. Louis, Mo., 1975.
16. NERVI, PIER LUIGI, *Structures.* New York: F. W. Dodge Corp., 1956.
17. O'ROURKE, M. J., and R. A. PARMELEE, "Serviceability Analysis for Wind Loads on Buildings," *Proceedings, Fourth International Conference on Wind Effects on Buildings and Structures.* London, September 1975.
18. REIHER, H., and F. J. MEISTER, "The Sensitivity of the Human Body to Vibrations" (in German), *Forschung auf dem Gebeite des Ingenieurwesens,* Vol. 2, No. 11 (1931).
19. SALVADORI, M. G., *The Message of the Structure.* New York: American Society of Civil Engineers (paper presented at the ASCE Spring Convention and Exhibit, Dallas, Tex., April 25–29, 1977).
20. SOMES, NORMAN F., *Abnormal Loading on Buildings and Progressive Collapse.* Washington, D.C.: National Bureau of Standards, May 1973.
21. WILSON, FOREST, *Emerging Form in Architecture, Conversations with Lev Zetlin.* Boston: Cahners Books, 1975.
22. WISS, J. F., and R. A. PARMELEE, "Human Perception of Transient Vibrations," *Journal of the Structural Division, ASCE,* Vol. 100 (April 1974), pp. 73–87.
23. ACI Committee 531, "Concrete Masonry Structures—Design and Construction," *Journal of the American Concrete Institute, Proceedings,* Vol. 67, No. 5 (May 1970).

3

LOADS

3–1 INTRODUCTION

As we read in Chapter 2, a basic functional requirement of structures is to *carry load.* Many other requirements may also be important to the mission of the structure, and once these structural requirements have been determined, the next step in the design process is the *determination of loads,* including the anticipated response of the structural system to these loads (Fig. 3–1). It is this *feeling* for the nature of loads and their effect on structural systems that is the beginning of our understanding of structural behavior. Of particular importance during the conceptual design phase, this feeling for loads and structural response allows the designer to match structural systems (with unique response characteristics) to specific types of loadings. For example, earthquake engineers, knowing that building structures must sustain large deformations without failure to survive intense earthquakes, select ductile structural frames and connections to provide this deformation capacity; designers of tall buildings, knowing the cost premium paid for carrying lateral loads by frame action at multistory heights, select shear walls and tubular systems instead.

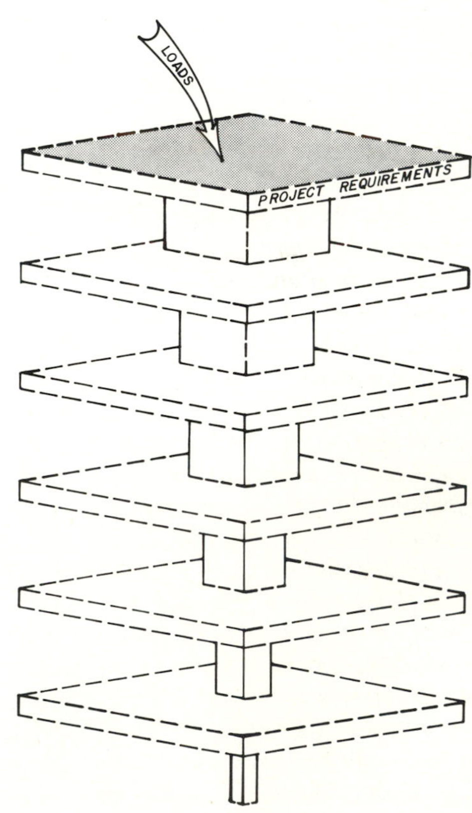

Figure 3–1. Systems Design Model

In addition to developing a feeling for structural loads and response, the structural engineer must make an accurate estimation of loads in order to assure a safe and economical design. Minimum design loads are often specified by building codes, but the designer must determine the specific loads to be used, which may be more (or less) than the code minimums. It is finally left to the judgment of the experienced engineer to determine the specific loads to be used for design of the structure.

This chapter examines the nature of structural loads: their origin, characteristics, and effect on structural systems. Procedures are presented for calculation of loads for use in both preliminary and final design, with the appendix material oriented toward final (more accurate) load determination. Our objective is to define further what the structure *must do*—before starting the process of structural assembly in Chapter 4.

3–2 CODES

Primary sources of information on loads are the various *codes* that have been established for the construction industry (Fig. 3–2). The structural engineer works with two broad classifications of codes: general building codes and design codes. *General building codes* are standards adopted by governmental bodies and regulatory agencies for the purpose of specifying minimum standards of construction. This type of code is often the source of so-called "minimum design loads" for structures. *Design codes* are more detailed technical standards for structural design, such as those published by the American Concrete Institute, American Institute of Steel Construction, and American Institute of Timber Construction. Quite often, design codes are incorporated (by reference) in the general building codes. Appendix 3A contains a listing of many of the general building codes and design-related codes likely to be encountered in practice.

Codes specify loads based on *type of use* and physical *dimensions* of the structure. The type of use obviously influences the nature and magnitude of expected loading—whether the structure supports automobiles or people, for example. The physical dimensions influence the magnitude and distribution of wind load and earthquake load and determine such factors as live load reductions.

Codes have the authority of law in many states, thereby establishing minimum design loads for most routine structures. However, most codes allow the engineer to use lower values if he or she can substantiate that the actual loads will be lower and can gain approval of the building officials. The following sections describe various types of code specified loads as well as several types of loads that are not normally included in building codes.

Figure 3–2. Codes

Classification of Buildings

The continuing trend toward rational methods of load calculation based on probability of occurrence and failure consequences has led ANSI A58.1–1982[1] to classify buildings in order of importance. Table 3–1 lists these classifications, which will be needed later for calculation of wind, snow, and earthquake loads.

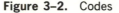

Table 3–1 Classification of Buildings and Other Structures for Wind, Snow, and Earthquake Loads[a]

Nature of occupancy	*Category*
All buildings and structures except those listed below	I
Building and structures where the primary occupancy is one in which more than 300 people congregate in one area	II
Buildings and structures designated as essential facilities including, but not limited to: hospitals and other medical facilities having surgery or emergency treatment areas; fire or rescue and police stations; primary communication facilities and disaster operation centers; power stations and other utilities required in an emergency; structures having critical national defense capabilities	III
Buildings and structures that represent a low hazard to human life or to property in the event of failure, such as agricultural buildings, certain temporary facilities, and minor storage facilities	IV

[a] *Minimum Design Loads for Buildings and Other Structures,* ANSI A58.1–1982 (New York: American National Standards Institute, 1982).

[1] *Building Code Requirements for Minimum Design Loads in Buildings and Other Structures,* A58.1–1982 (New York: American National Standards Institute, 1982).

Figure 3–3. Dead Load and Live Load

3–3 DEAD LOAD

All structures must support their own *dead load* as a minimum. The dead load of a structure is the weight of the structure itself and any permanent attachments (Fig. 3–3). Permanent attachments include fixed partitions, ceiling systems, floor coverings, roof coverings, air conditioning/heating ducts, lights, and so on. Movable partitions, furniture, and other nonpermanent loads are considered live loads. Equipment loads are normally considered a part of the uniform live load, or as a special loading case if the actual equipment loads exceed the effects of a uniform live load over the same floor area.

Less obvious dead loads are the *erection dead loads* that a structural member must carry during installation, some of which may "build in" permanent stresses in the member. For example, a composite beam (concrete slab and steel beam) that is shored until the wet concrete gains sufficient strength can carry all the dead load of the beam and the slab as a full composite member (Fig. 3–4). However, a composite beam that is unshored during the concreting operations must depend on just the flexual capacity of the steel member. The St. Louis Arch was erected as two cantilever members springing from each abutment and finally jacked apart at the top to insert the last closing segment at the crown. This obviously induced certain erection dead load stresses that had to be considered in the design. Still another illustration is the design of a masonry lintel (beam) to support wall loads based on assumed arching action of the masonry: that is, arching of masonry over the beam, resulting in a triangular load distribution (Fig. 3–5). This assumption is valid only if the lintel is shored until the masonry units attain adequate strength to act in this manner.

Precast and precast, prestressed concrete members (due to the inherent weight of the material) all experience erection dead load stresses which often significantly affect design. Termed *handling stresses,* these stresses depend on lifting sequence and points of "pickup" during removal from the forms, transportation to the job, and installation in the structure. Figure 3–6 illustrates several operations that induce handling stresses.

Dead loads are *static* loads: that is, they remain fixed and do not vary in intensity or location. Only when activated by an earthquake do dead loads take on a dynamic nature in the form of *inertial forces.* Appendix 3B contains a reference listing of dead loads.

Figure 3–4. Shored Composite Beam

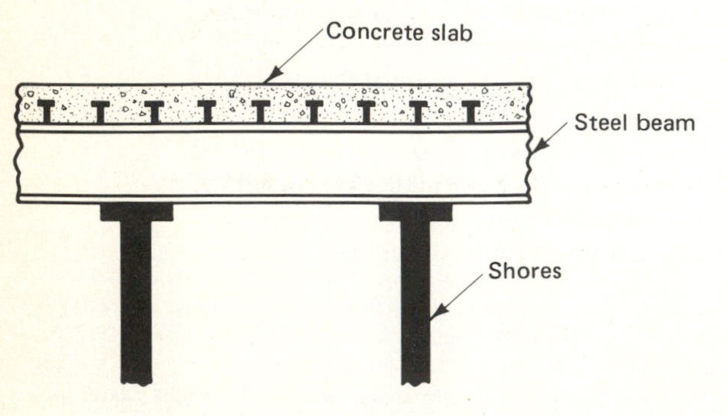

Figure 3–5. Masonry Lintel Load

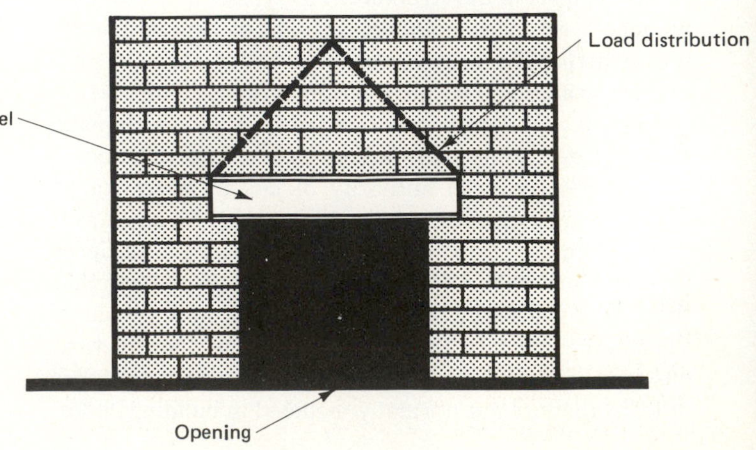

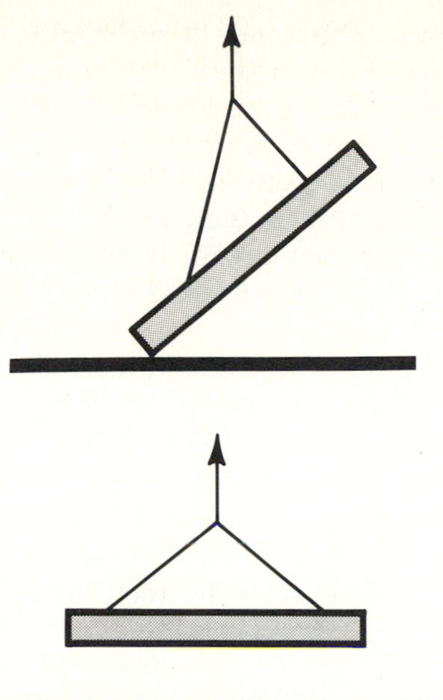

Figure 3–6. Handling Stresses

EXAMPLE PROBLEM 3–1: Dead Load

A floor spandrel (perimeter) beam supports a 4-ft (1.2-m) width of 6-in. (152-mm)-thick concrete slab, the same tributary width of acoustical tile ceiling system (suspended) and a 6-ft (1.2-m) height of exterior masonry wall composed of 4-in. (101-mm) brick and 8-in. (203-mm) concrete block (heavy weight). Calculate the dead load on the beam.

Solution Load tabulation (Use Appendix 3-B):

Concrete slab:
6 in. × $12\frac{1}{2}$ psf/in. × 4 ft = 300 lb/ft

Ceiling:
2 psf × 4 ft = 8 lb/ft

Brick wall:
39 psf × 6 ft = 234 lb/ft

Block wall:
55 psf × 6 ft = 330 lb/ft

Total load = 872 lb/ft (12,722 N/m)

3–4 LIVE LOAD

Live load is a nonpermanent load on the structure—other than wind, snow, or special types of equipment, which are usually treated separately (Fig. 3–3). Nonpermanent loads include people, furniture, storage, automobiles, minor equipment, and other items of a similar nature. *Construction loads* are a type of live load. These loads may consist of stacks of masonry or roofing material temporarily stored on portions of the structure during construction, construction equipment, formwork and supporting shores, and so on (Fig. 3–7).

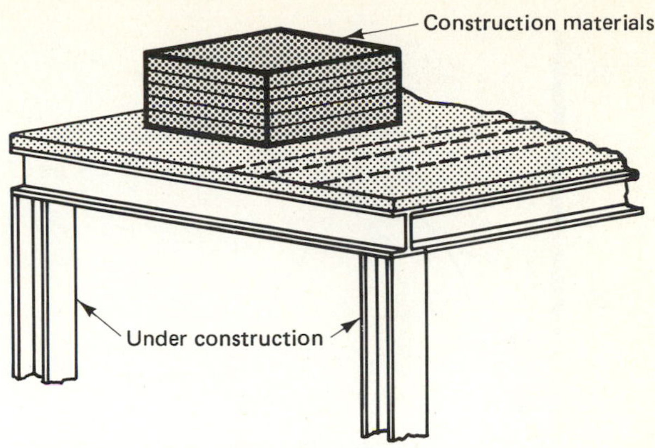

Figure 3–7. Construction Loads

Live loads are extremely variable by nature. They normally change, sometimes significantly, during a structure's lifetime as occupancy changes. Variation in the location of live load can create "patterns" of loading that accentuate bending stresses, as shown in Fig. 3–8. This *variability* in both magnitude and location is the basic characteristic of live loads that we must remember as designers.

Figure 3–8. Live Load Patterns

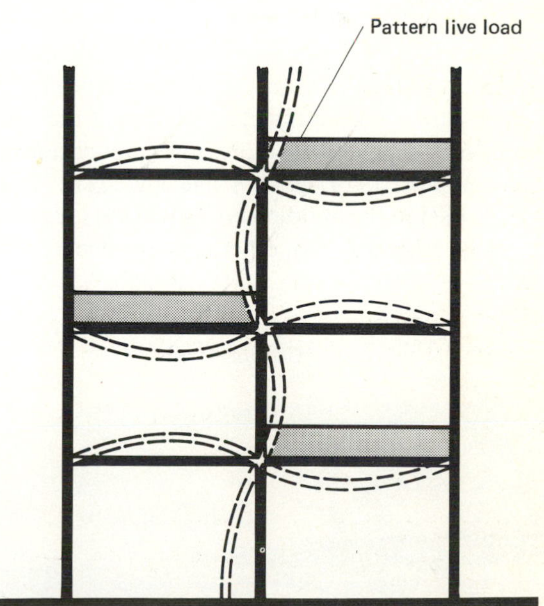

Minimum Live Loads

Code-specified *minimum loads* are based on a history of successful buildings. They incorporate some overload protection, allowance for construction loads, and serviceability considerations, such as vibration and deflection behavior. Actual live load surveys (Fig. 3–9) have demon-

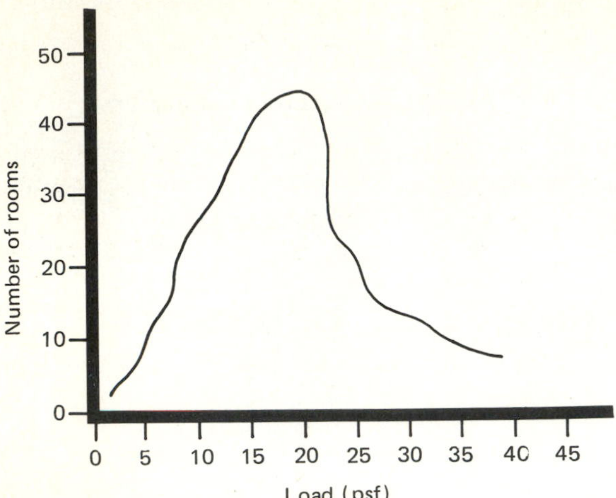

Figure 3–9. Live Load Survey

strated that actual live loads are in many cases considerably less than the minimum specified by the code. Much of this research has been done as part of an effort to establish probability based design loads and load factors—termed "probabilistic" design. As more work is done in this field, it may be expected that minimum live loads may be waived where adequate justification exists. In fact, this more rational approach to determination of live loads is already routinely done for the taller high-rise buildings by conducting wind tunnel tests to determine the design wind speed and pressure.

Building codes normally include *minimum uniform live loads* for design of roof and floor systems. These minimums are based on the classification of the structure by occupancy type: for example, *assembly* occupancy requires a 150-psf (7182-N/m²) live load. The floor live loads specified in the building codes include normal furniture, filing cabinets, and other similar items. Movable partitions may or may not be included in the uniform live load specified, with some codes specifying an additional partition load. *Concentrated live loads* (to be applied anywhere on the floor system) are frequently included to assure capacity to support safes, hand truck wheels, and automobiles. Both uniform and concentrated live loads are specified for automobile parking structures (Fig. 3–10). Appendix 3-C contains a tabulation of live loads.

Minimum roof live loads include allowance for minor snowfall and construction loads. Except in moderate climates, snow loads often exceed the minimum values and must be treated separately. Water load (due to ponding effects) may also exceed the code specified minimum roof live load.

The following minimum roof live load formula is specified by ANSI A58.1–1982:

$$L_r = 20R_1R_2 \geq 12 \qquad (3\text{–}1)$$

where L_r = design load on horizontal projection, psf

R_1 = reduction factor for tributary area:
- $R_1 = 1$ for $A_t < 200$ sq ft (18.6 m²)
- $R_1 = 1.2 - 0.001A_t$ for 200 sq ft (18.6 m²) $< A_t <$ 600 sq ft (55.7 m²)
- $R_1 = 0.6$ for $A_t \geq 600$ sq ft (55.7 m²)

R_2 = reduction for roof slope:
- $R_2 = 1$ for $F \leq 4$
- $R_2 = 1.2 - 0.05F$ for $4 < F < 12$
- $R_2 = 0.6$ for $F \geq 12$

A_t = tributary area

F = number of inches of rise per foot; for an arch or dome, the rise-to-span ratio multiplied by 32

Figure 3–10. Parking Structure Loads (Courtesy of Bethlehem Steel Corporation)

EXAMPLE PROBLEM 3–2: Live Load Determination

A hotel is to contain a restaurant and ballroom on the second floor, and the upper floors will contain guest rooms and corridors. Determine the live load to be used.

Solution Live load (use Appendix 3-C):

Restaurant	= 100 psf
Ballroom	= 100 psf
Second-floor corridor	= 100 psf
Guest rooms	= 40 psf
Guest corridors	= 40 psf

Live Load Reduction

Most codes allow a *reduction in live loads* for member design based on the total tributary area of floor supported. The basis for live load reduction is the unlikely possibility that full basic live load will occur simultaneously throughout the structure—and even if it did, the reduction formulas limit overstress to 30% on only a small portion of the total structure. Live load reduction is thus a *probability*-based approach.

According to ANSI A58.1–1982, members having an *influence area* of 400 sq ft or more may be designed for a reduced live load calculated by the following equation:

$$L = L_0\left(0.25 + \frac{15}{\sqrt{A_I}}\right) \qquad (3\text{–}2)$$

where L = reduced design live load per square foot of area supported by the member

L_0 = unreduced design live load per square foot of area supported by the member

A_I = influence area (square feet), equal to four times the tributary area for a column, to two times the tributary area for a beam, and to the panel area for a two-way slab

Limitations. The ANSI Code limits the reduced design live load to not less than 50% of the unreduced load for members supporting one floor, nor less than 40% for members supporting more than one floor.

For live loads that exceed 100 psf, and in garages for passenger cars only, design live loads on members supporting more than one floor are limited to a 20% reduction; no reduction is allowed otherwise. For live loads of 100 psf or less, no reduction is allowed for the following types of occupancy/use:

- Public assembly
- Garages
- One-way slabs
- Roofs (except as allowed by Eq. 3–1)

EXAMPLE PROBLEM 3–3: Live Load Reduction

A 10-story office building will consist of columns spaced 25 ft (7.6 m) on center in each direction. Assume a 20-psf (958-N/m²) roof live load and a 50-psf (2394-N/m²) floor live load. Calculate the reduced live load to be carried by a typical interior column at the ground level.

Solution

1. Roof live load (no reductions):

$$R_{LL} = 20.0 \text{ psf} \times 25 \text{ ft} \times 25 \text{ ft} = 12{,}500 \text{ lb } (55{,}600 \text{ N})$$

2. Floor live load:

$$\text{Area} = 9 \text{ floors} \times 25 \text{ ft} \times 25 \text{ ft} = 5627 \text{ sq ft}$$

$$A_I = 4 \times 5625 \text{ sq ft} = 22{,}500 \text{ sq ft}$$

By Eq. 3–1:

$$L = 50 \text{ psf}\left(0.25 + \frac{15}{\sqrt{22{,}500}}\right) = 17.5 \text{ psf}$$

$$\text{Maximum reduction} = 0.40(50) = 20 \text{ psf}$$

Therefore,

$$F_{LL} = 20.0 \text{ psf} \times 5625 \text{ sq ft} = 112{,}500 \text{ lb } (500{,}400 \text{ N})$$

3. Total live load:

$$\text{TL} = 12{,}500 + 112{,}500 = 125{,}000 \text{ lb } (556{,}000 \text{ N})$$

3–5 SNOW LOAD

When meteorological conditions of temperature and humidity cause condensation to fall in the form of snow, deposits accumulate on the roofs of structures and add gravity load (Fig. 3–11). The moisture content of the snow and the density of deposit can vary; thus no two snows are exactly alike in terms of the load imposed on a structure. Measurements in terms of equivalent inches of water are normally used to quantify the exact weight of a particular snow (Fig. 3–12). Wind often accompanies snowfall and can have a significant influence on the accumulated weight imposed on building structures by snowdrifts. Snow load normally increases or decreases depending on the following conditions:

- Decreased load due to slide-off of snow on roofs with slopes exceeding 30°

Figure 3–11. Snow Load (Courtesy of Robert Redfield and U.S. Army Cold Regions Research and Engineering Lab)

- Decreased load due to roofs having a clear exposure in windswept areas
- Increased load due to nonuniform accumulation on pitched or curved roofs
- Increased load in the valleys formed by multiple series roofs
- Increased load due to snow sliding off sloping roof areas onto adjacent roof areas
- Increased load on the lower levels of multilevel roofs and on roof areas adjacent to projections, such as penthouses, cooling towers, and parapet walls, due to drifting snow

Snow is a variable load and the designer must anticipate conditions that could cause drifts and other concentrated accumulations on structures. Figure 3–13 shows a roof collapse due to snow accumulation. The treatment of snow load as a uniform load will not suffice in many cases. As a result, ANSI A58.1–1982 specifies three snow loading conditions: *uniform load* imposed by the depth of snow, *unbalanced load,* and *drift load,* the latter two caused by wind conditions and particular building configurations.

MARCH 1978
GREENVILLE-SPARTANBURG AP
NATIONAL WEATHER SERVICE OFC
GREER, S. C.

Local Climatological Data

MONTHLY SUMMARY

LATITUDE 34° 54′ N LONGITUDE 82° 13′ W ELEVATION (GROUND) 957 FT. STANDARD TIME USED: EASTERN WBAN #03870

DATE	TEMPERATURE °F: MAXIMUM	MINIMUM	AVERAGE	DEPARTURE FROM NORMAL	AVERAGE DEW POINT	DEGREE DAYS BASE 65°: HEATING (SEASON BEGINS WITH JULY)	COOLING (SEASON BEGINS WITH JAN.)	WEATHER TYPES ON DATES OF OCCURRENCE (1 FOG, 2 HEAVY FOG, 3 THUNDERSTORM, 4 ICE PELLETS, 5 HAIL, 6 GLAZE, 7 DUSTSTORM, 8 SMOKE, HAZE, 9 BLOWING SNOW)	SNOW, ICE PELLETS OR ICE ON GROUND AT 07AM IN.	PRECIPITATION: WATER EQUIVALENT IN	SNOW, ICE PELLETS IN.	AVG. STATION PRESSURE IN. ELEV. 971 FEET M.S.L.	WIND: RESULTANT DIR.	RESULTANT SPEED M.P.H.	AVERAGE SPEED M.P.H.	FASTEST MILE: SPEED M.P.H.	DIRECTION	SUNSHINE: MINUTES	PERCENT OF POSSIBLE	SKY COVER TENTHS: SUNRISE TO SUNSET	MIDNIGHT TO MIDNIGHT	DATE
1	2	3	4	5	6	7A	7B	8	9	10	11	12	13	14	15	16	17	18	19	20	21	22
1	52	35	44	-3	33	21	0		0	0	0	28.93	35	6.0	8.9	19	N	514	75	4	5	1
2	35	29	32	-15	22	33	0	1 4 6	0	.37	2.8	29.13	06	7.9	8.2	12	E	0	0	10	9	2
3	36	29	33	-14	30	32	0	2 6	3	.38	0	28.70	36	4.2	7.2	22	NW	0	0	10	9	3
4	39	29	34	-13	17	31	0		1	0	0	28.90	35	12.4	12.7	19	N	574	83	4	4	4
5	47	17*	32*	-16	16	33	0		0	0	0	29.16	21	4.1	5.5	13	SW	694	100	0	0	5
6	60	30	45	-3	24	20	0		0	0	0	29.11	22	6.0	6.3	16	SW	696	100	2	3	6
7	53	38	46	-2	32	19	0	1	0	.18	0	29.14	03	6.7	7.3	13	NE	427	61	9	9	7
8	41	35	38	-10	34	27	0	1	0	.23	0	29.07	03	13.1	13.2	17	NE	0	0	10	10	8
9	40	35	38	-11	35	27	0	1 8	0	1.07	0	28.91	04	9.2	9.6	17	NE	0	0	10	10	9
10	58	38	48	-1	38	17	0	1	0	.23	0	28.64	32	5.4	8.6	23	NW	463	66	7	6	10
11	64	32	48	-1	34	17	0		0	0	0	28.91	21	6.3	6.6	17	SW	707	100	6	4	11
12	65	45	55	5	48	10	0	3	0	.37	0	28.91	22	6.6	8.2	19	SW	489	69	6	7	12
13	64	44	54	4	46	11	0		0	0	0	29.07	04	5.2	5.8	10	E	502	71	10	9	13
14	72	54	63*	13	52	2	0	1	0	.39	0	28.77	20	10.5	12.2	22	SW	525	71	4	5	14
15	73	44	59	9	40	6	0		0	0	0	28.90	22	4.5	5.0	18	SW	613	86	7	5	15
16	73	40	57	6	36	8	0		0	0	0	28.84	33	6.2	7.8	23	SW	653	91	3	4	16
17	49	29	39	-12	24	26	0		0	T	0	29.12	28	6.5	8.6	26	W	573	80	5	3	17
18	49	28	39	-12	20	26	0		0	0	0	29.33	28	3.7	6.8	14	SW	723	100	0	1	18
19	72	38	55	3	30	10	0		0	0	0	29.21	21	14.3	14.8	23	SW	725	100	2	1	19
20	75	42	59	7	38	6	0		0	0	0	29.18	21	7.6	8.1	15	SW	659	91	8	7	20
21	73	51	62	10	45	3	0		0	.19	0	29.02	23	10.0	12.7	27	SW	536	73	7	8	21
22	67	45	56	3	44	9	0	1	0	0	0	29.00	01	4.0	5.8	14	NE	731	100	0	0	22
23	75	43	59	6	41	6	0		0	0	0	29.00	23	.3	4.3	12	SW	712	93	2	1	23
24	76	46	61	8	50	4	0	1 3 8	0	.04	0	28.98	16	3.9	4.3	13	S	494	67	5	6	24
25	63	40	52	-2	46	13	0	1 3	0	2.19	0	29.05	03	13.1	13.4	23	NE	0	0	10	10	25
26	49	40	45	-9	41	20	0	1 3	0	.43	0	28.87	02	6.3	9.5	18	NE	268	36	10	10	26
27	54	43	49	-5	37	16	0	1	0	.02	0	28.85	30	5.2	7.9	12	N	286	39	7	6	27
28	67	34	51	-4	38	14	0		0	0	0	29.04	25	4.5	5.5	11	S	743	100	0	0	28
29	74	42	58	3	41	7	0		0	0	0	29.05	29	1.8	5.0	10	N	624	83	2	2	29
30	70	48	59	4	33	6	0		0	0	0	29.12	01	2.7	5.8	14	N	718	96	1	1	30
31	77*	40	59	3	39	6	0		0	0	0	29.08	19	3.1	6.6	13	SW	752	100	0	0	31
	SUM	SUM				TOTAL	TOTAL			TOTAL	TOTAL	FOR THE MONTH:						TOTAL	%	SUM	SUM	
	1862	1183				486	0	NUMBER OF DAYS		6.09	2.8	29.00	31	1.4	8.1	27	SW	15401	FOR	161	155	
	AVG.	AVG.	AVG.	DEP.	AVG.	DEP.	DEP.	PRECIPITATION		DEP.						DATE: 21		POSSIBLE	MONTH	AVG.	AVG.	
	60.1	38.2	49.2	-1.7	36	36	-13	≥ .01 INCH	13	0.76								22274	69	5.2	5.0	

NUMBER OF DAYS				SEASON TO DATE		
MAXIMUM TEMP.		MINIMUM TEMP.		TOTAL	TOTAL	
≥ 90°	≤ 32°	≤ 32°	≤ 0°	3378	0	
0	0	8	0	DEP.	DEP.	
				388	-13	

NUMBER OF DAYS	
SNOW, ICE PELLETS ≥ 1.0 INCH	1
THUNDERSTORMS	4
HEAVY FOG	1
CLEAR 11 PARTLY CLOUDY 11 CLOUDY 9	

GREATEST IN 24 HOURS AND DATES				GREATEST DEPTH ON GROUND OF SNOW, ICE PELLETS OR ICE AND DATE	
PRECIPITATION		SNOW, ICE PELLETS			
2.23	24–25	2.8	2	3	3+

Figure 3–12. Weather Bureau Data (Courtesy of National Climatic Center, National Oceanic and Atmospheric Administration)

Figure 3–13. Roof Collapse Under Snow Load (Courtesy of Wayne Tobiasson and U.S. Army Cold Regions Research and Engineering Lab)

Uniform Load

Minimum roof live loads specified by building codes may not be adequate to cover the anticipated snow loading conditions for a given area. The *uniform snow load* effect may be determined by reference to charts which show the maximum historical snow load for particular areas of the country (Fig. 3–14). The local building codes may also specify minimum snow loads. Freezing of the melted water around roof drains can prevent further migration of melted snow and result in additional accumulation when snowfall resumes (Fig. 3–15).

Flat Roofs. The uniform snow load on *flat roofs* in the contiguous United States may be determined by the following equation:

$$P_f = 0.7 C_e C_t I \rho_g \qquad (3\text{–}3)$$

where P_f = flat roof snow load, psf
C_e = exposure factor (Table 3–2)
C_t = thermal factor (Table 3–3)
I = importance factor (Table 3–4)
ρ_g = ground snow load, psf (from Fig. 3–14)

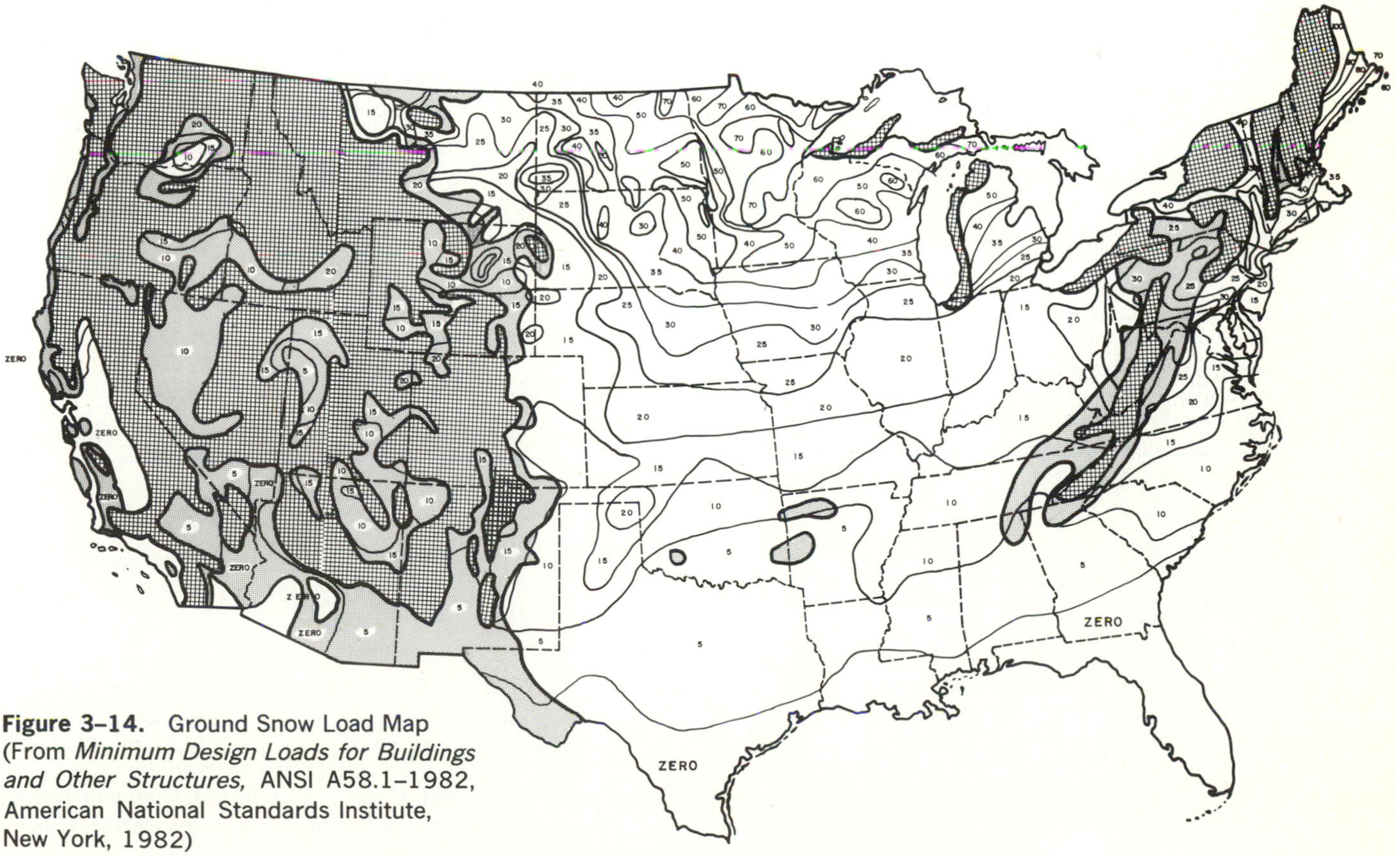

Figure 3–14. Ground Snow Load Map (From *Minimum Design Loads for Buildings and Other Structures,* ANSI A58.1–1982, American National Standards Institute, New York, 1982)

- Darker shading indicates extreme local variations that preclude mapping.
- Lighter shading indicates that zoned value is not appropriate for certain geographic settings, such as high country.

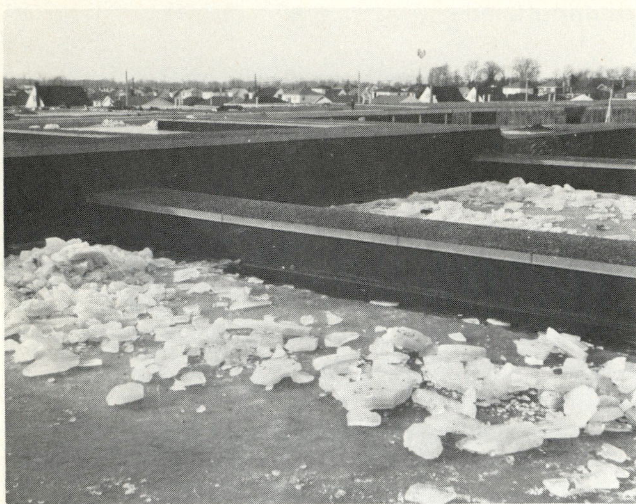

Figure 3–15. Ice Collection on Roof (Courtesy of Division of Building Research/National Research Council of Canada)

Table 3–2 Exposure Factor, C_e[a]

Siting of structure[b]	C_e
A. Windy areas with roof exposed on all sides with no shelter[c] afforded by terrain, higher structures, or trees	0.8
B. Windy areas with little shelter[c] available	0.9
C. *Normal siting.* Snow removed by wind cannot be relied on to reduce roof loads because of other higher structures or trees nearby	1.0
D. Areas that do not experience much wind and where terrain, other structures, or trees shelter the roof	1.1
E. Densely forested areas that experience little wind, with roof located tight in among conifers (cone bearing trees, mostly evergreens)	1.2

[a] *Minimum Design Loads for Buildings and Other Structures.* ANSI A58.1–1982 (New York: American National Standards Institute, 1982).

[b] These conditions should be representative of those that are likely to exist during the life of the structure. Roofs that contain several large pieces of mechanical equipment or other obstructions do not qualify for siting category A.

[c] Obstructions within a distance of $10h_0$ provide "shelter," where h_0 is the height of the obstruction above the roof level.

Table 3–3 Thermal Factor, C_t[a]

Thermal condition[b]	C_t
A. Heated structure	1.0
B. Structure kept just above freezing	1.1
C. Unheated structure	1.2

[a] *Minimum Design Loads for Buildings and Other Structures,* ANSI A58.1–1982 (New York: American National Standards Institute, 1982).

[b] These conditions should be representative of those that are likely to exist during the life of the structure.

Table 3–4 Importance Factor, *I*, for Snow Loads[a]

Category[b]	*Importance factor, I*
I	1.0
II	1.1
III	1.2
IV	0.8

[a] *Minimum Design Loads for Buildings and Other Structures,* ANSI A58.1–1982 (New York: American National Standards Institute, 1982).

[b] As defined in Table 3–1.

Minimum values. For locations where the ground snow load, ρ_g, is 20 psf (957 N/m)² or less, the flat roof snow load, P_f, shall not be less than the ground snow load multiplied by the importance factor. For locations where the ground snow load exceeds 20 psf (957 N/m)², the flat roof snow load shall not be less than 20 psf (957 N/m)² multiplied by the importance factor.

Sloped Roofs. The uniform snow load on sloping roofs may be determined by the following equation:

$$P_s = C_s P_f \tag{3–4}$$

where P_s = sloped roof snow load on the horizontal projection, psf

C_s = roof slope factor (from Fig. 3–16)

Portions of curved roofs having a slope exceeding 70° may be considered free from snow load. The roof slope factor for curved roofs may be determined by basing the slope on the vertical angle from the "eave" to the crown.

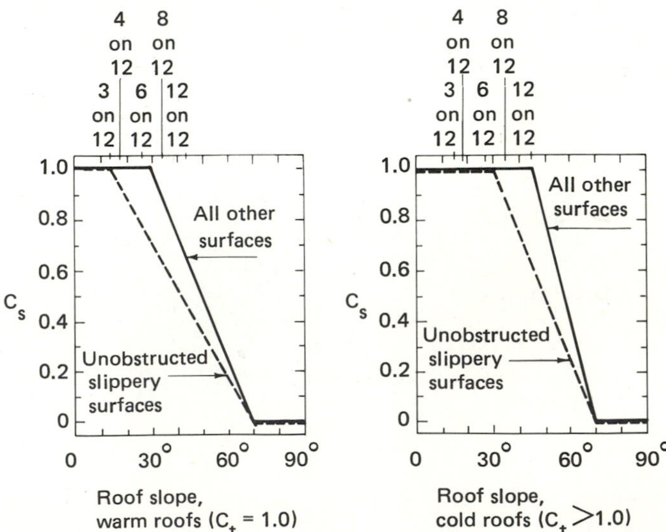

Figure 3–16. Roof Slope Factor (From *Minimum Design Loads for Buildings and Other Structures,* ANSI A58.1–1982, American National Standards Institute, New York, 1982)

Unbalanced Load

Unbalanced snow loading conditions may occur on flat roofs (Fig. 3–17) due to wind action and roof drainage patterns. "Pattern" placement of the uniform snow load (for calculating maximum moments) provides for this contingency.

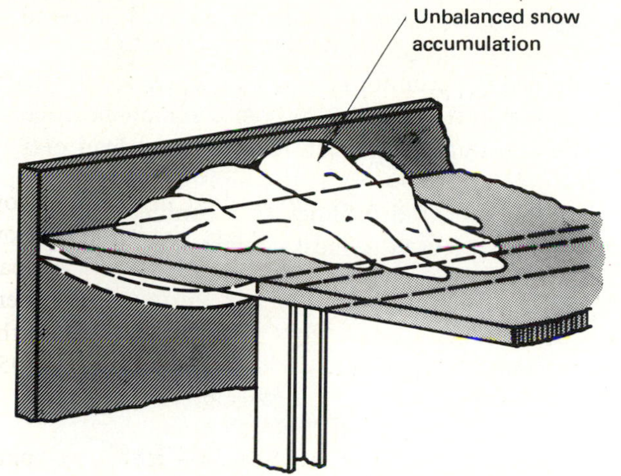

Figure 3–17. Unbalanced Snow Loading

Other unbalanced loads occur on the leeward slope of hip and gable roofs, curved roofs, and in the valleys of folded plate, sawtooth, and barrel vault roofs. Figure 3–18 illustrates the unbalanced loading parameters for hip and gable roofs. Similar procedures for other types of roofs are contained in the ANSI Standard.

Figure 3–18. Unbalanced Load on Hip and Cable Roof

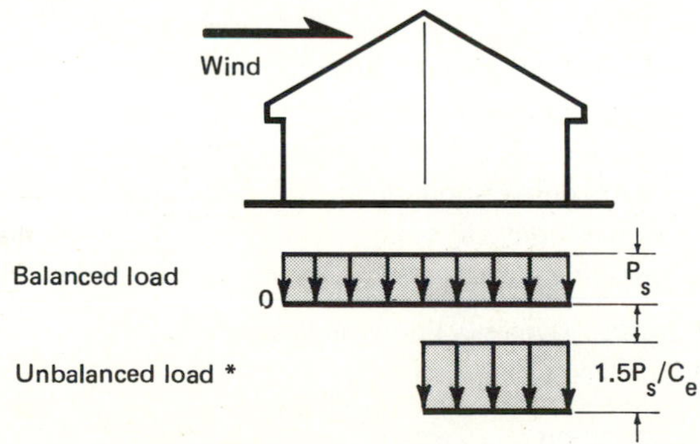

*If slope <15° or >70° unbalanced loads need not be considered.

Drift Loads

Drift loads caused by snow result when wind-driven snow accumulates next to vertical projections of the structure or adjacent buildings (Fig. 3–19). Size of the drifts depends on the ground snow load and the height and length

Figure 3–19. Snow Drift (Courtesy of Division of Building Research/National Research Council of Canada)

of the projection. Sliding snow from upper roof areas may add to the total drift load as well.

Figure 3–20 shows the geometry involved in calculating drift loads and the maximum intensity of the drift surcharge load is given by the following related equation:

$$\rho_d = h_d \gamma \qquad (3\text{–}5)$$

where ρ_d = maximum drift surcharge
γ = density of snow (Table 3–5)
h_d = drift height = $2I\rho_g/C_e\gamma$, but not greater than h_c
h_c = clear height from top of the balanced load on the lower roof to the closest point on the adjacent upper roof; it is assumed that all snow has blown off the upper roof near its eave
h_b = height of balanced snow load (i.e., balanced snow load, P_f or P_s, divided by appropriate density in Table 3–5)
l = length of drift (i.e., the common length of the upper and lower roofs)
w = width of drift

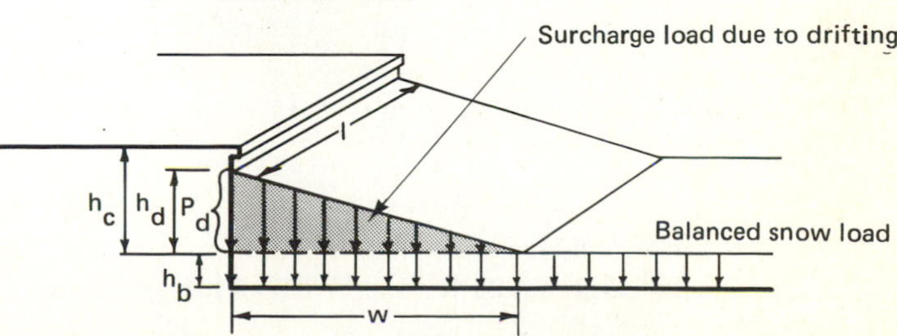

Figure 3–20. Drift Load Calculation (From *Minimum Design Loads for Buildings and Other Structures,* ANSI A58.1–1982. American National Standards Institute, New York, 1982)

- if $l \leq 50$ ft (15.2 m), $w = 3h_d$
- if $l > 50$ ft (15.2 m), $w = 4h_d$
- w is never less than 10 ft (3.05 m)
- if w exceeds the width of the lower roof, the drift shall be truncated at the far edge of the roof, not reduced to zero there

Table 3–5 Snow Densities[a]

Ground snow load, ρ_g (psf)	Drift density, γ (*pcf*)
1–10	Drifting not considered
11–30	15
31–60	20
Greater than 60	25

[a] *Minimum Design Loads for Buildings and Other Structures,* ANSI A58.1–1982 (New York: American National Standards Institute, 1982).

The drift equation is also applicable to higher roofs of adjacent buildings or terrain features that are within 20 ft (6.09 m) of the lower roof. A correction fraction equal to $(20 - s)/20$ shall be applied to the maximum drift load to account for the effect of spacing (where s is the spacing in feet).

EXAMPLE PROBLEM 3–4: Snow Load

Determine the snow load to be used in design of a roof structure in Winston-Salem, North Carolina. One edge of the roof is adjacent (for a distance > 50 ft) to a higher building (flat roof) which projects 15 ft (4.6 m) above the roof to be designed. Assume that normal siting and a thermal factor, C_t, of 1.1 and an importance factor, I, of 1.0.

Solution

1. From Fig. 3–14, ground snow in Winston-Salem is 20 psf (957 N/m²).
2. Uniform load for roof design:

$$P_f = 0.7\ C_e C_t I \rho_g$$
$$= 0.7(1.0)(1.1)(1.0)(20 \text{ psf}) = 15.4 \text{ psf } (737 \text{ N/m}^2)$$

3. Drift load: from Table 3–5, $\gamma = 15$ pcf (240 kg/m³).

$$h_b = \frac{15.4 \text{ psf}}{15 \text{ pcf}} = 1.03 \text{ ft } (0.314 \text{ m})$$

$$h_c = 15 \text{ ft} - 1.03 \text{ ft} = 13.97 \text{ ft } (4.3 \text{ m})$$

$$h_d = \frac{2I\rho_g}{C_e\gamma} = \frac{2(1.0)(20.0)}{(1.0)(15)} = 2.67 \text{ ft } (0.81 \text{ m})$$

$$\rho_d = h_d\gamma = 2.67(15) = 40.05 \text{ psf } (1917 \text{ N/m}^2)$$

$$w = 4h_d = 4(2.67 \text{ ft}) = 10.68 \text{ ft } (3.25 \text{ m})$$

3–6 EQUIPMENT LOAD

Equipment loads are exerted by items of equipment supported by the structure. Equipment loads fall into three categories: *static, dynamic,* and *moving.* Examples of *static* equipment loads include mechanical equipment, office copiers, vending machines, and so on. If significant in magnitude, equipment loads are generally specified as a separate loading case, although in many cases they can be absorbed in the uniform live load adopted for the structure as a whole (Fig. 3–21). It is common practice to include a "future equipment" concentrated load of 1 to 2 kips (4448 to 8896 N) at the midspan of beams and girders for industrial facilities as a contingency in case these beams are used to hoist equipment.

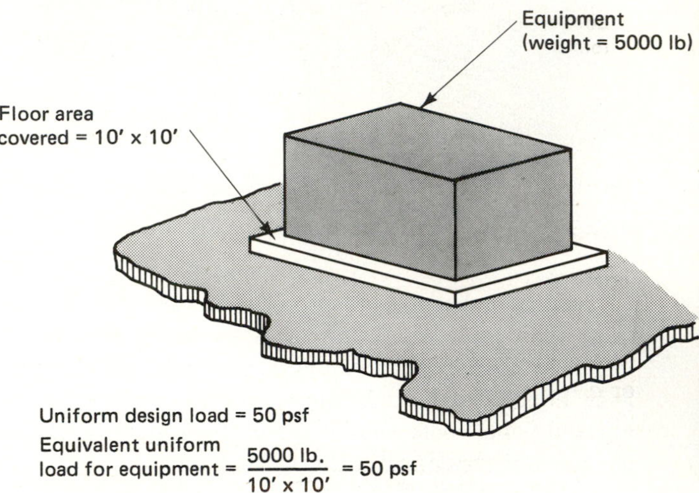

Figure 3–21. Equipment Load Compared to Design Live Load

Dynamic equipment loads are exerted by machinery with moving parts that transmit dynamic loads to the structure either through reciprocating (horizontal or vertical) motion or rotation of an eccentric mass (Fig. 3–22). Examples include weave looms, vibrating screens, and reciprocating compressors. The magnitude and fre-

Figure 3–22. Dynamic Equipment Loads

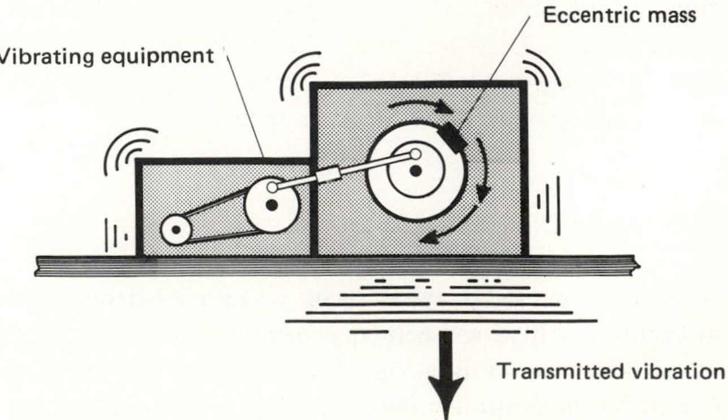

quency of the dynamic force and the natural frequency and damping of the structure are the critical factors involved in determining the effect of dynamic equipment loads on structural systems.

Moving equipment loads include cranes, monorails, and forklift trucks. Moving loads are characterized by the potential for application of the load at any point on the structure, thereby dictating a "worst-case" type of analysis. Moving loads associated with cranes and monorails normally have three directional components of force associated with them: gravity loading (including impact), longitudinal thrust, and horizontal thrust (Fig. 3–23).

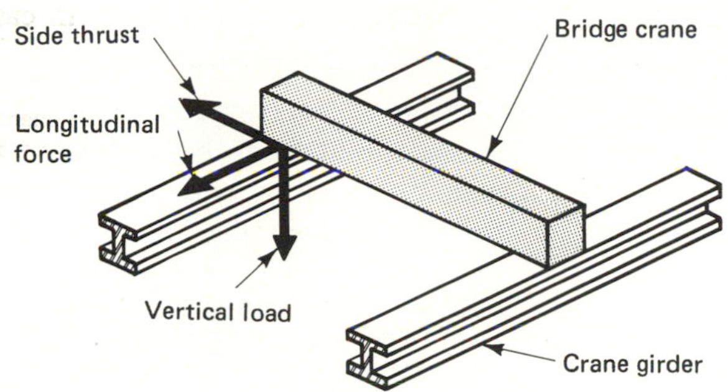

Figure 3–23. Crane Loading

The AISC Specification specifies the following loads for cranes:

1. Vertical impact
   a. 25% of maximum crane wheel loads for traveling cab-operated cranes
   b. 10% of maximum crane wheel loads for pendant-operated traveling cranes.
2. Side thrust: 20% of the sum of the weights of the lifted load and trolley
3. Longitudinal force: 10% of the maximum wheel loads

Several load combinations are normally involved with crane loadings—that combine the effects of dead load, live load, and wind load with the crane load.

Forklift trucks are characterized by the concentration of load on the front axle when loads are lifted and transported. The wheel load of a forklift truck can be significant and has led in the past to many cases of cracking and deterioration of concrete slabs.

3–7 VEHICULAR LOAD

Vehicular loading is primarily important for bridge-type structures. Although parking structures and ramps involve vehicular loading, its effect is normally covered by a specified uniform live load coupled with a concentrated live load (to be placed anywhere on the structure). Highway bridge loadings are specified by the American Association of State Highway and Transportation Officials (AASHTO) and railroad bridge loadings by the American Railway Engineering Association (AREA). These loads normally include a *gravity* component, a *longitudinal thrust* component, and *impact. Centrifugal* forces are also specified for curved bridge structures.

Highway Loading

Highway loading (for the gravity component) consists of either lane loads or truck loads. Specific code provisions define which type of loading will be used and under what circumstances. Figure 3–24 illustrates the lane loading and Fig. 3–25 illustrates the HS truck loading. Placement of the loads on the bridge structure for analysis must produce the maximum stress for each design consideration: maximum positive bending moment, maximum negative bending moment, maximum shear, and maximum support reaction. Influence lines for the structure are a useful design tool for determining the locations that produce these maximum stresses.

Figure 3–24. Highway Lane Loading (From *Standard Specifications for Highway Bridges,* 12th Ed., Washington, D.C., The American Association of State Highway and Transportation Officials, copyright 1977. Used by permission)

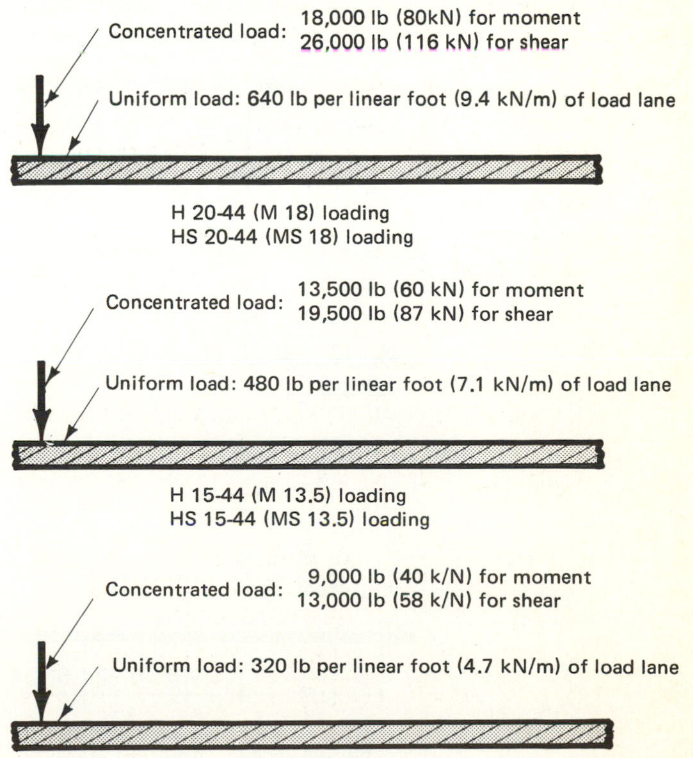

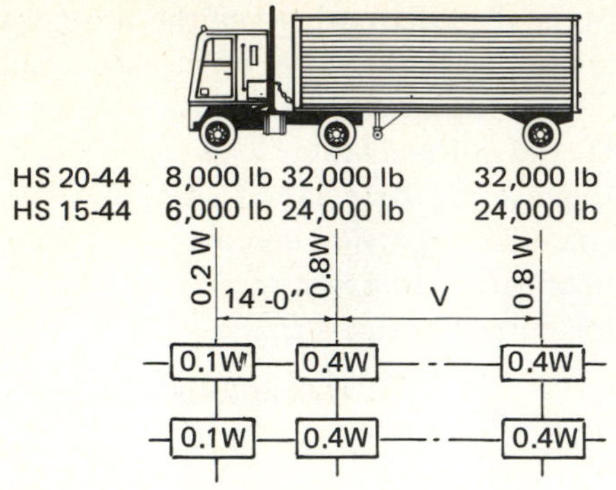

W = Combined weight on the first two axles, which is the same for the corresponding H truck.

V = Variable spacing — 14 to 30 ft. Spacing to be used is that which produces maximum stresses.

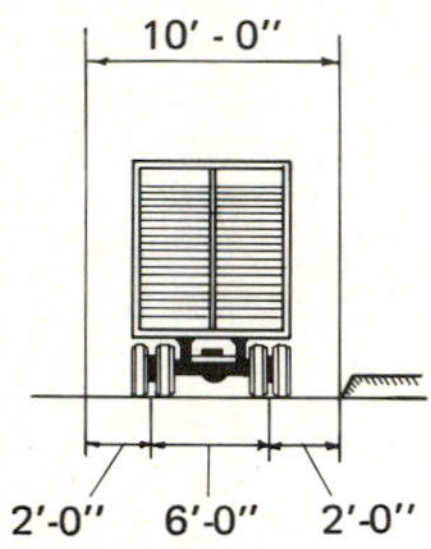

Figure 3–25. Highway Truck Loading (*Standard Specifications for Highway Bridges,* 12th Ed., Washington, D.C., The American Association of State Highway and Transportation Officials, copyright 1977. Used by permission)

(a) Influence line, positive moment at B

(b) Placement of truck load

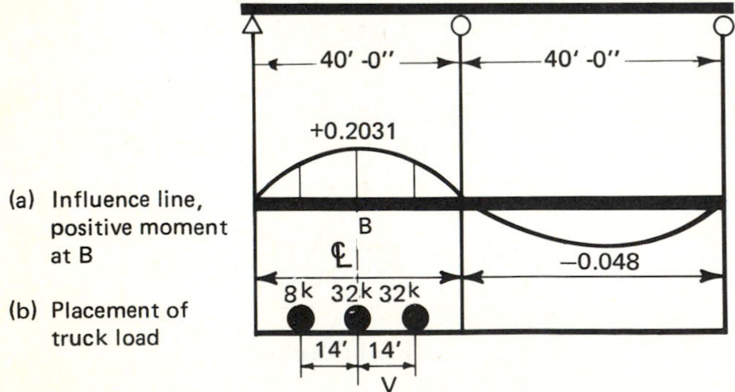

Figure 3–26. Example Problem 3–5

EXAMPLE PROBLEM 3–5: Highway Loading

Place an HS 20 truck on a two-span continuous highway bridge in the position to produce maximum positive moment at midspan of one of the girders. Use the influence line shown in Fig. 3–26(a).

Solution

1. Axle loads for HS 20 loading are 8 kips (35,584 N), 32 kips (1.4×10^5N) and 32 kips (1.4×10^5 N) (Fig. 3–24). These produce the maximum bending moment when positioned as close together as possible, $V = 14$ ft (4.26 m) with the middle axle at the midspan point.
2. Figure 3–26(b) illustrates the correct placement.

Longitudinal forces of 5% of the live load are required in addition to any forces in the longitudinal direction created by restraint at the support connections. The specified *impact* loading for highway bridge structures (to account for dynamic, vibrating, and impact loads) is given by the following formula:

$$I = \frac{50}{L + 125} \quad \text{or} \quad \frac{15.24}{L + 38} \tag{3–6}$$

where I = impact fraction as a fraction of the live load stress (maximum 30%)
L = length in feet (m) of the portion of the span that is loaded to produce the maximum stress in the member

Rail Loading

Rail loading is defined by a series of locomotive axle loads such as those shown in Fig. 3–27. As with highway loads, this rail loading must be placed on the structure to produce maximum stress. The specified *impact* for rail bridges constructed with steel members and carrying rolling equipment without hammer blow (diesels, electric locomotives, tenders, alone, etc.) is given by the following formula:

$$\text{For } L \text{ less than 80 ft: } I = \frac{100}{S} + 40 - \frac{3L^2}{1600} \tag{3–7}$$

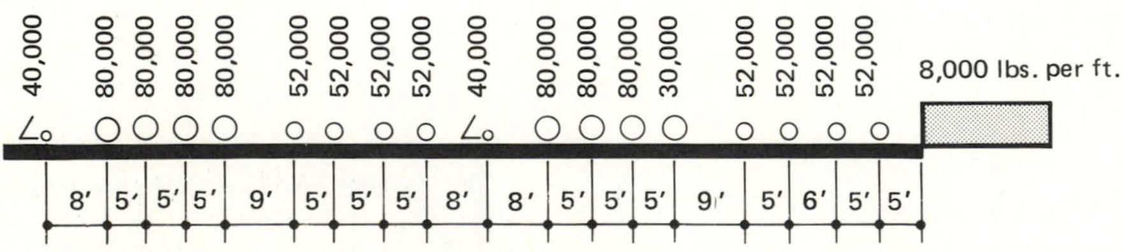

Figure 3–27. Railroad Loading (*Manual for Railway Engineering,* American Railway Engineering Association, Washington, D.C., 1981)

For L 80 ft or more: $I = \dfrac{100}{S} + 16 + \dfrac{600}{L - 30}$

where I = percentage of live load for impact
S = transverse distance, in feet, between longitudinal members (stringers, girders, etc.)
L = longitudinal distance, in feet, center to center of supports for stringers, girders, etc.

Similar formulas are given by AREA for different types of bridges, for multiple track installations, and for different construction materials.

3–8 IMPACT LOAD

Impact loads are dynamic loads exerted by equipment, cranes, vehicles, and so on. In addition to the impact loads already discussed in this chapter, the following impact loads are specified by the American Institute of Steel Construction:

For structures carrying live loads which induce impact, the assumed live load shall be increased as follows (if not otherwise specified):

For supports of elevators	100%
For traveling crane support girders and their connections	25%
Cab operated	25%
Pendant operated	10%
For supports of light machinery, shaft or motor driven, not less than	20%
For supports of reciprocating machinery or power-driven units, not less than	50%
For hangers supporting floors and balconies	33%

3–9 SOIL LOAD

Soil, as a compressible material, responds to confining pressure by consolidating and exerting a resisting force analogous to a spring. This force imposes a load on structures either by exerting *lateral earth pressure* (Fig. 3–28) or by exerting an upward *foundation bearing* or *friction pressure* on foundation elements (Fig. 3–29).

Lateral earth pressures range from 20 to 80 pcf (3128 to 12512 N/m³) depending on whether the pressure is considered *active, passive,* or *at rest. Active* pressures generate the lower numbers and are applicable when:

- The structure can rotate sufficiently for the soil to deform and reduce the angle of internal friction (Fig. 3–30).
- The wall height and soil condition allow the wall to be "earth formed" (assumes that soil moves before wall is poured).

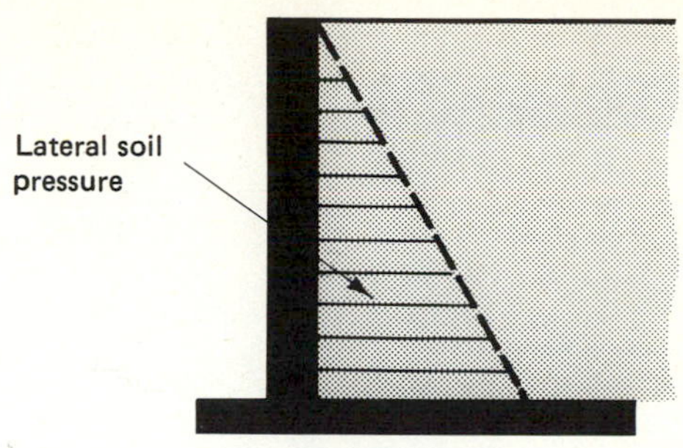

Figure 3–28. Lateral Soil Pressure

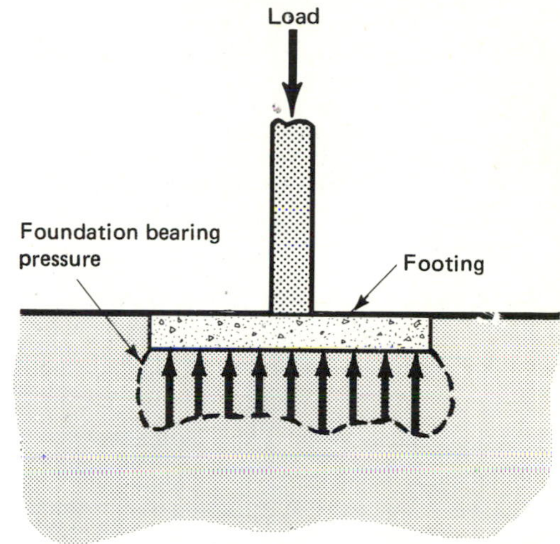

Figure 3–29. Foundation Soil Pressure

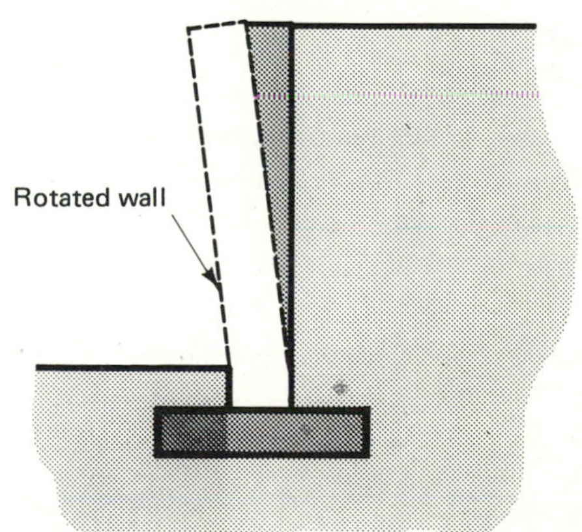

Figure 3–30. Active Pressure Due to Wall Rotation

- Conditions are such that a sloping excavation must be made and settlement of the backfill can be tolerated (Fig. 3–31) (assumes backfill is only lightly compacted).

Passive earth pressures generate the higher numbers and are applicable when the structural system actually presses against the soil. The toe of retaining wall footings, thrust blocks, water retaining structures, and similar structural elements pressing against the soil are designed for the passive condition.

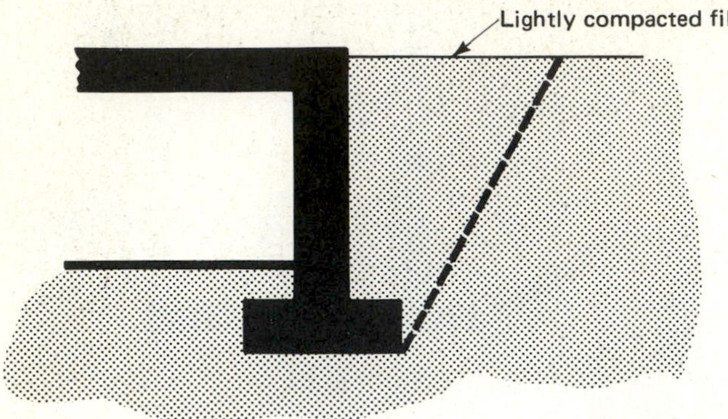

Figure 3–31. Active Pressure for Lightly Compacted Backfill

At-rest pressures are applicable when the structure undergoes no movement in either direction relative to the soil. It is also applicable when settlement cannot be tolerated for the condition shown in Fig. 3–31, and the backfill must be compacted to 95% of the Standard Proctor density.

The *soil bearing pressure* exerted on foundation elements depends directly on the relative stiffness of the soil and structural system. This "soil–structure interaction" can perhaps best be visualized by thinking of the soil as a series of springs supporting the foundation element (Fig. 3–32). It is obvious from the figure that the specific pressure at any point on the foundation will be related to the spring stiffness of the soil and to the flexural stiffness of the foundation element. It is quite common in foundation design to assume a uniform pressure distribution for isolated spread footings, and this is usually satisfactory. More complex foundations systems—such as mats, strap footings, and shear wall footings—often require a more accurate determination of the soil structure interaction and pressures. This is normally accomplished with a "beam on elastic foundation" analysis.

Figure 3–32. Foundation as "Beam-on-Elastic-Foundation"

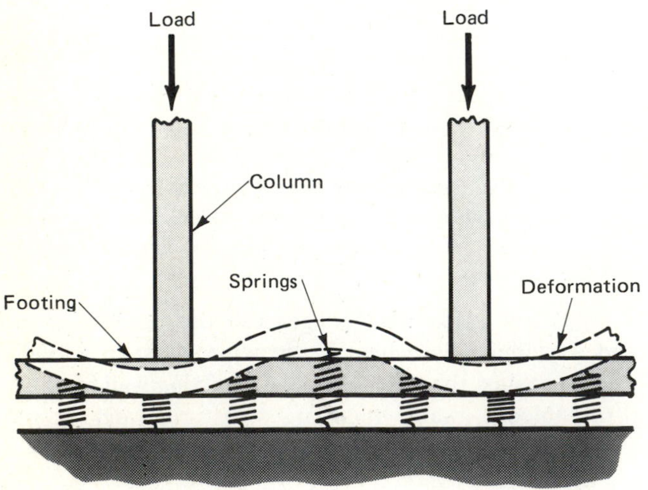

In addition to foundation bearing pressures, soils can exert other types of pressures on structures. These include heaving forces due to frost action or expansive clays. The forces on piles and other types of deep foundation elements are usually a combination of skin friction and end bearing forces, inducing primarily compressive stress in the element.

3–10 BLAST LOAD

Blast loads are created by conventional explosions, sonic booms, and detonation of military weapons. Blast loads are characterized by shock waves which exert a sudden overpressure (above atmospheric) on structures in their path—accompanied by very strong winds. The advancing shock wave is a moving wall of highly compressed air. As the expansion proceeds, the pressure distribution in the region behind the shock front gradually changes—and becomes negative (suction) with time. The winds shift direction at this point and blow inward toward the center of the explosion. Figure 3–33 shows the variation of overpressure with distance at various times.

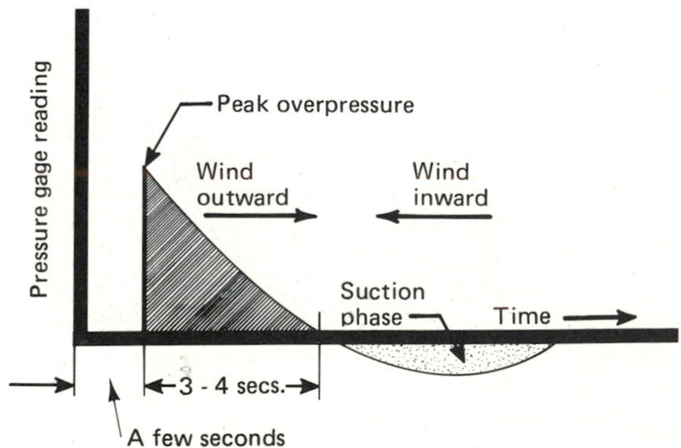

Figure 3–33. Variation of Overpressure with Time

The loading on aboveground structures from the air blast consists of a *diffraction phase* and a *drag phase.* The initial reflected pressures on the front of a structure (before envelopment by the shock wave) cause net lateral loads which constitute the diffraction phase. After envelopment of the structure, the mass and velocity of the air particles in the blast wave produce a drag loading, similar to wind loading, which acts on the structure together with the overpressure (Fig. 3–34). The peak overpressures are generated during the diffraction phase, and it is this sudden shock that causes much of the air blast damage.

The peak pressures created by *conventional explosions* are given by the following formula:

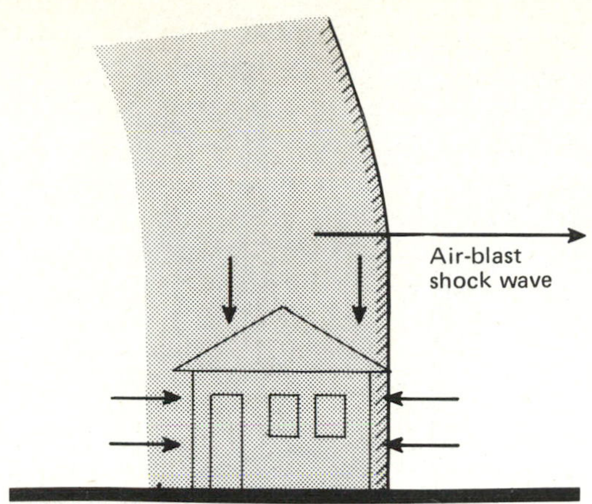

Figure 3–34. Blast Loading

$$P_s = \frac{4120}{Z^3} - \frac{105}{Z^2} + \frac{39.5}{Z} \text{ (lb/in.}^2\text{)} \quad (3\text{–}8)$$

$$P_s = \frac{180{,}780}{Z^3} - \frac{24{,}738}{Z^2} + \frac{50{,}718}{Z} \text{ (N/m}^2\text{)}$$

where P_s = peak pressure

$Z = \frac{r}{w^{1/3}}$, in which

r = distance from charge to gage, ft (m)

w = weight of charge, lb (N)

Nuclear weapons can produce peak blast overpressures as high as 200 or 300 psi (1.4 or 2.06 × 10^6 N/m^2) in close to the burst point, or they may be as low as 1 to 2 psi (6895 to 13,790 N/m^2) at places several miles away. A 5-megaton air burst will produce overpressures of 5 psi or more over an area of 168 sq mi (435 km^2). Most residential structures will collapse at 5 psi (34,475 N/m^2), 720 psf, which is normally accompanied by horizontal wind speeds of 160 mi/hr (257 km/hr).

The other major effect of weapons explosions is *ground shock* caused by air-induced shock and direct-transmitted shock from surface, near surface, or underground bursts. The shock-spectrum technique (described later in the discussion of seismic loads) is normally used to evaluate these effects.

3–11 MARINE LOAD

Marine loads are loads created by ocean currents and waves, termed *ocean forces,* and by ocean-going vessels, termed *mooring and docking impact forces.* These loads are imposed on structures such as piers, docks, bridge piers, and oil drilling platforms.

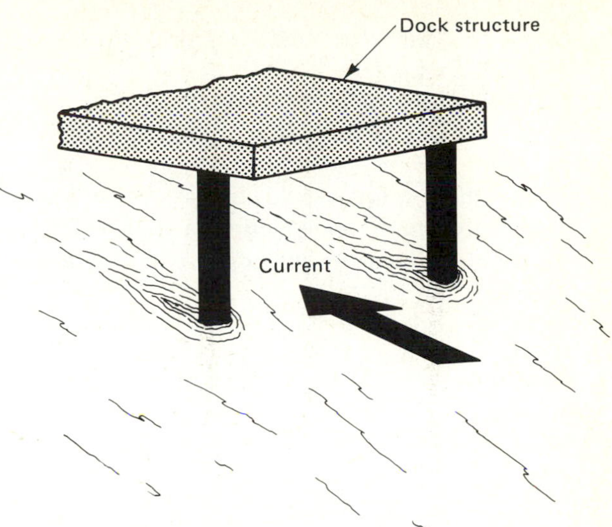

Figure 3–35. Ocean Current Drag Force

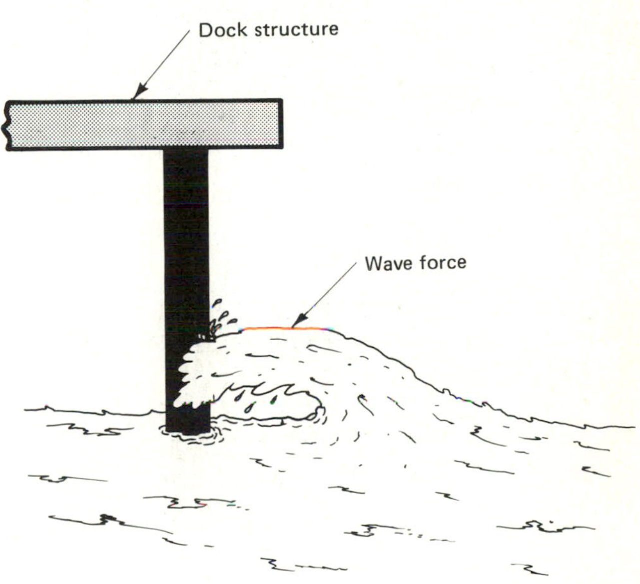

Figure 3–36. Ocean Wave Inertial Forces

Ocean Forces

Ocean forces include pressures created by *drag forces* (Fig. 3–35) and the dynamic action of waves, called *inertial forces* (Fig. 3–36).

Drag Forces. Drag forces for steady current flow are given by the following formula:

$$f_D = C_D \frac{\gamma}{2g} D u^2 \quad (3\text{–}9)$$

where f_D = drag force, pound per linear foot of pile, N/m

C_D = drag coefficient

γ = unit weight of fluid, 64.4 pcf (10,111 N/m³) for seawater
D = diameter of pile, ft (m)
u = velocity of fluid, ft/sec (m/sec)

When the flow is not steady but is constantly changing in velocity due to wave action, a modified version of the formula is used and integrated over the height of the structure:

$$F_D = C_D \frac{\gamma}{2g} DH^2K_D \qquad (3\text{–}10)$$

where

$$K_D = \frac{1}{H^2}\int_{-d}^{\eta} \pm u^2\, dz \qquad \text{(Fig. 3–37)}$$

and z = a vertical coordinate
d = depth of water below still-water level
η = distance of a surface particle above ($+\eta$) or below ($-\eta$) still-water level

Inertial Forces. Inertial forces are given by the following formula:

$$f_i = C_M \frac{\gamma\pi}{4g} D^2 \frac{du}{dt}$$

where f_i = inertial force, pounds per linear foot of pile (N/m)
C_M = coefficient of mass
du/dt = horizontal fluid acceleration, ft/sec per second (m/sec²)

The total inertial force on the pile at any moment can be found as

$$F_i = C_M \frac{\gamma}{2g} D^2HK_i \qquad (3\text{–}11)$$

where

$$K_i = \frac{\pi}{2H}\int_{-d}^{\eta} \frac{du}{dt}\, dz \qquad \text{(Fig. 3–37)}$$

This force is zero at crest and trough positions, with the maximum values occurring between these two positions. Therefore, the total maximum force is not the summation of the maximum values of F_D and F_i, since they occur at different times within the wave cycle. The drag force is predominant for high waves in shallow water, and the maximum force against the pile occurs at or near the crest of the waves. The inertial force is predominant for low waves in deep water, and the maximum force occurs at a phase angle of about 90° (one-quarter wavelength) or when the water surface at the pile is close to the still-water level. The coefficients K_D and K_i vary with the phase angle, and Fig. 3–37 shows values for K_D and K_i based on the solitary wave theory for waves near the breaking height.

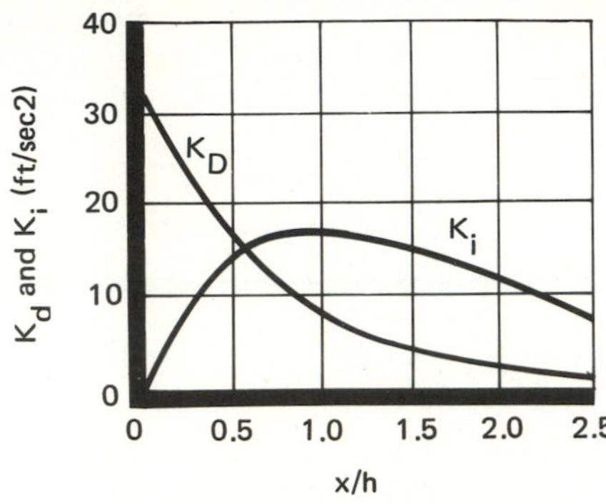

Figure 3–37. Coefficients for Drag and Inertial Forces (From R. O. Reid and C. L. Bretschneider, *Surface Waves and Offshore Structures: The Design Wave in Deep or Shallow Water, Storm Tide, and Forces on Vertical Piles and Large Submerged Objects,* Texas A & M Research Foundation, unpublished)

Mooring and Docking Forces

Mooring Forces. Mooring forces are transmitted to marine structures by the mooring lines that pull ships into or along the dock or hold them against the force of the wind and current (Fig. 3–38). Wind blowing against the side of the ship is the major force and is usually assumed to be between 10 and 20 psf (478 to 957 N/m²). The force of the current is given by the drag formula already discussed (Eq. 3–9) and will usually vary between 1 and 16 psf (47.8 to 7662 N/m²). This pressure is applied to the area of the ship below the water line when the ship is fully loaded.

Figure 3–38. Mooring Forces

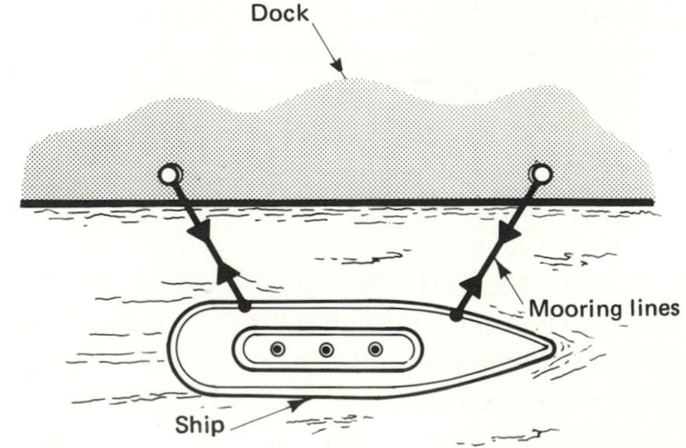

Docking Impact. Docking impact forces are caused by the ship striking the dock when berthing. The assumption usually made is that the maximum impact is that produced by a ship fully loaded striking the dock at an angle of 10° (with the face of the dock) and with a velocity of 0.25 to 0.5 ft/sec (0.076 to 0.15 m/sec) (Fig. 3–39).

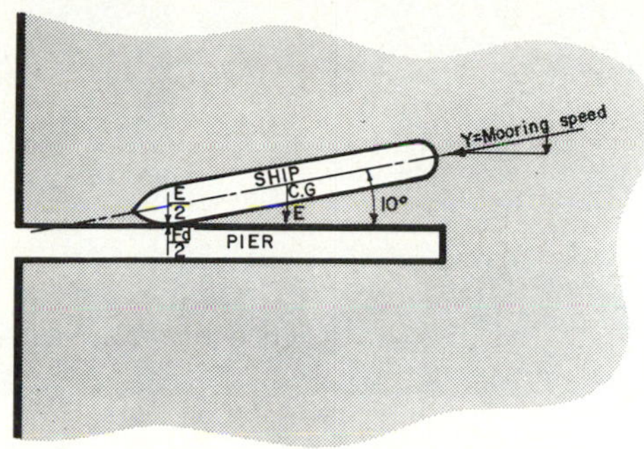

Figure 3–39. Docking Forces

The kinetic energy of impact is $E = \frac{1}{2}Mv^2$, and, substituting W/g for the mass M:

$$E = \frac{1}{2} \times \frac{W}{g} \times v^2 \qquad (3\text{–}12)$$

where E = energy, ft-tons (2240 lb) (N/m)
W = displaced weight of ship, long tons (N)
v = velocity of ship normal to dock, ft/sec (m/sec)
g = acceleration due to gravity, 32.2 ft/sec per second (9.81 m/sec²)

The energy to be absorbed by the fender system and dock is usually taken to be $\frac{1}{2}E$, and the remaining one-half is assumed to be absorbed by the ship and water.

3–12 WATER LOAD

Water exerts load on structures in a variety of ways. Uplift pressures on the bottom of underground tanks and basement structures due to the presence of ground water (under a hydraulic head) are called *hydrostatic forces.* These same hydrostatic forces can exert (together with lateral soil pressure) significant lateral pressures against retaining structures. Figure 3–40 illustrates these forces acting on a structure.

Water forces exerted on dam structures include normal hydrostatic uplift and lateral pressures plus inertial forces of the water mass during earthquakes. Figure 3–41 shows a loading diagram for a typical dam structure.

Water progressively impounded on roof structures due to excessive structural flexibility imposes *ponding loads* which must be considered in design or minimized by the design of the roof drainage system. Bridge structures crossing rivers experience another type of water force in the form of *stream forces* due to the water current and ice (Fig. 3–42). These forces exert drag pressures

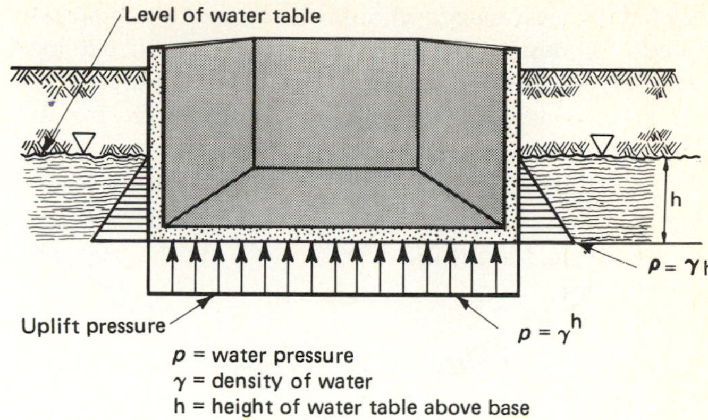

Figure 3–40. Hydrostatic Pressures

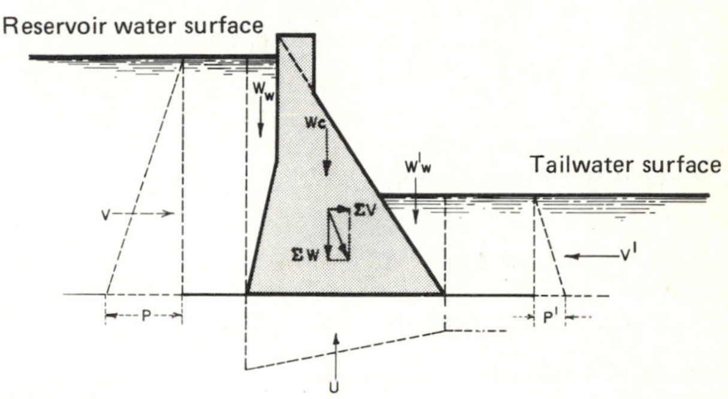

Figure 3–41. Dam Loading

Figure 3–42. Stream Forces

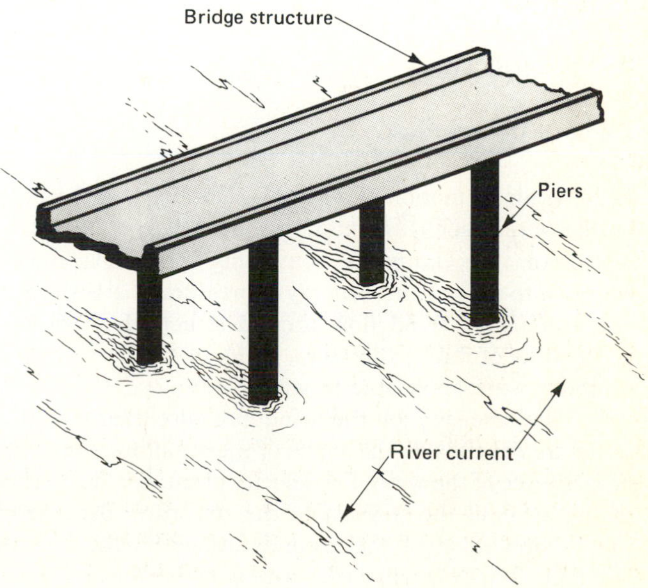

on bridge piers immersed in the water and are important loads for design of these structures. The current load can be calculated by Eq. 3–8. The pressure of ice on piers is normally assumed to be 400 psi (2.75×10^6 N/m^2).

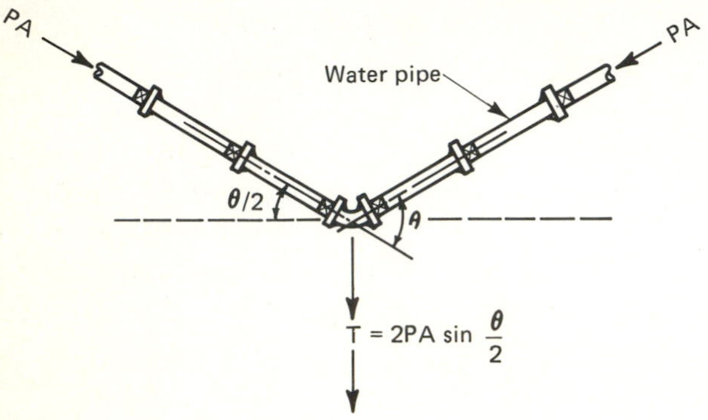

Figure 3–43. Water Thrust

Water transmitted under pressure in pipe distribution systems exerts still another type of pressure called *water thrust.* Water thrust is exerted against supports and pipe bends due to static and dynamic fluid action on the pipe (Fig. 3–43). This force can be calculated by the following formula:

$$T = PA \sin \frac{\theta}{2} \tag{3–13}$$

where T = thrust force, lb (N)
P = pressure in pipe, psi (N/m^2)
A = cross-sectional area of pipe, sq in. (m^2)
θ = angle of bend

3–13 WIND LOAD

General Description

Wind is air in motion (Fig. 3–44). Obstacles in the path of the wind—such as buildings and other structures—deflect or stop the wind, converting the winds kinetic energy into the potential energy of pressure, thereby creating *wind load.* Wind flow around structures is similar to the flow of fluids; therefore, many of the laws of fluid mechanics are applicable to wind action.

The intensity of the wind pressure exerted on a structure depends on the shape of the structure, the angle of incidence of the wind, the velocity of the air, the density of the air, and the stiffness of the structure. Because of the friction effect of the ground surface on the flow of wind, the wind velocity increases with height above the ground as the friction influence diminishes (Fig. 3–45). This "boundary layer" effect is analogous to the effect of pipe friction in fluid mechanics.

Figure 3–44. Smoke Visualization Showing Typical Flow around a Model of a Tall Building (Building Research Establishment Photograph, British Crown Copyright)

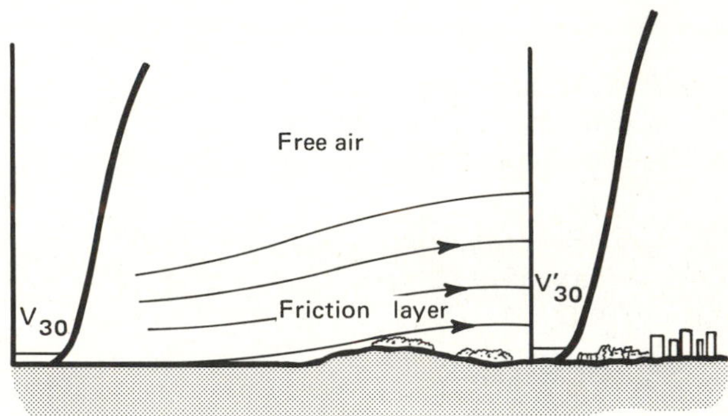

Figure 3–45. Effect of Ground Surface on Wind Velocity

As wind hits the structure and flows around it, several effects are possible (Fig. 3–46). Pressure on the windward face and suction on the leeward face creates *drag forces.* Unsymmetrical flow around the structure can create *lift forces* (analagous to flow around an airplane wing) which are perpendicular to the wind direction. Air turbulence at the leeward corners and edges can create *vortices,* which are high-velocity air currents that create circular updrafts and suction streams adjacent to the building. Periodic "shedding" of vortices creates dynamic effects and oscillation transverse to the direction of the wind.

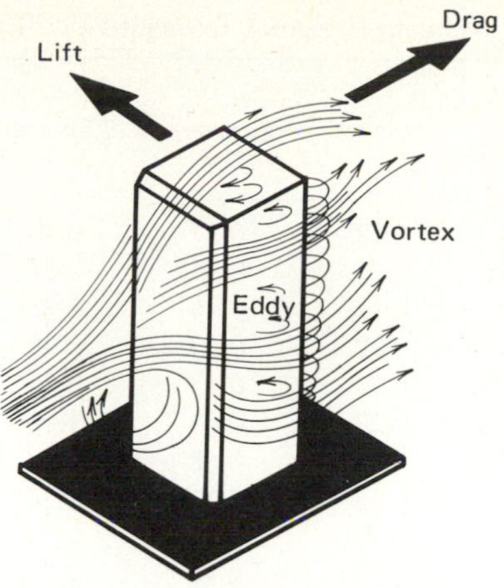

Figure 3–46. Wind Effects on Structures

Figure 3–47 illustrates the pattern of airflow around a tall building.

A constantly blowing wind will exert a fluctuating pressure on the windward and leeward faces of a building which can be approximated by a mean pressure (Fig. 3–48). In addition, wind *gusts* will buffet the structure with higher pressures, with the gust loading intensity varying with the stiffness of the structure. Wind load is very dynamic by nature and is closely related to the properties of the structure. Both gust buffeting and vortex shedding introduce dynamic effects. The famous "Galloping Gertie" bridge collapse (Fig. 3–49) is a vivid example of wind's dynamic nature.

Figure 3–47. Pattern of Airflow at Ground Level around a Tall Building, with a Low Building at a Small Distance to Windward (Building Research Establishment Photograph, British Crown Copyright)

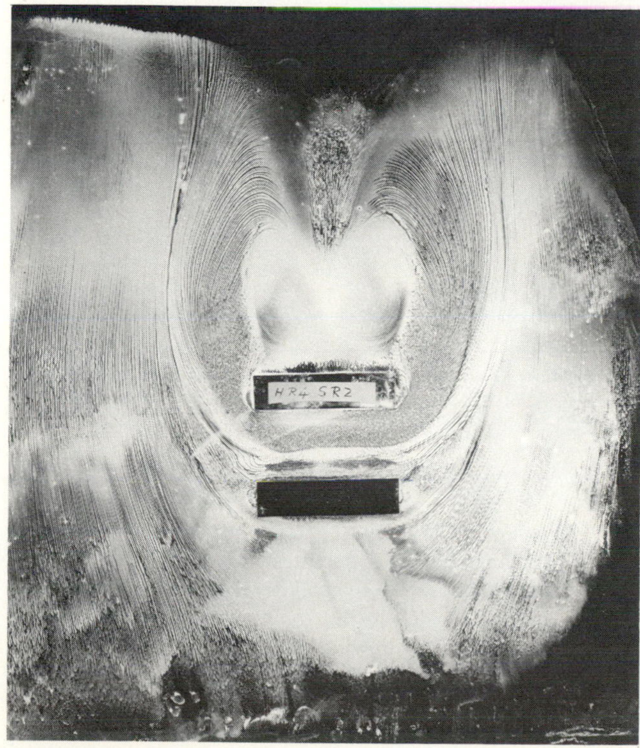

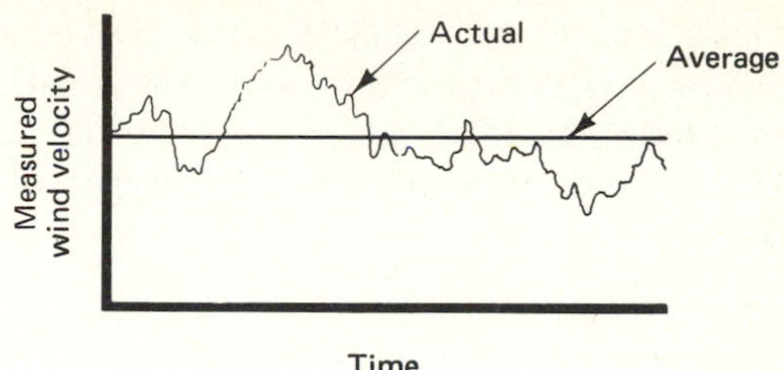

Figure 3–48. Mean Wind Pressure

Figure 3–49. Wind-Induced Oscillations of the Tacoma Narrows Bridge near Seattle, "Galloping Gertie" (Photography by F. B. Farquharson, Photography Collection, University of Washington Libraries)

Human tolerance to wind action has become increasingly important as building structures are built taller. The effects of the wind environment outside the building (due to downdrafts, etc.) must be considered as well as the effects of excessive lateral sway of the building's structural system. Downdrafts have been known to completely strip trees in plaza areas and to buffet pedestrians dangerously.

Design Approach

There are two approaches to design of structures for wind load effects. The first approach is the *static* approach, which approximates the wind force on a structure by an equivalent static load. This static load is calculated by the following formula, which considers the velocity of the wind, the shape of the structure, its exposure, and other factors:

$$p = 0.00256 C_p C_a G_h V_{30}^2 \left(\frac{h}{30}\right)^{2/a} \qquad (3\text{–}14)$$

where p = wind pressure, psf (N/m²)
C_p = shape coefficient
C_a = coefficient dependent on nearby topographic features
G_h = gust coefficient
a = exponent for velocity increase with height
V_{30} = basic wind speed at 30 ft (9.14 m) above ground, mi/hr
h = height at which wind pressure is to be calculated

Many codes simplify the calculation of wind pressures by providing tables that incorporate C_a and the height scaling factor for various exposure conditions.

The wind pressure formula for main wind-force resisting systems given by ANSI[2] is

$$p_z = q_z G_h C_p \qquad \text{(windward wall)} \qquad (3\text{–}15)$$

$$p_h = q_h G_h C_p \qquad \text{(leeward wall and roof)} \qquad (3\text{–}16)$$

where $p_{z,h}$ = wind pressure, psf (N/m²)
$q_{z,h}$ = effective velocity pressure, psf (N/m²), given by Eq. 3–17
G_h = gust coefficient (Table 3–8)
C_p = pressure coefficient (Fig. 3–51 and Table 3–9)

The *velocity pressures* (q_z and q_h) may be determined as follows:

$$q_z = 0.00256 k_z (IV)^2 \qquad (\text{for } q_h, \text{ set } z = h) \qquad (3\text{–}17)$$

where k_z = velocity pressure exposure coefficient given by Table 3–6
I = importance coefficient, Table 3–7 and Table 3–1
V = basic wind speed given by Fig. 3–50

The *gust factor*, G_h, is given in Table 3–8 for various exposure categories which are defined as follows:

[2] *Building Code Requirements for Minimum Design Loads in Buildings and Other Structures,* A58.1–1982 (New York: American National Standards Institute, 1982).

Table 3–6 Velocity Pressure Exposure Coefficient, K_z[a]

Height above ground level, z (ft)	*Exposure A*	*Exposure B*	*Exposure C*	*Exposure D*
0–15	0.12	0.37	0.80	1.20
20	0.15	0.42	0.87	1.27
25	0.17	0.46	0.93	1.32
30	0.19	0.50	0.98	1.37
40	0.23	0.57	1.06	1.46
50	0.27	0.63	1.13	1.52
60	0.30	0.68	1.19	1.58
70	0.33	0.73	1.24	1.63
80	0.37	0.77	1.29	1.67
90	0.40	0.82	1.34	1.71
100	0.42	0.86	1.38	1.75
120	0.48	0.93	1.45	1.81
140	0.53	0.99	1.52	1.87
160	0.58	1.05	1.58	1.92
180	0.63	1.11	1.63	1.97
200	0.67	1.16	1.68	2.01
250	0.78	1.28	1.79	2.10
300	0.88	1.39	1.88	2.18
350	0.98	1.49	1.97	2.25
400	1.07	1.58	2.05	2.31
450	1.16	1.67	2.12	2.36
500	1.24	1.75	2.18	2.41

[a] *Building Code Requirements for Minimum Design Loads in Buildings and Other Structures,* ANSI A58.1–1982 (New York: American National Standards Institute, 1982).

Table 3–7 Importance Coefficient (Wind)[a]

Structure classification category	*Importance coefficient, I* — *100 miles (160.9 km) from oceanline and other areas*	*Hurricane oceanline*
I	1.00	1.05
II	1.07	1.11
III	1.07	1.11
IV	0.95	1.00

[a] *Building Code Requirements for Minimum Design Loads in Buildings and Other Structures,* ANSI A58.1–1982 (New York: American National Standards Institute, 1982).

Exposure A: large city centers with at least 50% of the buildings having a height in excess of 70 ft (21.3 m). Use of this exposure category shall be limited to those areas for which terrain representative of Exposure A prevails in the upwind direction for a distance of at least ½ mile or 10 times the height of the building or structure, whichever is greater. Possible channeling effects or increased velocity pressures due to the building

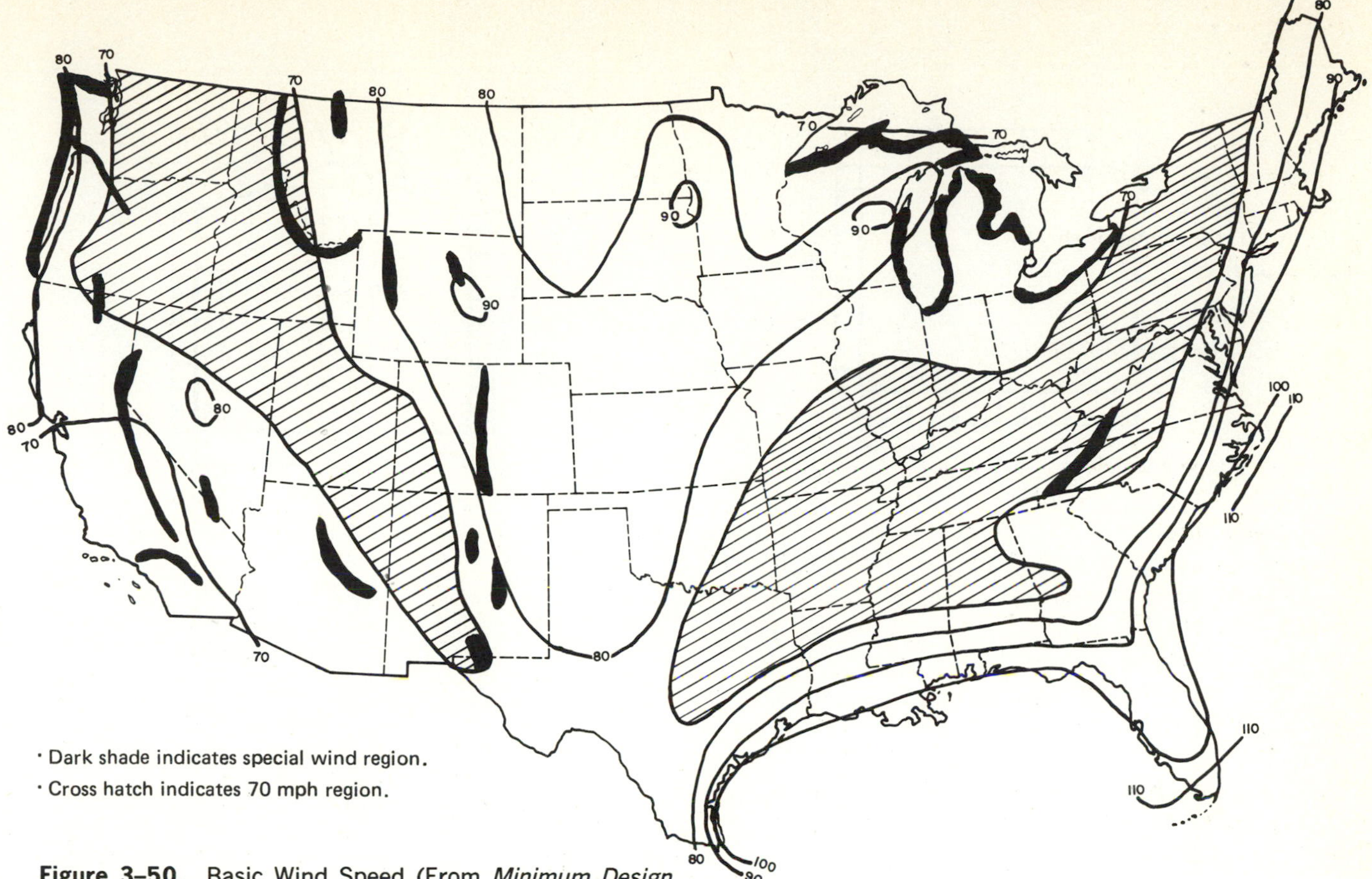
· Dark shade indicates special wind region.
· Cross hatch indicates 70 mph region.

Figure 3–50. Basic Wind Speed (From *Minimum Design Loads for Buildings and Other Structures,* ANSI A 58.1–1982, American National Standards Institute, New York, 1982)

or structure being located in the wake of adjacent buildings shall be taken into account.

Exposure B: urban and suburban areas, well-wooded areas, or other terrain with numerous closely spaced obstructions having the size of single-family dwellings or larger. Use of this exposure category shall be limited to those areas for which terrain representative of Exposure B prevails in the upwind direction for a distance of at least 1500 ft (457 m) or 10 times the height of the building or structure, whichever is greater.

Exposure C: open terrain with scattered obstructions having heights generally less than 30 ft (9.14 m). This category includes flat, open country, and grasslands.

Exposure D: flat, unobstructed coastal areas directly exposed to wind blowing over large bodies of water. This exposure shall be used for those areas representative of Exposure D extending inland from the shoreline a distance of 1500 ft (457 m) or 10 times the height of the building or structure, whichever is greater.

For *flexible buildings* and *structures* (having a height exceeding five times the least horizontal dimension or a fundamental frequency less than 1 Hz), G_h must be determined by a rational analysis that incorporates

Table 3–8 Gust Response Factors, G_h and G_z[a]

Height above ground level, z (ft)	*Exposure A*	*Exposure B*	*Exposure C*	*Exposure D*
0–15	2.36	1.65	1.32	1.15
20	2.20	1.59	1.29	1.14
25	2.09	1.54	1.27	1.13
30	2.01	1.51	1.26	1.12
40	1.88	1.46	1.23	1.11
50	1.79	1.42	1.21	1.10
60	1.73	1.39	1.20	1.09
70	1.67	1.36	1.19	1.08
80	1.63	1.34	1.18	1.08
90	1.59	1.32	1.17	1.07
100	1.56	1.31	1.16	1.07
120	1.50	1.28	1.15	1.06
140	1.46	1.26	1.14	1.05
160	1.43	1.24	1.13	1.05
180	1.40	1.23	1.12	1.04
200	1.37	1.21	1.11	1.04
250	1.32	1.19	1.10	1.03
300	1.28	1.16	1.09	1.02
350	1.25	1.15	1.08	1.02
400	1.22	1.13	1.07	1.01
450	1.20	1.12	1.06	1.01
500	1.18	1.11	1.06	1.00

[a] *Building Code Requirements for Minimum Design Loads in Buildings and Other Structures,* ANSI A58.1–1982 (New York: American National Standards Institute, 1982).

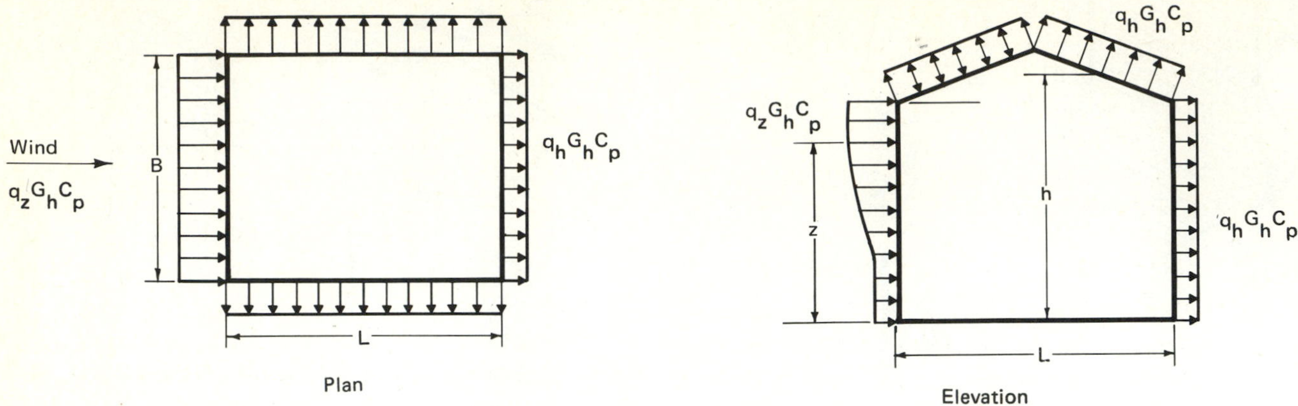

Figure 3–51. Pressure Coefficients

Table 3–9 External Pressure Coefficients for Average Loads on Main Wind-Force Resisting Systems[a,b]

Roof Pressure Coefficients C_p for Use with q_h

Wind direction	h/L	*Windward* *Angle, θ (deg)* 0	10–15	20	30	40	50	>60	*Leeward*
Normal to ridge	≤0.3	−0.7	0.2[c] −0.9[c]	0.2	0.3	0.4	0.5	0.01θ	−0.7 for all values of h/L
	0.5	−0.7	−0.9	−0.75	−0.2	0.3	0.5	0.01θ	
	1.0	−0.7	−0.9	−0.75	−0.2	0.3	0.5	0.01θ	
	≥1.5	−0.7	−0.9	−0.9	−0.9	−0.35	0.2	0.01θ	
Parallel to ridge	h/B or h/L ≤2.5				−0.7				−0.7
	h/B or h/L ≥2.5				−0.8				−0.8

Wall Pressure Coefficients C_p

Surface	L/B	C_p	*For use with:*
Windward wall	All values	0.8	q_z
Leeward wall	0–1	−0.5	q_h
	2	−0.3	
	>4	−0.2	
Side walls	All values	−0.7	q_h

[a] *Minimum Design Loads for Buildings and Other Structures,* ANSI A58.1–1982 (New York: American National Standards Institute, 1982).

[b] *h,* mean roof height; *B,* horizontal dimension measured normal to wind; *L,* horizontal dimension measured parallel to wind.

[c] Both values of C_p shall be used in assessing load effects.

the dynamic properties of the structure (ANSI Appendix A6 contains one method).

The *pressure coefficients* (C_p) are illustrated in Fig. 3-51 and listed in Table 3-9.

The following example illustrates the procedure for calculating wind loads.

EXAMPLE PROBLEM 3-6: Wind Load Determination

Calculate the design wind loads for a 70-ft (21.3-m) office building 120 ft × 120 ft (36.5 m × 36.5 m) in plan in Miami, Florida. Assume Exposure B and 20-ft (6.1-m) bays. Resolve the pressure into lateral forces for each frame applied at the floor levels.

Solution

1. Determine design pressure for the windward wall.
   a. From Fig. 3-50, basic wind speed:

$$V = 100 \text{ mi/hr (160 km/hr)}$$

   b. From Table 3-7, importance coefficient:

$$I = 1.05$$

   c. From Table 3-8, gust response factor:

$$G_h = 1.36$$

   d. From Table 3-9, external pressure coefficient:

$$C_p = 0.8$$

$$p_z = q_z G_h C_p = 0.00256 k_z (IV)^2 G_h C_p$$
$$= 0.00256 k_z [1.05(100)]^2 (1.36)(0.8)$$

   e. k_z values are obtained from Table 3-6 and calculation results for each height are shown in Table 3-10.

2. Determine design pressure for leeward wall.

$$p_h = q_h G_h C_p$$
$$= (20.6)(1.36)(-0.5) = -14.0 \text{ psf } (-670.3 \text{ N/m}^2)$$

Table 3-10 Velocity Pressures for Example Problem 3-6

Height (ft)	k_z	q_z *(psf)*	p_z *(psf)*	p_h *(psf)*
0–15	0.37	10.4	11.3	14.0
20	0.42	11.8	12.8	14.0
25	0.46	13.0	14.1	14.0
30	0.50	14.1	15.3	14.0
40	0.57	16.1	17.5	14.0
50	0.63	17.8	19.4	14.0
60	0.68	19.2	20.9	14.0
70	0.73	20.6	22.4	14.0

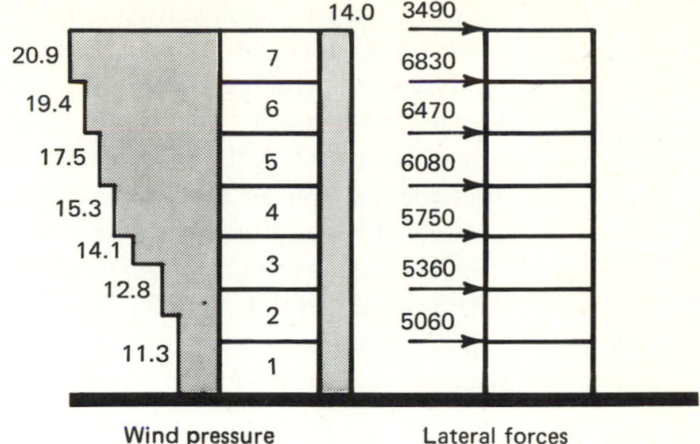

Figure 3-52. Example Problem 3-6

3. Determine lateral loads at each floor (see Fig. 3-52).

$p_2 = (11.3 \text{ psf} + 14.0 \text{ psf})(20 \text{ ft})(10 \text{ ft}) = 5060$ lb (22,506 N)

$p_3 = (12.8 \text{ psf} + 14.0 \text{ psf})(20 \text{ ft})(10 \text{ ft}) = 5360$ lb (23,841 N)

$p_4 = (14.1 \text{ psf} + 14.0 \text{ psf} + 15.3 \text{ psf} + 14.0 \text{ psf})(20 \text{ ft})(5 \text{ ft}) = 5750$ lb (25,531 N)

$p_5 = (15.3 \text{ psf} + 14.0 \text{ psf} + 17.5 \text{ psf} + 14.0 \text{ psf})(20 \text{ ft})(5 \text{ ft}) = 6080$ lb (27,043 N)

$p_6 = (17.3 \text{ psf} + 14.0 \text{ psf} + 19.4 \text{ psf} + 14.0 \text{ psf})(20 \text{ ft})(5 \text{ ft}) = 6470$ lb (28,778 N)

$p_7 = (19.4 \text{ psf} + 14.0 \text{ psf} + 20.9 \text{ psf} + 14.0 \text{ psf})(20 \text{ ft})(5 \text{ ft}) = 6830$ lb (30,379 N)

$p_8 = (20.9 \text{ psf} + 14.0 \text{ psf})(20 \text{ ft})(5 \text{ ft}) = 3490$ lb (15,523 N)

Similar procedures are outlined by the ANSI Code for calculating design pressures for components and cladding, several types of roof systems and for structures such as chimneys and tanks.

The *static method* takes into consideration the load magnification effect caused by gusts in resonance with along-wind vibrations of the structure but does not include allowances for across-wind or torsional loading, vortex shedding, or instability due to galloping or flutter.

Shielding or the effect of adjacent structures is considered in the exposure coefficient used in the static method. However, it is a mistake to assume that because a building is in a congested area, it is protected from the full force of the wind. Depending on the orientation and configuration of the adjacent structures, greater wind pressures can be generated than those imposed on a structure sitting in an open field.

The second design approach to wind loading is a

dynamic analysis which involves a determination of the wind loading through wind tunnel tests and a true dynamic structural analysis. Whereas the effective velocity pressure used in the static approach takes into account the dynamic response of ordinary buildings and structures to gusts in the direction parallel to the wind, it does not provide for the effects of vortex shedding or instability due to galloping. For buildings whose height exceeds five times the least horizontal dimension, and for buildings whose dynamic properties tend to make them wind sensitive, a detailed dynamic analysis should be performed.

Figure 3–53. Wind Tunnel Tests (Courtesy of A. G. Davenport, University of Western Ontario)

The dynamic analysis will include wind tunnel tests (Fig. 3–53) which will simulate (to scale) the building structure, the surrounding structures and environment, and wind velocities. Normally, one scale model of the structure is instrumented with pressure transducers and is used to measure pressure effects. Another scale model (with stiffness properties equivalent to the proposed structure) is used to determine dynamic effects and deflections.

Extreme Wind Conditions

Extreme winds, such as *thunderstorms, hurricanes,* and *tornadoes,* buffet structures with forces that may exceed those assumed for normal design. Design for such winds is usually not directly considered by building codes, although the ANSI method does incorporate hurricane wind speeds for a specified probability of occurrence. The following paragraphs describe the characteristics of these extreme winds and their effect on structures.

Thunderstorms (Fig. 3–54) are caused by the interaction of masses of cold and hot air with temperature, pressure, and humidity factors determining the potential for formation of the storm. Wind speeds in thunderstorms have typical wind speeds of 20 to 70 mi/hr (32 to 112 km/hr), with the swirling wind action exerting high suction forces on roofing and cladding elements.

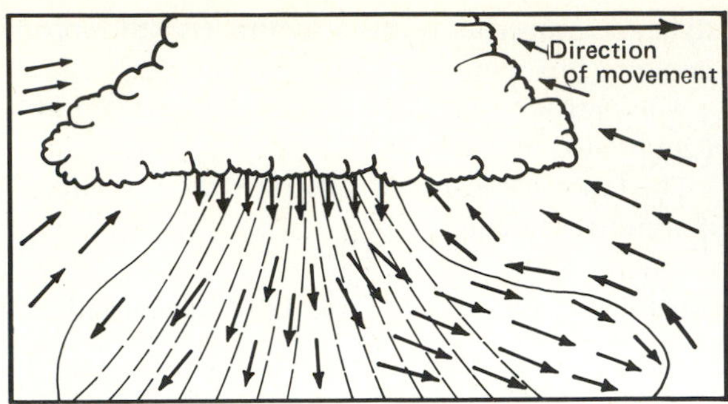

Figure 3–54. Thunderstorms

Hurricanes are formed when regions of low air pressure develop over the ocean and create violent patterns of wind currents (Fig. 3–55). These wind currents whirl around the "eye" of the hurricane, which is a relatively calm area with little wind. The typical hurricane system has a diameter of about 300 mi (482 km); the diameter of the eye is normally 30 to 40 mi (48 to 64 km), and vertically the circulation may extend to about 9 mi (14.5 km). The entire storm system moves at the speed of the steering current, generally less than 55 mi/hr (88.5 km/hr). The highest wind speeds in hurricanes can approach 200 mi/hr (321 km/hr), which exceeds the maximum (specified by Code) basic wind velocity of 120 mi/hr (193 km/hr) for any area of the United States. Except in rare instances (such as defense installations), a structure would not be designed for 200-mi/hr (321-km/hr) winds. Usually, the only modification to the basic design procedure to accommodate these extreme hurricane winds is a greater attention to details and to anchoring elements (windows, canopies, etc.) that can become damaging missiles during a 200-mi/hr wind (321 km/hr).

Figure 3–55. Hurricanes [Official Photograph, National Oceanic and Atmosphere Administration (NOAA)]

Figure 3–56. Tornados [Official Photograph, National Oceanic and Atmosphere Administration (NOAA)]

Tornadoes develop within severe thunderstorms and consist of a vortex of air that appears as a funnel-shaped cloud (Fig. 3–56). The funnel, typically 200 to 800 ft (60 to 243 m) in diameter, moves with respect to the ground with speeds of 20 to 60 mi/hr (32 to 97 km/hr) in a path approximately 9 miles (14.5 km) long and directed predominately toward the northeast. High wind speeds and low interior air pressures account for the destructive nature of the tornado. Sudden changes in air pressure, dropping as low as 10% of the total atmospheric pressure [or 2000 psf (95,760 N/m²)] may cause buildings to explode. Tornadoes contain the most powerful of all winds, causing damage estimated at $100 million a year in the United States alone. Figure 3–57 shows the tornado strike probability in the United States and Figure 3–58 shows regional intensities and pressure drops.

Wind speeds in tornadoes can vary from approximately 100 mi/hr (160 km/hr) to as high as 300 to 400 mi/hr (483 to 644 km/hr). Buildings are not usually designed to sustain a direct hit from a tornado, but buildings in areas with a high tornado frequency should be carefully detailed and extra attention given to anchorage of roof decks, and so on. As tornadoes seemingly become more frequent, more information is becoming available for design of tornado-resistant structures, and it is likely that building codes will be changed to reflect the increasing frequency of tornadoes.

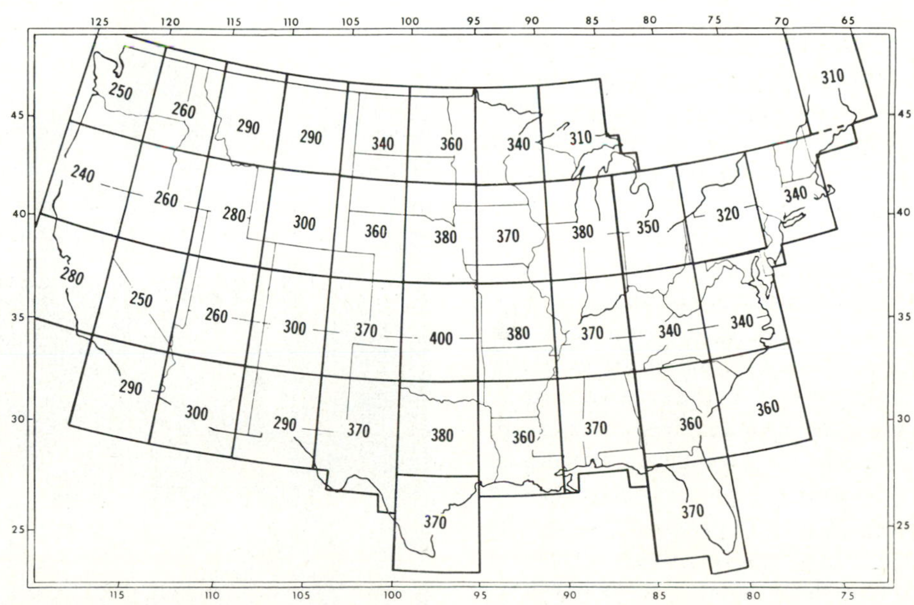

CALCULATED TORNADO WINDSPEED BY FIVE-DEGREE SQUARES FOR 10^7 PROBABILITY PER YEAR

Figure 3–57. Tornado Windspread for 10^{-7} Probability per Year [From E. Markee, J. Beckerly, and K. Sanders, *Technical Basis for Interim Regional Tornado Criteria, Wash-1300* (UC-11), U.S. Atomic Energy Commission, Office of Regulation, Washington, D.C.]

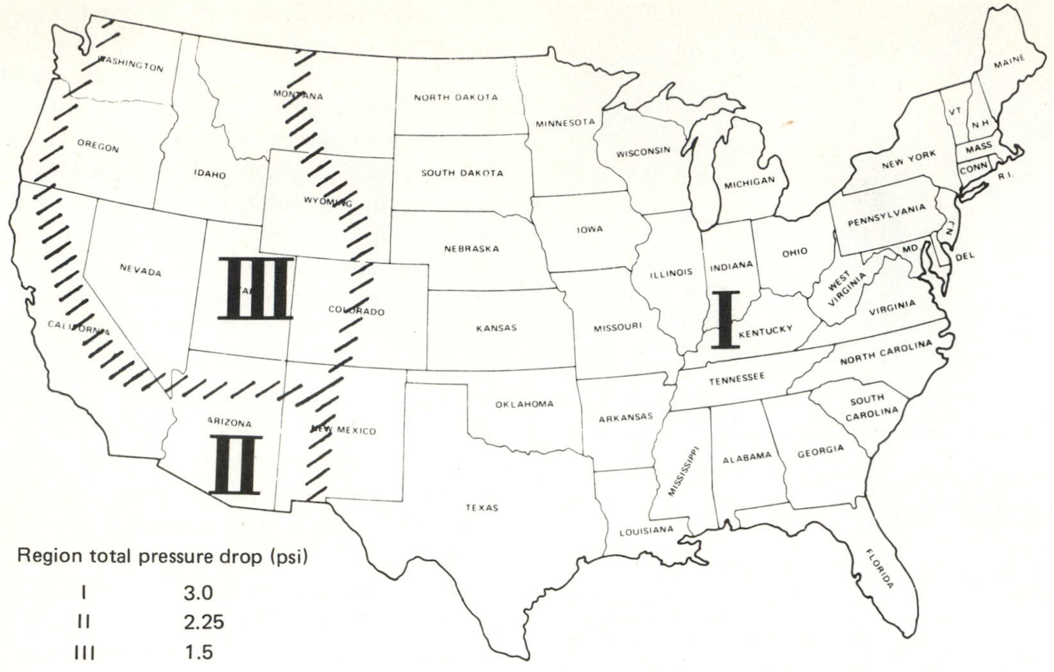

Figure 3–58. Tornado Intensity Regions and Pressure Drops [From E. Markee, J. Beckerly, and K. Sanders, *Technical Basis for Interim Regional Tornado Criteria,* Wash-1300 (UC-11), U.S. Atomic Energy Commission, Office of Regulation, Washington, D.C.]

3–14 EARTHQUAKE (SEISMIC) LOADING

General Description

The loading produced by *earthquakes* is the result of a complex interaction between ground motion and the response characteristics of the structure. Because these loads are active—changing with time—they are termed *dynamic* loads and differ from wind loads in one important aspect: whereas wind loads are externally applied loads, and hence are proportional to the exposed surface of the structure, earthquake loads are *inertia* forces. The inertia forces result from the distortion produced by both the earthquake motion and the lateral resistance of the structure. Their magnitude is a function of the *mass* and *stiffness* of the structure, rather than the exposed surface.

Strong ground motion and amplified response in the structure can produce loads of significant magnitude. Figures 3–59 and 3–60 show the effects of one particular earthquake on building structures. These loads can be so severe that it becomes impractical to design structures to resist them solely in the elastic range of material behavior. Some amount of inelastic deformation is necessary to absorb the energy generated by the larger earthquakes.

The following sections on ground motion and structural response provide further insight into the nature of earthquake loads and establish the basis for the design approach.

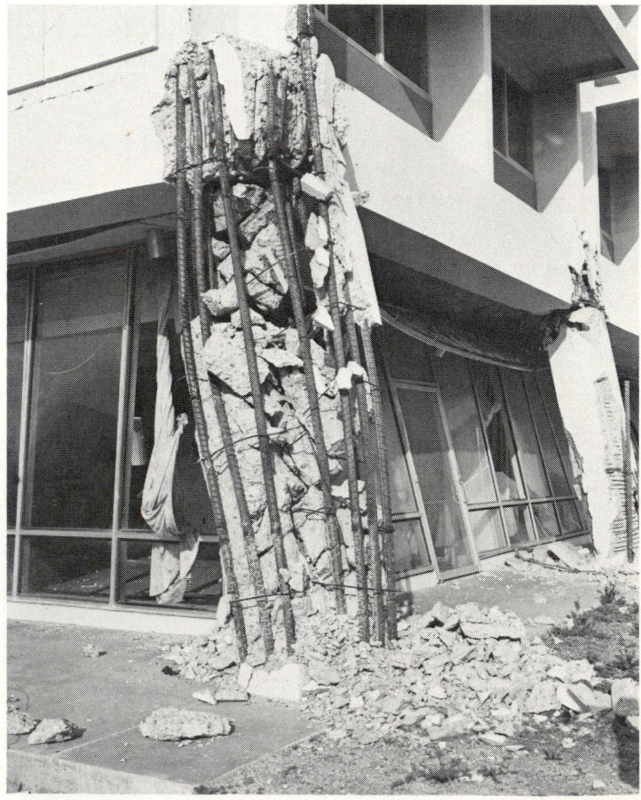

Figure 3–59. Earthquake Damage (Courtesy of Earthquake Engineering Research Library, California Institute of Technology)

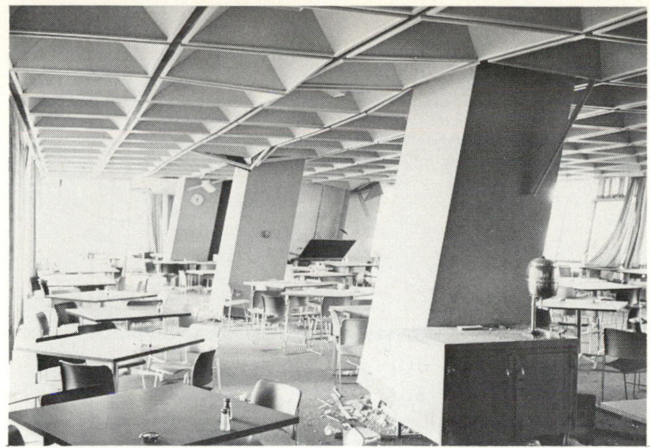

Figure 3-60. Earthquake Damage (Courtesy of Earthquake Engineering Research Library, California Institute of Technology)

Ground Motion

Earthquakes are generated by extreme pressures in the earth's crust causing slippage of plates along fault lines (Fig. 3-61), called tectonic activity. This slippage generates pressure waves in the earth's crust which travel at speeds of 2.5 to 5.0 mi/sec (4 to 8 km/sec). The pressure waves cause acceleration and displacement of the ground surface in three directional components: two horizontal and one vertical. The motion of the ground surface can be recorded with an accelograph and graphical plots produced showing the variation (with time) of velocity, acceleration, and displacement (see Fig. 3-62).

The ground motion parameter most commonly used in structural analysis is the variation of the ground *acceleration* with time in the immediate vicinity of the structure. The character and the variation of the acceleration is primarily dependent on the magnitude and focal depth of the earthquake, the epicentral distance of the site considered (distance from the epicenter or origin of the first

Figure 3-61. Earthquake Fault (Photo by Karl V. Steinbrugge)

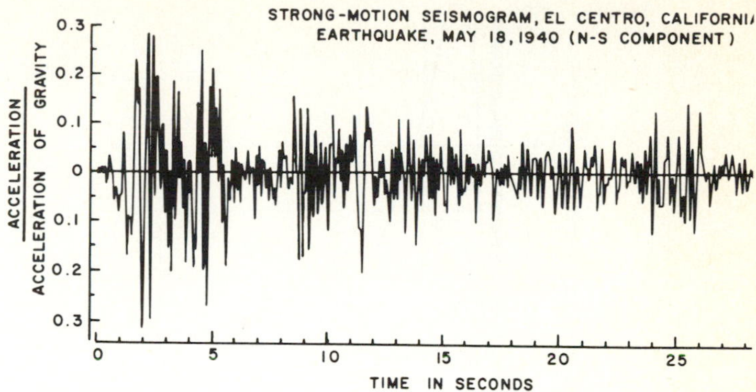

Figure 3-62. Accelogram (From *Earthquake Forces on Tall Structures* by Henry J. Degenkolb)

seismic waves), and the properties of the intervening and surrounding ground. The most significant characteristics describing the ground acceleration (which influence structural response) are the *maximum acceleration, frequency* characteristics, and *duration* of the large-amplitude pulses. Earthquake motions will generally be random in character, and there is no way to anticipate the exact nature of any future earthquake at a particular site. However, continuing studies of the seismic history and geology of a region can yield valuable estimates of the expected range of the significant ground acceleration parameters.

Structural Response

Linear Elastic Model. A simple structural model can be used to illustrate the response of structural systems to earthquake motion. Figure 3-63 shows a single mass attached to a column element—modeling, perhaps, a single-story structure—with the lumped mass representing the mass of the roof system, and the column representing the lumped stiffness of all the columns. This model is also called a single-degree-of-freedom (SDF) model. If this structural model has a stiff column and a light mass (and, therefore, a short period of vibration) and is subjected to earthquake base motion, it will tend to accelerate at the same rate as the ground. This response can be plotted graphically as shown in Fig. 3-64, which

Figure 3-63. Single-Degree-of-Freedom Model

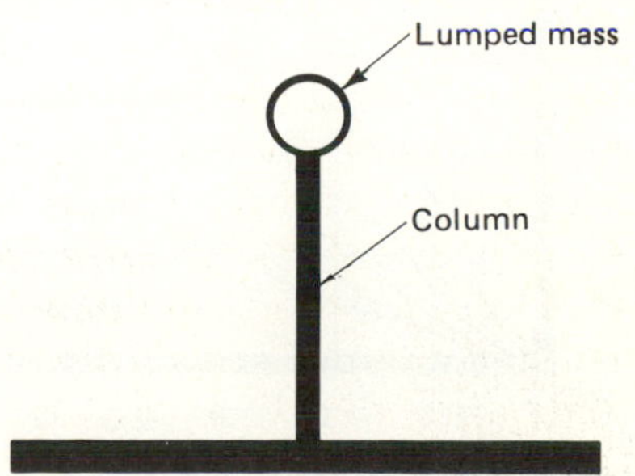

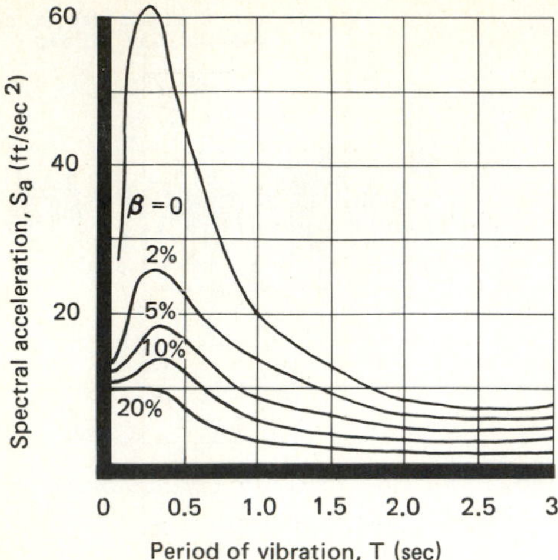

Figure 3–64. Acceleration Response

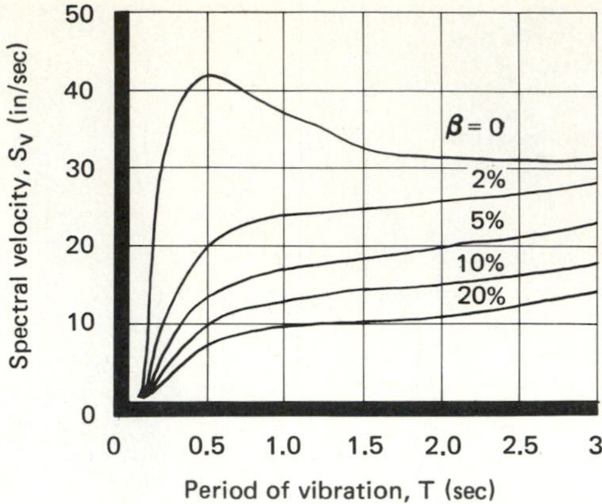

Figure 3–66. Velocity Response

shows *spectral* (maximum) acceleration versus period of vibration. The stiffer structures (with short periods) experience the higher accelerations and the more flexible structures (with longer periods) experience lower accelerations. The various curves represent varying degrees of *damping*—the frictional resistance to motion provided by the structural system and its attachments.

If the structural model has a flexible column and a heavy mass (and, therefore, a long period of vibration) and is subjected to earthquake base motion, the mass does not have time to move, and the displacement in the column is equal to the displacement of the ground. Figure 3–65 shows plots of spectral displacement versus period of vibration. Note that the peak displacements occur for the more flexible structures (with longer periods).

Other curves showing velocity (Fig. 3–66) can also be plotted as a function of the period of vibration. All three of these curves can conveniently be plotted together

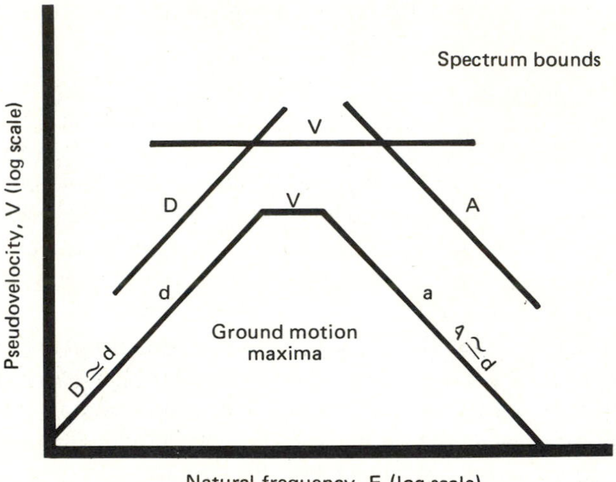

Figure 3–67. Plot of Acceleration, Displacement, and Velocity

on a logarithmic scale as shown in Fig. 3–67. Because of the linear relationship between the logarithms of both the spectral displacement (S_d) and acceleration (S_a) with the logarithms of the spectral velocity (S_v) and circular frequency (ω), it is possible to have a single plot showing the variation of all three response quantities with the frequency (or period). These *response spectrum* plots can be used to represent the response of single-degree-of-freedom (SDF) systems to a specific ground motion. Response spectrum graphs (such as Fig. 3–68) are generated by numerical integration of actual earthquake acceleration records to determine the maximum values for each period of vibration. These maximum values can be interpreted as the amplified response of the structure when compared to the curve representing the ground motion.

Figure 3–65. Displacement Response

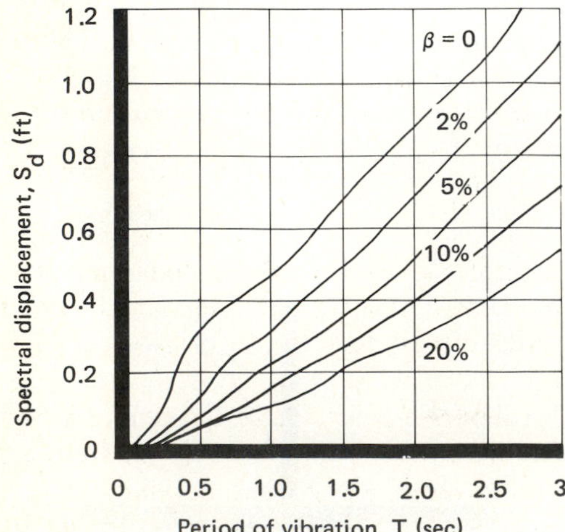

Actual building structures, of course, contain more than a single degree of freedom. A three-story structure, for example, contains three lumped masses as shown in Fig. 3–69, and the structure will consequently have three

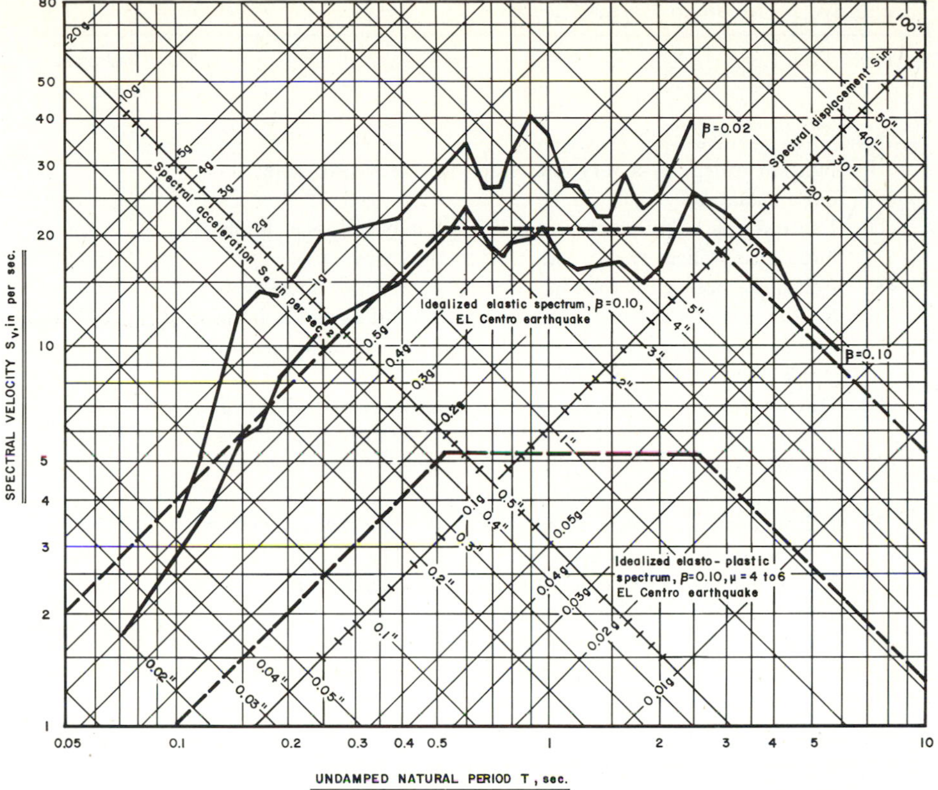

Figure 3–68. Response Spectrum (From John A Blume, Nathan M. Newmark, and Leo H. Corning, *Design of Multi-story Reinforced Concrete Buildings for Earthquake Motions, Portland Cement Association,* Chicago, 1961)

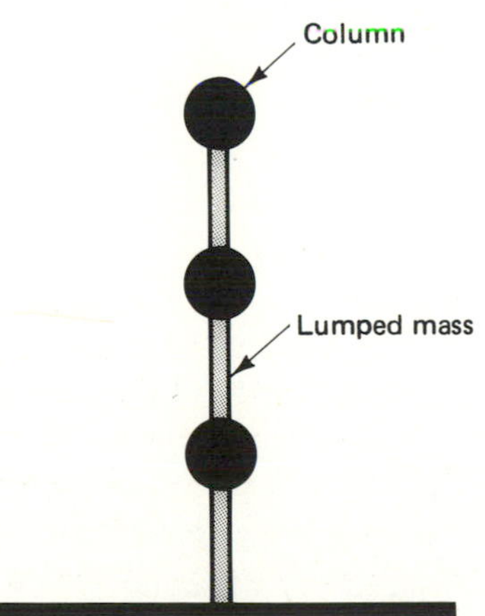

Figure 3–69. Three Degrees of Freedom

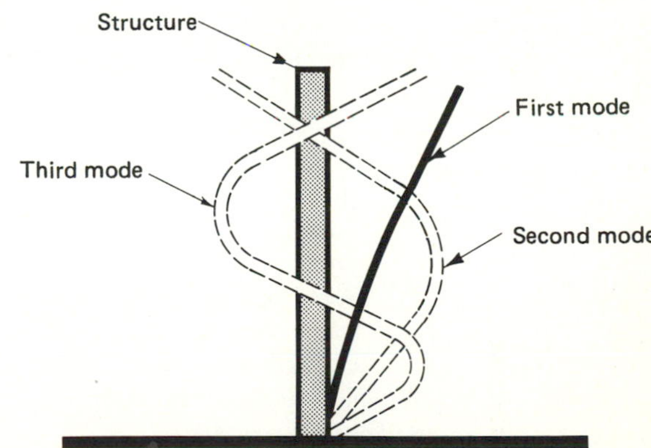

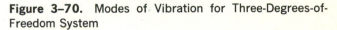

Figure 3–70. Modes of Vibration for Three-Degrees-of-Freedom System

modes of vibration (Fig. 3–70). Each component mode of vibration, however, behaves as an independent SDF system with a characteristic frequency; therefore, the maximum modal response values may be obtained by combining appropriate response spectra for SDF systems. The individual modal maxima generally do not occur simultaneously. The maximum values for all modes are, therefore, normally combined by taking the "square root of the sum of the squares" of the modal maxima. Values of acceleration, velocity, and displacement can thus be determined at each floor level.

Inelastic Behavior. Earthquake forces determined from the linear elastic model would be far in excess of the levels required by present building codes. Actual structures undergo nonlinear inelastic deformations

Table 3–11 Modified Mercalli Scale

		Ground acceleration, a
I.	Not felt except by a very few under especially favorable circumstances.	
II.	Felt only by a few persons at rest, especially on upper floors of buildings. Delicately suspended objects may swing.	0.002–0.003g
III.	Felt quite noticeably indoors, especially on upper floors of buildings, but many people do not recognize it as an earthquake. Standing motor cars may rock slightly. Vibration like passing truck. Duration estimated.	0.003–0.007g
IV.	During the day felt indoors by many, outdoors by few. At night some awakened. Dishes, windows, doors disturbed; walls make creaking sound. Sensation like heavy truck striking building. Standing motor cars rocked noticeably.	0.007–0.015g
V.	Felt by nearly everyone; many awakened. Some dishes, windows, etc., broken; a few instances of cracked plaster; unstable objects overturned. Disturbances of trees, poles, and other tall objects sometimes noticed. Pendulum clocks may stop.	0.015–0.03g
VI.	Felt by all; many frightened and run outdoors. Some heavy furniture moved; a few instances of fallen plaster or damaged chimneys. Damage slight.	0.03–0.07g
VII.	Everybody runs outdoors. Damage negligible in buildings of good design and construction; slight to moderate in well-built ordinary structures; considerable in poorly built or badly designed structures; some chimneys broken. Noticed by persons driving motor cars.	0.07–0.15g
VIII.	Damage slight in specially designed structures; considerable in ordinary substantial buildings, with partial collapse; great in poorly built structures. Panel walls thrown out of frame structures. Fall of chimneys, factory stacks, columns, monuments, walls. Heavy furniture overturned. Sand and mud ejected in small amounts. Changes in well water. Disturbs persons driving motor cars.	0.15–0.3g
IX.	Damage considerable in specially designed structures; well-designed frame structures thrown out of plumb; great in substantial buildings, with partial collapse. Buildings shifted off foundations. Ground cracked conspicuously. Underground pipes broken.	0.3–0.07g
X.	Some well-built, wooden structures destroyed; most masonry and frame structures destroyed with foundations; ground badly cracked. Rails bent. Landslides considerable from river banks and steep slopes. Shifted sand and mud. Water splashed over banks.	0.7–1.5g
XI.	Few, if any, (masonry) structures remain standing. Bridges destroyed. Broad fissures in ground. Underground pipelines completely out of service. Earth slumps and land slips in soft ground. Rails bent greatly.	1.5–3.0g
XII.	Damage total. Waves seen on ground surfaces. Lines of sight and level distorted. Objects thrown upward into the air.	3.0–

which effectively reduce the magnitude of the inertia forces (Fig. 3–71). Yielding at points of maximum moment in a framed structure produces increased rotations and deflections, which allow greater energy absorption of the earthquake forces (Fig. 3–72). Structural design must obviously ensure that this assumed ductility is provided in the member and connection design—and without jeopardizing overall frame stability.

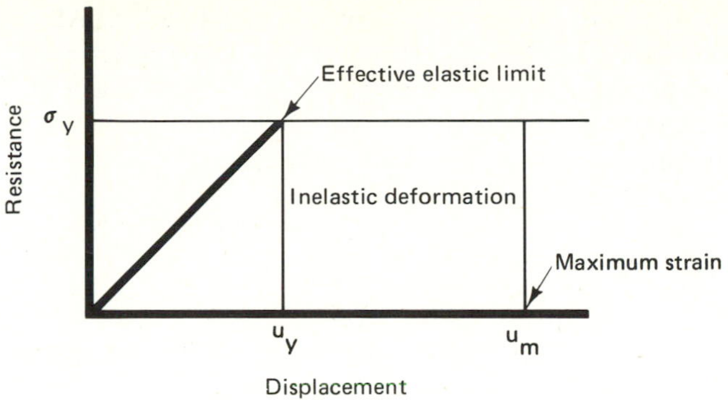

Figure 3–71. Inelastic Behavior

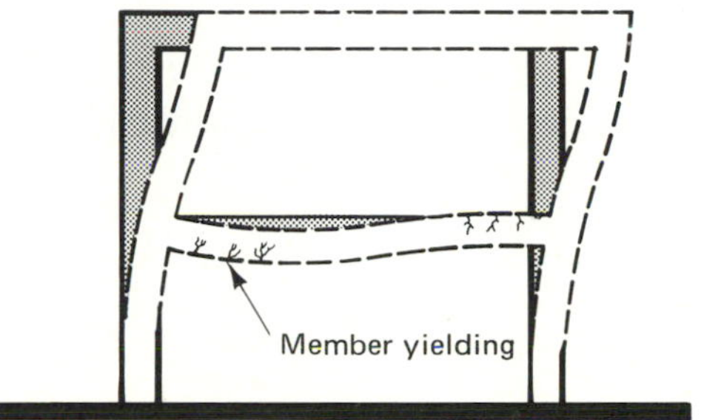

Figure 3–72. Member Yielding

Damage Scale. While the specific characteristics of ground motion are the important parameters for engineering analysis, a single parameter scale has been adopted by most of the industry as an indicator of earthquake intensity and damage effects. The commonly accepted industry standard for measuring the intensity and effects of earthquakes is the Modified Mercalli (MM) scale, illustrated in Table 3–11.

Design Approach

The minimum requirements for seismic design are specified in most building codes. Many codes leave the decision as to whether buildings shall be designed for earthquakes to local authorities. However, seismic design is generally required by federal agencies, such as the Veterans' Administration, GSA, Corps of Engineers, and NAVFAC. In addition, seismic design may be required or should be considered in the design of the following structures: nuclear power plants, high-rise buildings, and emergency facilities.

Where seismic design is performed, two design methods are available: *static analysis* and *dynamic analysis.* The static method utilizes an *equivalent lateral load* procedure which approximates the dynamic loads with a set of externally applied forces (similar to wind load). The dynamic method utilizes a structural dynamic analysis to determine the inertia forces. Both methods must somehow account for the *ground motion,* the *natural frequency* of the structure, *damping,* and *inelastic action* (ductility).

Static Design Method. The static method recommended by ANSI 58.1–1982 establishes a percentage of the total dead load of the structure as a horizontal force applied at the base, termed "base shear." This method uses an estimate of the fundamental period of vibration based on simple formulas that involve only a general description of the building type and overall dimensions. Ground motion and the other parameters are incorporated in a set of coefficients which convert the dead load into an equivalent lateral force.

Base shear. The equation for base shear is

$$V = ZIKCSW \tag{3-18}$$

where V = total lateral shear at the base
Z = numerical coefficient depending on earthquake zone: $\frac{1}{8}$ for Zone 0; $\frac{3}{16}$ for Zone 1; $\frac{3}{8}$ for Zone 2; $\frac{3}{4}$ for Zone 3; 1.0 for Zone 4 (see Fig. 3–73)
I = occupancy importance factor (Tables 3–12 and 3–1)
K = numerical coefficient depending on framing type (Table 3–13)
C = numerical coefficient depending on natural frequency
S = numerical coefficient for soil profile (Table 3–14)
W = total dead load

The product CS need not exceed 0.14 or, for Soil Profile 3 in Seismic Zones No. 3 and No. 4, the product need not exceed 0.11. W is taken equal to the total weight of the structure and applicable portions of other components including, but not limited to, the following:

1. Partitions and permanent equipment, including operating contents.
2. For storage and warehouse structures, a minimum of 25% of the floor live load.
3. The effective snow load (snow loads up to 30 psf are not considered; for loads over 30 psf, the effective snow load is taken as the full snow load unless reduced by the local regulatory authority).

The coefficient I (Table 3–12) modifies the force level based on the *importance* of the building, such as essential facilities and buildings housing large numbers of people. The K coefficient (Table 3–13) accounts for the *ductility* of various framing systems and their ability to absorb energy in the inelastic range. The coefficients Z, C, and S relate directly to *ground motion:* the coeffi-

Table 3–12 Occupancy Importance Factor for Seismic Design[a]

Building classification	*I*
I	1.0
II	1.25
III	1.5
IV	NA[b]

[a] *Minimum Design Loads for Buildings and Other Structures,* ANSI A58.1–1982 (New York: American National Standards Institute, 1982).

[b] Not applicable.

cient C approximates the shape of the response spectrum curve for a selected ground motion; Z (Fig. 3–73) and S (Table 3–14) modify the curve to relate to the seismic history and soil profile of the building location. All of the coefficients except C may be found by referring to the appropriate tables and maps. C must be calculated based on an estimate of the natural period of the building as follows:

$$C = \frac{1}{15\sqrt{T}} \tag{3–19}$$

where T is the fundamental period of the building. *The value of C need not exceed 0.12.*

Natural frequency. The fundamental period of the building, T, in formula 3–19 may be established based on the properties of the seismic resisting system in the direction being analyzed and the use of established methods of mechanics. The period calculated by this method,

Table 3–13 Horizontal Force Factor *K* for Buildings and Other Structures[a,b]

Arrangement of lateral force resisting elements	*Value of* K
1. *Bearing wall system:* A structural system with bearing walls providing support for all, or major portions of, the vertical loads. Seismic force resistance is provided utilizing:	
Unreinforced masonry walls	4.00[c]
Reinforced concrete or reinforced masonry walls or braced frames	1.33
One-, two-, or three-story light wood or metal frame wall systems	1.00
2. *Building frame system:* A structural system with an essentially complete space frame providing support for vertical loads. Seismic force resistance is provided by shear walls or braced frames.	1.00
3. *Moment-resisting frame system:* A structural system with an essentially complete space frame providing support for vertical loads. Seismic force resistance is provided by one of the following moment resisting frame systems:	
Ordinary concrete frames	2.50[c,d]
Ordinary steel frames	1.00
Special frames	0.67[e]
4. *Dual system:* A structural system with an essentially complete space frame providing support for vertical loads. Seismic force resistance is provided by a combination of a special moment resisting frame system and shear walls or braced frames.	0.80
5. *Elevated tanks:* Tanks plus full contents where tanks are supported on four or more cross-braced legs and not supported on a building.	2.50[f]
6. *Structures other than buildings:* Structures other than buildings or elements of buildings.	2.00

[a] *Minimum Design Loads for Buildings and Other Structures,* ANSI 58.1–1982 (New York: American National Standards Institute, 1982).

[b] Where wind load would produce higher stresses, this load shall be used instead of the loads resulting from earthquake forces.

[c] Not permitted in Seismic Zones 2, 3, and 4.

[d] For systems not specially detailed for cyclic inelastic straining.

[e] Special moment-resisting space frames conforming to the requirements of "Specification for the Design, Fabrication, and Erection of Standard Steel for Buildings," Part II, Sections 2.7, 2.8, and 2.9 or "Building Code Requirements for Reinforced Concrete, Appendix A."

[f] The minimum value of KC shall be 0.12 and the maximum value of KCS need not exceed 0.29 or 0.23 for soil profile 3 in Seismic Zones 3 and 4. The tower shall be designed for an accidental torsion of 5%. Elevated tanks that are supported by buildings or do not conform to type or arrangement of supporting elements as described above shall be designed using $C_p = 0.3$.

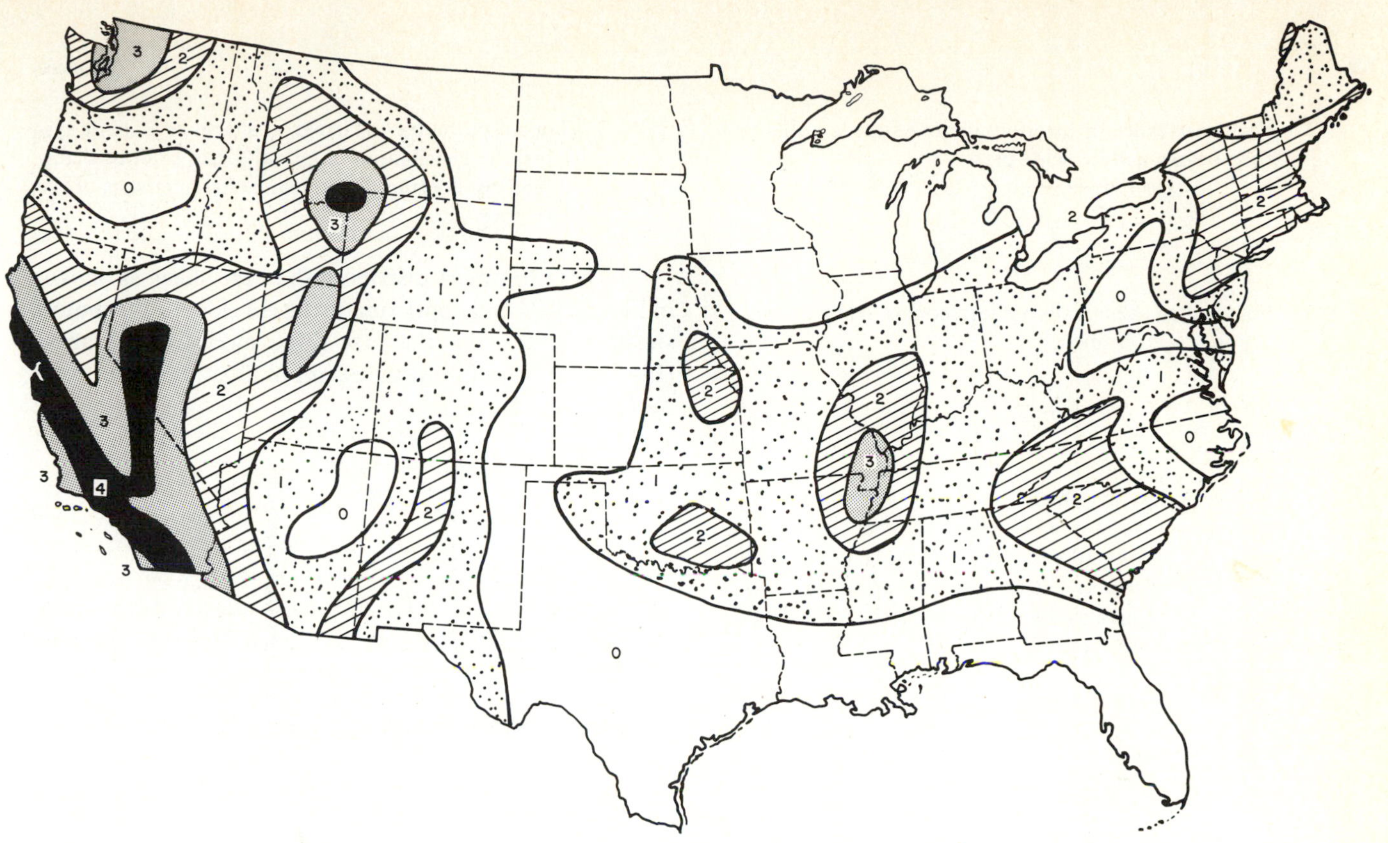

Figure 3–73. Map for Seismic Zones (From *Minimum Design Loads for Buildings and Other Structures,* ANSI A 58.1-1982, American National Standards Institute, New York, 1982)

Table 3–14 Soil Profile Coefficients[a,b]

Type	*Description*	*Coefficient*
S_1	Rock of any characteristic, either shalelike or crystalline in nature; such material may be characterized by a shear wave velocity greater than 2500 ft/sec *or* Stiff soil conditions where the soil depth is less than 200 ft and the soil types overlying rock are stable deposits of sands, gravels, or stiff clays	1.0
S_2	Deep cohesionless or stiff clay conditions, including sites where the soil depth exceeds 200 ft and the soil types overlying rock are stable deposits of sands, gravel, or stiff clays	1.2
S_3	Soft to medium-stiff clays and sands, characterized by 30 ft or more of soft to medium-stiff clays with or without intervening layers of sand or other cohesionless soils	1.5

[a] *Minimum Design Loads for Buildings and Other Structures,* ANSI A58.1-1982 (New York: American National Standards Institute, 1982).

[b] In locations where the soil properties are not known in sufficient detail to determine the soil profile type or where the profile does not fit any of the three types, soil profile S_2 or soil profile S_3 shall be used, whichever gives the larger value for *CS.*

however, must not exceed by more than 20% the one calculated by the approximate formulas that follow. Established methods of mechanics include the *Rayleigh method,* described in Appendix 3D.

The value for T may be determined by the following approximate formulas:

1. For shear walls, or exterior concrete frames utilizing deep beams and/or wide piers:

$$T = \frac{0.05 h_n}{\sqrt{D}} \tag{3–20}$$

2. For isolated shear walls not interconnected by frames, or for braced frames:

$$T = \frac{0.05 h_n}{\sqrt{D_S}} \tag{3–21}$$

3. For buildings in which the lateral forces resisting system consists of moment-resisting space frames capable of resisting 100% of the required lateral forces and such system is not enclosed by or adjoined by more rigid elements tending to prevent the frame from resisting lateral forces:

$$T = C_T h_n^{3/4} \tag{3–22}$$

where h_n = height above the base, ft
D = dimension of the structure in direction parallel to applied force, ft
D_S = longest dimension of a shear wall or braced frame in a direction parallel to applied forces
C_T = 0.035 for steel frames and 0.025 for concrete frames

Vertical distribution. After calculation of the fundamental period and base shear, the total lateral force V must be distributed over the height of the structure. The following formulas are used for *vertical distribution:*

1. For structures having regular shapes of framing systems:

$$V = F_t + \sum_{i=1}^{N} F_i \tag{3–23}$$

where F_t is the concentrated force at the top given by the following formula, but need not exceed $0.25V$ and may be considered as 0, where T is 0.7 sec or less:

$$F_t = 0.07TV \tag{3–24}$$

and $\Sigma_{i=1}^{N} F_i = \Sigma F_x$, where F_x is the force at each level designated as x.

$$F_x = \frac{(V - F_t) w_x h_x}{\sum_{i=1}^{N} w_i h_i} \tag{3–25}$$

where w_i, w_x = that portion of w which is assigned level i or x
h_i, h_x = height above the base to level i or x, ft

2. For structures having irregular shapes or framing systems, the distribution of lateral forces shall be determined considering the dynamic characteristics of the structure.

These formulas distribute the base shear as shown in Figure 3–74 and approximate the envelope of maximum shears produced by the first three modes of vibration (Fig. 3–75). The following example illustrates the *static design method.*

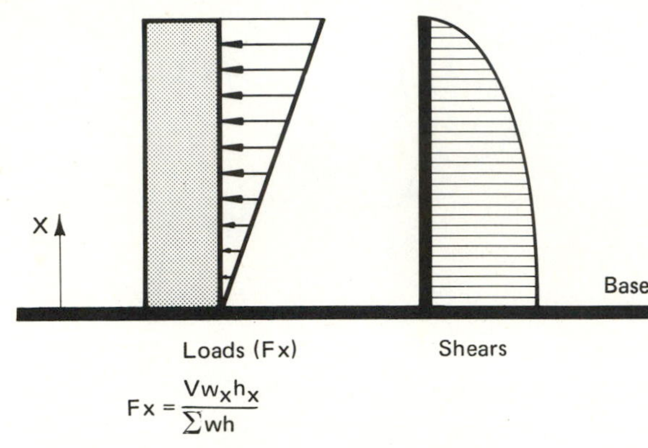

Figure 3–74. Distribution of Base Shear

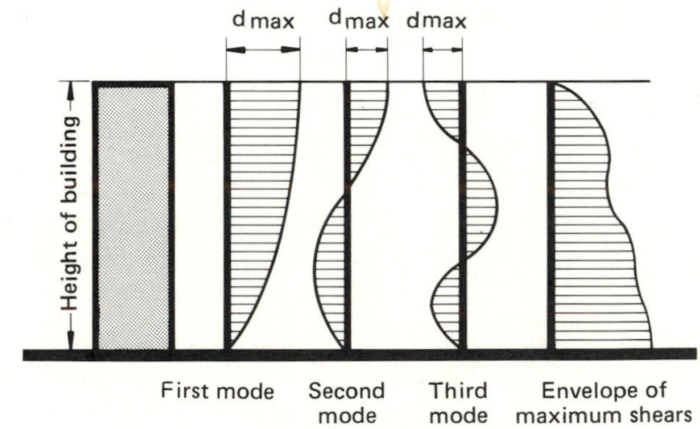

Figure 3–75. Envelope of Maximum Shears

EXAMPLE PROBLEM 3–7: Earthquake Load by Equivalent Lateral Force Procedure

Calculate the earthquake forces on a braced steel frame building to be built in Charleston, South Carolina. The building will be 60 ft x 60 ft (18.3 m x 18.3 m) in plan and will be four stories high, 15 ft (4.6 m) between floors. The roof dead load is 20 psf (958 N/m²), and the floor dead load is 100 psf (4788 N/m²). Wall weight is 10 psf (479 N/m²); assume no snow load or partition load.

Solution

1. Calculate W.

Floors = (60 ft)(60 ft)(100 psf)(3) = 1,080,000 lb

$$\text{Roof} = (60\text{ ft})(60\text{ ft})(20\text{ psf}) = 72{,}000\text{ lb}$$
$$\text{Walls} = (60\text{ ft})(4)(15\text{ ft})(4)(10\text{ psf}) = 144{,}000\text{ lb}$$
$$W = 1{,}296{,}000\text{ lb} = 1{,}296\text{ kips}\ (5{,}764{,}608\text{ N})$$

2. Determine coefficients:

$Z = \frac{3}{8}$ (building is in Zone 2 by Fig. 3–73)

$I = 1.5$ (assume Class III building and consult Table 3–12)

$K = 1.0$ (Table 3–13)

$$C = \frac{1}{15\sqrt{T}}, \quad T = \frac{0.05h_n}{D_S} = \frac{0.05(60\text{ ft})}{60\text{ ft}} = 0.387\text{ sec}$$

$$C = \frac{1}{15\sqrt{0.387}} = 0.107$$

$S = 1.5$ (Table 3–14)

(*Note: CS* need not exceed 0.14, but ignore for this example.)

3. Calculate V:

$$V = (3/8)(1.5)(1.0)(0.107)(1.5)(1296) = 117\text{ kips }(520{,}416\text{ N})$$

4. Calculate force at each level:

$$F_t = 0 \text{ since } T < 0.7\text{ sec}$$

$$F_x = \frac{(V-0)w_x h_x}{\sum_{i=1}^{N} w_i h_i}$$

See Table A.

Table A

Level	*Floor/Roof weight*	*Wall weight*	w_x	h_x	$w_x h_x$	F_x[a]
1	360	36	396	15	5,940	16.9
2	360	36	396	30	11,880	33.9
3	360	36	396	45	17,820	50.8
4	72	18	90	60	5,400	15.4
					$\Sigma\, w_i h_i = 41{,}040$	117

[a] $F_x = (w_x h_x/41040)\ (117\text{ kips}) = 0.00285 w_x h_x$.

Dynamic Design

A *dynamic analysis* is the second method for performing seismic analysis. This method of seismic analysis is required for certain types of structures and is the only true representation of the response of the building to earthquake motion. Computer programs are available for performing a linear or nonlinear response history analysis which is a solution of the equations of motion at succeeding discrete steps of an accelograph record. This procedure is complicated, but the nonlinear analysis does directly account for the actual inelastic behavior of the structure. The most commonly used method of dynamic analysis is *modal analysis,* which utilizes the technique of combining the response of the independent (SDF) modes of vibration based on the *response spectrum* approach to ground motion. This method is described in Appendix 3D and involves three steps: (1) determination of the natural periods of vibration and mode shapes for the structure; (2) calculation of modal shears, displacements, and so on, based on a selected design spectrum; and (3) calculation of total response values by combining the individual modal values.

Choice of Design Approach

Both the *equivalent lateral force* and the *modal analysis* procedures are applicable to structures that have generally uniform ductility throughout their frames and minimum eccentricity between centers of mass and centers of resistance. The latter characteristic allows uncoupling of the motions in the two lateral directions and the torsional direction and permits independent analysis for each lateral component, with torsion effects considered only indirectly by summing story shears times eccentricities. Structures with strongly coupled lateral-torsional motions may be analyzed by extending the modal method to three degrees of freedom per floor: two translational motions and one torsional motion. Structures with significant ductility variance may be analyzed with nonlinear response history computer programs.

The *equivalent lateral force* procedure is most applicable for structures with uniform distribution of mass and stiffness, since the simple formulas only estimate the fundamental period and distribution of forces over the height. The *modal analysis* procedure accounts for several periods and mode shapes and is more accurate in determining total base shear, as well as vertical distribution.

3–15 SAFETY FACTORS AND COMBINATION OF LOADS

The term *safety factor* has been used for years by structural engineers to describe the margins of safety incorporated in design to account for uncertainties. Uncertainty is associated with material properties, residual stress, accidental imperfections, erection tolerances, accidental overloads, and other factors. *Allowable stress* design methods lump all of these factors into a single safety factor. However, since uncertainty can be treated with probability theory, there is a clearly established trend to think of structural safety in terms of factors that are based on the probability of certain events occurring, and to separate consideration of load uncertainty from material uncertainty. Called *strength design* or LRFD (for load and resistance factor design), this method assigns "load factors" to loads and combinations of loads to account for uncertainty associated with loads. ANSI A58.1–1982 recommends loads and load combinations for both allowable stress design and strength design. The following sections describe the ANSI approach.

Loads

The following loads (and associated symbols) shall be considered:

D = dead load consisting of:
- the weight of the member itself
- the weight of all materials of construction incorporated into the building to be permanently supported by the member, including built-in partitions
- the weight of permanent equipment

E = earthquake load
F = loads due to fluids with well-defined pressures and maximum heights
L = live loads due to intended use and occupancy, including loads due to movable objects and partitions, traveling cranes, and loads temporarily supported by the structure during maintenance; if resistance to impact loads is taken into account in design, such effects shall be included with live load L
L_r = roof live loads
S = snow loads
R = rain loads
H = loads due to the weight and lateral pressure of soil and water in soil
P = loads, forces, and effects due to ponding
T = self-straining forces and effects arising from contraction or expansion resulting from temperature changes, shrinkage, moisture changes, creep in component materials, movement due to differential settlement, or combinations thereof
W = wind load

Load Combinations: Allowable Stress Design

All loads shall be considered to act in the following combinations:

(1) D

(2) $D + L + (L_r \text{ or } S \text{ or } R)$

(3) $D + (W \text{ or } E)$

(4) $D + L + (L_r \text{ or } S \text{ or } R) + (W \text{ or } E)$

When structural effects of F, H, P, or T are significant, they shall be considered in design. The following load combination factors may be used.

- 0.75 for combinations including in addition to D:

$$L + (L_r \text{ or } S \text{ or } R) + (W + E)$$
$$L + (L_r \text{ or } S \text{ or } R) + T$$
$$(W \text{ or } E) + T$$

- 0.66 for combinations including in addition to D:

$$L + (L_r \text{ or } S \text{ or } R) + (W \text{ or } E) + T$$

Load Combinations: Strength Design

Design strength shall exceed the effects of the factorial loads in the following combinations:

(1) $1.4D$

(2) $1.2D + 1.6L + 0.5(L_r \text{ or } S \text{ or } R)$

(3) $1.2D + 1.6(L_r \text{ or } S \text{ or } R) + (0.5L \text{ or } 0.8W)$

(4) $1.2D + 1.3W + 0.5L + 0.5(L_r \text{ or } S \text{ or } R)$

(5) $1.2D + 1.5E + (0.5L \text{ or } 0.2S)$

(6) $0.9D - (1.3W \text{ or } 1.5E)$

The load factor on L in combinations 3, 4, and 5 shall equal 1.0 for garages, areas occupied as places of public assembly, and all areas where the live load is greater than 100 psf. When structural effects of F, H, P, or T are significant, they shall be considered in design as the following factored loads: $1.3F$, $1.6H$, $1.2P$, and $1.2T$.

Dead Load Counteracting Live Load

When loads other than dead counteract dead loads in a structural member or joint, special care shall be exercised by the designer to ensure adequate safety for possible stress reversals.

3–16 SUMMARY

In this chapter we described the basic types of structural loads. Our emphasis was on understanding the nature of these loads as well as methods to calculate their magnitude.

As we have seen in this chapter, structural design loads are often based on standards adopted by building codes, sometimes leading us to a false sense of security regarding our designs. Codes cannot replace the individual judgment and experience of the engineer responsible for the design. We must have an understanding of the nature of loads, the response of the structure, and load path concepts in order to provide safe and economical designs.

Chapter 4 begins the process of assembling materials and elements into structural systems capable of supporting the loads and meeting the other project requirements. We will later return to structural loads as we develop preliminary designs in Chapters 5, 6, 7, and 8.

PROBLEMS

3–1. The beam designated B-1 in Figure 3–76 is part of the floor framing system for an office building. The building floor is a 4-in. (102-mm) concrete slab and the ceiling is acoustical fiber tile supported by a ceiling suspension system. The building is lighted by ceiling fixtures and heated and cooled with ceiling supplied air diffusers. Draw a sketch of beam B-1 showing the mathematical model, including the design dead and live loads. Show a tabulation of the loads to include how you arrived at the total loads.

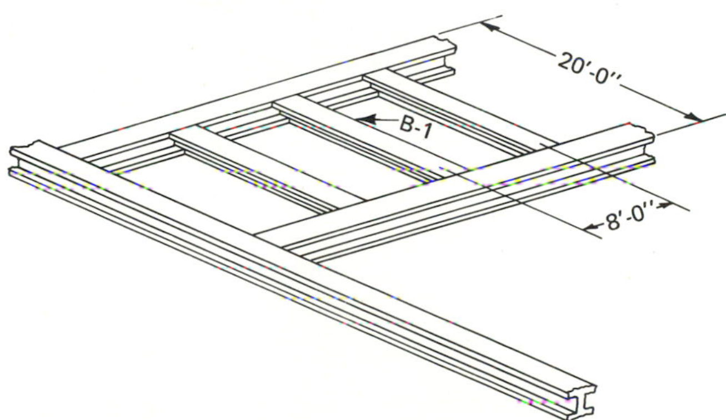

Figure 3–76. Problem 3–1

3–2. The roof system shown in the Fig. 3–77 is to be designed for snow load. The building is located in Portland, Maine, and is adjacent to a higher structure as shown. Draw a sketch showing the expected snow configuration on the roof and calculate and show the loads.

Figure 3–77. Problem 3–2

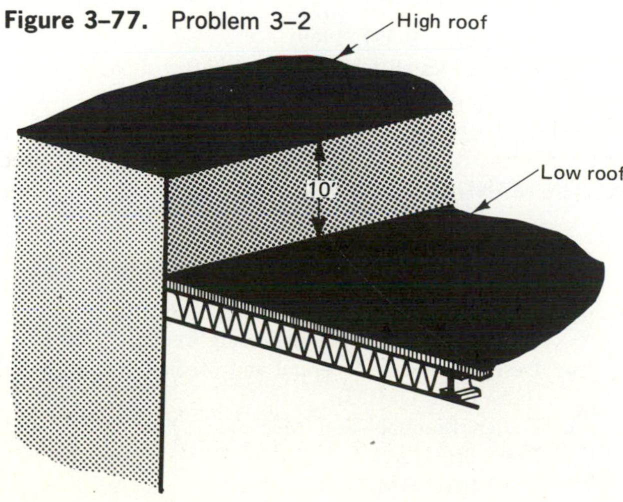

3–3. Figure 3–78 shows the influence line for negative moment at the interior support of a two-span bridge. Draw a sketch showing where you would place an HS 20 truck loading to produce the maximum negative moment.

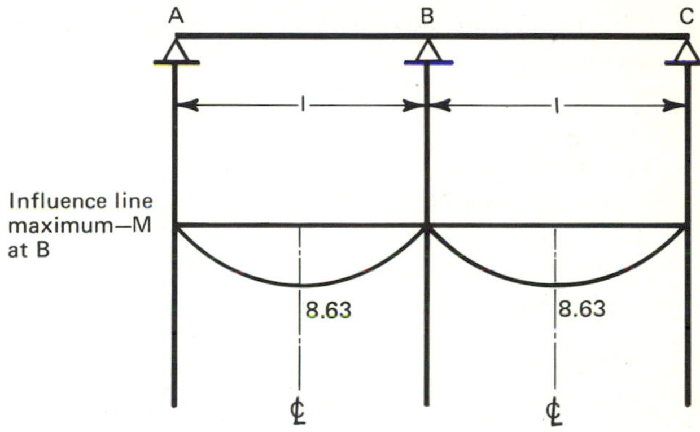

Figure 3–78. Problem 3–3

3–4. Figure 3–79 shows an underground rectangular concrete structure. The walls are supported by the base slab (assume a uniform distribution of soil pressure on the slab) and the water table is assumed to be at the ground surface. Draw a sketch showing the structure and the loads for which the base slab must be designed.

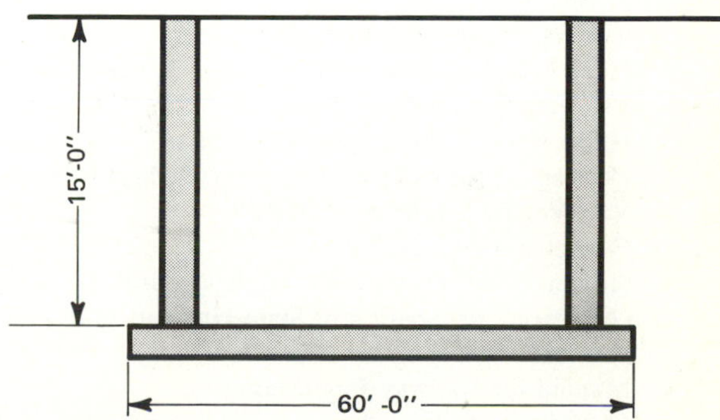

Figure 3–79. Problem 3–4

3–5. A 10-story hotel, 100 ft × 100 ft (30.5 m × 30.5 m) in plan, is to be built on the coast of South Carolina. Assume a 100-ft (30.5-m) total height. Draw a sketch showing the design wind loads over the height of the building structure.

Resolve the wind pressure into concentrated lateral loads applied at each floor (and roof) level and show these forces.

3–6. A three-story building, 100 ft × 100 ft (30.5 m × 30.5 m) in plan, is to be built on the coast of California and the structural framing system will be a moment-resisting steel frame. Assume that $S = 1.2$ and $I = 1.0$. The roof dead load is 30 psf (1436 N/m²) and the floor dead load is 100 psf (4788 N/m²). The masonry walls, 4 in. (102 mm) thick, are supported at each floor level and the columns weigh 30 plf (438 N/m). Draw a sketch showing the lateral earthquake loads for which you would design the structure. Show appropriate calculations to indicate how you arrived at the design loads.

APPENDIX 3A: Codes

GENERAL BUILDING CODES

- Standard Building Code (Southern Building Code Congress, 1982)
- Uniform Building Code (International Conference of Building Officials, 1979)
- National Building Code (American Insurance Association)
- Basic Building Code (Building Officials and Code Administrators International, Chicago)
- American National Standard Building Code (American National Standard Institute, ANSI A58.1–1982)
- South Florida Building Code (1982)
- New York City Building Code (1970)
- National Fire Protection Association (NFPA)

DESIGN-RELATED CODES

- Design Manual—Structural Engineering (NAVFAC DM-2 1970)
- Seismic Design for Buildings (TM 5–809–10; NAVDOCKS P. 355; AFM 88–3, Chap. 13; 1966)
- Air Force Design Manual—Principles and Practices for Design of Hardened Structures (AFSWC-TDR-62-138), 1962 (AFWL-TR-74-102; pending release)
- HUD (Housing and Urban Development) Minimum Property Standards (1972)
- HEW (Health, Education and Welfare) (Various Handbooks for Design of Medical Facilities)
- VA (Veteran's Administration) (Various Design Handbooks)
- Standard Specifications for Highway Bridges (American Association of State Highway and Transportation Officials)
- Manual for Railway Engineering (American Railway Engineering Association, 1980)
- Building Code Requirements for Reinforced Concrete (American Concrete Institute, ACI 318–77)
- Specification for the Design Fabrication and Erection of Structural Steel for Buildings (American Institute of Steel Construction, 1978)
- Specification for the Design of Cold-Formed Steel Structural Members (American Iron and Steel Institute, 1980)
- Aluminum Construction Manual (Aluminum Association, 1971)
- Design Manual for Structural Tubing (Committee of Steel Pipe Producers, 1974)
- Metal Building Systems Manual (Metal Building Manufacturer's Association, 1971)
- Building Code Requirements for Engineered Brick Masonry (BIA, 1969)
- Specifications for the Design and Construction of Load Bearing Concrete Masonry (National Concrete Masonry Association, 1974)
- Design Specifications for Light Metal Plate Connected Wood Trusses (Truss Plate Institute, 1970)
- PCI Design Handbook (Prestressed Concrete Institute)
- PCI Post-tensioning Manual (Prestressed Concrete Institute)
- Timber Construction Manual (American Institute of Timber Construction, 1974)
- Standard Specifications and Load Tables (Steel Joist Institute)
- Steel Roof Deck Design Manual (Metal Deck Institute, 1975)
- Minimum Performance Standard for Single-Wall Air-Supported Structures (Canvas Products Association, 1971)
- Structures to Resist the Effects of Accidental Explosions (NAVFAC P-397, 1969)
- Seismic Design for Buildings (NAVFAC P-355, 1973)

DESIGN MANUALS

- Harbor and Coastal Facilities (NAVFAC-DM-26)
- Shore Protection Manual (Vols. 1, 2, 3) (U.S. Army Coastal Engineering Center, 1973)
- Design Manual—Hospital and Medical Facilities (NAVFAC DM-33)
- Design Manual—Soil Mechanics, Foundations and Earth Structures (NAVFAC DM-7)

APPENDIX 3B: Minimum Dead Loads

	psf
Walls	
4-in. clay brick, medium absorption	39
4-in. concrete brick, heavy aggregate	46
4-in. concrete brick, light aggregate	33
8-in. clay brick, medium absorption	79
8-in. concrete brick, heavy aggregate	89
8-in. concrete brick, light aggregate	68
$12\frac{1}{2}$-in. clay brick, medium absorption	115
$12\frac{1}{2}$-in. concrete brick, heavy aggregate	130
$12\frac{1}{2}$-in. concrete brick, light aggregate	98
17-in. clay brick, medium absorption	155
17-in. concrete brick, heavy aggregate	174
17-in. concrete brick, light aggregate	130
22-in. clay brick, medium absorption	194
22-in. concrete brick, heavy aggregate	216
22-in. concrete brick, light aggregate	160
4-in. brick, 4-in. load-bearing structural clay tile backing	60
4-in. brick, 8-in. load-bearing structural clay tile backing	75
8-in. brick, 4-in. load-bearing structural clay tile backing	102
8-in. load-bearing structural clay tile	42
12-in. load-bearing structural clay tile	58
4-in. concrete block, heavy aggregate	30
6-in. concrete block, heavy aggregate	43
8-in. concrete block, heavy aggregate	55
12-in. concrete block, heavy aggregate	85
4-in. concrete block, light aggregate	21
6-in. concrete block, light aggregate	30
8-in. concrete block, light aggregate	35
12-in. concrete block, light aggregate	55
2-in. furring tile, one side of masonry wall, add to above figures	12
4-in. stone	55
4-in. glass block	18
Windows: glass, frame, and sash	8
Curtain walls	See manufacturer
1-in. structural glass	15
$\frac{1}{4}$-in. corrugated cement asbestos	3
Exterior stud walls	
2×4 at 16 in., $\frac{5}{8}$-in. gyp. insulation, $\frac{3}{8}$-in. siding	11
2×6 at 16 in., $\frac{5}{8}$-in. gyp. insulation, $\frac{3}{8}$-in. siding	12
With brick veneer	48
Partitions	
4-in. clay tile	18
6-in. clay tile	28
8-in. clay tile	34
4-in. concrete block, heavy aggregate	30
6-in. concrete block, heavy aggregate	42
8-in. concrete block, heavy aggregate	55
12-in. concrete block, heavy aggregate	85
4-in. concrete block, light aggregate	20
6-in. concrete block, light aggregate	28
8-in. concrete block, light aggregate	38
12-in. concrete block, light aggregate	55
Wood studs 2×4, unplastered	4

APPENDIX 3B: Minimum Dead Loads (*Continued*)

	psf
Movable steel partitions	4
1-in. plaster, cement	10
1-in. plaster, gypsum	5
Lathing, metal	$\frac{1}{2}$
Lathing, $\frac{1}{2}$-in. gypsum board	2
Wood studs 2 × 4, unplastered	4
Wood studs 2 × 4, plastered one side	12
Wood studs 2 × 4, plastered two sides	20
Concrete slabs	
Concrete, reinforced-stone, per inch of thickness	$12\frac{1}{2}$
Concrete, reinforced-lightweight sand, per inch of thickness	$9\frac{1}{2}$
Concrete, reinforced, lightweight, per inch of thickness	9
Concrete, plain stone, per inch of thickness	12
Concrete, plain, lightweight, per inch of thickness	$8\frac{1}{2}$
Floor finish	
3-in. wood block on mastic, no fill	10
1-in. cement finish on stone-concrete fill	32
1-in. terrazzo on stone-concrete fill	32
Marble and mortar on stone-concrete fill	33
3-in. wood block on $\frac{1}{2}$-in. mortar base	16
Solid flat tile on 1-in. mortar base	23
2-in. asphalt block, $\frac{1}{2}$-in. mortar	30
1-in. terrazzo, 2-in. stone concrete	32
$1\frac{1}{2}$-in. terrazzo floor finish directly on slab	19
$\frac{3}{4}$-in. ceramic or quarry tile on $\frac{1}{2}$-in. mortar bed	16
$\frac{3}{4}$-in. ceramic or quarry tile on 1-in. mortar bed	23
$\frac{1}{4}$-in. linoleum or asphalt tile	1
$\frac{7}{8}$-in. hardwood flooring	4
$\frac{3}{4}$-in. subflooring	3
Slate, per inch of thickness	15
Floor fill	
Cinder concrete, per inch	9
Lightweight concrete, per inch	8
Sand, per inch	8
Stone concrete, per inch	12

	12-in. spacing	*16-in. spacing*	*24-in. spacing*
Wood-joist floors (no plaster): double wood floor joist sizes (in.)			
2 × 6	6	5	5
2 × 8	6	6	5
2 × 10	7	6	6
2 × 12	8	7	6
Ceilings			
Acoustical fiber tile			1
Gypsum board, per $\frac{1}{8}$ in. of thickness			0.55
Plaster on tile or concrete			5
Suspended metal lath and gypsum plaster			10
Suspended metal lath and cement plaster			15
Suspended steel channel system			2
Mechanical duct allowance			4
Roof and wall coverings			
Asbestos-cement shingles			4

APPENDIX 3B: Minimum Dead Loads (*Continued*)

	psf
Asphalt shingles	2
Cement tile	16
Clay tile (for mortar add 10 lb)	
2-in. book tile	12
3-in. book tile	20
Roman	12
Spanish	19
Ludowici	10
Composition	
Three-ply ready roofing	1
Four-ply felt and gravel	$5\frac{1}{2}$
Five-ply felt and gravel	6
Copper or tin	1
Corrugated asbestos-cement roofing	4
Corrugated iron	2
Gypsum sheating, $\frac{1}{2}$-in.	2
Insulation, roof boards (per inch of thickness)	
Polystyrene foam	0.2
Urethane foam with skin	0.5
Cellular glass	0.7
Perlite	0.8
Fibrous glass	1.1
Fiberboard	1.5
Plywood, per $\frac{1}{8}$ in. of thickness	0.4
Skylight, metal frame, $\frac{3}{8}$-in. wire glass	8
Slate, $\frac{3}{16}$-in.	7
Slate, $\frac{1}{4}$-in.	10
Wood sheathing, per inch thickness	3
Wood shingles	3
Waterproofing, 5-ply membrane	5
2-in. wood decking, Douglas fir	5
3-in. wood decking, Douglas fir	8
Metal deck, 20 gage	2.5
Metal deck, 18 gage	3

	pcf
Materials	
Cast-stone masonry (cement, stone, sand)	144
Cinder fill	57
Concrete, plain	
Cinder	108
Expanded-slag aggregate	100
Haydite (burned-clay aggregate)	90
Slag	132
Stone (including gravel)	144
Vermiculite and perlite aggregate, non-load-bearing	25–50
Other light aggregate, load-bearing	70–105
Concrete, reinforced	
Cinder	111
Slag	138
Stone (including gravel)	150
Masonry, ashlar	
Granite	165
Limestone, crystalline	165
Limestone, oolitic	135

APPENDIX 3B: Minimum Dead Loads (*Continued*)

	pcf
Marble	173
Sandstone	144
Masonry, brick	
Hard (low absorption)	130
Medium (medium absorption)	115
Soft (high absorption)	100
Masonry, rubble mortar	
Granite	153
Limestone, crystalline	147
Limestone, oolitic	138
Marble	156
Sandstone	137
Terra-cotta, architectural	
Voids filled	120
Voids unfilled	72
Wood, seasoned	
Ash, commercial white	41
Cypress, southern	34
Fir, Douglas, coast region	34
Oak, commercial reds and whites	47
Pine, southern yellow	37
Redwood	28
Spruce, red, white, and Sitka	29
Western hemlock	32

APPENDIX 3C: Minimum Live Loads[a]

Occupancy or use	*Live load (psf)*
Apartments (*see* Residential)	
Armories and drill rooms	150
Assembly areas and theaters	
Fixed seats	60
Movable seats	100
Lobbies	100
Platforms (assembly)	100
Stage	150
Balcony (exterior)	100
On one- and two-family residences only and not exceeding 100 sq ft	60
Bowling alleys, poolrooms, and similar recreational areas	75
Corridors	
First floor	100
Other floors, same as occupancy served except as indicated	
Dance halls and ballrooms	100
Decks (patio and roof), same as area served or for the type of occupancy accommodated	
Dining rooms and restaurants	100
Dwellings (*see* Residential)	
Fire escapes	100
On single-family buildings only	40

[a] *Minimum Design Loads for Buildings and Other Structures*, ANSI A58.1–1982 (New York: American National Standards Institute, 1982).

APPENDIX 3C: Minimum Live Loads[a] (*Continued*)

Occupancy or use	*Live load (psf)*
Garages (passenger cars only)—for trucks and buses, use AASHO lane loads (see below for concentrated load requirements)	50
Grandstands (*see* Stadium and arena bleachers)	
Gymnasiums, main floors and balconies	100
Hospitals	
Operating rooms, laboratories	60
Private rooms	40
Wards	40
Corridors, above first floor	80
Hotels (*see* Residential)	
Libraries	
Reading rooms	60
Stack rooms (books and shelving at 65 pcf) but not less than	150
Corridors, above first floor	80
Manufacturing	
Light	125
Heavy	250
Marquees and canopies	75
Office buildings (file and computer rooms shall be designed for heavier loads based on anticipated occupancy)	
Offices	50
Lobbies	100
Penal institutions	
Cell blocks	40
Corridors	100
Residential	
Dwellings (one- and two-family)	
Uninhabitable attics without storage	10
Uninhabitable attics with storage	20
Habitable attics and sleeping areas	30
All other areas	40
Hotels and multifamily houses	
Private rooms and corridors serving them	40
Public rooms and corridors serving them	100
Stadium and arena bleachers	100
Schools	
Classrooms	40
Corridors above first floor	80
Sidewalks, vehicular driveways, and yards, subject to trucking	250
Stairs and exitways	100
Storage warehouse	
Light	125
Heavy	250
Stores	
Retail	
First floor	100
Upper floors	75
Wholesale, all floors	125
Walkways and elevated platforms (other than exitways)	60
Yards and terraces, pedestrians	100

APPENDIX 3C: Minimum Live Loads[a] (*Continued*)

Minimum Concentrated Loads

Location	Load (lb)
Elevator machine room grating (on area of 4 sq in.)	300
Finish light floor plate construction (on area of 1 sq in.)	200
Garages (passenger cars with not more than 9 passengers; on area of 20 sq in.)	2000
Office floors	2000
Scuttles, skylight ribs, and accessible ceilings	200
Sidewalks	8000
Stair treads (on area of 4 sq in. at center of tread)	300

APPENDIX 3D: Dynamic Analysis by Modal Superposition

GENERAL

The modal superposition method, or *modal analysis,* models the structure as a lumped-mass model, with a single degree of freedom at each floor level. Each mode of vibration is treated separately as a single-degree-of-freedom (SDF) system, and the dynamic response determined with a selected response spectrum. Analysis is performed for each of two mutually perpendicular axes. Single-bay frames may be used to represent the structure provided that all frames parallel to the earthquake motion are similar, and floors act as rigid diaphragms to distribute forces equally. If these assumptions cannot be made, an analysis that considers horizontal distribution to all frames must be used. Total response is determined by combining the response of the individual modes: normally, the lowest three modes of vibration or all modes of vibration for structures with periods greater than 0.4 (ATC recommendation, Ref. 4).

The following sections describe the steps involved in a modal analysis and Example Problem 3–8 illustrates the procedure.

NATURAL FREQUENCY

Dynamic analysis computer programs may be used to calculate the periods of vibration and mode shapes, or any elastic frame analysis program may be used with an approximate iterative procedure attributed to Rayleigh and others. The Rayleigh procedure is described in the following paragraphs.

The structural system is first modeled as a lumped-mass system (mass concentrated at the floor levels). There will be as many modes of vibration as there are degrees of freedom, one for each lumped mass. The displacement at the ith mass vibrating in the mth mode, assuming steady-state harmonic vibration, can be expressed as follows:

$$\text{displacement} = \phi_{im} \sin \omega_m t$$

where ϕ_{im} = maximum displacement of mass at ith level when vibrating in the mth mode, in.
ω = natural circular frequency ($2\pi f$), rad/sec
t = time, sec

The velocity of the mass is expressed as the first derivative of the displacement:

$$\text{velocity} = \omega_m \phi_{im} \cos \omega t$$

The acceleration of the mass is expressed by the second derivative of the displacement:

$$\text{acceleration} = \omega_m^2 \phi_{im} \sin \omega_m t$$

The negative value of this acceleration can be considered a reversed effective force (or inertial force) when multiplied by the mass m_i. The maximum inertial force at any level of the structure can then be expressed as

$$\text{inertia force } (F_{im}) = m_i \omega_m^2 \phi_{im} = \frac{w_i}{g} \omega_m^2 \phi_{im}$$

where w_i = weight at level i
g = acceleration of gravity

To find the value of ω, it is necessary to find a set of displacements (ϕ_{im}) at each mass point that when multiplied by the weight, w, and circular frequency, ω^2, give inertial forces, F, which will produce these same deflections when applied to the structure. A series of successive approximations can be made as follows:

1. Assume a deflection shape ϕ for the structure. A linear approximation is usually good for the first mode shape. Express the shape ϕ as a "normalized" quantity ϕ_{ia}, the lowest value set equal to 1.0.
2. Calculate the inertia forces at each floor level by multiplying the weight by the assumed shape.
3. Compute the deflections of the structure at each level due to lateral forces equal to the inertia forces determined in step 2.
4. Recompute the deflected shape ϕ' (normalized) and compare with the assumed shape in step 1. Two or three cycles may be necessary to get approximate agreement.
5. Determine the natural frequency by equating the work done on the system (inertia force times deflection) to the kinetic energy at zero displacement [mass times (velocity)^2]. Since the velocity is expressed by the first derivative of the displacement:

$$K \text{ (kinetic energy)} = \frac{1}{2} m_i V^2 = \frac{1}{2}\frac{w_i}{g} V^2$$

$$= \frac{1}{2}\frac{w_i}{g}(A\phi'_{im}\omega_m)^2$$

$$U \text{ (work done)} = \frac{1}{2} F_i \phi_{im}$$

$$= \frac{1}{2}\left(\frac{w_i}{g} A\omega^2\phi_{ia}\right) A\phi'_{im}$$

The $A\omega^2/g$ term can be treated as a constant and eliminated from the expression for F_i, with an equal effect on the ϕ_{im} term in the expressions for K and U. Setting K equal to U, we get

$$\omega^2 = \frac{g \sum_{i=1}^{N} w_i \phi_{ia} \phi'_{im}}{A \sum_{i=1}^{N} w_i (\phi'_{im})^2} \qquad (3\text{–}26)$$

where ϕ_{ia} = assumed normalized shape
ϕ'_{im} = calculated normalized shape
A = normalized constant = ϕ_{im}/ϕ'_{im}

Table 3–15 illustrates this procedure for the first mode of the structure shown in Figure 3–80.

Higher modes of vibration also contribute to structural response. The actual deflected shape of the structure can be imagined to consist of linear combinations of the various modal shapes. This combination can be expressed mathematically as follows:

$$\phi_{ia} = \sum_{i=1}^{N} \psi_m \phi_{im}$$

Table 3–15 Calculation of First Mode Frequency for Example Problem 3–8

Cycle	Level	*Weight,* w_i	*Assumed shape,* ϕ_{ia}	*Inertia force,* $F_i = w_i\phi_{ia}$	*Computed deflection,* ϕ_{im}	*Computed shape,* ϕ'_{im}	$F_i\phi'_{im}$	$w_i(\phi'_{im})^2$
1	1	117.7	1.0	117.7	8.074	1.000	117.7	117.7
	2	117.7	2.0	235.4	17.382	2.153	506.8	545.6
	3	117.7	3.0	353.1	23.946	2.966	1047.3	1035.4
	4	37.0	4.0	148.0	27.428	3.397	502.8	426.9
							2174.6	2125.6
2	1	117.7	1.000	117.70	7.992	1.000	117.7	117.7
	2	117.7	2.153	253.41	17.158	2.147	544.1	542.6
	3	117.7	2.966	349.10	22.952	2.872	1002.6	970.8
	4	37.0	3.397	125.69	26.483	3.314	416.5	406.4
							2080.9	2037.5
3	1	117.7	1.000	117.70	7.848	1.000	117.7	117.7
	2	117.7	2.147	252.70	16.823	2.144	541.8	541.0
	3	117.7	2.872	338.03	22.896	2.917	986.0	1001.5
	4	37.0	3.314	122.62	25.897	3.300	404.6	402.9
							2050.1	2063.1

$A = 7.848$

$$\omega_1^2 = \frac{387(2050.1)}{7.848(2063.1)} = 49.0$$

$\omega_1 = 7.0$ rad/sec
$F_1 = 1.11$ Hz
$T_1 = 0.898$ sec

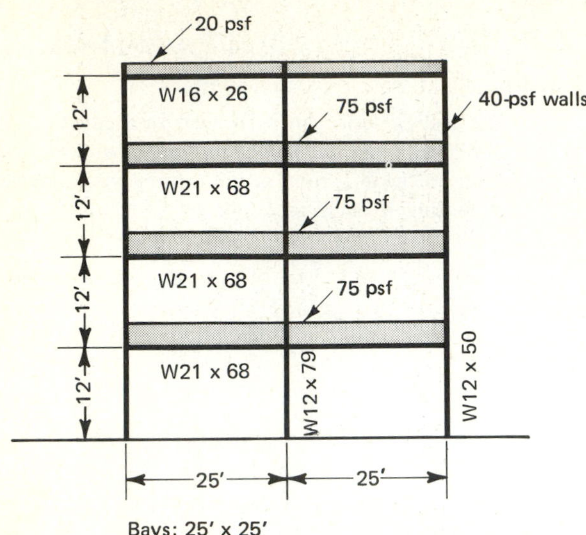

Figure 3–80. Example Problem 3–8

where ϕ_{ia} = assumed deflection of mass i
ϕ_{im} = deflection coordinate for mth mode
ψ_m = *participation factor* for mth mode
N = number of modes

The *participation factor* can be calculated as follows:

$$\psi_m = \frac{\sum_{i=1}^{N} w_i \phi_{ia} \phi_{im}}{\sum_{i=1}^{N} w_i \phi_{im}^2} \tag{3–27}$$

and the participation of the mth mode in the assumed deflection at mass i is $\psi_m \phi_{im}$. This participation concept allows higher modes to be calculated by sweeping out previously determined mode shapes as follows:

$$\phi_{is} = \phi_{ia} - \psi_m \phi_{im}$$

where ϕ_{is} is the assumed deflection "swept" of previously determined modes.

Tables 3–16 and 3–17 illustrate this procedure for the second and third modes of the structure shown in Fig. 3–80.

BASE SHEAR

The total inertia force (base shear) for a given mode is given by the product of the participation factor, the modal inertia force, and spectral displacement as follows:

$$V_m = \sum_{i=1}^{N} \psi_m F_{im} S_D$$

The participation factor was defined earlier (Eq. 3–27) and with the ϕ_{ia} term in the participation factor set equal to 1.0:

$$V_m = \frac{\sum_{i=1}^{N} w_i \phi_{im}}{\sum_{i=1}^{N} w_i \phi_{im}^2} \left(\frac{w_i}{g} \phi_{im} \omega^2\right) S_D$$

Since $S_D = S_a/\omega^2$,

$$V_m = \frac{S_a \sum_{i=1}^{N} (w_i \phi_{im})^2}{g \sum_{i=1}^{N} w_i \phi_{im}{}^2} = \frac{S_a}{g} \overline{W}_m \tag{3–28}$$

where V_m = base shear for mth mode
ψ_m = modal participation factor
F_{im} = modal inertia force at mass i
S_D = spectral displacement
S_a = spectral acceleration
w_i = weight assumed concentrated at level i
ϕ_{im} = modal displacement at mass i
$\overline{W}_m$ = effective weight for mode m

With the response spectrum approach, S_a will normally be taken from the inelastic spectrum curve to account for ductility effects. The formula for base shear may also be expressed in terms of spectral velocity as follows:

$$V_m = \left(\frac{\omega S_v}{g}\right) \overline{W}_m = \left(\frac{2\pi S_v}{g T_m}\right) \overline{W}_m$$

The term in parentheses has been replaced by an approximate coefficient, C_{sm}, in the ATC-3 recommendations as follows:

$$V_m = C_{sm} \overline{W}_m \tag{3–29}$$

where

$$C_{sm} = \frac{1.2 A_v S}{R T_m^{2/3}}$$

and A_v = effective peak velocity given by seismic zoning maps (A_v is proportional to S_v)
S = coefficient for the soil profile characteristics of the site
R = response modification factor to account for inelastic action
T_m = period of the building for the mth mode

Table 3–16 Calculation of Second Mode Frequency for Example Problem 3–8

Cycle	Level	Weight, w_i	First mode shape, ϕ_{i_1}	Assumed shape, ϕ_{ia}	$w_i\phi_{ia}\phi_{i_1}$	$\psi_1\phi_{i_1}$	Swept shape, $\psi_{is} = \phi_{ia} - \psi_1\phi_{i_1}$	Inertia force, $F_i = w_i\phi_{is}$	Computed deflection, ϕ_{im}	Computed shape, ϕ'_{im}	$F_i\phi'_{im}$	$w_i(\phi'_{im})^2$
1	1	117.7	1.000	1.00	117.7	0.198	0.802	94.40	0.605	1.000	94.4	117.7
	2	117.7	2.144	0.800	201.9	0.425	0.375	44.14	0.465	0.769	33.9	69.6
	3	117.7	2.917	0.400	137.3	0.578	−0.178	−20.95	−0.220	−0.364	7.6	15.6
	4	37.0	3.300	−0.400	−48.8	0.653	−1.053	−38.96	−0.948	−1.567	61.1	90.9
					408.1						197.0	293.8

$$\psi_1 = \frac{w_i\phi_{ia}\phi_{i_1}}{w_i\phi_{i_1}^2} = \frac{408.1}{2063.1} = 0.198$$

Cycle	Level	Weight, w_i	First mode shape, ϕ_{i_1}	Assumed shape, ϕ_{ia}	$w_i\phi_{ia}\phi_{i_1}$	$\psi_1\phi_{i_1}$	Swept shape, $\psi_{is} = \phi_{ia} - \psi_1\phi_{i_1}$	Inertia force, $F_i = w_i\phi_{is}$	Computed deflection, ϕ_{im}	Computed shape, ϕ'_{im}	$F_i\phi'_{im}$	$w_i(\phi'_{im})^2$
2	1	117.7	1.000	1.000	117.7	−0.00218	1.002	117.94	0.858	1.000	117.9	117.7
	2	117.7	2.144	0.769	194.1	−0.00467	0.774	91.10	0.762	0.888	80.9	92.8
	3	117.7	2.917	−0.364	−125.0	−0.00635	−0.358	−42.14	−0.339	−0.395	16.6	18.4
	4	37.0	3.300	−1.567	−191.3	−0.00719	−1.560	−57.72	−1.436	−1.674	96.6	103.7
					−4.5						312.0	332.6

$$\psi_1 = \frac{w_i\phi_{ia}\phi_{i_1}}{w_i\phi_{i_1}^2} = \frac{-4.5}{2063.1} = 0.00218$$

Cycle	Level	Weight, w_i	First mode shape, ϕ_{i_1}	Assumed shape, ϕ_{ia}	$w_i\phi_{ia}\phi_{i_1}$	$\psi_1\phi_{i_1}$	Swept shape, $\psi_{is} = \phi_{ia} - \psi_1\phi_{i_1}$	Inertia force, $F_i = w_i\phi_{is}$	Computed deflection, ϕ_{im}	Computed shape, ϕ'_{im}	$F_i\phi'_{im}$	$w_i(\phi'_{im})^2$
3	1	117.7	1.000	1.000	117.7	0.0009	0.999	117.58	0.895	1.000	117.6	117.7
	2	117.7	2.144	0.888	224.1	0.0019	0.886	104.28	0.826	0.923	96.3	100.3
	3	117.7	2.917	−0.395	−135.6	0.0025	−0.398	−46.84	−0.364	−0.407	19.1	19.5
	4	37.0	3.300	−1.674	−204.4	0.0029	−1.677	−62.05	−1.545	−1.726	107.1	110.2
					1.8						340.1	347.7

$$\psi_1 = \frac{1.8}{2063.1} = 0.00218 \qquad A = \frac{0.895}{1.0} = 0.895$$

$\omega_2 = 20.6$
$f_2 = 3.27$ Hz
$T_2 = 0.305$ sec

$$\omega_2^2 = \frac{387(340.1)}{0.895(347.7)} = 423$$

MODAL FORCES, DEFLECTIONS, AND DRIFTS

The modal force at each level, F_{xm}, is determined as follows:

$$F_{xm} = \frac{w_x\phi_x}{\sum_{i=1}^{N} w_i\phi_{im}} V_m \tag{3–30}$$

DESIGN VALUES

The design values for base shear, forces, and deflections can be calculated by taking the "square root of the sum of the squares" of each of the modal values. The equation for base shear, V_t, becomes

$$V_t = (V_1^2 + V_2^2 + V_3^2 + \cdots + V_N)^{1/2} \tag{3–31}$$

ANSI recommends that V_t be not less than 90% of that computed by the *equivalent lateral force procedure.*

EXAMPLE PROBLEM 3–8: Dynamic Analysis

Determine the earthquake forces for the four-story building shown in Fig. 3–80 by the modal analysis method. Use the El Centro inelastic response spectrum (Fig. 3–68) for the ground motion parameters. Assume 10% damping and a ductility factor $\mu = 4$ to 6.

Solution The first step is to determine the *natural frequencies* and *mode shapes* for the first three modes. This calculation may be performed with a dynamic analysis computer program or by the Rayleigh procedure using standard frame analysis programs. Assume that each of the building frames parallel to the earthquake motion have similar stiffness and mass characteristics and that the floor systems act as rigid diaphragms, thereby providing for approximately equal distribution of story forces to each frame. These assumptions allow the use of a single-bay model for analysis. The Rayleigh solutions for this problem are shown in Tables 3–15, 3–16, and 3–17.

The mode shapes can then be used to calculate the *effective weight* $\overline{W}_m$ for each mode as shown in Table 3–18. The *base shear* and force at each level for each mode is then calculated as follows:

Table 3–17 Calculation of Third Mode Frequency for Example Problem 3–8

Cycle	Level	Weight, w_i	First mode shape, ϕ_{i_1}	Second mode shape, ϕ_{i_2}	Assumed shape, ϕ_{ia}	$w_i\phi_{ia}\phi_{i_1}$	$w_i\phi_{ia}\phi_{i_2}$	$\psi_1\phi_{i_1}$ (1)	$\psi_2\phi_{i_2}$ (2)	Swept shape, $\phi_{is}=\phi_{ia}$, $-(1)-(2)$	Inertia force, $F_i=w_i\phi_{is}$	Comp. deflection, ϕ_{im}	Comp. shape, ϕ'_{im}	$F_i\phi'_{im}$	$w_i(\phi'_{im})^2$
1	1	117.7	1.000	1.000	1.000	117.7	117.7	0.0272	−0.212	1.185	139.47	0.383	1.000	139.5	117.7
	2	117.7	2.144	0.923	−1.000	−252.3	−117.8	0.0583	−0.196	−0.862	−101.46	−0.215	−0.561	56.9	37.0
	3	117.7	2.917	−0.407	0.200	68.7	−9.6	0.0793	0.086	0.035	4.12	−0.057	−0.149	−0.6	2.6
	4	37.0	3.300	−1.726	1.000	122.1	−63.9	0.0898	0.366	0.544	20.13	0.292	0.762	151.3	21.5
						56.2	−73.6							211.1	178.8

$$\psi_1=\frac{56.2}{2063.1}=0.0272 \qquad \psi_2=\frac{-73.6}{347.7}=-0.212 \qquad A=0.383$$

$$\omega_3^2=\frac{387(211.1)}{0.383(178.8)}=1192$$

Cycle	Level	Weight, w_i	First mode shape, ϕ_{i_1}	Second mode shape, ϕ_{i_2}	Assumed shape, ϕ_{ia}	$w_i\phi_{ia}\phi_{i_1}$	$w_i\phi_{ia}\phi_{i_2}$	$\psi_1\phi_{i_1}$ (1)	$\psi_2\phi_{i_2}$ (2)	Swept shape, $\phi_{is}=\phi_{ia}$, $-(1)-(2)$	Inertia force, $F_i=w_i\phi_{is}$	Comp. deflection, ϕ_{im}	Comp. shape, ϕ'_{im}	$F_i\phi'_{im}$	$w_i(\phi'_{im})^2$
2	1	117.7	1.000	1.000	1.000	117.7	117.7	0.0087	0.0437	0.948	111.58	0.325	1.000	111.6	117.7
	2	117.7	2.144	0.923	−0.561	−141.6	−60.9	0.0187	0.0403	−0.620	−72.97	−0.141	−0.434	31.7	22.2
	3	117.7	2.917	−0.407	−0.149	−51.2	7.1	0.0253	−0.0178	−0.142	−16.71	−0.040	−0.123	2.1	1.8
	4	37.0	3.300	−1.726	0.762	93.0	−48.7	0.0287	−0.0754	0.809	29.93	0.493	1.351	40.4	67.5
						17.9	15.2							185.8	209.2

$$\psi_1=\frac{17.9}{2063.1}=0.0087 \qquad \psi_2=\frac{15.2}{347.7}=0.0437 \qquad A=0.325$$

$$\omega_3^2=\frac{387(185.8)}{0.325(209.2)}=1057$$

Cycle	Level	Weight, w_i	First mode shape, ϕ_{i_1}	Second mode shape, ϕ_{i_2}	Assumed shape, ϕ_{ia}	$w_i\phi_{ia}\phi_{i_1}$	$w_i\phi_{ia}\phi_{i_2}$	$\psi_1\phi_{i_1}$ (1)	$\psi_2\phi_{i_2}$ (2)	Swept shape, $\phi_{is}=\phi_{ia}$, $-(1)-(2)$	Inertia force, $F_i=w_i\phi_{is}$	Comp. deflection, ϕ_{im}	Comp. shape, ϕ'_{im}	$F_i\phi'_{im}$	$w_i(\phi'_{im})^2$
3	1	117.7	1.000	1.000	1.000	117.7	117.7	0.0630	−0.0280	0.965	113.58	0.330	1.000	113.6	117.7
	2	117.7	2.144	0.923	−0.434	−109.5	−47.1	0.1350	−0.0258	−0.543	−63.91	−0.158	−0.479	30.6	27.0
	3	117.7	2.917	−0.407	−0.123	−42.2	5.9	0.1838	0.0114	−0.318	−37.43	−0.138	−0.418	15.6	20.6
	4	37.0	3.300	−1.726	1.351	164.9	−86.3	0.2079	0.0483	1.095	40.52	0.484	1.467	59.4	79.6
						130.9	−9.8							219.2	244.9

$$\psi_1=\frac{130.9}{2063.1}=0.063 \qquad \psi_2=\frac{-9.8}{347.7}=-0.028 \qquad A=0.330$$

$\omega_3=32.4$
$f_3=5.15$ Hz
$T_3=0.19$ sec

$$\omega_3^2=\frac{387(219.2)}{0.33(244.9)}=1049$$

Table 3–18 Calculation of Effective Weights for Example Problem 3–8

Mode	*Level*	w_i	ϕ_{im}	$w_i\phi_{im}$	$(w_i\phi_{im})^2$	$w_i\phi_{im}^2$
1	1	117.7	1.000	117.7	13,853	117.7
	2	117.7	2.144	252.3	63,679	541.0
	3	117.7	2.917	343.3	117,876	1,001.5
	4	37.0	3.300	122.1	14,908	402.9
				835.4	210,316	2,063.1
	$\overline{W}_1 = \frac{210,316}{2063.1} = 101.9$ kips					
2	1	117.7	1.000	117.7	13,853	117.7
	2	117.7	0.923	108.6	11,802	100.3
	3	117.7	−0.407	−47.9	2,295	19.5
	4	37.0	−1.726	−63.9	4,078	110.2
				114.5	32,028	347.7
	$\overline{W}_2 = \frac{32,028}{347.7} = 92.1$ kips					
3	1	117.7	1.000	117.7	13,853	117.7
	2	117.7	−0.479	−56.4	3,178	27.0
	3	117.7	−0.418	−49.2	2,420	20.6
	4	37.0	1.467	54.2	2,946	79.6
				66.3	22,397	244.9
	$\overline{W}_3 = \frac{22,397}{244.9} = 91.4$ kips					

Mode 1: $T_1 = 0.898$ sec, $\overline{W}_1 = 101.9$ kips. From Fig. 3–68, $S_a = 0.1g$. By Eq. 3–28,

$$V_1 = \frac{0.1g}{g}(101.9) = 10.2 \text{ kips}$$

$$F_{11} = \frac{117.7}{835.4}(10.2) = 1.44 \text{ kips}$$

$$F_{21} = \frac{252.3}{835.4}(10.2) = 3.08 \text{ kips}$$

$$F_{31} = \frac{343.3}{835.4}(10.2) = 4.19 \text{ kips}$$

$$F_{41} = \frac{122.1}{835.4}(10.2) = 1.49 \text{ kips}$$

Mode 2: $T_2 = 0.305$ sec, $\overline{W}_2 = 92.1$ kips. From Fig. 3–68, $S_a = 0.17g$. By Eq. 3–28,

$$V_2 = \frac{0.17g}{g}(92.1) = 15.7 \text{ kips}$$

$$F_{12} = \frac{117.7}{114.5}(15.7) = 16.14 \text{ kips}$$

$$F_{22} = \frac{108.6}{114.5}(15.7) = 14.89 \text{ kips}$$

$$F_{32} = \frac{-47.9}{114.5}(15.7) = -6.57 \text{ kips}$$

$$F_{42} = \frac{-63.9}{114.5}(15.7) = -8.76 \text{ kips}$$

Mode 3: $T_3 = 0.19$ sec, $\overline{W}_3 = 91.4$ kips. From Fig. 3–68, $S_a = 0.17\ g$. By Eq. 3–28,

$$V_3 = \frac{0.17g}{g}(91.4) = 15.5 \text{ kips}$$

$$F_{13} = \frac{117.7}{66.3}(15.5) = 27.52 \text{ kips}$$

$$F_{23} = \frac{-56.4}{66.3}(15.5) = -13.19 \text{ kips}$$

$$F_{33} = \frac{-49.2}{66.3}(15.5) = -11.50 \text{ kips}$$

$$F_{43} = \frac{54.2}{66.3}(15.5) = 12.67 \text{ kips}$$

Total *design* values are then computed by taking the "square root of the sum of the squares" as follows:

$$V_T = (10.2^2 + 15.7^2 + 15.5^2)^{1/2} = 24.3 \text{ kips}$$

Complete results are tabulated in Table 3–19.

Table 3–19 Calculation of Total-Story Shear Forces for Example Problem 3–8

	Mode 1		*Mode 2*		*Mode 3*		
Level	*Force*	*Shear*	*Force*	*Shear*	*Force*	*Shear*	*Total shear*[a]
4	1.49		−8.76		12.67		
		1.49		−8.76		12.67	15.47
3	4.19		−6.57		−11.50		
		5.68		−15.33		1.17	16.39
2	3.08		14.89		−13.19		
		8.76		−0.44		−12.02	14.88
1	1.44		16.14		27.52		
		10.20		15.70		15.50	24.30

[a] Calculated by "square root of the sum of the squares" method.

REFERENCES

1. ACI Committee 442, *Response of Multi-story Concrete Structures to Lateral Forces,* SP-36. Detroit, Mich.: American Concrete Institute, 1973.
2. Air Force Design Manual, *Principles and Practices for Design of Hardened Structures,* AFSWC–TDR–62–138, 1962.
3. AMRHEIN, JAMES, *Reinforced Masonry Engineering Handbook.* Los Angeles, Calif.: Masonry Institute of America, 1972.
4. Applied Technology Council, *Tentative Provisions for the Development of Seismic Regulations for Buildings,* ATC 3–06, NSF 78–8, NBS Special Publication 510. Washington, D.C.: U.S. Government Printing Office, 1978.
5. BEAUFAIT, FRED, ed., *Tall Buildings, Planning, Design and Construction.* Nashville, Tenn.: Civil Engineering Program, Vanderbilt University, 1974.
6. BEEDLE, LYNN, et al., *Structural Steel Design.* New York: The Ronald Press Company, 1964.
7. BIGGS, JOHN M., *Introduction to Structural Dynamics.* New York: McGraw-Hill Book Company, 1964.
8. BLUME, JOHN, NATHAN NEWMARK, and LEO CORNING, *Design of Multi-story Concrete Buildings for Earthquake Motions.* Chicago: Portland Cement Association, 1961.
9. *Building Code Requirements for Minimum Design Loads in Buildings and Other Structures.* New York: American National Standards Institute, 1982.
10. CLOUGH, RAY W., and JOSEPH PENZIEN, *Dynamics of Structures.* New York: McGraw-Hill Book Company, 1975.
11. *Design of Structures to Resist the Effects of Atomic Weapons.* EM 1110, U.S. Army Corps of Engineers, July 1959.
12. *Design Manual—Structural Engineering.* NAVFAC DM-2, October 1970.
13. Enwright Associates, Inc., "Design Reference Notes: Earth Pressure Considerations" (unpublished).
14. Enwright Associates, Inc., "Design Reference Notes: Vibration Considerations" (unpublished).
15. FINTEL, MARK, ed., *Handbook of Concrete Engineering.* New York: Van Nostrand Reinhold Company, 1974.
16. *Harbor and Coastal Facilities.* NAVFAC DM-26.
17. LRFD Committee (ASCE), "LRFD Papers," *Journal of the Structural Division, ASCE,* Vol. 104, No. ST-9 (September 1978).
18. Massachusetts Institute of Technology, *Live Load Effects in Office Buildings.* National Science Foundation, PB-224 636 (60/1279), July 1975.
19. MEHTA, KISHER C., and JOSEPH E. MINOR, and JAMES R. MCDONALD, "The Tornadoes of April 3–4, 1974: Windspeed Analysis," 1975 ASCE National Structural Engineering Convention—Meeting Preprint 2490.
20. MERRITT, FREDRICK, ed., *Structural Steel Designers Handbook.* New York: McGraw-Hill Book Company, 1972.
21. NEWMARK, NATHAN M., and EMILIO ROSENBLUETH, *Fundamentals of Earthquake Engineering.* Englewood Cliffs, N.J.: Prentice-Hall, Inc., 1971.
22. NORRIS, et al., *Structural Design for Dynamic Loads.* New York: McGraw-Hill Book Company, 1959.
23. PETERKA, J. A., and J. E. CERMAK, "Adverse Wind Loading Induced by Adjacent Buildings," 1975 ASCE National Structural Engineering Convention—Meeting Preprint 2456.
24. *Proceedings of the Symposium on Soil-Structure Interaction,* University of Arizona, 1964.
25. QUINN, ALONZO DEF., *Design and Construction of Ports and Marine Structures,* 2nd ed. New York: McGraw-Hill Book Company, 1972.
26. RADZIMINSKI, JAMES, and JOSEPH BIEDENBACH, eds., *The Proceedings, Earthquake Engineering Conference, January 23–24, 1975.* Columbia, S.C.: College of Engineering, University of South Carolina, June 1975.
27. RICHART, F. E., JR., et al., *Vibration of Soils and Foundations.* Englewood Cliffs, N.J.: Prentice-Hall, Inc., 1970.
28. ROSENBLUETH, EMILIO, *Design of Earthquake Resistant Structures.* New York: John Wiley & Sons, Inc., 1980.
29. SALVADORI, MARIO, and ROBERT HELLER, *Structure in Architecture—The Building of Buildings.* Englewood Cliffs, N.J.: Prentice-Hall, Inc., 1963.
30. SCHUELLER, WOLFGANG, *Highrise Building Structures.* New York: John Wiley & Sons, Inc., 1977.
31. *Shore Protection Manual,* Vols. 1–3. U.S. Army Coastal Engineering Research Center, 1973. Washington, D.C.
32. SIMIU, EMIL, and ROBERT SCANLAN, *Wind Effects on Structures, An Introduction to Wind Engineering.* New York: John Wiley & Sons, Inc., 1978.
33. *Structures to Resist the Effects of Accidental Explosion,* NAVFAC P-397. Washington, D.C.: Department of the Navy, 1969.
34. *STRUDL DYNAL,* McDonnell Douglas Automation Company, St. Louis, October 1980.
35. Task Committee on Wind Forces, "Wind Forces on Structures," *Transactions of the ASCE,* Paper No. 3269, Vol. 126, Part 2 (1961), pp. 1124–1198.
36. WHITMAN, BIGGS, et al., "Seismic Design Decision Analysis," *Journal of the Structural Division, ASCE,* Vol. 101, No. ST-5 (May 1975).
37. WILSON, FORREST, *Emerging Form in Architecture—Conversations with Lev Zetlin.* Boston: Cahners Books, 1975.
38. *Wind Loads on Buildings and Structures.* Washington, D.C.: U.S. Department of Commerce, National Bureau of Standards, 1970.

4

ELEMENTS

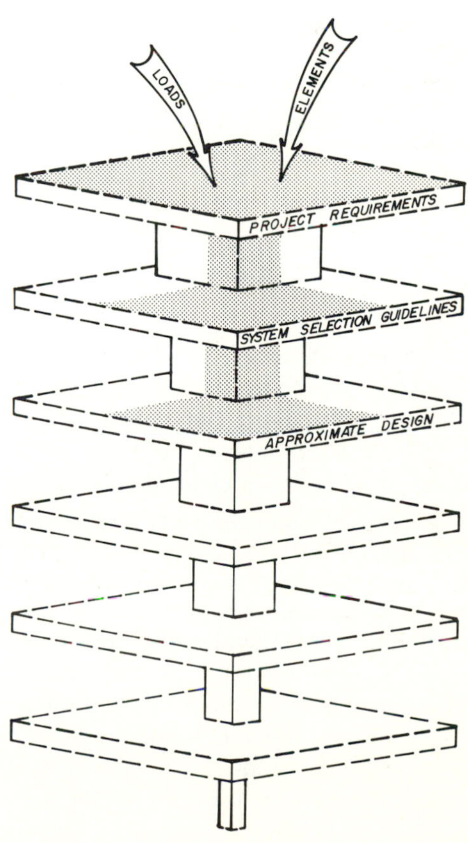

Figure 4–1. Systems Design Model

4–1 INTRODUCTION

In this chapter we begin the process of structural assembly with the *element* (Fig. 4–1). Elements are the basic component of structural assembly—the smallest unit into which structures can be divided—and are dimensionally described primarily by the single dimension of *length*. As one-dimensional building blocks, they represent the simplest form of structural material and, together with other elements, form subsystems and systems.

Analysis (for moments, shears, and axial loads) and *design* (member proportioning) of elements are fundamental course subjects covered in almost all civil engineering curricula, a rational approach based on the concept that understanding of structures needs to progress from the simple to the complex. In Chapter 1 we placed *member design* as the *final step* in an iterative sequence of activities leading to a design solution. This apparent conflict between our educational approach and design practice is, of course, the basic reason for a systems design study. Rather than depart radically from this well-established educational approach, we will build on the knowledge of elements and continue the "pyramiding" of knowledge, while developing the "systems design model" that establishes a framework for "telescoping" knowledge back to the member design level. This requires a constant emphasis on behavior and conceptual understanding: as elements of various sizes and types are interconnected to form a structural system, and as we understand how the assembled elements behave, only then can we complete the cycle and design individual elements compatible with the requirements of the total structure.

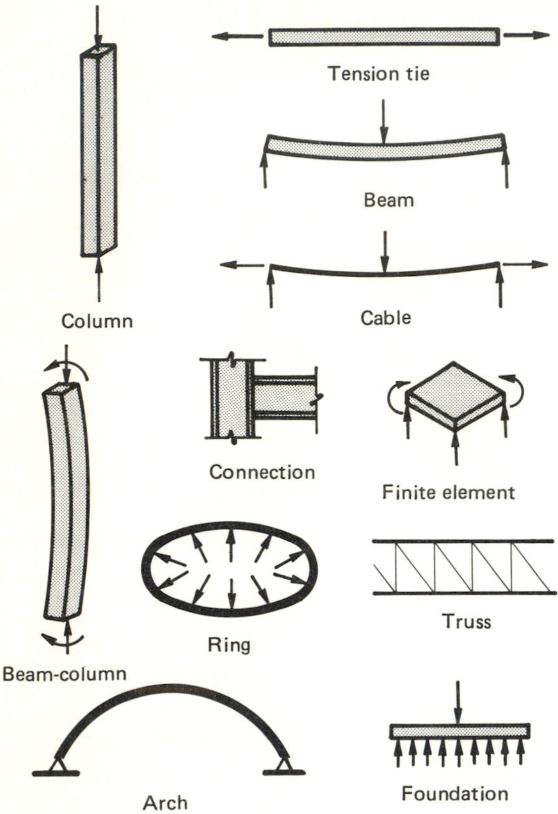

Figure 4–2. Structural Elements

This chapter describes 11 basic elements (Fig. 4–2). The description includes the history and development of the element form, types, and uses for the element, behavior, approximate analysis, and design techniques. Our purpose is to build an element "vocabulary," complete with a knowledge of behavior, application concepts, and preliminary design skills.

The *approximate analysis* and *approximate design* methods provide a basis for conceptual mathematical understanding and are valuable aids during preliminary studies and system evaluations. We will develop approximate methods for member design based on interpolation between *established limits* of usable stress (Fig. 4–3), which channels our structural estimates into reasonable boundaries. With these tools we can make quick decisions during creative team design sessions without the far too customary retreat into detail design formulas. The approximate methods are based on code equations adopted within the United States, but their approximate nature provides the flexibility for international use.

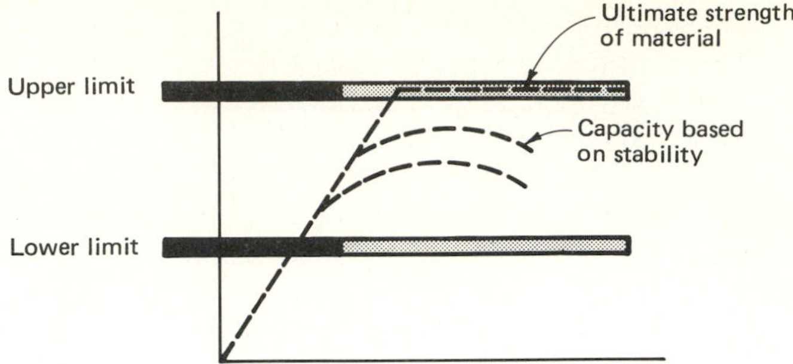

Figure 4–3. Member Stress Limits

4–2 COLUMNS

Columns are *compression* elements that carry forces primarily in direct axial compression of the member material (Fig. 4–4). (For classification purposes, columns that also carry moment are termed beam-columns and are discussed later in this chapter.) Normally considered to be vertical members (oriented with the long axis vertical),

Figure 4–4. Column Element (Courtesy of Bethlehem Steel Corporation)

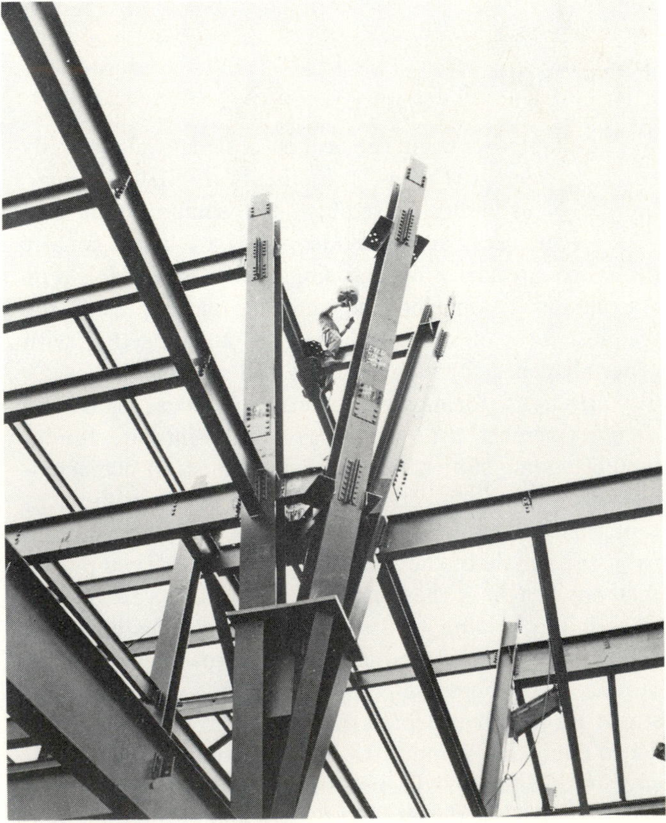

columns can actually be positioned in any orientation. They are defined by their length dimension running between points of support and can be very short (e.g., footing piers) or very long (e.g., bridge piers) and are major elements in trusses, building frames, and substructure supports for bridges. Commonly used names for columns include *studs, struts, posts, piers, piles,* and *shafts.* They are normally constructed with symmetrical cross sections (round, square, or wide-flange shapes) but may be constructed in any geometrical fashion. Virtually every common construction material is used for column construction, including steel, concrete, masonry, timber, and plastic.

History and Development

The first column element was probably the wood branch used in construction of early hut-type shelters. Thrusting the individual pieces into the ground achieved resistance to overturning and transferred imposed loads directly into the soil. Pole construction later developed from this concept, which saw timber poles driven into the ground without the use of footings or foundations. Only later did problems with rot and deterioration of the wood promote the use of separate footings of a more durable material.

Large slabs of stone, such as those found at Stonehenge, probably were an independent development of the column form. The stone columns were initially made very massive so that they would be stabilized by their own weight. This development continued with the use of massive stone columns in Greek and Roman structures.

The use of slender sections for columns grew out of the development of wrought iron, cast iron, steel, and concrete materials. The first engineering analysis method was prescribed by Van Musschenbrock in a paper published in 1729. He presented an empirical column curve which is similar to those in use today:

$$P = \frac{kBD^2}{L^2} \tag{4–1}$$

where P = column strength
k = empirical factor
B and D = breadth and width of the section
L = length of column

Later Euler gave us the classical formula for column strength:

$$P_{cr} = \frac{\pi^2 EI}{L^2} \tag{4–2}$$

Later researchers have progressively refined our understanding of column behavior, and present-day codes generally reflect the state of the art of our current knowledge.

Types and Uses

Common types of *metal compression members* are illustrated in Fig. 4–5. Single angles are rarely used, except in light roof trusses, because of eccentricities at the connections, and the tee shape is often used in roof trusses. Column sections for buildings are normally wide-flange sections, although the pipe section is the most efficient and suitable for medium loads and simple connection details.

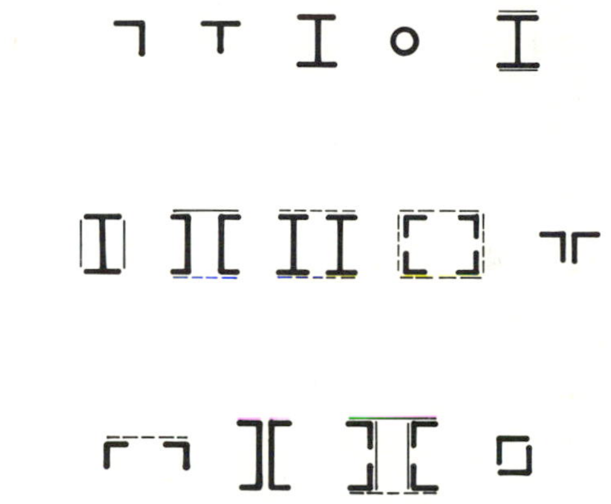

Figure 4–5. Metal Compression Members

Compression members in wood, masonry, and concrete are normally square or rectangular sections due to the forming process. *Wood members* are used extensively in residential construction and are referred to as "studs." *Masonry columns* are used for moderate loads and usually in conjunction with canopies, walkways, and other light structural systems. *Concrete columns* are used as widely as metal members, both in the cast-in-place form and as precast sections. Although the cost/strength ratio of concrete in compression is superior to steel, the weight of concrete (and its effect on the foundation system) limits the practical height of concrete columns. The tallest concrete building constructed to date with concrete columns is the 859-ft (262-m) Water Tower Place building in Chicago. In contrast, the Sears Tower in Chicago uses steel columns over a height of 1454 ft (443 m).

Behavior

Column *strength* is governed by material strength or buckling. Figure 4–6 shows the limits of column strength, with limits of elastic buckling behavior defined by the Euler curve as a function of *effective length.* The effective length factor (k) corrects the classical Euler column (pinned at both ends) for actual connection conditions

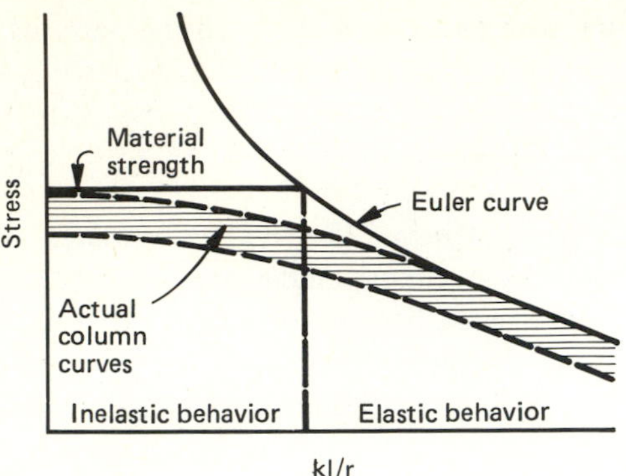

Figure 4–6. Column Strength

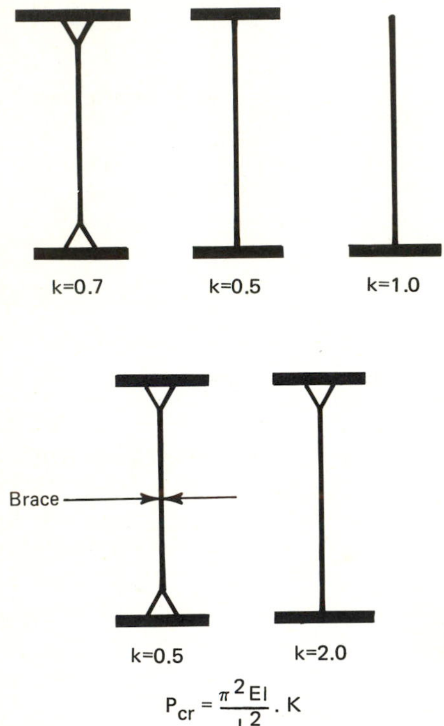

Figure 4–7. Buckling Load for Various Support Conditions

(fixed, free, etc.). Figure 4–7 illustrates the effect of varying support conditions on the buckling load.

Actual columns vary from the theoretical limits defined in Fig. 4–6 because of initial imperfections, crookedness, and residual stress effects. Design curves fall in a range below the theoretical limits and are described by mathematical formulas developed to correlate experimental column tests with analytical techniques.

Approximate Analysis

Columns (by our definition) carry only axial loads. Bending moments introduced by eccentric load paths (applied off center of the column) are usually small and therefore of no significance during preliminary analysis. Regardless of the horizontal framing system used for floors and roofs, the load applied over the *tributary area* of the column supported at each level will approximate the column design loads. The tributary area is defined as that area of a floor or roof which supports all of the loads whose *load path* leads to the column. This area can be visualized as a series of floor and roof areas "stacked" on the column (Fig. 4–8). This concept of "column stacks" allows rapid determination of column dead and live loads without the necessity for detailed member load analysis at the floor or roof subsystem levels. Appropriate live load reduction formulas can also be used very conveniently with this method.

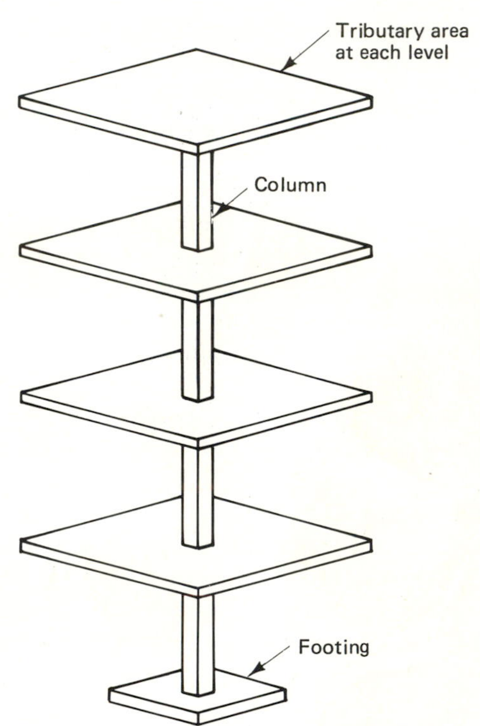

Figure 4–8. Tributary Area to Columns

A preliminary determination must also be made of the support conditions (bracing) for the column in order to determine the effective length. Most columns will be parts of systems that are braced in each direction (by X-bracing, shear walls, etc.) and the assumption of an effective length factor of 1.0 is valid. Columns that are parts of other systems are really beam-columns and will be discussed later.

EXAMPLE PROBLEM 4–1: Determination of Column Load

Determine the total load for an interior column of a three-story concrete building with bays of 25 ft × 25 ft (7.6 m × 7.6 m) (see Fig. 4–9). Dead loads are 75 psf (3591

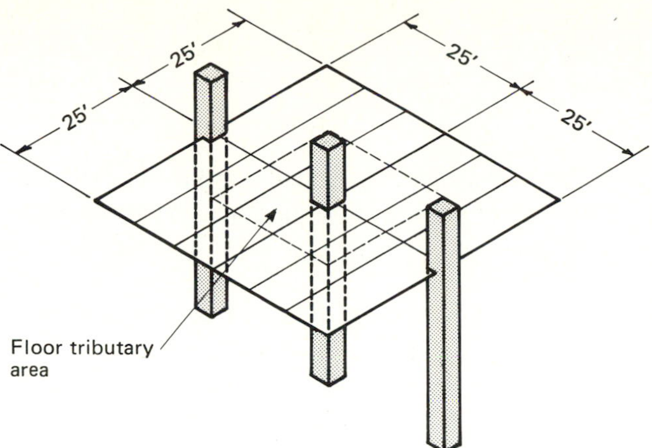

Figure 4-9. Example Problem 4-1

N/m²) for the roof and floors and live loads are 20 psf (957 N/m²) for the roof and 50 psf (2394 N/m²) for the floors. Assume no live load reduction.

Solution

1. Tributary area at each level = 25 ft × 25 ft = 625 sq ft.
2. Roof load = area × total load
   = 625 sq ft × (75 psf + 20 psf)
   = 59,375 lb
3. Floor load = area × total load
   = 2(625 sq ft) × (75 psf + 50 psf)
   = 156,250 lb

   Total column load
   = 215,625 lb (959,100 N)

Approximate Design

Many handbooks and tables are available for rapid selection of preliminary column sizes. When the right table is not available or we are involved in conceptual sessions that require both a *feeling* for the mathematical possibilities and rapid response, it is often desirable to have an alternative method.

The simplest approach to approximate column design is to work with the basic axial stress formula:

$$A = \frac{P}{F_a} \tag{4-3}$$

where A = area of column, square inches (m²)
P = axial load, kips (N): ultimate or working load
F_a = equivalent stress allowed by design method, ksi (Pa)

The upper limit of F_a is the compressive strength of the material (with a factor of safety). The lower limit is the stress at which buckling occurs at maximum permitted kl/r ratios, where k is the effective length factor, l is the unbraced column length, and r is the radius of gyration. These limits can be established on a simple graph for each major column material as described in the following sections.

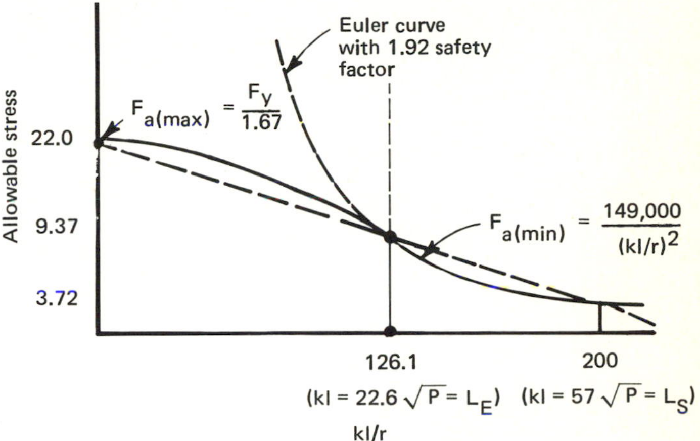

Figure 4-10. Structural Steel (A36) Column Curve

Steel Columns. Figure 4–10 shows a column curve for A36 structural steel. The upper limit, $F_{a\,(max)}$, is the yield strength divided by a factor of safety of 1.67. An intermediate point, $F_{a\,(int)}$, is the Euler buckling stress (factor of safety of 1.92) which controls beyond kl/r equal to C_c (126.1 for A36 steel), where C_c is the column slenderness ratio separating elastic and inelastic buckling. The lower limit, $F_{a\,(min)}$, occurs at the maximum kl/r ratio of 200. The intermediate and lower limits can also be expressed in terms of unbraced lengths L_E and L_S, respectively. Straight lines connecting the three points provide a linear approximation to the column curve which can be used for approximate design.

The radius of gyration, r, for steel columns can be approximated by the following expression:

$$r_x = 0.90\sqrt{A}, \qquad r_y = 0.55\sqrt{A} \tag{4-4}$$

Substituting the value for r_y into AISC Eq. 1.5–2 (since the y axis usually controls), the formula for F_a in the elastic range becomes

$$\frac{P}{A} = F_a = \frac{45,072(A)}{(kl)^2} \tag{4-5}$$

or

$$A = \frac{P\sqrt{kl}}{212} \tag{4-6}$$

For A36 steel:

$$F_{a\,(\text{int})} = 9.37 \text{ ksi } (64.6 \text{ MPa})$$

$$\text{at } kl/r = 126.1 \text{ and } kl = L_E$$

$$\frac{kl}{r} = \frac{kl}{0.55\sqrt{A}} = 126.1, \qquad (kl)^2 = 4810(A)$$

From Eq. 4–6,

$$(kl)^2 = 4810\left(\frac{\sqrt{P}\,kl}{212}\right)$$

$$kl = 22.6\sqrt{P} \text{ inches } (0.57\sqrt{P} \text{ meters}) \qquad (4\text{–}7)$$

$$F_{a\,(\text{min})} = 3.73 \text{ ksi } (25.7 \text{ MPa})$$

$$\text{at } kl/r = 200 \text{ and } kl = L_S$$

$$\frac{kl}{r} = \frac{kl}{0.55\sqrt{A}} = 200, \qquad (kl)^2 = 12{,}100(A)$$

From Eq. 4–6,

$$(kl)^2 = 12{,}100\left(\frac{\sqrt{P}\,kl}{212}\right)$$

$$kl = 57\sqrt{P} \text{ inches } (1.45\sqrt{P} \text{ meters}) \qquad (4\text{–}8)$$

Equations 4–7 and 4–8 are approximate formulas for the column length (kl) associated with the intermediate and lower stress limits. The following example illustrates the approximate design method.

Example Problem 4–2: Preliminary Steel Column Design

Select a preliminary steel column (A36 steel) for an axial load of 100 kips (4.4×10^6 N) and an effective length (kl) of 15 ft (4.51 m).

Solution

1. Establish stress limits based on Fig. 4–10.

$$\text{Max. allowable stress} = 22 \text{ ksi } (151.7 \text{ MPa})$$

$$\text{Int. allowable stress} = 9.37 \text{ ksi } (64.6 \text{ MPa})$$

$$\text{at } kl = 22.6\sqrt{P}$$

$$kl = 22.6\sqrt{P} = 22.6\sqrt{100}$$

$$= 18.8 \text{ ft } (5.73 \text{ m})$$

2. Select allowable stress by interpolation.

$$F_a = 9.37 + \frac{3.8}{18.8}(22.0 - 9.37) = 11.92 \text{ ksi}$$

Assume that $F_a = 12.0$ ksi.

$$A = \frac{P}{F_a} = \frac{100}{12.0} = 8.33 \text{ in.}^2 \ (5373 \text{ mm}^2)$$

3. An alternative check with the *AISC Steel Handbook* yields the following selection.

   W8 × 28: Allow $P = 95$ kips, $A = 8.25$ in.^{2}.

   W8 × 31: Allow $P = 131$ kips, $A = 9.13$ in.^{2}.

Thus the approximate method gives a reasonable answer.

After a few trials with this method, we can estimate the allowable stress without performing the detailed interpolation or use Appendix 4A, which contains values for L_E and L_S. The following example illustrates this refinement.

Example Problem 4–3: Preliminary Steel Column Design

Select a preliminary steel column (A36 steel) for an axial load of 80 kips (355,840 N) and an effective length (kl) of 12 ft (3.66 m).

Solution

1. Establish stress limits based on Fig. 4–10.

$$\text{Max. allowable stress} = 22.0 \text{ ksi } (151.7 \text{ MPa})$$

$$\text{Int. allowable stress} = 9.37 \text{ ksi } (64.6 \text{ MPa})$$

$$\text{at } kl = 22.6\sqrt{P}$$

$$kl = 22.6\sqrt{P} = 22.6\sqrt{80}$$

$$= 202 \text{ in.}$$

$$= 16.8 \text{ ft } (5.12 \text{ m})$$

$$\text{Min. allowable stress} = 3.72 \text{ ksi } (25.6 \text{ MPa})$$

$$\text{at } kl = 57\sqrt{P}$$

$$kl = 57\sqrt{P} = 57\sqrt{80} = 509 \text{ in.}$$

$$= 42.4 \text{ ft } (12.9 \text{ m})$$

2. Select allowable stress by estimate.

$$(kl = 16.8 \text{ ft})\ 9.37 < F_a < 22.0\ (kl = 0 \text{ ft})$$

   Estimate F_a (kl = 12 ft) = 12.0 ksi.

$$A = \frac{P}{F_a} = \frac{80}{12.0} = 6.67 \text{ in.}^2 \ (4300 \text{ mm}^2)$$

3. Use W8 × 24 (A = 7.08 in.^{2}).
4. Appendix 4A can also be used directly to select the lightest section with $L_E \geq 12$ ft and an area A' estimated between the following limits:

$kl = 0$	$kl = 12$	$kl = L_E$
$A = 3.63$	$A = A'$	$A = 8.54$

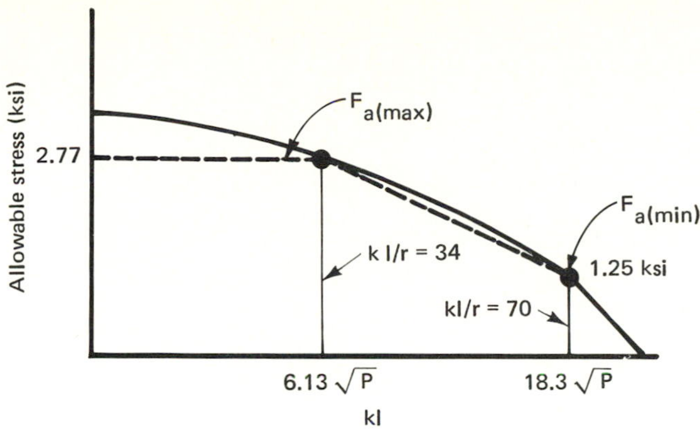

Figure 4–11. Reinforced Concrete Column Curve (3000 psi)

Concrete Columns. Figure 4–11 shows a column curve for reinforced concrete columns based on the ultimate strength method. The upper limit, $F_{a\,(\max)}$, is the ultimate stress determined by the column capacity formula:

$$P_u = 0.80\phi[0.85 f_c' A_g + f_y A_{st}] \qquad (4\text{–}9)$$

where P_u = ultimate capacity of column (kips or N)
ϕ = capacity reduction factor
f_c' = strength of concrete (ksi or MPa)
A_g = area of column (in.^{2} or mm^2)
f_y = yield strength of steel (ksi or MPa)
A_{st} = area of steel (in.^{2} or mm^2)

For 3.0 ksi (20.7 MPa) concrete and Grade 60 steel, assume 4% reinforcement for preliminary design. (The ACI Code allows 1 to 8%.)

$$P_u = 0.80(0.70)[0.85(3)A_g + 60(0.04A_g)]$$
$$= 2.77A_g$$

or

$$\frac{P_u}{A_g} = F_{a\,(\max)} = 2.77 \text{ ksi (19.1 MPa)} \qquad (4\text{–}10)$$

This stress will approximate the column size required based on ultimate loads up to a kl/r value of 34, where moment magnification (to account for column slenderness effects) must be considered. The corresponding approximate value for kl can be derived as follows:

Since $r = 0.3h$ (ACI Code recommendation), where h is the depth of column in direction stability is being considered,

$$\frac{kl}{0.3h} = 34, \qquad kl = 10.2h = 10.2\sqrt{A}$$

From Eq. 4–10,

$$kl = 10.2\sqrt{\frac{P}{2.77}} = 6.13\sqrt{P} \qquad (4\text{–}11)$$

$$\left(\text{for } \frac{kl}{r} = 34,\ F_a = 2.77 \text{ ksi}\right)$$

The upper practical limit for the slenderness ratio is about 70; the minimum moment to be magnified is equal to an eccentricity (e/t) of 0.1. The minimum moments (due to eccentric loads) must be magnified above kl/r = 34 by the following formula:

$$\delta = \frac{C_m}{1 - P_u/\phi P_c} \qquad (4\text{–}12)$$

where δ = moment magnifier
C_m = factor to account for ratio of end moments = 1.0 for braced frames
P_u = ultimate axial load (kips or N)
P_c = elastic buckling load (kips or N)
ϕ = capacity reduction factor

The lower limit of stress, $F_{a\,(\min)}$, can be determined as follows:

$$P_u = A_g F_a = bhF_a = b(3.33r)F_a \qquad (4\text{–}13)$$

$$P_c = \frac{\pi^2 EI}{l^2} \qquad (4\text{–}14)$$

where b = width of column, in.
h = depth of column in direction stability is being considered, in.
EI = effective stiffness ($\simeq 0.4EIg$ for approximate design)
E = modulus of elasticity, ksi (3.12×10^3 ksi for 3.0-ksi concrete)
I = moment of inertia, in.^{4}
I_g = moment of inertia based on gross section, in.^{4}
l = height of column, in.

Therefore,

$$P_c = \frac{(9.85)(3.12 \times 10^3)(0.4bh^3)}{12l^2}$$

$$P_c = \frac{1034bh^3}{l^2} = \frac{1034b(r/0.3)^3}{l^2} = \frac{38{,}296br^3}{l^2} \qquad (4\text{–}15)$$

Substituting Eqs. 4–15 and 4–13 into Eq. 4–12:

$$\delta = \frac{1.0}{1 - \dfrac{b(3.33r)F_a(l^2)}{0.7(38{,}296br^3)}} \qquad (4\text{–}16)$$

$$= \frac{1.0}{1 - 0.000124F_a(l/r)^2}$$

Setting δ equal to an arbitrary maximum limit of 5.0 (e/t = 0.5) and l/r equal to 70 yields

$$5.0 = \frac{1.0}{1 - 0.000124F_a(70)^2}$$

$$F_{a\,(\min)} = 1.31 \text{ ksi } (9.03 \text{ MPa})$$

which will occur at $kl/r = 70$, $kl/0.3h = 70$, $kl = 21h = 21\sqrt{A} = 21\sqrt{P/1.31}$.

$$kl = 18.3\sqrt{P} \quad \left(\text{for } \frac{kl}{r} = 70,\ F_a = 1.31 \text{ ksi}\right) \qquad (4\text{–}17)$$

For light loads and/or long columns, kl calculated by Eq. 4–17 may be less than the actual column length, in which case use kl/r = 70 directly to size the column. The following example illustrates the approximate design method.

Example Problem 4–4: Preliminary Concrete Column Design

Select a preliminary concrete column to support an axial dead load of 114 kips (507,072 N) and an effective length of 15 ft (4.57 m).

Solution

1. Determine P_u:

$$P_u = 1.4(114) = 160 \text{ kips } (711{,}680 \text{ N}).$$

2. Establish stress limits by Fig. 4–11.

$$\text{Max. allowable stress} = 2.77 \text{ ksi } (19.1 \text{ MPa})$$
$$\text{at } kl = 6.13\sqrt{P}$$
$$6.13\sqrt{P} = 6.13\sqrt{160} = 77.5 \text{ in. } (6.46 \text{ ft}) (1.97 \text{ m})$$
$$\text{Min. allowable stress} = 1.31 \text{ ksi } (8.03 \text{ MPa})$$
$$\text{at } kl = 18.3\sqrt{P}$$
$$18.3\sqrt{P} = 18.3\sqrt{160} = 231 \text{ in. } (19.3 \text{ ft}) (5.9 \text{ m})$$

3. Select allowable stress by interpolation.

$$F_a = 1.31 + \frac{4.8}{13.3}(2.77 - 1.31) = 1.84 \text{ (assume 1.85)}$$

$$A = \frac{P_u}{F_a} = \frac{160}{1.85} = 86.5 \text{ in.}^2 \ (55{,}793 \text{ mm}^2)$$

4. Select column size = 10 in. × 10 in. (4% steel).
5. An alternative check with the *ACI Handbook* yields the following selection: 10 in. × 10 in. column with 4% reinforcement.

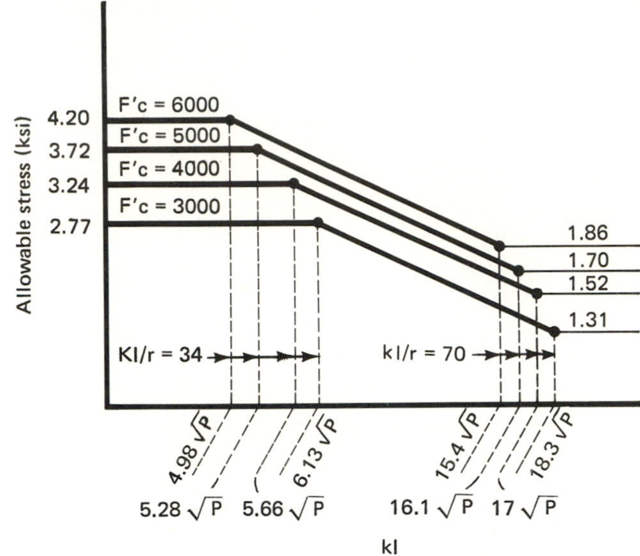

Figure 4–12. Reinforced Concrete Column Curves for Various Concrete Strengths and 4% Steel Reinforcement

As we learned previously, the detailed interpolation can be replaced with a simple estimate after we obtain a feeling for the numbers. Curves for other concrete strengths and reinforcement percentages are shown in Fig. 4–12.

Masonry Columns. Figure 4–13 shows column curves for reinforced masonry which can be applied in a similar manner to those previously developed for steel and concrete. The upper limit, $F_{a\,(\max)}$, is the stress determined by the following formulas:

$$\text{For brick:} \quad P = C_eC_s(f_m + 0.80p_gf_s)A_g \qquad (4\text{–}18)$$

$$\text{For block:} \quad P = C_eC_s(f_m + 0.65p_gf_s)A_g \qquad (4\text{–}19)$$

where C_e = eccentricity coefficient = 1.0 for no moment
C_s = correction for slenderness effects
f_m = allowable masonry stress, ksi = $0.20f'_m$ (brick), $0.225f'_m$ (block)
f'_m = masonry strength, ksi
p_g = reinforcement, %
f_s = allowable steel stress, ksi or MPa
A_g = gross area of column, in.² or mm²

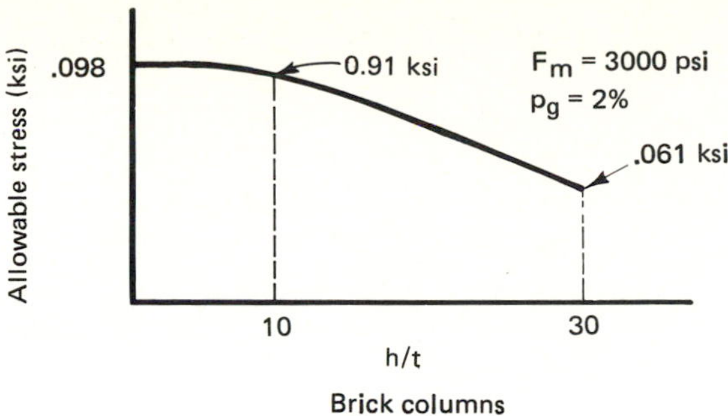

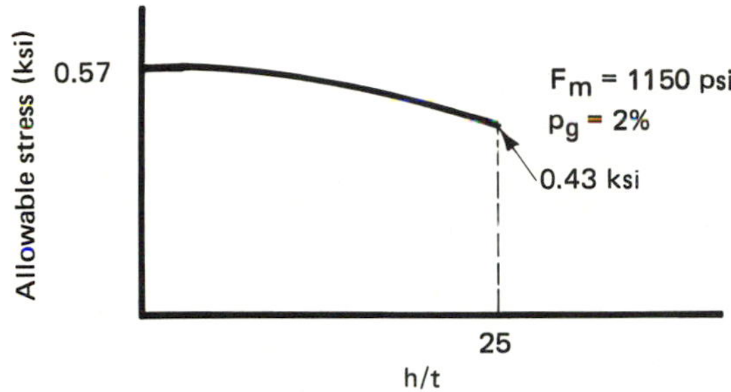

Figure 4–13. Masonry Column Curves

Steel reinforcement in masonry is limited to the range 1 to 4%. Assume 2% reinforcement for preliminary design. For *brick masonry* of 3.0 ksi (20.7 MPa) (f'_m) strength and 2% steel reinforcement (Grade 60), Eq. 4–18 gives

$$P = 1.0(C_s)[0.20(3.0) + 0.80(0.02)(24.0)]A_g$$

$$F_{a\,(\max)} = \frac{P}{A_g} = 0.98C_s \text{ (ksi)} \tag{4–20}$$

$$= 6.75C_s \text{ (MPa)}$$

For *concrete masonry* of 1.15 ksi (7.93 MPa) (f'_m) strength and 2% steel reinforcement, Eq. 4–19 gives

$$P = 1.0(C_s)[0.225(1.15) + 0.65(0.02)(24.0)]A_g$$

$$F_{a\,(\max)} = \frac{P}{A_g} = 0.572C_s \text{ (ksi)} \tag{4–21}$$

$$= 3.94C_s \text{ (MPa)}$$

The slenderness corrections for h/t ratios are given by the following formulas:

$$\text{For brick:}\quad C_s = 1.20 - \frac{h}{52t} \tag{4–22}$$

$$\text{For block:}\quad C_s = 1.0 - \left(\frac{h}{40t}\right)^3 \tag{4–23}$$

Maximum h/t is limited to 25 for concrete masonry columns.

Wood Columns. Figure 4–14 shows column curves for wood. The upper limit, $F_{a\,(\max)}$, is established by the maximum allowable compressive stress for the specific grade and species of wood. The lower limit is given by the following formulas:

For round columns:

$$F_{a\,(\min)} = \frac{3.619E}{(l/r)^2} \tag{4–24}$$

For square or rectangular columns:

$$F_{a\,(\min)} = \frac{0.30E}{(l/d)^2} \tag{4–25}$$

The maximum l/d ratio is limited to 50. For $l/d = 50$ and $E = 1600$ ksi (11.0 GPa), Eq. 4–25 gives

$$F_{a\,(\min)} = \frac{0.30(1600)}{(50)^2} = 0.192 \text{ ksi} \tag{4–26}$$

$$= 1.32 \text{ MPa}$$

Column Stress Summary. Allowable stress ranges established for column elements are:

Steel:	9.37 to 22.0 ksi (64.6 to 151.7 MPa) for A36 steel, based on working loads.
Concrete:	1.31 to 2.77 ksi (9.0 to 19.1 MPa) for 3.0-ksi concrete, based on ultimate loads.
Masonry:	0.61 to 0.98 ksi (4.2 to 6.8 MPa) for brick based on working loads.
	0.43 to 0.57 ksi (3.0 to 3.9 MPa) for block based on working loads.
Wood:	0.192 to 1.5 ksi (1.3 to 10.3 MPa) based on working loads.

Figure 4–14. Wood Column Curves

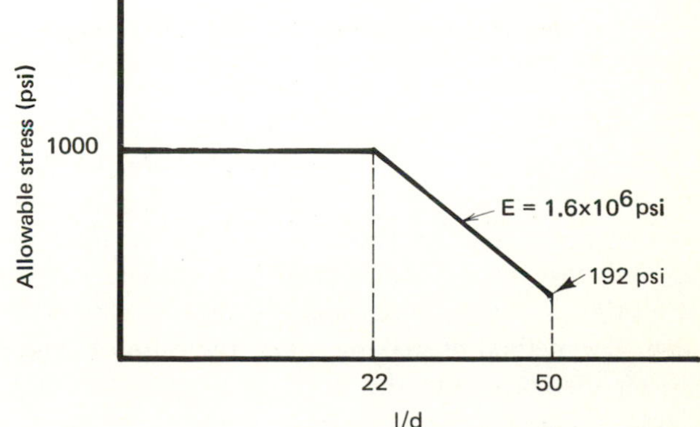

4–3 TENSION TIES

Tension ties are elements that carry load in axial tension, or stretching of the member material. Ties serve as chord and web members in trusses, as hangers to carry suspended loads, and as bracing members. The requirement for high tensile strength normally dictates the use of steel for these members. Other names for tension ties include *struts, tie rods,* and *hangers.*

History and Development

Ties were used very early to resist the horizontal thrust exerted by dome and arch structures and were also used in earthquake areas to tie building walls together. The development of wrought iron provided the material for these early applications, and later developments in steel spawned wider usage, including applications in bridge trusses and roof construction. The major uses today are for tension members in truss systems (plane and space), for bracing (vertical and horizontal systems), and for tie rods between arch or rigid frame foundations.

Types and Uses

Common types of tension tie members are illustrated in Fig. 4–15. Rods are used extensively for tie rods and for bracing. Single angles are probably the most common member used, although heavier loads and connection difficulties favor the use of double angles. The tee, wide-flange, and pipe shapes are used for truss systems.

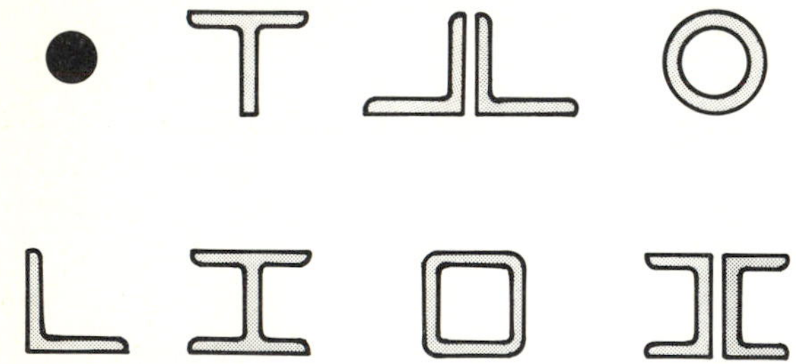

Figure 4–15. Tension Tie Elements

Behavior

The tension capacity of tension tie members is simply dictated by the tensile strength of the material and the efficiency of connections. Elimination of material at connections for bolt holes results in a "net section" area less than the gross member area. The efficiency of this net section is further influenced by the ductility of the metal, the method of making holes, the ratio of gage to fastener diameter, and the ratio of net area in tension to area in bearing.

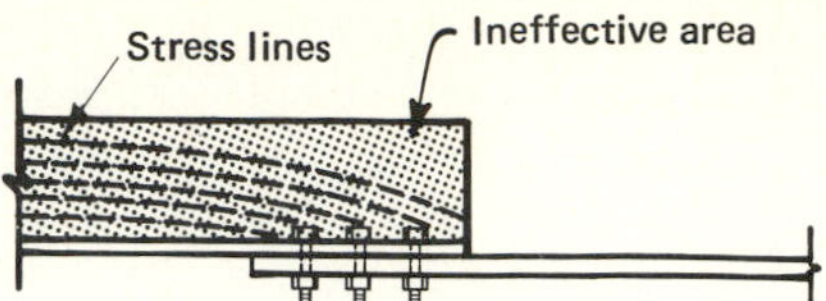

Figure 4–16. Shear Lag Effect

The distribution of cross-sectional material relative to connection plates determines still another efficiency factor called *shear lag.* If only one leg of an angle tension member is connected as shown in Fig. 4–16, the *flow* of stress from the full member cross section into the narrow section of material connected to the bolts may not be 100% effective. For this reason, both AREA and AASHTO specify the net area as equal to the net area of the connected leg plus one-half the area of the unconnected leg. AISC specifies a reduction coefficient between 0.75 and 0.90 depending on the specific connection type.

Tension members are not subject to buckling and, therefore, there is no theoretical limit on the slenderness ratios (l/r) that could be used. However, these members may be subject to load reversal during shipping and erection, or from wind or some other loading. Most specifications require that slenderness be kept below certain maximum values in order to guarantee some minimum compressive strength in the member.

Approximate Analysis

Loads on *hangers* can be readily calculated by the tributary area concept described earlier for columns. *Tie rods* carry loads in support of other structural elements or subsystems, and these loads must come from an analysis of system behavior. Tension members of *truss* and *bracing* systems are covered later in this chapter and in Chapter 5.

Approximate Design

Preliminary design of tension members requires an estimate of the net area available for resisting the loads and the practical capacity of potential connections. The full gross area of members with welded connections is effective, unless the welding process lowers the base metal strength, such is the case with aluminum members.

Bolted Metal Members. The approximate net area will be 80 to 90% of the member gross area. The required area is then

$$A_T = \frac{P}{0.80F_t} \qquad (4\text{–}27)$$

Allowable metal tension stresses (F_t) are shown in Table 4–1.

Table 4–1 Tension Strength of Metals

Metal	Ultimate tension strength ksi	MPa	Allowable stress[a] ksi	MPa
Structural steel				
A36	80	551.6	22.0	151.7
A242	70	482.7	30.0	206.9
A440	70	482.7	30.0	206.9
A441	70	482.7	30.0	206.9
A572	60	413.7	25.0	172.4
Aluminum				
6061-T6	42	289.6	19(11)	131(75.8)
Cold-formed steel				
A570	53	365.4	22	151.7

[a] Reduced allowable stress for welded connections shown in parentheses.

Example Problem 4–5: Preliminary Steel Tension Member Design

Determine the required area for a steel tension member (A36 steel) carrying an axial load of 20 kips (88,960 N).

Solution

1. Determine allowable tension stress from Table 4–1.

$$F_t = 22.0 \text{ ksi } (151.7 \text{ MPa})$$

2. Calculate required area based on Eq. 4–27.

$$A_t = \frac{P}{0.80F_t} = \frac{20.0}{0.80(22)} = 1.14 \text{ in.}^2 \ (735 \text{ mm}^2)$$

3. Select two angles: 2 in. $\times$ 2 in. $\times$ $\frac{3}{16}$ in. (A = 1.43 in.^{2}).

Bolted Wood or Laminated Timber Members. The capacity of a wood member may be limited by the ability to physically locate sufficient bolts, with proper edge distances, and so on, to carry the allowable load of the net section due to the relative low strength of wood in bearing at various angles to the grain. For this reason, it is often advisable to first establish an idea of the number and size of bolts that can be potentially used at the connection, and then use this estimate for determining member size and capacity. Appendix 4E gives approximate bolt capacities for various size bolts and species of wood.

Concrete Members. When precast construction is used, concrete members present some of the same limitations that were discussed for wood construction. The bearing capacity of the bolts (embedded studs or through bolts) may control the usable tension member capacity. Cast-in-place tension members have monolithic connections, but are usually controlled by available room for tension lap splices of the rebar or direct butt-welded splices. These considerations usually control available tension capacity. Practical tension stresses in concrete range from 0.75 to 1.5 ksi (5.2 to 10.3 MPa).

Tension Tie Summary. Allowable stresses established for tension tie elements are:

Steel:	11.0 to 22.0 ksi (75.8 to 151.7 MPa)
Concrete:	0.75 to 1.5 ksi (5.2 to 10.3 MPa)
Wood:	0.5 to 1.5 ksi (3.4 to 10.3 MPa)

4–4 CABLES

Cables are tension elements, but unlike tension ties that carry loads axially, cable loads are usually applied at an angle to the cable axis (Fig. 4–17). Cables are characterized by a geometry that incorporates a definite sag, from which the cable derives its load-carrying ability. Depending on the application, cables may also be called *guys, strand,* and *catenaries.* High-tensile strength wire is usually the construction material used for cables.

Figure 4–17. Loads Applied to Cables

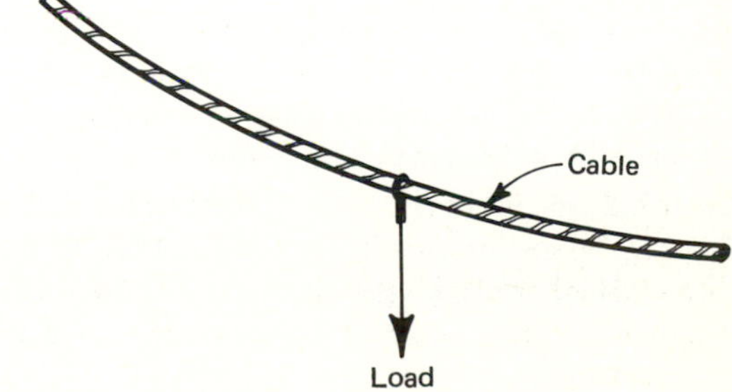

History and Development

The cable element was used early for the construction of rope suspension bridges across rivers and streams and later provided support for tents and similar shelters. In Rome, the catenary form of the cable element was used for temporary roofing of theaters and amphitheaters. The Roman Coliseum still contains attachments for masts to carry the ropes, and there are stone bollards at the base for anchorage. Despite this early history of roof applications, the cable found its first major use in suspension bridge construction. Only as recently as the mid-1960s have cables reemerged as elements for building construction, with projects such as the arena at Raleigh and Madison Square Garden.

Types and Uses

Cables are used to support roof structures, tie-back systems, and bridges. Cables used in buildings can be classified into four groups:

1. Suspension roofs
2. Hangers
3. Auxiliary structural components
4. Main structural system

The suspension roof and suspension bridge applications are the most common uses, with the cable element employed to span large distances. The cable has several advantages over the plate girder, truss, and arch when spans exceed about 150 ft (45.7 m): shear bending and bending instability are involved with an increase in span in the plate girder; the truss suffers from connection costs and depth of construction as spans increase; the arch becomes increasingly vulnerable to compression buckling. Cable limitations are only in dead weight and anchorage. Common cable materials are shown in Table 4–2 and cable anchorage methods are shown in Fig. 4–18.

Behavior

Three types of behavior characterize the cable element: *changing geometry, nonlinear behavior,* and *dynamic action.* Since cables have very little bending stiffness, they must change configuration in order to continue to carry loads in tension. As the cable geometry changes to accommodate the loads (Fig. 4–19), this same change in geometry creates differences in the cable forces. Cable deflection also changes the geometry with increasing load producing deflections which change the geometry—which change the cable forces. This behavior results in a nonlinear load

Table 4–2 Cable Properties (Bridge Strand: Single Strand, Multiple Wire)

Diam (in.)	*Weight per foot (approx. lb)*	*Metallic area (approx. sq in.)*	*Breaking strength (tons)*[a]
1/2	0.52	0.150	15
9/16	0.66	0.190	19
5/8	0.82	0.234	24
11/16	0.99	0.284	29
3/4	1.18	0.338	34
13/16	1.39	0.396	40
7/8	1.61	0.459	46
15/16	1.85	0.527	54
1	2.10	0.600	61
1 1/16	2.37	0.677	69
1 1/8	2.66	0.759	78
1 3/16	2.96	0.846	86
1 1/4	3.28	0.938	96
1 5/16	3.62	1.03	106
1 3/8	3.97	1.13	116
1 7/16	4.34	1.24	126
1 1/2	4.73	1.35	138
1 9/16	5.13	1.47	150
1 5/8	5.55	1.59	162
1 11/16	5.98	1.71	176
1 3/4	6.43	1.84	188
1 13/16	6.90	1.97	202
1 7/8	7.39	2.11	216
1 15/16	7.89	2.25	230
2	8.40	2.40	245
2 1/16	8.94	2.55	261
2 1/8	9.49	2.71	277
2 3/16	10.1	2.87	293
2 1/4	10.5	3.04	310
2 5/16	11.2	3.21	327
2 3/8	11.7	3.38	344
2 7/16	12.5	3.57	360
2 1/2	12.8	3.75	376
2 9/16	13.6	3.94	392
2 5/8	14.5	4.13	417
2 11/16	15.2	4.33	432
2 3/4	15.9	4.54	452
2 7/8	17.4	4.96	494
3	18.9	5.40	538
3 1/8	20.5	5.86	584
3 1/4	22.2	6.34	625
3 3/8	23.9	6.83	673
3 1/2	25.7	7.35	724
3 5/8	27.6	7.88	768
3 3/4	29.5	8.43	822
3 7/8	31.5	9.00	878
4	33.6	9.60	925

[a] Class A coating.

OPEN SOCKET CONNECTION

Cable-to-column flange

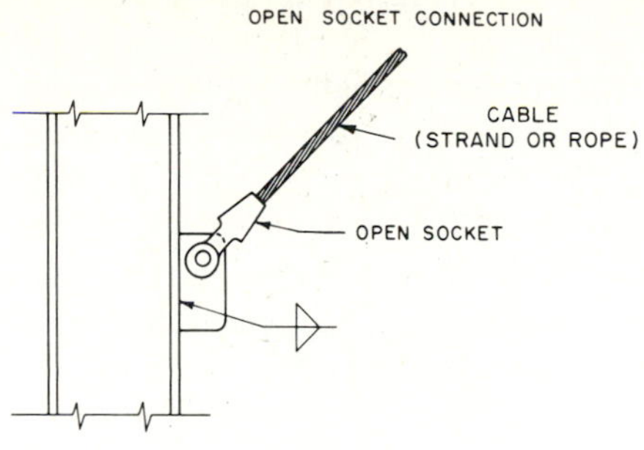

OPEN-SOCKET APPLICATION

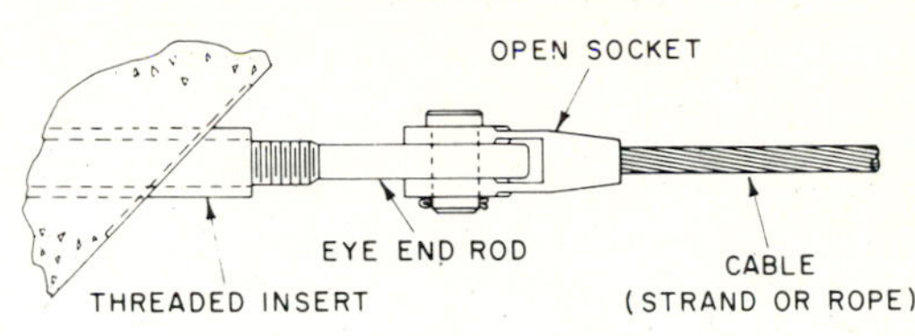

OPEN SOCKET CONNECTION

Beam-to-cable hanger

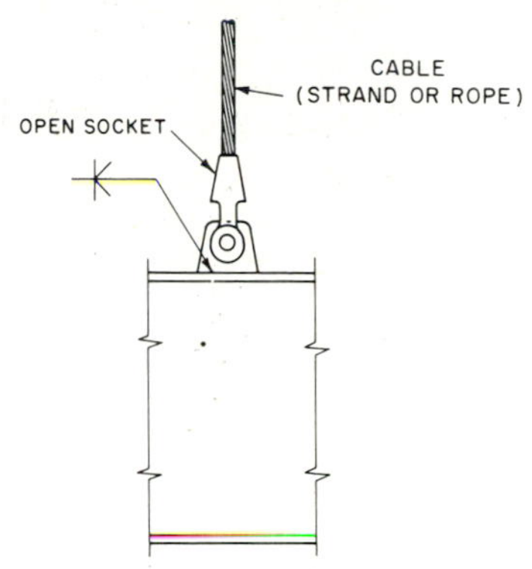

CLOSED-SOCKET APPLICATION

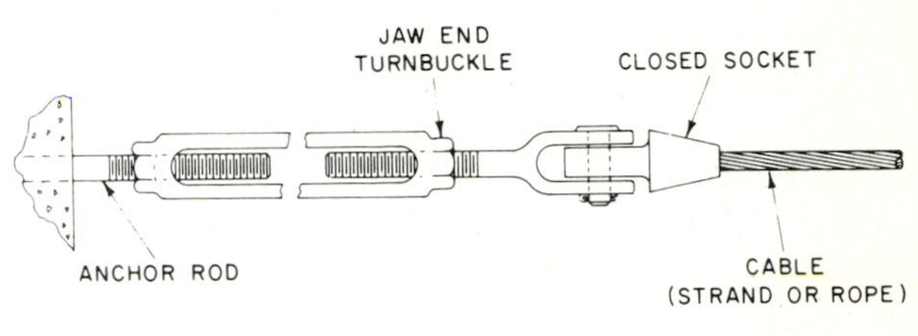

CLOSED BRIDGE SOCKET APPLICATION

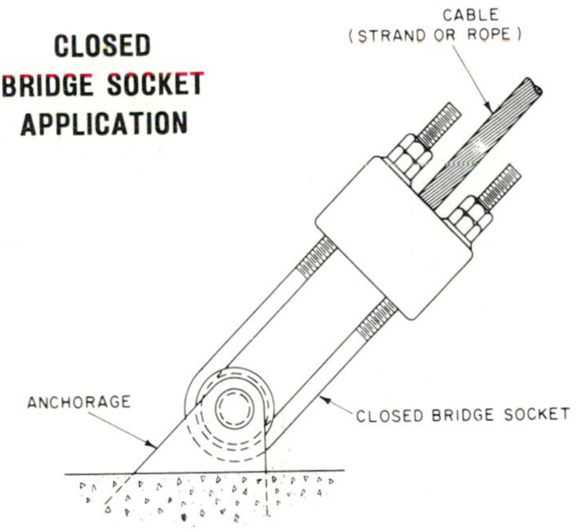

TYPE 6 ANCHOR SOCKET APPLICATION

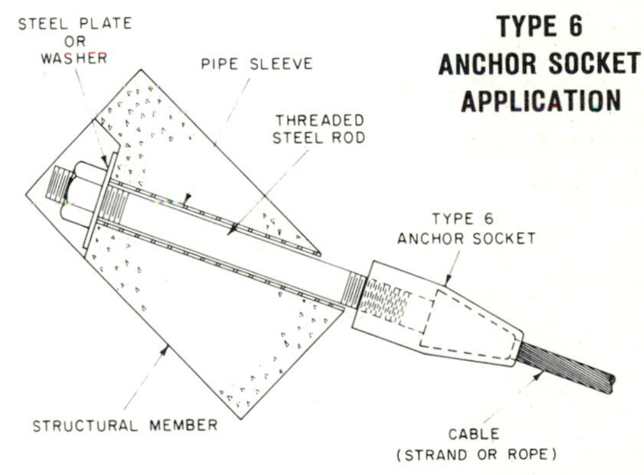

TYPE 7 ANCHOR SOCKET APPLICATION

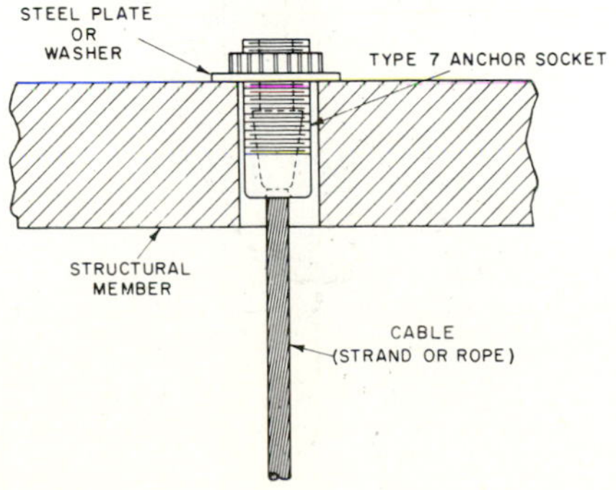

TYPE 2 OR 5 ANCHOR SOCKET APPLICATION

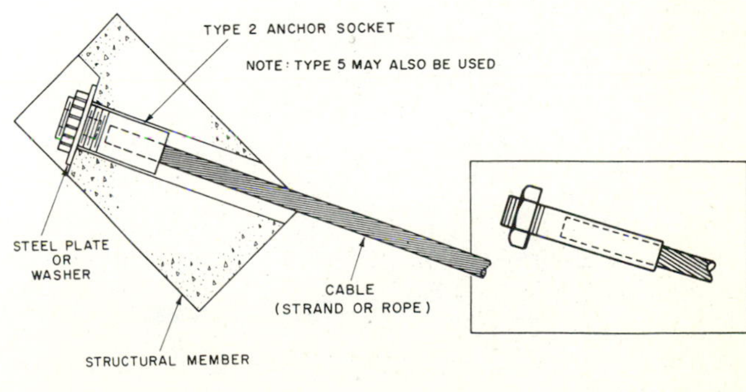

Figure 4–18. Cable Anchorage (From *Cable Roof Structures,* Bethlehem Steel Corporation, 1968)

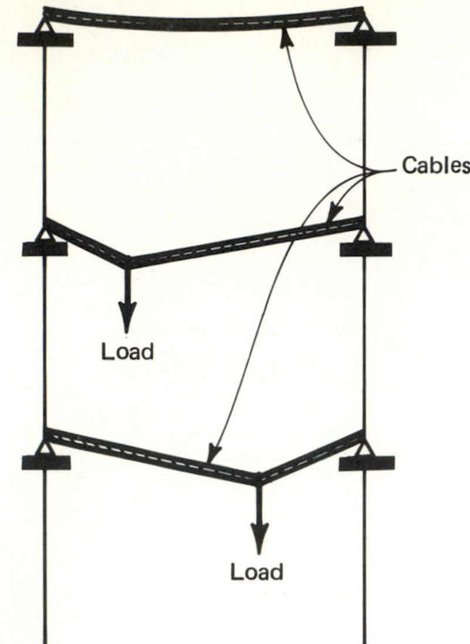

Figure 4–19. Cable Geometry Changes

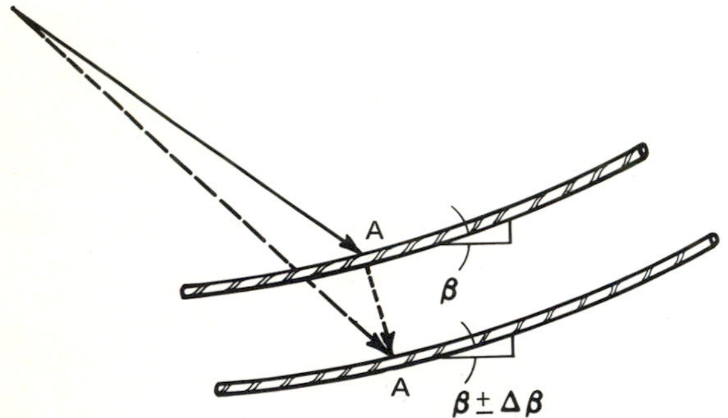

Figure 4–20. Load–Displacement Relationships for Cable Elements

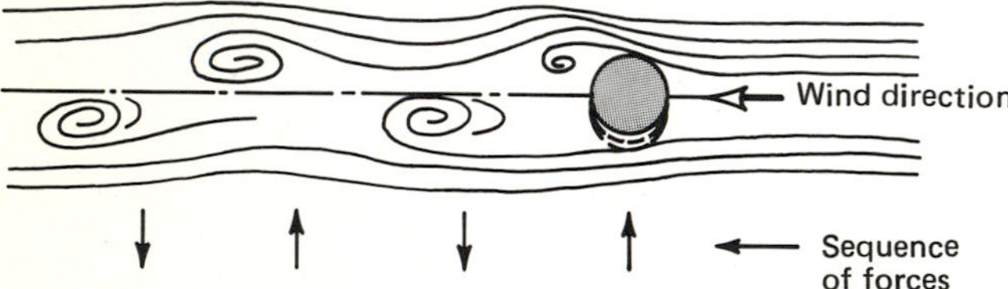

Figure 4–21. Wind Action on Cables

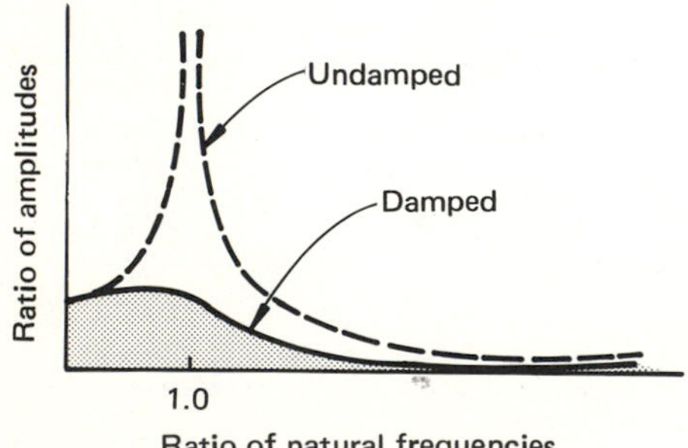

Figure 4–22. Dynamic Response of Cables

displacement curve (Fig. 4–20), an important characteristic of cable elements.

Because cables are light and flexible members, they are particularly sensitive to vibration, flapping, rippling, fluttering, and galloping, all forms of dynamic behavior. Wind is the principal cause of much of this dynamic behavior, although movement of vehicles, machinery, and sound waves may also contribute. Lift and vortex shedding created by the wind (Fig. 4–21) impose dynamic forces that create behavior patterns in relation to the ratio of the frequency of the exciting force to the natural frequency of the cable (Fig. 4–22).

Approximate Analysis

The analysis of cables differs from that of other elements (in which the deformations are at first neglected) and consists of two main steps:

1. Determination of the cable shape
2. Computation of the forces and stresses

Figure 4–23 shows cable geometry and corresponding force and shape parameters. The deflection of the cable at any point is denoted by $d(x)$, the sag of the cable is denoted by f, and the cable-force horizontal component is denoted by H. The analysis of the cable subjected to a load $q(x)$ is based on the following equation:

$$Hd(x) = M(x) \tag{4–28}$$

M is the bending moment due to the load acting on a simple beam of the same span. The cable force is determined by

$$S = \frac{H}{\cos \beta} \tag{4–29}$$

As an example, the cable shown in Fig. 4–24 is loaded by a concentrated load W applied at a point al from the left support. The bending moment M due to the load on a simple beam is

$$M = R_A(al)$$

From Eq. 4–28,

$$H = \frac{M}{d} = \frac{R_A(al)}{h}$$

and

$$S = \frac{H}{\cos \beta} = \frac{H}{al}\sqrt{(al)^2 + h^2}$$

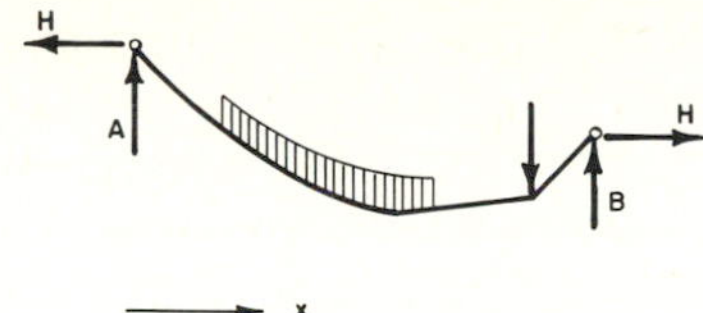

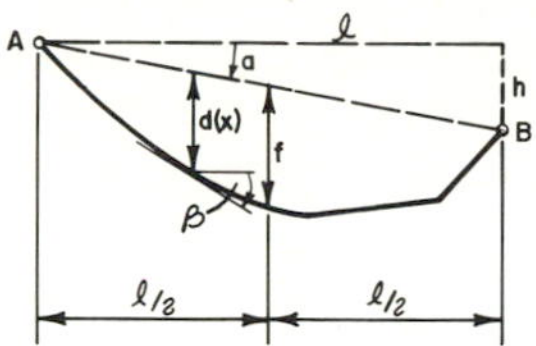

Figure 4–23. Cable Geometry

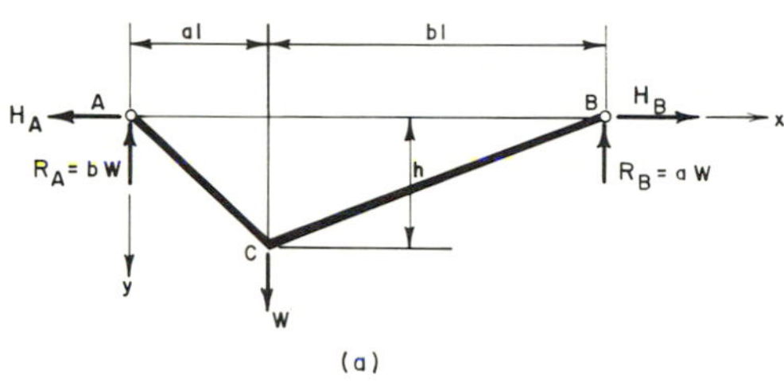

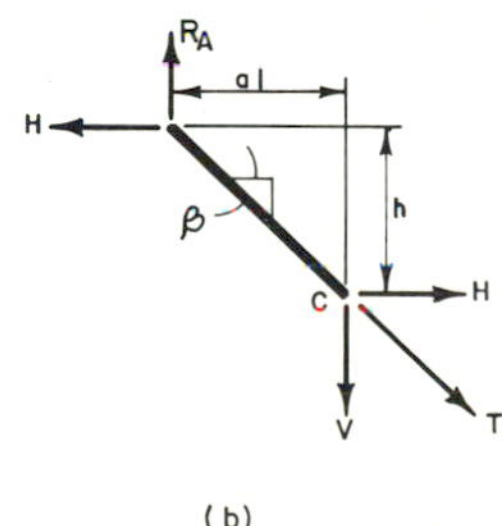

Figure 4–24. Concentrated Load on Cable

The cable length (L) is given by

$$L = \sqrt{(al)^2 + h^2} + \sqrt{(bl)^2 + h^2}$$

If the deflection (h) is preestablished, the horizontal tension is calculated directly and the cable length must be determined. On the other hand, if the cable length is preestablished, h must be calculated and used to determine the horizontal tension H.

Additional loads coming on the cable will change the shape and create higher stress, which will also change the shape due to extension of the cable. These effects can be studied with nonlinear analysis computer programs; however, for many applications, the additional load will be of the same magnitude and shape as the original load and the effect on geometry will be minimum. Cable extension will increase the deflection and decrease the cable tension and preliminary analysis usually neglects this effect. Approximate formulas for calculating cable forces are given in the following sections.

Uniform Load. For a uniform load q on the horizontal projection,

$$H = \frac{ql}{8n} \qquad \text{where } n = \frac{h}{l} \text{ (sag ratio)} \tag{4–30}$$

$$S = \frac{H}{l}\sqrt{l^2 + (4f + h)^2} \tag{4–31}$$

For this case, the cable shape is a quadratic parabola.

Triangular Load (Peak at Midspan). For a triangular load with maximum value of q at midspan,

$$H = \frac{ql}{12n} \tag{4–32}$$

$$S = \frac{H}{l}\sqrt{l^2 + (3f + h)^2} \tag{4–33}$$

Triangular Load (Minimum at Midspan). For a triangular load with maximum value of q at the supports (0 at midspan),

$$H = \frac{ql}{24n} \tag{4–34}$$

$$S = \frac{H}{l}\sqrt{l^2 + (6f + h)^2} \tag{4–35}$$

Cable Length. The cable length for the above cases is given by

$$L = l\left(l + \frac{8}{3}n^2 + \frac{h^2}{2l^2}\right) \tag{4–36}$$

The following example illustrates the approximate method.

EXAMPLE PROBLEM 4–6: Preliminary Cable Analysis

Determine the maximum cable force and cable length for a uniformly loaded cable with a span of 60 ft (18.3 m) and an initial sag of 6 ft (1.83 m). The uniform load is 600 lb/ft (8755 N/m).

Solution

1. Determine cable force by Eq. 4–31.

$$S = \frac{H}{l}\sqrt{l^2 + (4f + h)^2}$$

From Eq. 4–30,

$$H = \frac{ql}{8n} = \frac{600(60)}{8(0.10)} = 45{,}000 \text{ lb } (200{,}160 \text{ N})$$

Therefore,

$$S = \frac{45{,}000}{60}\sqrt{(60)^2 + (4 \times 6 + 0)^2}$$

$$= 48{,}466 \text{ lb } (215{,}576 \text{ N})$$

2. Determine cable length by Eq. 4–36:

$$L = l\left(l + \frac{8}{3}n^2 + \frac{h^2}{2l^2}\right)$$

$$= 60\left[l + \frac{8}{3}(0.10)^2 + \frac{0}{2(60)^2}\right]$$

$$= 61.6 \text{ ft } (18.8 \text{ m})$$

Approximate Design

Optimum sag–span ratios for cables are in the range of 1:10. Once cable geometry and forces have been determined, the required cable area can be calculated as follows:

$$A = \frac{T}{F_t} \qquad (4\text{–}37)$$

where T = tension force
F_t = allowable stress in tension

Table 4–2 gives ultimate stress values for several cable steels.

Cable anchorage forces must be resisted by either rings (tension or compression) or by direct anchorage into a foundation element.

4–5 BEAMS

Beams are bending elements—they carry loads primarily normal to their longitudinal axis and transfer these loads from support to support by flexural bending of the member material. Generally considered to be elements of horizontal subsystems, beams can actually be oriented in any direction: for example, as parts of walls and sloping assemblies. Other names for beams include *purlins, girts, stringers, girders, rafters, joists, spandrels, cantilevers, brackets,* and *lintels.*

History and Development

Early beams were constructed of trunks of trees and other timber materials or slabs of stone. The Greeks and Romans used stone extensively to span between closely spaced columns to form the roof system for magnificent temples. Although iron was discovered much earlier, the iron beam did not become a structural element until the end of the eighteenth century as a series of disastrous fires in timber frame buildings and the general availability of good cast iron led to the substitution of iron beams for timber beams. The next 50 years saw an intensive development of fireproof floors and of the iron beam as a major element of these floors.

Parallel developments in reinforced concrete and laminated timber construction, and still later in prestressed concrete, provided an increasingly wider selection of beam materials. Today, there are almost an infinite number of materials of various strengths and behavior that can be used in beam construction.

Types and Uses

Beams are primarily used for floor and roof construction, bridges, and platforms. Steel and other metal beams generally take the form of web and flange sections, while concrete, timber, and masonry beams are generally rectangular in shape, although concrete beams often are T-shaped (Fig. 4–25).

Members that carry the direct loads of deck construction are often termed *secondary members,* while the heavier members that support several secondary members are termed *primary members* or girders. In bridge con-

Figure 4–25. Beam Types

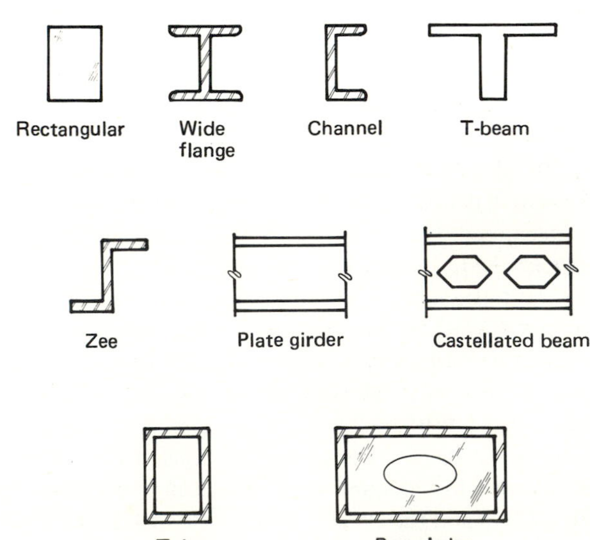

struction, the terms *floor beams* and *stringer beams* are used to denote the same secondary and primary functions.

Composite beams are made of a combination of steel and concrete materials consisting of a concrete deck bonded to supporting steel beams by studs welded to the beam flanges at regular intervals. Other forms of composite beams are shown in Fig. 4–26. These include *flitch plates* (a combination of wood and steel plate popular in residential construction), *plywood box beams,* and *stub-girder* construction.

Figure 4–26. Composite Beams

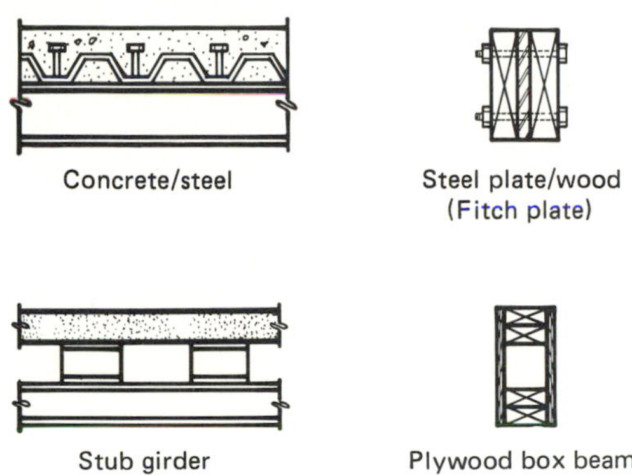

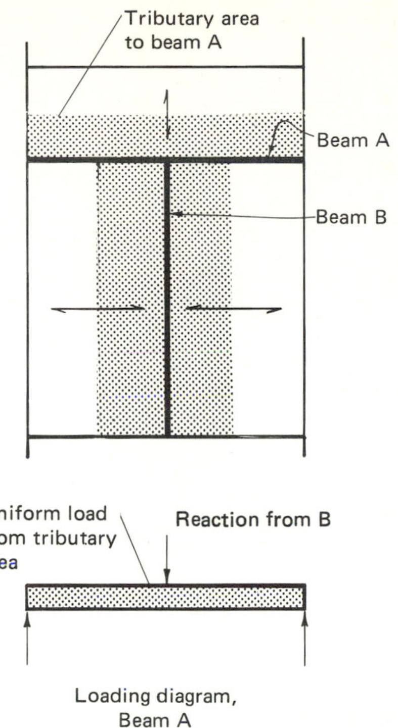

Figure 4–27. Load Distribution to Beams

Behavior

The flexural (bending) behavior of beams leads to failure either by rupture of the material in tension or crushing of the material in compression. In addition, cross sections constructed of thin elements can fail by various forms of instability: local flange buckling, web buckling and lateral-torsional buckling (similar to column elements). Shear stresses are also particularly important for concrete and timber beams.

Behavior of individual beams is then a function of material strength in flexure and shear, dimensions of thin plate components, and overall bracing of the member to prevent lateral-torsional buckling. Deflection behavior is related to the member properties, material modulus of elasticity, and to the time-dependent effects of creep.

Approximate Analysis

Approximate analysis of beams includes (1) load and equivalent load determination, (2) mathematical model assumptions for indeterminate beams, and (3) shear and moment calculation.

Loads. Uniform floor and roof loads can be assumed to travel to supporting beams in proportion to the *tributary area* supported (Fig. 4–27). More intricate and complex load distribution patterns are not justified for preliminary analysis. *Concentrated* loads result from either direct application of equipment, or wheel loads, or from end reactions imposed by intersecting secondary members. Two or three concentrated loads are easy to handle in preliminary analysis, but several loads—particularly of varying locations and magnitude—can complicate the analysis. Several concentrated loads can be converted to an equivalent uniform load by summing the total load, dividing by the span length, and multiplying the result by a correction factor. Figure 4–28 shows the correction factor for one, two, and three concentrated loads. By the time three loads are applied, the correction factor is at 1.33. This procedure is most helpful for three or more loads and a correction factor of 1.10 to 1.25 is suggested.

Indeterminate Beams. Continuous beams and beams that rigidly frame into columns as part of a moment-resisting frame require simplifying assumptions for preliminary analysis. One method is to use tabulated moment and shear coefficients for braced frames, such as those contained in the ACI Code (Table 4–3). A modified version of these coefficients (based on limit strength design) is also contained in Table 4–3 and provides greater flexibility, including allowances for frame action.

The second method of preliminary analysis is to assume inflection points (points of zero moment) at

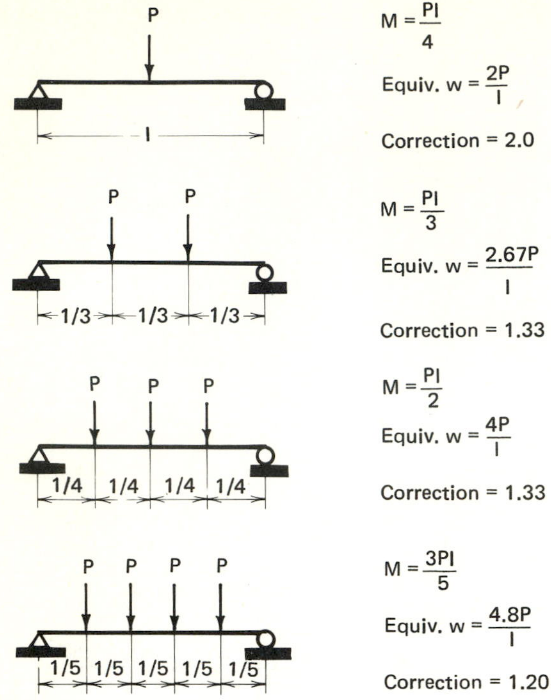

Figure 4–28. Equivalent Uniform Loads

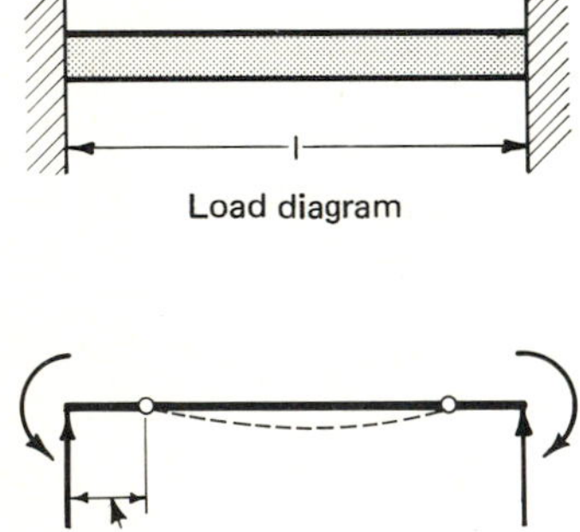

Figure 4–29. Location of Inflection Points

enough points to make the beam determinate. Generally, this requires two inflection points per span, except in the case of the exterior spans of continuous beams resting on simple supports. A fixed-end beam (Fig. 4–29) has inflection points under uniform load at a point $0.21l$ from each support. Beams with less restraint may have inflection points as close as $0.10l$ and with more restraint as far as $0.25l$. Rough sketching of the load patterns and anticipated deformed shape can provide a clue to the approximate location of inflection points (Fig. 4–30).

Rigid frames can generally be assumed to have inflection points at $0.21l$, corresponding to the fixed-end condition. Continuous beam inflection points should be established as follows:

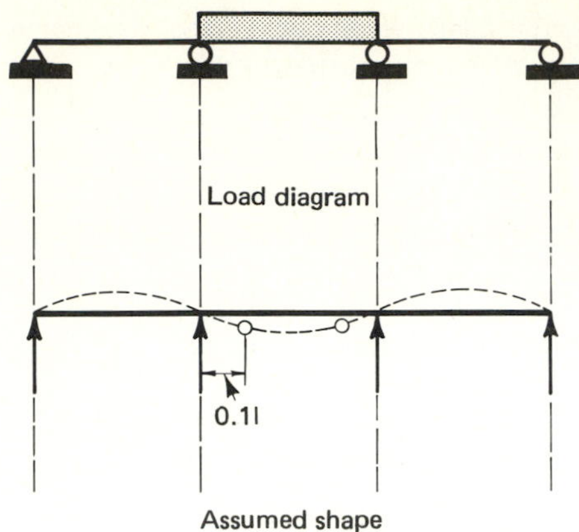

Figure 4–30. Sketching of Deformation to Locate Inflection Points

Uniform load:	$0.25l$
Pattern live load:	
Positive moment:	$0.10l$
Negative moment:	$0.25l$ at center support (point of maximum moment calculation)
	$0.10l$ at far support

Shear and Moment Calculation. Once loads have been determined and appropriate assumptions made for indeterminate beams, shears and moments can be calculated using the simple beam formulas (Fig. 4–31):

$$+M = \frac{wl_1^2}{8} \tag{4–38}$$

$$V = \frac{wl}{2} \tag{4–39}$$

$$-M = \frac{wl_1}{2}(l_2) + \frac{wl_2^2}{2} \tag{4–40}$$

Figure 4–31. Shear and Moment Calculation

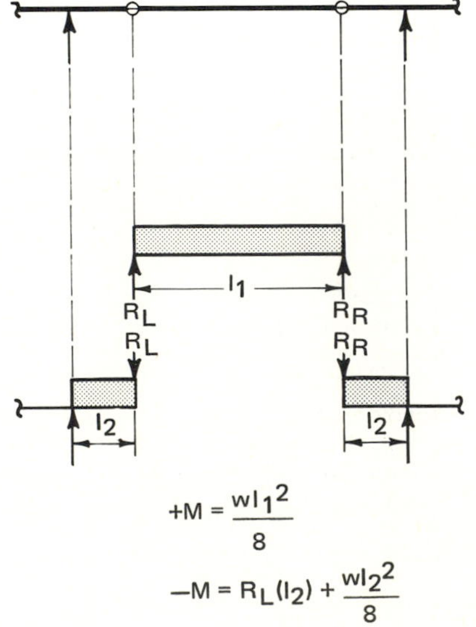

Table 4–3 Analysis Coefficients[a,b]

	ACI[c] code	Alternate[d] coefficients
Moments in Beams		
Positive moment		
End spans		
Discontinuous end unrestrained	$w_u l_n^2/11$	$w_u l_n^2/15.8$
Discontinuous end integral with support	$w_u l_n^2/14$	$w_u l_n^2/19.6$
Interior spans	$w_u l_n^2/16$	$w_u l_n^2/19.6$
Negative moment at exterior face of first interior support		
Two spans	$w_u l_n^2/9$	$w_u l_n^2/12.7$
More than two spans	$w_u l_n^2/10$	
Negative moment at other faces of interior supports	$w_u l_n^2/11$	$w_u l_n^2/13.2$
Negative moment at face of all supports for slabs with spans not exceeding 10 ft; and beams where ratio of sum of column stiffnesses to beam stiffnesses exceeds 8 at each end of the span	$w_u l_n^2/12$	
Negative moment at interior face of exterior support for members built integrally with supports		
Where support is a spandrel beam	$w_u l_n^2/24$	$w_u l_n^2/25$
Where support is a column	$w_u l_n^2/16$	$w_u l_n^2/14.3$
Shear in Beams		
Shear in end members at face of first interior support	$1.15 w_u l_n/2$	$(0.40 + l_o/5l_n) w_u l_n \geq 1.2 w_u l_n/2$
Shear at face of all other supports	$w_u l_n/2$	$(0.55 + 0.02 w_l/w_d) w_u l_n \geq (0.50 + l_o/12 l_n) w_u l_n$
Shear at exterior end of exterior span	$w_u l_n/2$	
Moments in Columns[d]		
Moment at exterior column		
Top of joint rotating toward span	$M_1 = 0.07 w_{ul} l_{nl}^2 - 0.50 w_{dc} l_{nc}^2$	
Top of joint rotating toward cantilever	$M_c = 0.50 w_{uc} l_{nc}^2 - 0.175 w_{dl} l_{nl}^2$	
(Use M_1 or M_c, whichever is larger)		
Moment at interior column		
Top of joint rotating toward span	$M_i = 0.06(w_{ui} l_{ni}^2 - w_{dj} l_{nj}^2)$	
Top of joint rotating toward span	$M_j = 0.06(w_{uj} l_{nj}^2 - w_{di} l_{ni}^2)$	
(Use M_i or M_j, whichever is larger)		

[a] For *braced frames* only.

[b] w_u = design ultimate load
w_d = unit dead load
w_l = unit live load
w_{ul} = design ultimate load on span l
w_{uc} = design ultimate load on cantilever c
w_{ui} = design ultimate load on span i
w_{uj} = design ultimate load on span j
w_{dc} = unit dead load on cantilever c
w_{dl} = unit dead load on span l
w_{di} = unit dead load on span i
w_{dj} = unit dead load on span j
l_n = distance between face of supports
l_{nl} = l_n for span l
l_{nc} = l_n for cantilever span c
l_{ni} = l_n for span i
l_{nj} = l_n for span j
l_o = length of span on opposite face of column

[c] ACI Committee 318, *Building Code Requirements for Reinforced Concrete,* ACI 318–77 (Detroit, Mich.: American Concrete Institute, 1977). Permitted for two or more approximately equal spans (the larger of two adjacent spans not exceeding the shorter by more than 20%) with loads uniformly distributed, where the unit live load does not exceed three times the unit dead load.

[d] Richard W. Furlong and Carlos Rezende, "Alternate to ACI Analysis Coefficients," *Journal of the Structural Division,* Vol. 105, No. ST11 (November 1979) (New York: American Society of Civil Engineers). General conditions require loads to be uniformly distributed; ratio between unit live load and unit dead load must not exceed 5; exterior spans are at least half as long as the adjacent interior span.

The following example illustrates the approximate analysis methods.

EXAMPLE PROBLEM 4–7: Approximate Beam Analysis

Determine the maximum positive moment for span A of the four span beam shown in Fig 4–32(a).

Solution

1. Calculate the equivalent uniform load.

$$w = \frac{\Sigma P}{l} \times \text{correction factor}$$

$$= \frac{8 + 12 + 15 + 10}{32}(1.20) = 1.69 \text{ kips/ft (8028 N/m)}$$

2. Sketch the deformed shape as shown in Fig. 4–32(b).
3. Assume inflection points at 4 ft (1.2 m) from the supports (as shown) based on evaluation of the sketch.
4. Calculate maximum positive moment.

$$+M = \frac{wl_1^2}{8} = \frac{1.69(24)^2}{8} = 121.7 \text{ ft-kips (165,025 N/m)}$$

Figure 4–32. Example Problem 4–7

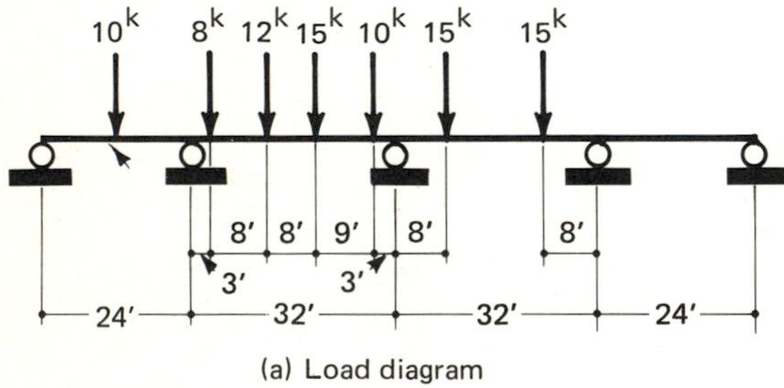

(a) Load diagram

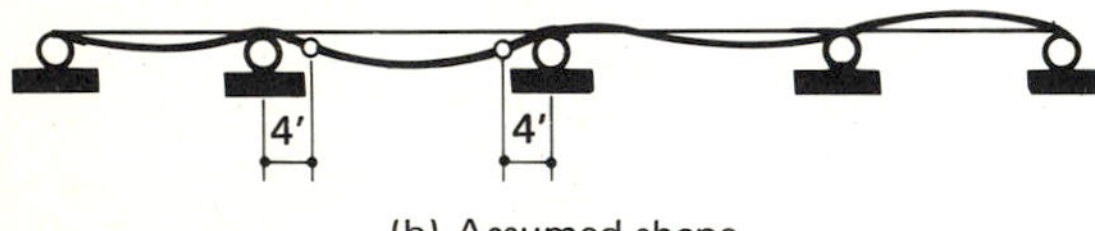

(b) Assumed shape

Approximate Design

Approximate member design for beams involves (1) selection of a minimum depth to control deflection, (2) determining the lateral bracing to be provided, and (3) then use of the simple bending formula:

$$F_b = \frac{Mc}{I} \tag{4–41}$$

where I = moment of inertia of beam section
c = distance from neutral axis to extreme fiber
M = applied moment
F_b = allowable bending stress

Steel Beams. The *minimum depth* (d) for simple span beams can be established assuming a maximum total load deflection of 1/240 and an "average" allowable stress (A36 steel) of 18,000 psi (124.1 MPa). Set

$$\Delta = \frac{5wl^4}{384EI} = \frac{l}{240}$$

$$w = \frac{EI}{3.13l^3} = \frac{(30 \times 10^6)I}{3.13l^3} = \frac{(9.58 \times 10^6)I}{l^3} \tag{4–42}$$

and

$$f = \frac{Mc}{I} = \frac{wl^2c}{8I} = 18{,}000 \text{ psi}$$

$$c = \frac{144{,}000I}{wl^2} \tag{4–43}$$

Substituting Eq. 4–42 into Eq. 4–43 yields

$$c = \frac{144{,}000I}{wl^2} = \frac{144{,}000I(l^3)}{(9.58 \times 10^6)Il^2} = 0.015l$$

Since $d = 2c$,

$$d = 2(0.015l) = 0.03l(l/33) \qquad (l \text{ and } d \text{ in inches}) \tag{4–44}$$

or

$$d = 0.03(l \times 12) = 0.36l\ (l/2.77) \qquad (l \text{ in feet and } d \text{ in inches}) \tag{4–45}$$

This derivation results in the commonly used rule of thumb that steel beam depths (inches) should be from *one-third to one-half of the span* (feet).

The *allowable bending stress* for steel depends on the unbraced length. Figure 4–33 shows an approximate

Figure 4–33. Allowable Bending Stress for Steel (A36)

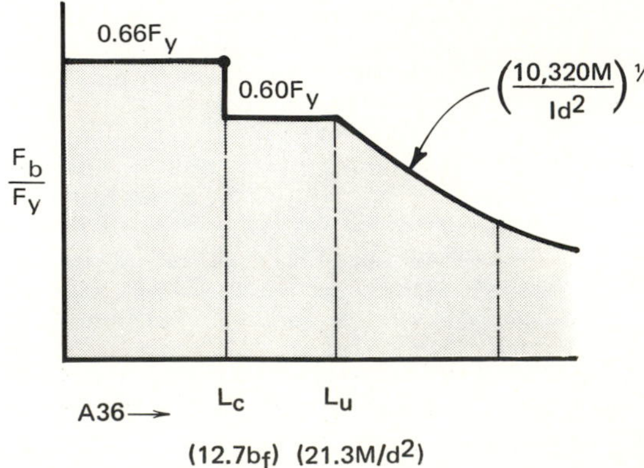

curve which describes the relationship between allowable bending stress and unbraced length for A36 steel. The approximate relationships among allowable stress, beam depth, and unbraced length can be developed as follows.

According to AISC, the *maximum length* (L_c) for which a bending stress of $0.66F_y$ can be used is

$$L_c \leq \frac{76b_f}{\sqrt{F_y}} \tag{4–46a}$$

or

$$L_c \leq \frac{20{,}000}{(d/A_f)F_y} \tag{4–46b}$$

Equation 4–46a will result in the lowest value for most optimum weight steel sections. For A36 steel,

$$L_c = \frac{76b_f}{\sqrt{F_y}} = \frac{76b_f}{\sqrt{36}} = 12.7b_f \tag{4–47}$$

Values for L_c for various beam sections are tabulated in Appendix 4B.

The *maximum length* (L_u) for which a bending stress of $0.60F_y$ can be used is (for most cases) given by AISC Eq. 1.5–7 as follows:

$$L_u = \frac{12{,}000}{(d/A_f)F_b}$$

If A_f is approximated as $0.86M/dF_b$ and F_b is set equal to 22 ksi (0.60 F_y),

$$L_u \simeq \frac{21.3M}{d^2} \tag{4–48}$$

For unbraced lengths beyond L_u, AISC Eq. 1.5–7 can be used directly to solve for F_b:

$$F_b = \frac{12{,}000}{ld/A_f}, \qquad A_f \simeq \frac{0.86M}{dF_b}$$

$$F_b = \sqrt{\frac{10{,}320M}{ld^2}} \leq 0.60F_y \tag{4–49}$$

The following example illustrates the approximate method.

EXAMPLE PROBLEM 4–8: Preliminary Steel Beam Design

Select a preliminary steel beam member to span 25 ft (7.6 m). The moment has already been determined to be 100 ft-kips (135,600 N/m) and bracing will be provided at 8-ft (2.4-m) intervals.

Solution

1. Determine approximate depth to control deflection.

$$d \text{ (in.)} = \tfrac{1}{2} \text{ span in feet}$$

$$d = \frac{25}{2} = 12.5 \text{ in. (317 mm)}$$

Use a 14-in. beam depth.

2. Determine allowable bending stress by Eq. 4–49.

$$F_b = \sqrt{\frac{10{,}320(100)}{8(14)^2}} = 25.6 \text{ ksi} > 22.0 \text{ max.}$$

Use F_b = 22.0 ksi (151.7 MPa).

3. Alternatively, determine L_u by Eq. 4–48:

$$L_u = \frac{21.3(100)}{(14)^2} = 10.8 \text{ ft}$$

Since l = 8 ft, F_b = 22.0 ksi

4. Select member size.

$$s = \frac{M}{F_b} = \frac{100(12)}{22.0} = 54.5 \text{ in.}^3 \text{ (893,800 mm}^3\text{)}$$

From Appendix 4B, use W 14 × 38 (s = 54.6 in.³).

Concrete Beams. The *minimum depth* for simple span concrete beams specified by the 1977 ACI Code is as follows (see Table 4–4):

$$\text{Min. } h = \frac{l}{16} \text{ (} h \text{ and } l \text{ in inches or mm)} \tag{4–50}$$

or

$$h = \frac{3}{4l} \text{ (} h \text{ in inches and } l \text{ in feet)} \tag{4–51}$$

The *ultimate resisting capacity* of a reinforced concrete section is shown in Fig. 4–34.

$$M_u = \phi f'_c bd^2\omega(1 - 0.59\omega) \tag{4–52}$$

where M_u = ultimate moment, ft-k
ϕ = capacity reduction factor
f'_c = concrete strength, ksi
b = width of beam, in.
d = depth to tensile steel, in.
$\omega = pf_y/f'_c$
p = reinforcing percentage
f_y = yield strength of reinforcing steel, ksi

Reinforcing steel percentage varies between 0.35 and 1.6%. If 0.85% is adopted for preliminary design:

Table 4–4 Minimum Thickness of Beams or One-Way Slabs Unless Deflections Are Computed[a,b]

Member	Minimum thickness, h			
	Simply supported	*One end continuous*	*Both ends continuous*	*Canti-lever*
Solid one-way slabs	$l/20$	$l/24$	$l/28$	$l/10$
Beams or ribbed one-way slabs	$l/16$	$l/18.5$	$l/21$	$l/8$

[a] ACI Committee 318, *Building Code Requirements for Reinforced Concrete,* ACI 318–77 (Detroit, Mich.: American Concrete Institute, 1977).

[b] Members not supporting or attached to partitions or other construction likely to be damaged by large deflections. The span length l is in inches.

$$\omega = \frac{\rho f_y}{f'_c} = 0.0085 f_y/f'_c \qquad (4\text{–}53)$$

For 3000 psi (20.7 MPa) concrete and Grade 60 steel:

$$\omega = \frac{0.0085(60)}{3.0} = 0.17$$

$$M_u = 0.9(3.0)bd^2(0.17)[1 - 0.59(0.17)]$$

$$= 0.41bd^2 \qquad (4\text{–}54)$$

Setting Eq. 4–54 equal to Eq. 4–41, we get

$$M_u = 0.41bd^2 = F_b\left(\frac{I}{c}\right)$$

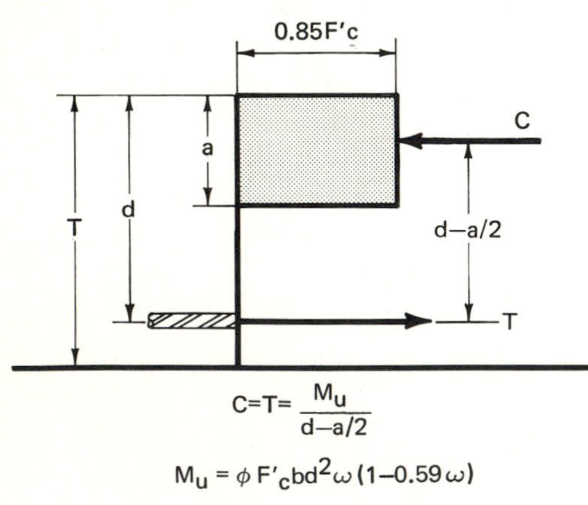

Figure 4–34. Ultimate Strength of Concrete Beam

Assume $d = 0.8h$ as the approximate depth to steel:

$$I = \frac{bh^3}{12} = \frac{b(d/0.8)^3}{12} = \frac{bd^3}{6.14}$$

$$F_b = \frac{0.41bd^2(6.14)(d/2)}{bd^3}$$

$$= 1.26 \text{ ksi } (8.69 \text{ MPa}) \qquad (4\text{–}55)$$

This is the equivalent allowable bending stress for 3000-psi (20.7-MPa) concrete and 0.85% reinforcement. Other values for different concrete strengths and reinforcement percentages are given in Table 4–5.

Table 4–5 Equivalent Allowable Bending Stress (Ultimate) for Concrete Beams[a]

Concrete strength		*Reinforcement*		*Allowable stress (ultimate)*	
psi	*MPa*	*(%)*	ω	*ksi*	*MPa*
3000	20.7	0.85	0.17	1.26	8.7
3000	20.7	1.60[b]	0.32	2.15	14.8
4000	27.6	1.00	0.15	1.50	10.3
4000	27.6	2.14[b]	0.32	2.85	19.7
5000	34.5	1.25	0.15	1.89	13.0
5000	34.5	2.52[b]	0.30	3.41	23.5
6000	41.4	1.40	0.14	2.13	14.7
6000	41.4	2.83[b]	0.28	3.87	26.7

[a] Based on Grade 60 steel.

[b] Maximum reinforcement = 0.75ρb.

Tests have shown that laterally unbraced reinforced concrete beams of any reasonable dimensions, even when very deep and narrow, will not fail prematurely by lateral buckling provided the beams are loaded without lateral eccentricity that could cause torsion. The ACI Code limits the unbraced length to 50 times the least width b.

Shear capacity of preliminary beam sizes should be checked by the following equation, which limits allowable shear stress to a reasonable value:

$$v_u = \frac{V_u}{\phi b_w d} \leq 6\sqrt{f'_c} \text{ ksi} \leq 0.5\sqrt{f'_c} \text{ N/m}^2 \quad (4\text{–}56)$$

where f'_c = concrete strength, ksi or N/m^2
b_w = width of beam, in. or m
d = depth to tension steel, in. or m
V_u = shear force, kips or N
ϕ = capacity reduction factor (0.85 for shear)

The following example illustrates the approximate design method.

EXAMPLE PROBLEM 4–9: Preliminary Concrete Beam Design

Select a preliminary concrete beam to simple span 25 ft (7.6 m). The ultimate moment (M_u) is 140 ft-kips (189,840 N/m) and the maximum ultimate shear is 14 kips (62,272 N). Use 3000 psi (20.7 MPa) concrete and Grade 60 steel.

Solution

1. Determine approximate depth (h) to control deflection.

$$h \text{ (in.)} = \tfrac{3}{4} \text{ span in feet}$$

$$\text{Min. } h = \frac{3}{4}(25) = 18.75 \text{ in. (476 mm)}$$

2. Determine allowable stress from Table 4–5. For 0.85% reinforcement,

$$F_b = 1.26 \text{ ksi (8.7 MPa)}$$

3. Select member size:

$$s = \frac{M}{F_b} = \frac{140(12)}{1.26} = 1333 \text{ in.}^3 \text{ (21,861,200 mm}^3\text{)}$$

Try b = 12 in. (305 mm).

$$h = \sqrt{\frac{6s}{b}} = \sqrt{\frac{6(1333)}{12}} = 25.8 \text{ in. (655 mm)}$$

Use a 12 in. wide × 26 in. deep (305 mm × 660 mm) beam.

$$\begin{aligned} \text{Reinforcement} &= 0.0085bd \\ &= 0.0085(12)(0.8 \times 26) \\ &= 2.12 \text{ in.}^2 \text{ (1367 mm}^2\text{)} \end{aligned}$$

4. Check shear capacity by Eq. 4–56.

$$v_u = \frac{14}{0.85(12)(0.8 \times 26)} = 0.065 \text{ ksi (448 kPa)}$$

$$0.065 \text{ ksi} < 6\sqrt{f'_c} \text{ (328 ksi)} \qquad \text{O.K.}$$

Wood Beams. The *minimum depth* for wood beams can be established with a procedure similar to the one used previously for steel beams. This procedure establishes the following:

$$\text{Min. } d = 0.02l$$

or

$$d = \frac{3}{5}l \text{ (} l \text{ in feet and } d \text{ in inches)} \quad (4\text{–}57)$$

Lateral support requirements for wood beams do not usually present problems. Therefore, the bending formula can be used directly with the allowable bending stress for the grade and species:

$$S = \frac{M}{F_b}$$

Prestressed Concrete Beams. The *minimum depth* for simple span prestressed concrete beams is

$$\text{Min. } h = \frac{l}{40} \text{ (} h \text{ and } l \text{ in in. or m)}$$

or

$$h = 0.3l \text{ (} h \text{ in in., } l \text{ in ft)} \quad (4\text{–}58)$$

For depressed strand construction the *allowable stresses* at midspan control design and the section properties for the final stress condition (after all loads are in place) can be expressed as follows:[1]

$$S_T = \frac{(1-R)M_o + M_d + M_l}{Rf_{ti} + f_{cs}}$$

$$S_B = \frac{(1-R)M_o + M_d + M_l}{f_{ts} + Rf_{ci}}$$

For 20% losses, and setting $F_T = Rf_{ti} + f_{cs}$ and $F_B = f_{ts} + Rf_{ci}$ yields

$$S_T = \frac{0.20M_o + M_d + M_l}{F_t} \quad (4\text{–}59)$$

$$S_B = \frac{0.20M_o + M_d + M_l}{F_B} \quad (4\text{–}60)$$

$$f_{cci} = \frac{S_B}{S_T + S_B}(f_{ti} + f_{ci}) - f_{ti} \quad (4\text{–}61)$$

$$P_i = A_c f_{cci} \quad (4\text{–}62)$$

[1] Arthur N. Nilson, *Design of Prestressed Concrete* (New York: John Wiley & Sons, Inc., 1978).

Table 4–6 Allowable Stresses for Prestressed Concrete

f'_c	f'_{ci}	f_{ti}	f_{ts}	f_{ci}[a]	f_{cs}	F_T	F_B[a]
				Concrete			
5000	2500	150	424	1500	2250	2370	1624
5000	3000	164	424	1800	2250	2381	1864
5000	3500	177	424	2100	2250	2392	2104
6000	3000	164	464	1800	2700	2831	1904
6000	3500	177	464	2100	2700	2841	2144
6000	4000	190	464	2400	2700	2852	2384

Prestressing Steel
Seven-Wire Strand, f_{pu} = 270 ksi

Nominal diameter (in.)	$\frac{3}{8}$	$\frac{7}{16}$	$\frac{1}{2}$	$\frac{9}{16}$	0.600
Area (sq in.)	0.085	0.115	0.153	0.192	0.215
Weight (plf)	0.29	0.40	0.53	0.65	0.74
$0.7f_{pu}A_{ps}$ (kips)	16.1	21.7	28.9	36.3	40.7
$0.8f_{pu}A_{ps}$ (kips)	18.4	24.8	33.0	41.4	46.5
$f_{pu}A_{ps}$ (kips)	23.0	31.0	41.3	51.8	58.1

[a] Addition of nonprestressed reinforcement in the initial tension zone can be used to increase these allowable values in the range of 800 psi.

where S_T, S_B = section modulus required
R = ratios expressing effective prestress force after losses
M_o = moment due to weight of member
M_d = moment due to dead load
M_l = moment due to live load
f_{ti} = allowable initial tension stress
f_{ts} = allowable service tension stress
f_{ci} = allowable initial compressive stress
f_{cs} = allowable service compressive stress
f_{cci} = initial concrete centroidal stress
P_i = initial prestress force

Table 4–6 lists the allowable stress for various strengths of concrete. Although based on depressed strand construction, Eqs. 4–59 and 4–60 can be used for straight strand provided that provisions are made to minimize excessive stress at the ends of the member, such as sheathing. The following example illustrates the approximate design method.

EXAMPLE PROBLEM 4–10: Preliminary Prestressed Beam Design

Select a prestressed double-tee member to span 30 ft (9.1 m) and carry a dead load of 50 psf (2394 N/m²) and a live load of 75 psf (3591 N/m²).

Solution

1. Determine minimum depth by Eq. 4–58.

$$h = 0.3l = 0.3(30) = 9 \text{ in. (229 mm)}$$

Use a 12-in. deep member (305 mm).

2. Determine moments. Assume an 8-ft (2.44-m)-wide double-tee member and that weight of member = 35 psf (1675 N/m²).

$$M_o = \frac{wl^2}{8} = \frac{(8 \times 0.035)(30)^2}{8} = 31.5 \text{ ft-kips (42,714 Nm)}$$

$$M_d = \frac{wl^2}{8} = \frac{(8 \times 0.05)(30)^2}{8} = 45.0 \text{ ft-kips (61,020 Nm)}$$

$$M_l = \frac{wl^2}{8} = \frac{(8 \times 0.075)(30)^2}{8} = 67.5 \text{ ft-kips (91,530 Nm)}$$

3. Determine allowable stresses. Assume that f'_c = 5000 psi (34.5 MPa) and f'_{ci} = 3500 psi (24.1 MPa). By Table 4–7,

$$F_T = 2392 \text{ psi (16.5 MPa)}, \qquad F_B = 2104 \text{ psi (14.5 MPa)}$$

4. Calculate required section properties. Assume 20% losses.

$$S_T = \frac{0.20M_o + M_d + M_l}{F_T}$$

$$= \frac{(0.20 \times 31.5 + 45 + 67.5)12}{2.39}$$

$$= 596.5 \text{ in.}^3 \text{ (9,782,600 mm}^3\text{)}$$

$$S_B = \frac{0.20M_o + M_d + M_l}{F_B}$$

Table 4–7 Allowable Horizontal Shear Load for One Connector, q (kips)[a,b]

Connector[c]	Specified compressive strength of concrete f'_c[d] (ksi)		
	3.0	*3.5*	*≥4.0*
$\frac{1}{2}$-in. diam. × 2-in. hooked or headed stud	5.1	5.5	5.9
$\frac{5}{8}$-in. diam. × $2\frac{1}{2}$-in. hooked or headed stud	8.0	8.6	9.2
$\frac{3}{4}$-in. diam. × 3-in. hooked or headed stud	11.5	12.5	13.3
$\frac{7}{8}$-in. diam. × $3\frac{1}{2}$-in. hooked or headed stud	15.6	16.8	18.0
Channel C3 × 4.1	4.3ω	4.7ω	5.0ω
Channel C4 × 5.4	4.6ω	5.0ω	5.3ω
Channel C5 × 6.7	4.9ω	5.3ω	5.6ω

[a] *Specification for the Design, Fabrication, and Erection of Structural Steel for Buildings* (Chicago: American Institute of Steel Construction, 1978).

[b] Applicable only to concrete made with ASTM C33 aggregates.

[c] The allowable horizontal loads tabulated may also be used for studs longer than shown.

[d] ω = length of channel, inches.

$$= \frac{(0.20 \times 31.5 + 45 + 67.5)12}{2.1}$$

$$= 678.9 \text{ in.}^3 \ (11{,}133{,}960 \text{ mm}^3)$$

5. Select member. From Appendix 4B, use a 8 ft wide × 18 in. deep double-tee (S_T = 1966, S_B = 701, A = 344, weight = 45 psf).

   Use actual section properties to calculate required prestress:

$$f_{cci} = \frac{S_B}{S_T + S_B}(f_{ti} + f_{ci}) - f_{ti}$$

$$= \frac{701}{1926 + 701}(0.177 + 2.1) - 0.177$$

$$= 0.42 \text{ ksi } (2.89 \text{ MPa})$$

$$P_i = A_c f_{cci} = 344(0.42) = 144 \text{ kips } (640{,}512 \text{ N})$$

$$\text{No. } \tfrac{1}{2}\text{-in. strand req'd} = \frac{144}{28.9} = 4.98$$

Composite Steel–Concrete Beams. The minimum depth for composite steel beams can be estimated as 0.80 × the depth required for a non-composite section. Likewise, the required section modulus can be estimated as 0.70 × the non-composite requirement.

The total horizontal shear to be resisted by the studs (each side of the point of maximum positive moment) is given by AISC Eq. 1.11–4 as follows:

$$V_h = \frac{A_s F_y}{2} \qquad (4\text{–}63)$$

where f'_c = specified compression strength of concrete, kips/sq in.

A_c = actual area of effective concrete flange, sq in.

A_s = area of steel beam, sq in.

The total number of studs per beam is equal to 2 × V_h. The allowable shear load per connector is given in Table 4–7.

Beam Summary. Allowable stresses established for beam elements are:

Steel:	12.0–24.0 ksi (82.7–165.5 MPa)
Concrete:	1.26–3.87 ksi (8.7–26.7 MPa) (ultimate loads)
Wood:	1.0–2.0 ksi (6.9–13.8 MPa)
Prestressed:	1.6–2.8 ksi (11.0–19.3 MPa)

4–6 BEAM-COLUMNS

Beam-columns carry loads in combined axial load and bending (Fig. 4–35). As primary members in rigid frames, they carry moments as a result of gravity load on intersecting girders and may also carry moments induced by lateral wind or earthquake loads on the frame. These moments are a result of frame action; other moments may originate from eccentric connections, bending load applied directly to the member (top chords of trusses, for example), and from initial imperfections in column alignment.

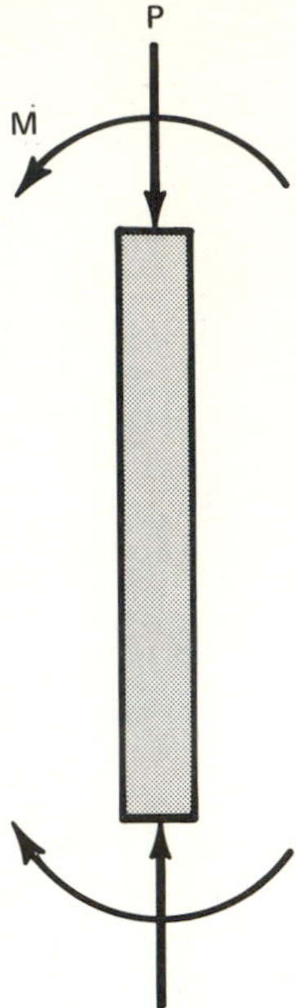

Figure 4–35. Beam Column Loads

Reinforced concrete construction generally includes beam-columns because of the inherent rigidity at the monolithic connections. The ACI Code specifies a minimum moment for every concrete column, a requirement that we incorporated earlier in the approximate design procedure for concrete columns.

History and Development

The first full use of beam-columns in the United States was probably in the Home Insurance Building built in Chicago in 1883, built with a steel frame and one of the first to develop continuity between the floor girders and the supporting columns. The Ingalls Building in Cincinnati was one of the first concrete frame buildings to achieve this continuity and was constructed in 1902. Since this early development, beam-columns have continued to play an increasing role in the development of structural form. The development of indeterminate analysis methods has provided us with the tools to calculate the moments in these more complex structural elements. An even greater contribution has been made by the numerous researchers who have painstakingly studied the behavior of beam-columns in the laboratory. This development is continuing, and each new edition of a major design code refines the beam-column design procedures.

Types and Uses

Most beam-column elements have shapes similar to column elements. Single-story rigid frames, however, quite often make use of tapered sections for more efficient use

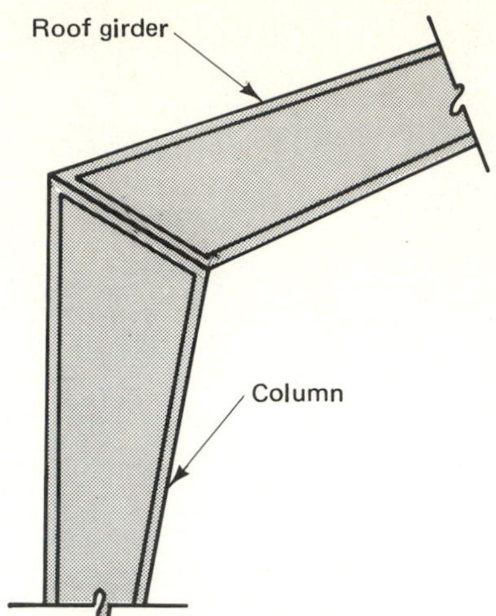

Figure 4–36. Tapered Beam Columns

of material (Fig. 4–36). This is particularly true in the preengineered metal building industry.

Beam-columns are used primarily to provide lateral stability and to reduce the size and depth of floor members in building structures. They are also used to carry loads from panel point to panel point in trusses. Vierendeel trusses (Fig. 4–37) use the beam-column as the primary load resisting element. Bridge piers usually act as beam-columns in carrying longitudinal and transverse loads from the superstructure. Even pile foundation systems are beam-columns when subjected to lateral loads.

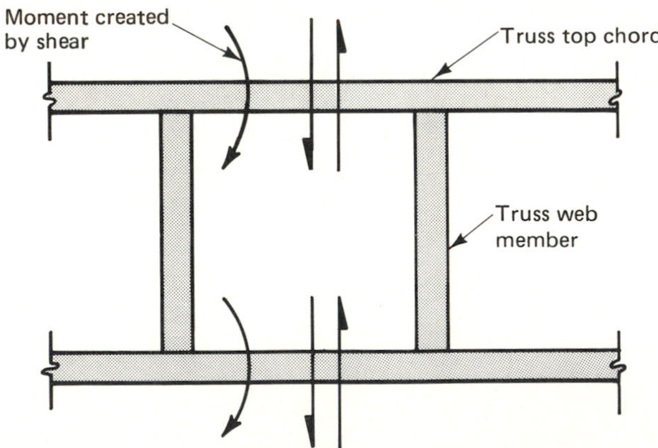

Figure 4–37. Vierendeel Truss

Concrete beam-columns usually are called on to carry moments about both principal axes of bending, a result of continuity in each direction. *Steel* beam-columns, on the other hand, seldom rely on weak-axis being for either lateral resistance or gravity load resistance. Bending about both axes is termed "biaxial bending," a common occurrence for corner columns and for pattern loads on interior columns (Fig. 4–38).

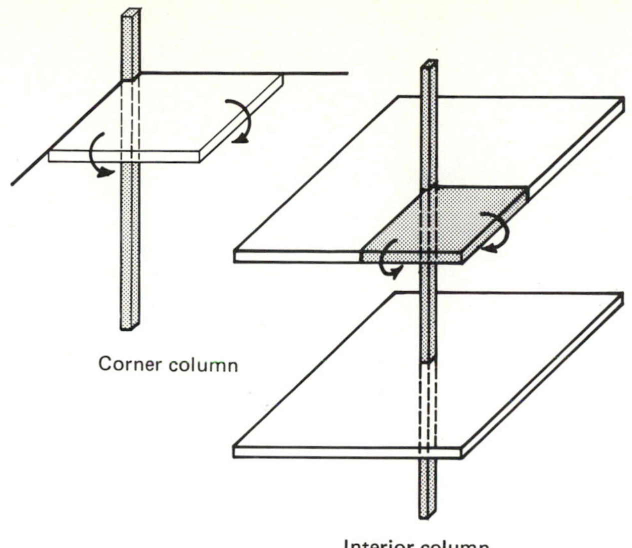

Figure 4–38. Biaxial Bending

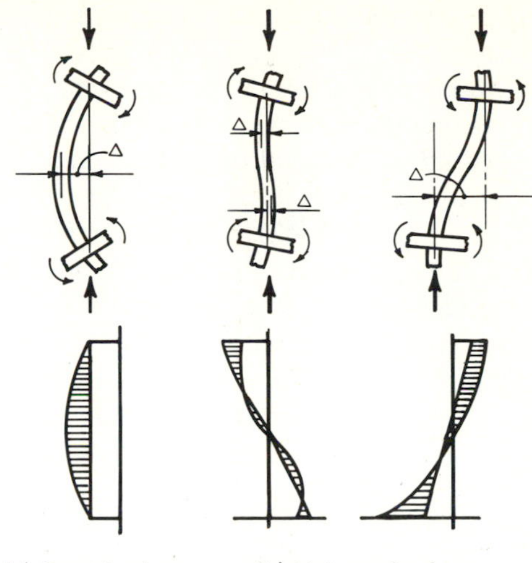

Figure 4–40. Braced and Unbraced Frame Column Moment Patterns

Behavior

The loads and corresponding moments on beam-columns (Fig. 4–39) cause deflections which, when multiplied by the axial load, result in additional moment. This *moment magnification* effect is a characteristic of all beam-columns and is the major design problem for strong-axis bending. A beam-column element of a braced frame will have its maximum moment at a point somewhere between midheight and the column end [Fig. 4–40(a)]. An unbraced frame beam-column (in the sway mode) will have its maximum moment at the column ends [Fig. 4–40(b)]. The pure column aspects of behavior are also present and may cause buckling in the weak direction. Finally, the buckling behavior that is characteristic of beams—lateral torsional buckling—may cause twisting and buckling transverse to the plane of bending.

Figure 4–39. Moment Magnification in Beam-Columns

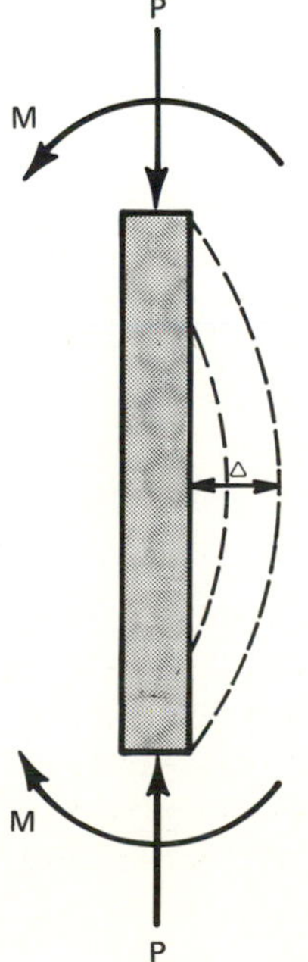

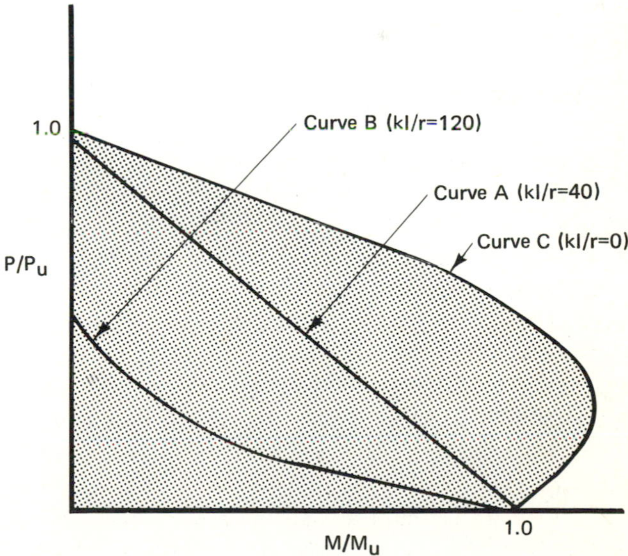

Figure 4–41. Beam-Column Interaction Diagrams

Convenient graphical forms for dealing with this behavior are the "interaction diagrams" shown in Fig. 4–41. The theoretical maximum strength is shown by curve A for a column with no instability. For kl/r greater than 40, increasing slenderness decreases the beam-column capacity (curve B for $kl/r = 120$, for example). Curve C represents a concrete column in which the lower axial loads result in less moment capacity in the tension control range.

As a result of this behavior, *short and torsionally strong* beam-columns will bend essentially in the plane of the applied forces and will develop as much strength as if they were restrained from deflecting in the weak direction (curve 1 in Fig. 4–42). *Long slender box columns,* with small bending moments in the strong plane, will buckle in a direction normal to the plane of bending and will behave essentially the same as if loaded concentrically (curve 2, Fig. 4–42). For *intermediate-length* columns, with varying lateral-torsional stiffness, some combination of the two failure modes can be expected as the moment is magnified by the axial load. Curve 3 represents such a column, with $kl/r = 60$, whose maximum capacity can be related to the $kl/r = 0$ curve by magnifying the moment of curve 1 by $P\Delta$. A *torsionally weak* beam-column is apt to twist as well as bend during buckling failure and may follow curve 4.

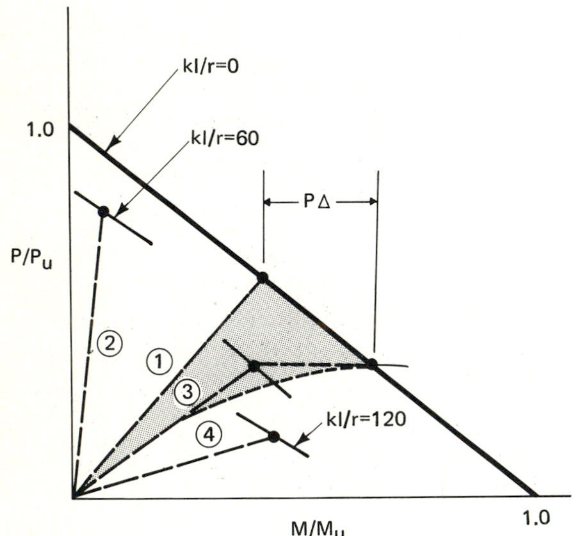

Figure 4–42. Beam-Column Behavior Related to Interaction Diagrams

Column strength can be expressed by the following interaction equation:

$$\frac{f_a}{F_a} + \frac{\delta f_b}{F_b} \leq 1.0 \tag{4–64}$$

where f_a = axial compressive stress (P/A)
F_a = allowable compressive stress for column section
f_b = bending stress (Mc/I)
F_b = allowable bending stress
δ = moment magnification factor

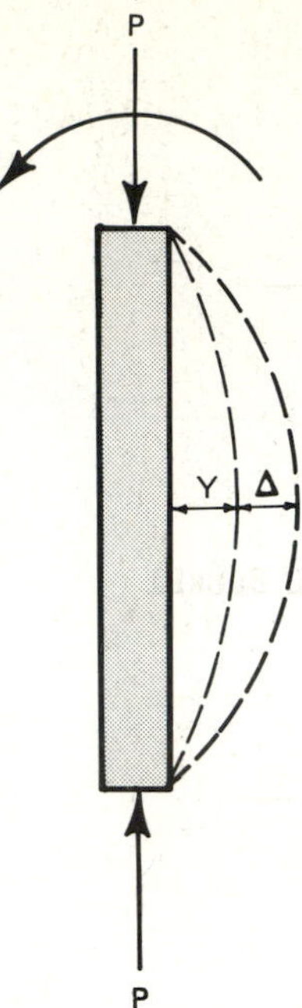

Figure 4–43. Moment Magnification Due to $P\Delta$ Effect

Moment Magnification. The moment magnification factor (δ) can be derived (Fig. 4–43) by computing the secondary bending moment resulting from the $P\Delta$ effect. By moment–area principles and assuming a sine curve to represent the secondary bending moment curve,

$$\Delta = \frac{P}{EI}(\Delta + Y)\frac{L}{2}\frac{2}{\pi}\frac{L}{\pi} = (\Delta + Y)\frac{PL^2}{\pi^2 EI} \tag{4–65}$$

or

$$\Delta = (\Delta + Y)\frac{P}{P_e}$$

where P_e is the Euler buckling load $= \pi^2 EI/L^2$.

$$\Delta = Y\left(\frac{P/P_e}{1 - P/P_e}\right) = Y\left(\frac{\alpha}{1-\alpha}\right)$$

where $\alpha = P/P_e$.

$$Y_{\max} = Y + \Delta = Y + Y\left(\frac{\alpha}{1-\alpha}\right) = \frac{Y}{1-\alpha} \tag{4–66}$$

The maximum bending moment including the axial effect is then

$$M_{\max} = M_o + P_{y\,(\max)}$$

By Eq. 4–66,

Table 4–8 Values of C_m

Category	Loading conditions	f_b[a]	C_m	Remarks
A	Computed moments maximum at end; joint translation not prevented	$\frac{M_2}{S}$	0.85	$M_1 < M_2$; $\frac{M_1}{M_2}$ negative
B	Computed moments maximum at end; no transverse loading; joint translation prevented	$\frac{M_2}{S}$	$\left(0.6\text{–}0.4\frac{M_1}{M_2}\right)$ but not less than 0.4	
C	Transverse loading; joint translation prevented	$\frac{M_2}{S}$		
	Ends restrained		0.85	
	Ends unrestrained	$\frac{M_3}{S}$	1.0	

[a] Calculate bending stress based on M_1, smaller end moment; M_2, larger end moment, positive clockwise; M_3, maximum interior moment.

$$M_{\max} = M_o + P\left(\frac{Y}{1-\alpha}\right) \tag{4–67}$$

$$M_{\max} = M_o + \left(\frac{\alpha\pi^2 EI}{L^2}\right)\frac{Y}{1-\alpha}$$

$$\left(\frac{1-\alpha}{M_o}\right)M_{\max} = 1 + \left(\frac{\pi^2 EIY}{M_o L^2} - 1\right)\alpha$$

$$M_{\max} = M_o\left(\frac{C_m}{1-\alpha}\right) \tag{4–68}$$

where

$$C_m = 1 + \left(\frac{\pi^2 EIY}{M_o L^2} - 1\right)\alpha \tag{4–69}$$

The term $C_m/(1 - \alpha)$ is the magnification factor. C_m recommendations for several conditions of loading are shown in Table 4–8.

The magnification factor (δ) can be separated into two parts[2]: δ_b to account for the magnification of *gravity load moments* in a nonsway (braced) condition, and δ_s to account for the magnification of *lateral load moments* in a sway (unbraced) condition. In accordance with the previous development, δ_b can then be expressed as

$$\delta_b = \frac{C_m}{1 - P/P_e} \geq 1.0 \tag{4–70}$$

The effective length factor (κ) for use in determining P_e is theoretically less than 1.0 for braced conditions due to the restraint provided by the girders. However, most codes specify a value of 1.0 for use in this case.

The moment magnification for unbraced frames (δ_s) can be expressed as follows:

$$\delta_s = \frac{1.0}{1 - \Sigma P/\Sigma P_e} \geqslant 1.0 \tag{4–71}$$

where C_m has been conservatively assumed to equal 1.0.

The summation of both column load (P) and buckling load (P_e) is based on the concept that all columns in a given story must sway equally, and thus the moment magnification will depend on the total story stiffness. The effective length factor (κ) depends on the relative stiffness at each end of the member, but can be set equal to 1.0 if the $P\Delta$ procedure discussed in the next section is used.

Approximate Analysis

Axial loads in beam-columns can be calculated using the tributary area concepts discussed in Section 4–2. Bending moments come from analysis of frame action, and several approximate methods are available to assist in this analysis, including Table 4–3 for braced frames. Further discussion is contained in the Chapter 5 section on rigid frames.

Moment magnification factors (δ_b and δ_s) can be approximated from graphs and approximate formulas. Such a formula for δ_s is[3,4]

[2] J. S. Ford, D. C. Chang, and J. E. Breen, "Behavior of Unbraced Multipanel Concrete Frames," *Journal of the American Concrete Institute,* Vol. 78, No. 2 (March–April 1981), p. 99.

[3] ACI Committee 318, Commentary on *Building Code Requirements for Reinforced Concrete,* ACI 318–77 (Detroit, Mich.: American Concrete Institute, 1977), p. 48.

[4] James G. MacGregor and Sven E. Hage, "Stability Analysis and Design of Concrete Frames," *Journal of the Structural Division, ASCE,* Vol. 103, No. ST-10 (October 1977).

$$\delta_s = \frac{1}{1-Q} = \frac{1}{1 - \Sigma P\Delta/Hh_s} \quad (4\text{–}72)$$

where Q = stability index (Q_u for concrete design) which should not exceed 0.2 maximum

ΣP = summation of loads in a given story (P_u for concrete design)

Δ = elastically computed first-order lateral deflection due to H (Δ_u for concrete design based on reduced effective stiffness, EI)[5]

H = total lateral force within the story (H_u for concrete design)

h_s = height of story

Further, if the ratio, Q, is equal to or less than 0.04, the frame may be considered braced and $\delta_s = 1.0$. Approximate procedures for calculating Δ are also discussed in Chapter 5.

Approximate Design

Once axial loads, bending moments, and moment magnifiers have been determined through analysis, beam-columns can be designed using forms of the interaction equation developed earlier. A convenient form of the interaction equation for direct design is the following:

$$A = \frac{P}{F_a} + \frac{\delta_b M_G}{F_b}\left(\frac{A}{S}\right) + \frac{\delta_s M_w}{F_b}\left(\frac{A}{S}\right) \quad (4\text{–}73)$$

where A = area, in.^{2} or mm^2

F_a = allowable compressive stress if axial force alone existed

F_b = allowable bending stress if bending alone existed

M_G = moment due to gravity load

$M_w(M_E)$ = moment due to wind (or earthquake)

δ_b = moment magnifier for braced conditions

δ_s = moment magnifier for sway conditions

S = section modulus, in.^{3} or mm^3

The term A/S is called the bending factor and relates bending stresses to axial stresses. Since the moment magnifier (δ) depends on member properties, an initial estimate of δ must be made to start the preliminary design process.

Steel Beam-Columns. Figure 4–44 shows plots of the interaction equations for steel beam-columns based on kl/r values. Maximum beam-column capacity is available when $kl/r < 40$, as shown in Fig. 4–44. Moment magnification effects are small and can be neglected.

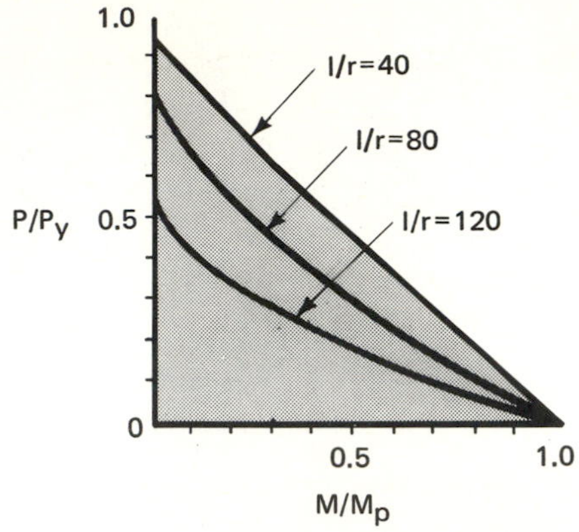

Figure 4–44. Steel Beam-Column Interaction Diagrams

Assuming that $A/S \simeq 2.6/d$, substitution into Eq. 4–74 gives

$$A_{\text{req'd}} = \frac{P}{F_a} + \frac{(2.6)}{d}\frac{M}{F_b} \quad (4\text{–}74)$$

When $kl/r > 40$, the moment magnification term must be incorporated and the equation becomes

$$A_{\text{req'd}} = \frac{P}{F_a} + \frac{\delta_b M_G}{F_b}\left(\frac{2.6}{d}\right) + \frac{\delta_s M_w}{F_b}\left(\frac{2.6}{d}\right) \quad (4\text{–}75)$$

Figure 4–45. Moment Magnification Values for Steel Beam-Columns

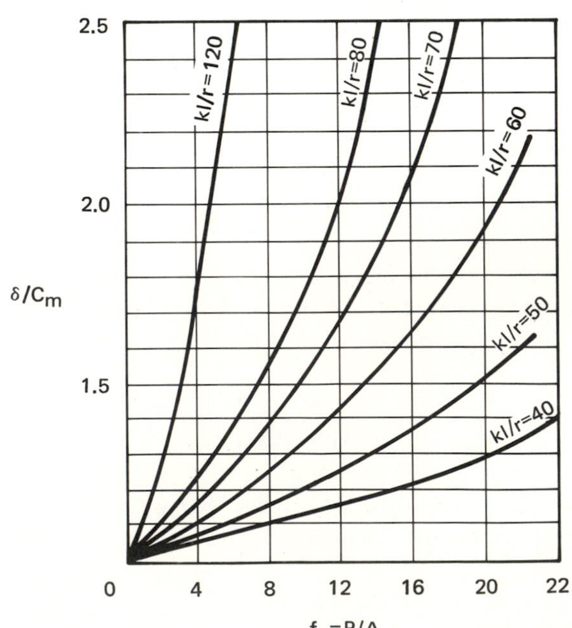

[5] MacGregor and Hage recommend $I = 0.4I_g$ for beams and $I = 0.8I_g$ for columns.

Figure 4–45 shows curves for δ/C_m based on the level of compressive stress (F_a) and these can be used for an estimate of the magnification factor, δ_b. The following example illustrates the approximate method.

EXAMPLE PROBLEM 4–11: Preliminary Steel Beam-Column Design

A beam-column, as part of a moment-resisting frame, must be designed to support an axial load of 100 kips (444,800 N) and a gravity-load moment of 50 ft-kips (67,800 Nm), single curvature bending, and a wind moment of 60 ft-kips (81,360 Nm). Total lateral load on the frame at the story is 30 kips (133,440 N) and total story vertical load is 300 kips (1,334,000 N). Story height is 12 ft (3.65 m). Select a preliminary steel section (A36 steel).

Solution

1. Determine allowable compressive stress F_a (Fig. 4–10).

Max. allowable $F_a = 22.0$ ksi (151.7 MPa)

Int. allowable $F_a = 9.37$ ksi (64.6 MPa) at $kl = 22.6\sqrt{P}$

$$kl = 22.6\sqrt{P} = 22.6\sqrt{100}$$
$$= 226 \text{ in. (18.8 ft, 5.73 m)}$$

Select F_a = 14.0 ksi (96.5 MPa).

2. Determine allowable bending stress, F_b. Assume lateral bracing provided by the floor system (l = 12 ft) and assume an 8-in.-deep column. By Eq. 4–49,

$$F_b = \sqrt{\frac{10{,}320M}{ld^2}} \leq 0.60F_y$$

$$F_b = \sqrt{\frac{10{,}320(110)}{12(8)2}} = 34.4$$

Use max. F_b = 22 ksi (151.7 MPa).

3. Determine moment magnifier for gravity loads. Estimate minimum A with no magnification.

$$A_{\min} = \frac{100}{14.0} + \frac{2.6(50 \times 12)}{22.0(8)} = 16.0 \text{ in.}^2$$

By Eq. 4–4,

$$r_x \simeq 0.9\sqrt{A} = 0.9\sqrt{16.0} = 3.60$$

$$\frac{kl}{r} = \frac{1.0(12 \times 12)}{3.60} = 40.0$$

Since $kl/r \leq 40$, δ_b = 1.0. (If kl/r is > 40, Fig. 4–45 can be used to determine δ_b.)

4. Determine moment magnifier for lateral loads. Assume drift limited to 0.0025. From Eq. 4–72,

$$Q = \frac{\Sigma P\Delta}{Hh_s} = \frac{300(0.0025)}{30} = 0.0025 < 0.04$$

(Lateral moments are not magnified, δ_s = 1.0. If Q is > 0.04, use Eq. 4–72 to solve for δ_s.)

5. Select member size. Check A required for wind moment plus gravity moment (33% stress increase allowed).

$$A = \frac{100}{14.0 \times 1.33} + \frac{2.6(110 \times 12)}{22.0(8)1.33}$$
$$= 20.0 \text{ in.}^2 \text{ (12,900 mm}^2\text{)}$$

From Appendix 4A, use W8 × 67 (A = 19.7 in.²).

Concrete Beam-Columns. Maximum capacity for concrete beam-columns is available when $kl/r < 34$ (Fig. 4–46) and no magnification is required. The interaction equation must contain a modified value for F_b and F_a due to the influence of axial load on the moment capacity of reinforced concrete members. Surveys of actual concrete structures have concluded that 98% of the columns in braced frames had h/t less than 12.5 ($l/r < 41$) and e/t less than 0.64, while 98% of the columns in unbraced frames had h/t less than 18 ($l/r < 60$) and e/t less than 0.84. As a result, for approximate beam-column design, we can assume a member in the compression controlling range only and use a straight-line interaction equation.

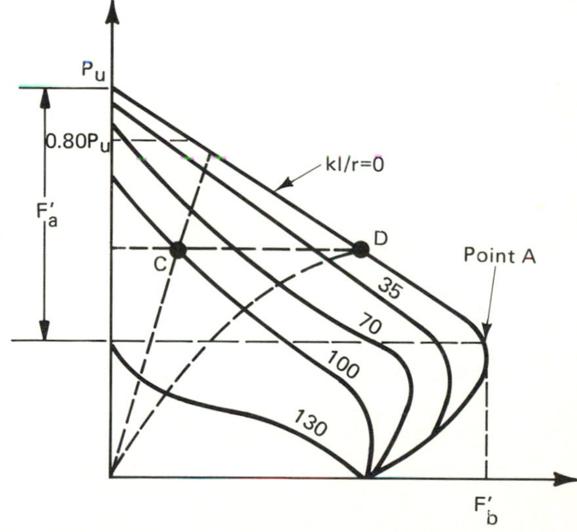

Figure 4–46. Concrete Beam-Column Interaction Curves

The equivalent F'_b (point A in Fig. 4–46) can be determined by using the equation for the balanced condition—the dividing point between the compression controls and tension controls regions:

$$\phi P_b e_b = \phi\left[0.85f'_c ab\left(0.6d - \frac{0.51d}{2}\right) + A'_s f'_y(0.6d - 0.15d) + A_s f_y(0.4d)\right]$$

For 4% steel and 3000 psi (20.7 MPa) concrete:

$$\phi P_b e_b = \phi[0.85(3.0)(0.51d)b(0.345d) + 0.02bd(60.0)(0.45d) + 0.02bd(60.0)(0.4d)]$$

$$\phi P_b e_b = \phi[1.468bd^2] = 0.7(1.468bd^2) = 1.02bd^2 \quad (4\text{–}76)$$

This compares to a value of $0.41bd^2$ established for M_u in Eq. 4–56.

$$I = \frac{bh^3}{12} = \frac{b(d/0.8)^3}{12} = \frac{bd^3}{6.14}$$

$$F_b' = \frac{Mc}{I} = \frac{1.02bd^2(6.14)(d/2)}{bd^3} \quad (4\text{–}77)$$

$$= 3.13 \text{ ksi } (21.6 \text{ MPa})$$

When $kl/r > 34$, the moment magnifier can be used to magnify the moment and relate the actual beam-column back to the $kl/r = 0$ curve. For example, point C on the $kl/r = 100$ of Fig. 4–46, when magnified, moves to point D on the $kl/r = 0$ curve. Thus one curve can be used for beam-column capacity; only different concrete strengths or reinforcement percentages will necessitate additional curves.

Modified values for F_a (F_a') and P_u (P_u') are necessary to correct the vertical scale of the interaction diagram:

$$F_a' = \frac{P_u - P_b}{A}$$

where $P_b = \phi(0.85)f_c'ab$
$= 0.7(0.85)(f_c')(0.51d)b = 0.30bdf_c'$

and by Eq. 4–10, omitting the maximum capacity factor of 0.80 and substituting $bd/0.8$ for Ag:

For 3000-psi concrete:

$$P_u = 4.325bd$$

Therefore,

$$P_u - P_b = 4.325bd - 0.30bd(3.0) = 3.42bd = 0.79P_u$$

Because $F_a = 0.80P_u/A$,

$$F_a' = \left(\frac{0.79}{0.80}\right)F_a = \left(\frac{0.79}{0.80}\right)2.77 = 2.74 \text{ ksi}$$

and

$$P_u' = P_u - P_b = P_u - 0.3bdf_c' \quad (4\text{–}78)$$

F_a in this equation is $F_{a\,(max)}$ from Eq. 4–10 and is based on kl/r in the plane of bending.

The modified interaction equation for concrete is then, with

$$\frac{A}{S} = \frac{bh(6)}{bh^2} = \frac{6}{h}$$

$$A_{req'd} = \frac{P_u'}{F_a'} + \frac{\delta(6)}{h}\frac{M_u}{F_b'} \quad (4\text{–}79)$$

where all values are for the plane of bending. (*Note:* It may occasionally be necessary to check columns for stability and/or bending in the other plane as well.)

The ACI Code specifies a minimum moment, M_2, equal to $(0.6 + 0.03h)$ and a factor β_d (ratio of maximum factored dead load moment to maximum factored total load moment, always positive) for computation of C_m values. Figure 4–47 shows moment magnifier curves based on β_d and various kl/r ratios.

Table 4–9 lists F_b' values for other strengths of concrete. The following example illustrates the approximate method.

Table 4–9 Allowable Stress for Concrete Beam-Columns

f_c'		F_a'		F_b' ($kl/r = 0$)	
psi	*MPa*	*ksi*	*MPa*	*ksi*	*MPa*
3000	20.7	2.74	18.9	3.13	21.6
4000	27.6	3.08	21.2	3.48	23.9
5000	34.5	3.44	23.7	3.80	26.2
6000	41.4	3.81	26.3	4.12	28.4

EXAMPLE PROBLEM 4–12: Preliminary Concrete Beam-Column Design

Select a preliminary column to carry the following loads:

Axial load:

DL = 100 kips = (444,800 N) $f_c' = 3000$ psi = (20.7 MPa)

LL = 50 kips = (222,400 N) $f_y = 60{,}000$ psi = (41.4 MPa)

Moment:

$M_{DL} = 25$ ft-kips (33,900 Nm)

$M_{LL} = 15$ ft-kips (20,340 Nm)

$M_{WL} = 75$ ft-kips (101,700 Nm)

Story height is 15 ft (4.57 m) and the total lateral load is 40 kips (177,920 N) and there are a total of four columns, each with the same load.

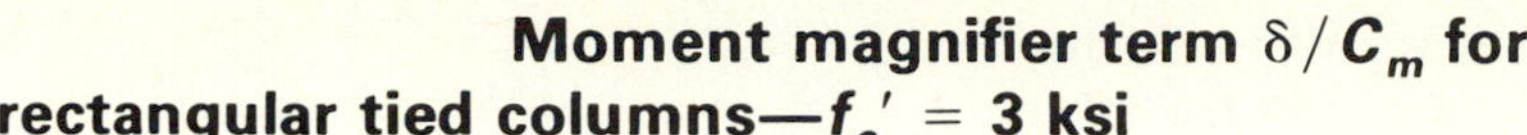

Moment magnifier term δ / C_m for rectangular tied columns—$f_c' = 3$ ksi

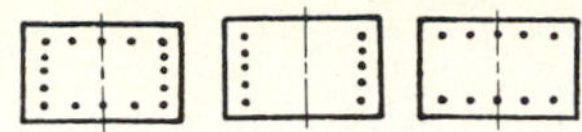

$$\frac{\delta}{C_m} = \frac{1}{1 - (0.00137)\left(\dfrac{P_u(1+\beta_d)}{A_g}\right)\left(\dfrac{kl_u}{h_e}\right)^2}$$

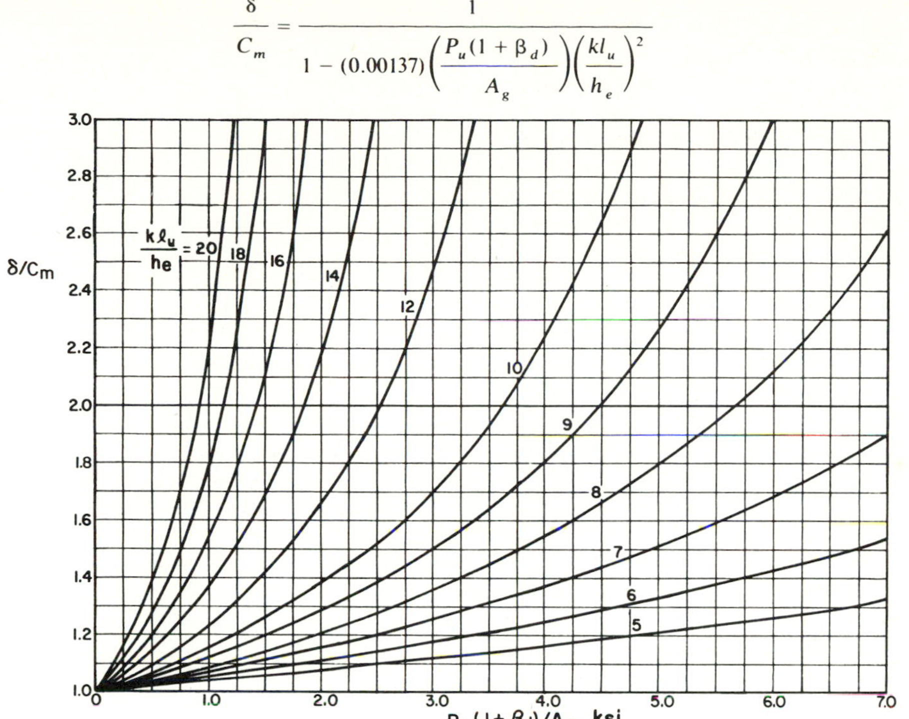

(a)

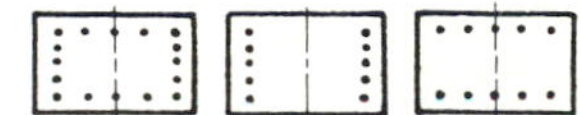

Moment magnifier term δ / C_m for rectangular tied columns—$f_c' = 4$ ksi

$$\frac{\delta}{C_m} = \frac{1}{1 - (0.00119)\left(\dfrac{P_u(1+\beta_d)}{A_g}\right)\left(\dfrac{kl_u}{h_e}\right)^2}$$

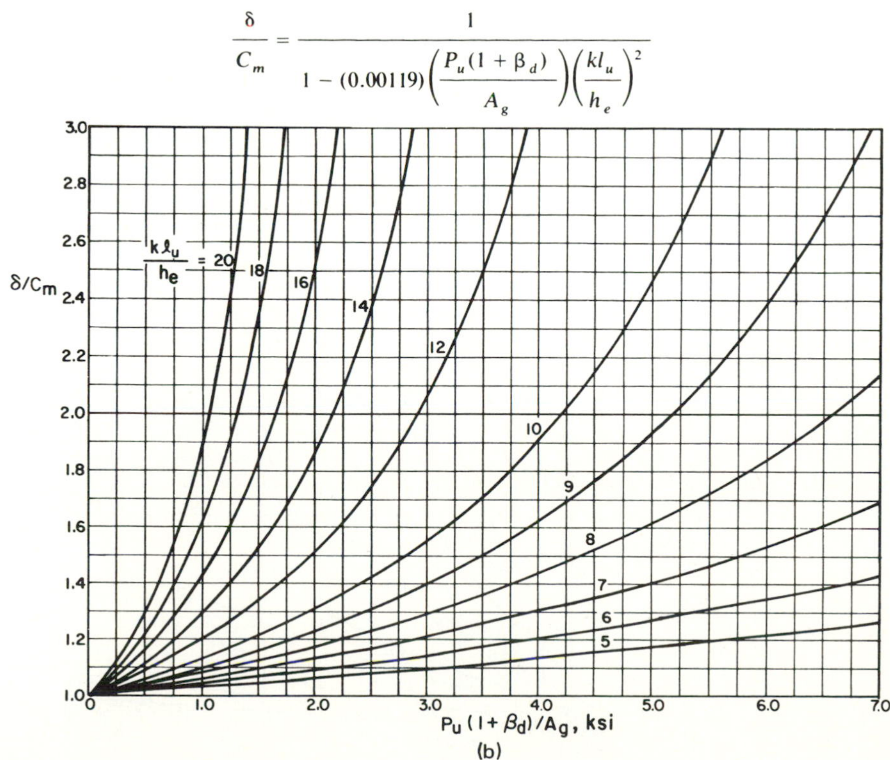

(b)

Figure 4–47. Moment Magnification Values for Concrete Beam-Columns (From ACI Committee 340, *Design Handbook*, Vol. *2: Columns,* American Concrete Institute, Detroit, Mich., 1978)

Moment magnifier term δ / C_m for rectangular tied columns—$f_c' = 5$ ksi

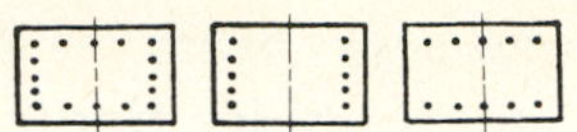

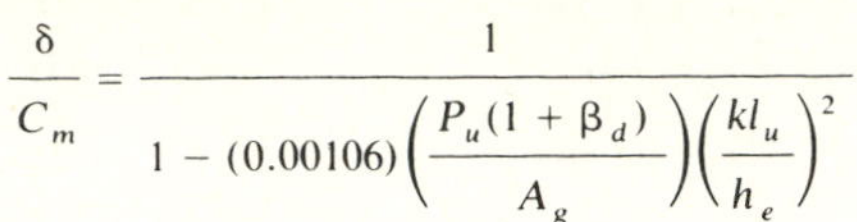

$$\frac{\delta}{C_m} = \frac{1}{1 - (0.00106)\left(\dfrac{P_u(1 + \beta_d)}{A_g}\right)\left(\dfrac{kl_u}{h_e}\right)^2}$$

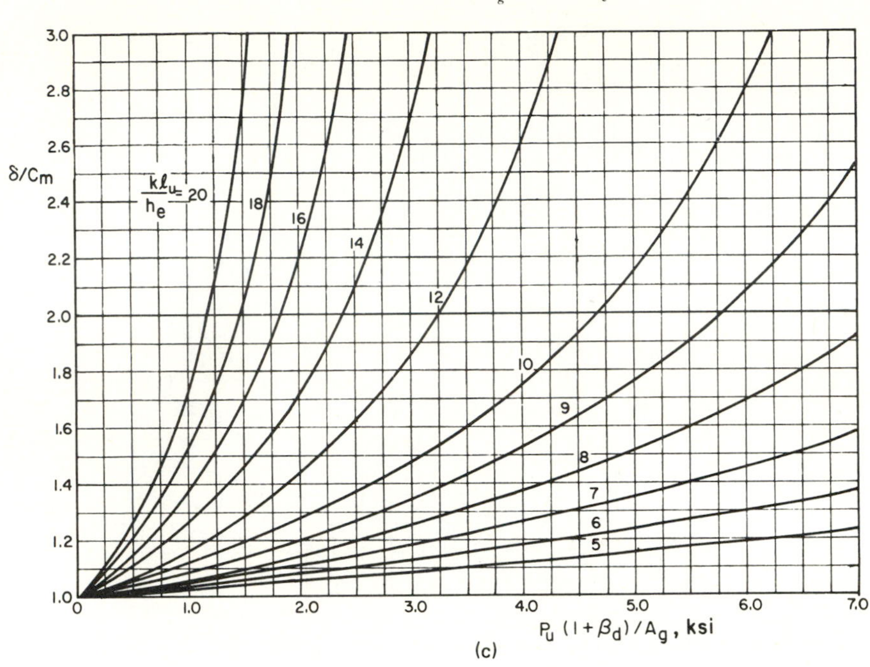

(c)

Moment magnifier term δ / C_m for rectangular tied columns—$f_c' = 6$ ksi

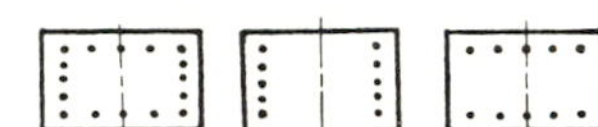

$$\frac{\delta}{C_m} = \frac{1}{1 - (0.000971)\left(\dfrac{P_u(1 + \beta_d)}{A_g}\right)\left(\dfrac{kl_u}{h_e}\right)^2}$$

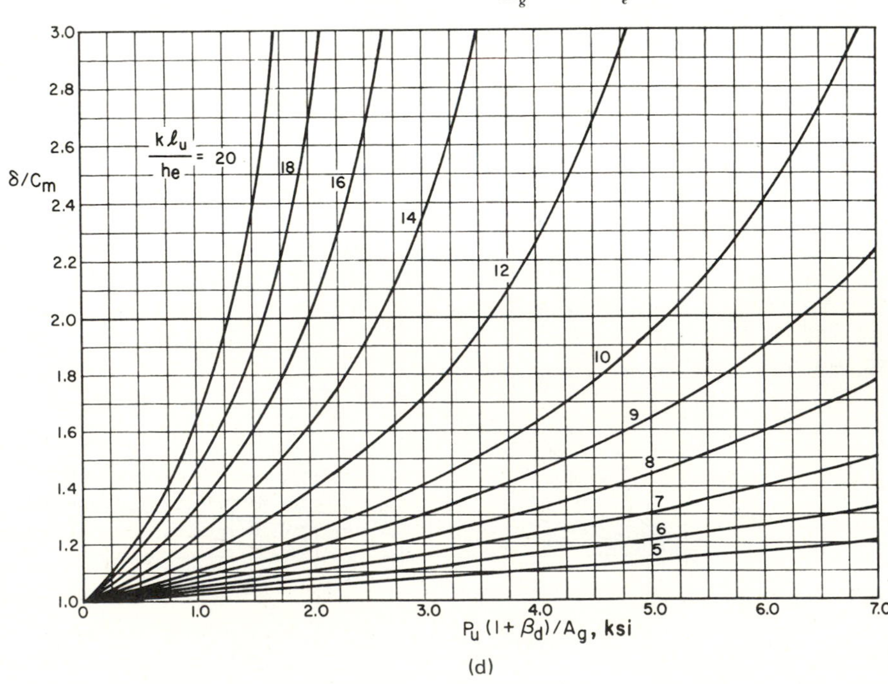

(d)

Figure 4–47 (*Continued*)

Solution

1. Calculate ultimate loads.

$$P_u = 1.4(100) + 1.7(50)$$
$$= 225 \text{ kips } (1{,}000{,}800 \text{ N})$$
$$M_u \text{ (gravity)} = 1.4(25) + 1.7(15)$$
$$= 60 \text{ ft-kips } (81{,}360 \text{ Nm})$$
$$M_u \text{ (wind)} = 1.7(75) = 127 \text{ ft-kips } (172{,}212 \text{ Nm})$$

2. Determine minimum column size before magnification. Try h = 12 in. (305 mm) (use 0.75 reduction for wind combination). Assume that F_a = 2.77 ksi (19.1 MPa). By Eq. 4–78,

$$P'_u = 225 - 0.30(12)(0.8 \times 12)3.0$$
$$= 121 \text{ kips } (538{,}208 \text{ N})$$

Assume that F'_a = 2.74 ksi (18.9 MPa), and F'_b = 3.13 ksi (21.6 MPa).

$$A_{\min} = \frac{P'_u}{F'_a} + \frac{6M_u}{F'_b h}$$
$$= \frac{121(0.75)}{2.74} + \frac{6(187 \times 12)(0.75)}{3.13(12)}$$
$$= 301 \text{ in.}^2 \ (194{,}145 \text{ mm}^2)$$
$$\sqrt{301} = 17.3 \text{ in. } (439 \text{ mm}) > 12 \text{ in. } (305 \text{ mm}) \quad \text{N.G.}$$

Try h = 16 in. (406 mm).

$$P'_u = 225 - 0.30(12)(0.8 \times 16)3$$
$$= 87 \text{ kips } (386{,}976 \text{ N})$$
$$A_{\min} = \frac{87(0.75)}{2.74} + \frac{6(187 \times 12)(0.75)}{3.13(16)}$$
$$= 226 \text{ in.}^2 \ (145{,}770 \text{ mm}^2)$$
$$b = \frac{226}{16} = 14.1 \text{ in. } (358 \text{ mm})$$

Assume a 14 in. × 16 in. (356 mm × 406 mm) column = 224 in.² (14,450 mm²).

3. Check M_{DL} and M_{LL} magnification.

$$\frac{kl}{r} = \frac{1.0 \times 15 \times 12}{0.3(16)} = 37.5 > 34$$

Magnification must be considered.

4. Calculate magnifier for gravity moment, δ_b.

$$\text{Moment ratio } \beta_d = \frac{M_D}{M_T} = \frac{1.4(25)}{60} = 0.58$$
$$\frac{P_u(1+\beta_d)}{A_g} = \frac{225(1+0.58)}{14 \times 16}$$
$$= 1.58 \text{ ksi } (10.89 \text{ MPa})$$

From Fig. 4–47, δ/C_m = 1.32, δ_b = 1.32.

5. Calculate magnifier for wind moment, δ_s. Assume that drift limitation = 0.0025 and use 0.75 reduction for combined loads.

$$Q = \frac{\Sigma P_u \Delta}{H_u h_s} = \frac{0.75(225)(4)(0.0025)}{0.75(40)(1.7)} = 0.0033 < 0.04$$
$$\delta_s = 1.0$$

6. Calculate required area for gravity load.

$$P'_u = P_u - P_b$$
$$P'_u = 225 - 0.30(14)(16 \times 0.8)3 = 64 \text{ kips } (284{,}672 \text{ N})$$
$$A = \frac{P'_u}{F_a} + (\delta_b)\frac{6M_u}{F'_b h}$$
$$A = \frac{64}{2.74} + \frac{(1.32)6(60 \times 12)}{3.13(16)} = 137 \text{ in.}^2 \ (88{,}365 \text{ mm}^2)$$

7. Calculate the required area for gravity plus wind.

$$A = \frac{P'_u}{F'_a}(0.75) + (\delta_b)\frac{6M_{uG}}{F'_b h}(0.75) + \frac{(\delta_s)6M_{uw}(0.75)}{F'_b h}$$
$$= \frac{64(0.75)}{2.74} + \frac{(1.32)(6)(60 \times 12)0.75}{3.13(16)}$$
$$+ \frac{1.0(6)(127 \times 12)0.75}{3.13(16)}$$

A = 240 in.² (154,800 m²) > 224 in.² for 16 in. × 14 in. column.

8. However, use 14 in. × 16 in. since reinforcement is based on 4% and can be increased to cover difference.
9. Check with *ACI Design Handbook* (Vol. 2: Columns, p. 89): by columns 7.10.3, the section is O.K.

4–7 CONNECTIONS

Connections are "linkage" elements that join other elements together to form subassemblies—joining beams to columns, columns to footings, bracing to columns, and diaphragms to edge beams. Their function is to transfer shear, moment, and axial forces from one element to another.

Much of structural analysis theory assumes that connections are either fully fixed (allow no relative rotation between the connected parts) or are fully free (hinged joint with no moment resistance). Structural design, however, must deal with connections that fit neither of these mathematical model assumptions. Most connections exhibit some degree of moment resistance and varying degrees of rotation without deterioration. Rotation capacity (ductility) is an extremely important factor in rigid frame behavior, especially for ultimate strength design methods and resistance to earthquake forces, requiring large inelastic deformations without failure.

The behavior of the connection (i.e., moment–rotation characteristics) affects the strength and stability of the connected elements (buckling load for columns, for example)—and the stability of the entire structural system—the "weak link in the chain" concept. Correct assumptions for design must be based on load path continuity through the connections, assuring that assumed stability forces are actually provided by the physical construction of the connection.

History and Development

The connection element became an important structural component when the first frame buildings were erected during the early 1900s. Prior to that time, connections were primarily either simple bearing types or the types used in bridge truss construction. Neither of these early types required any moment resistance, and an understanding of their behavior was not critical to structural design.

Since the early 1900s, evolution of connection methods and research into their basic behavior have paved the way for major development of the beam-column frame (both rigid and simple), until today this method of construction represents by far the most commonly used structural system. In contrast to early columns, which were stabilized by their own weight, columns today achieve stability through connections. As the trend toward stronger, lighter, and more slender structural elements continues, the critical development of connections will play a major role in the success or failure of structural systems.

Types and Uses

The various types of connections are usually associated with the framing method employed. The following major types are commonly used:

- Bearing connections
- Truss connections
- Moment-resisting connections (rigid)
- Shear connections
- Hinged connections (simple)
- Semirigid connections
- Tension connections
- Friction connections
- Shear-friction connections
- Elastic spring
- Sliding connections

These types are shown in Fig. 4–48 together with the mathematical model for each.

Bearing connections are used for post and lintel frames and for bearing wall systems. Their primary function is to provide a support reaction to resist gravity load without exceeding the bearing strength of the connected elements. *Truss connections* are used to connect chord members and web members in trusses. Their primary function is to transmit tension and compression forces between elements. *Moment-resisting connections* are essential elements in any type of continuous or rigid-frame construction. This includes many types of brackets, cantilevers, portal frames, wind frames, and continuous girder framing. Their function is to provide moment resistance as connecting elements seek to rotate relative to each other. Normally shear and axial forces must also be transmitted through this type of connection. This type of connection is classified Type 1 by AISC.

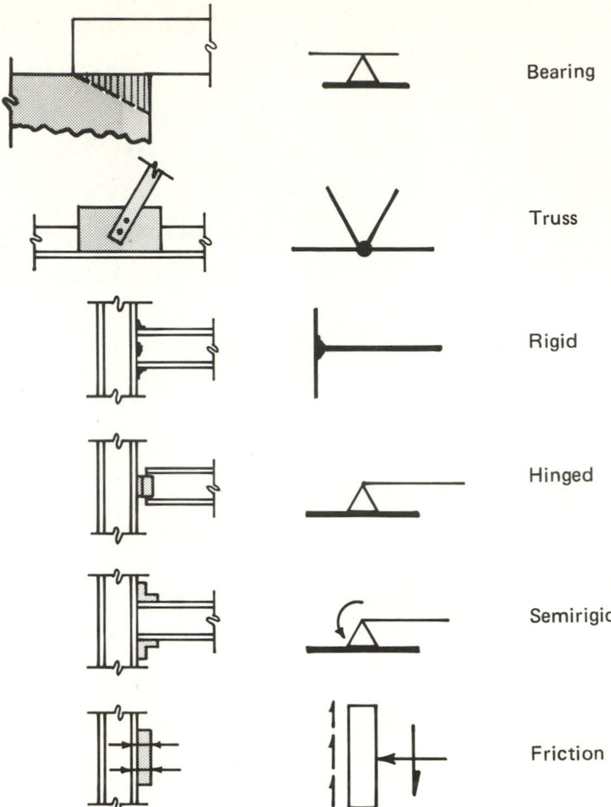

Figure 4–48. Connection Elements

Shear connections are used in braced frames to connect beams and columns, at inflection points in continuous framing, at construction joints in concrete systems, and for connecting secondary beams with primary beams or girders. In addition to transmitting shear forces, these connections often are required to transmit axial forces. When adequate rotation capacity is also provided, they become *hinged connections.* Hinged connections may take the form of pure hinges, which are used for longer spans of bridges and other major structures, or they may simply remain "disguised" in the form of either bearing connections or shear connections. In either case, the primary function is to provide moment-free rotation consistent with the deformations of the connected elements. Hinged connections are AISC Type 2.

Semirigid connections are used to resist wind moments in frames, provide partial restraint at column-footing joints, and provide partial continuity in concrete and masonry construction. They must possess sufficient ductility to deform beyond their assumed "design" rotation position under the influence of heavier loads, but yet maintain a reserve moment capacity equal to the semirigid assumption. This type of connection is classified Type 3 by AISC.

Tension connections are the opposite of bearing connections. Their function is to transmit tension forces between members by welds, bolts, sockets, pins, or other devices. These connections are often subjected to impact and fatigue loads and are the most critical of all the connection types, because connection failure often leads to total system collapse. *Friction connections* resist translational or sliding forces by virtue of the activation of friction resistance between materials. Typical uses include retaining wall footings, high-strength bolted friction connections to transmit shear, and precast concrete/masonry wall connections. *Shear-friction connections* resist translational shear forces (one element translating relative to another) by friction, but with the normal forces provided by tension reinforcement. This connection is primarily used in concrete construction for brackets, corbels, and haunches. *Sliding connections* provide no restraint to translational forces and are primarily used to provide for volume-change expansion in bridges and at expansion joints in building structures.

Elastic springs provide support reactions parallel to the spring axis and with accompanying elastic deformation. They are used to dampen vibrating machinery, dampen earthquake forces, and serve as a mathematical model for soil–structure interaction.

Behavior

The stress–strain behavior of connections is often complex and difficult to predict without experimental testing. In *metals,* behavior includes shear deformation of bolts and web plates, bearing deformation of material adjacent to bolt holes and opposite compressive flange forces, tension elongation of flange connecting plates and bolts in tension, bending deformation of clip angles, and friction strain between contact surfaces. This behavior is almost impossible to analyze by exact mathematical procedures. The ductility of metals, however, allows local yielding and redistribution of stresses—permitting simplifying assumptions and approximate design techniques.

In *concrete,* connection behavior involves mechanics of stress transfer between concrete and reinforcing steel, development of reinforcement, confinement of concrete shear areas, and redistribution of stress as higher loads cause local yielding. Experimental and approximate methods must also be used to provide a basis for design.

Tension Transfer. The forces transferred through connections include tension, compression, shear, and moment. Tension forces are transferred by direct joining of material (i.e., butt welded joints), bolted lap splices which rely on the shear strength of bolts for the force transfer, direct tension applications of bolts, embedment of reinforcing steel in concrete, and lap splices of reinforcing steel. With the exception of the direct tension applications, these methods rely on local yielding and redistribution of stress along the length of the joint for effectiveness (Fig. 4–49). The usual design assumption is that the entire joint is effective in achieving the tension transfer.

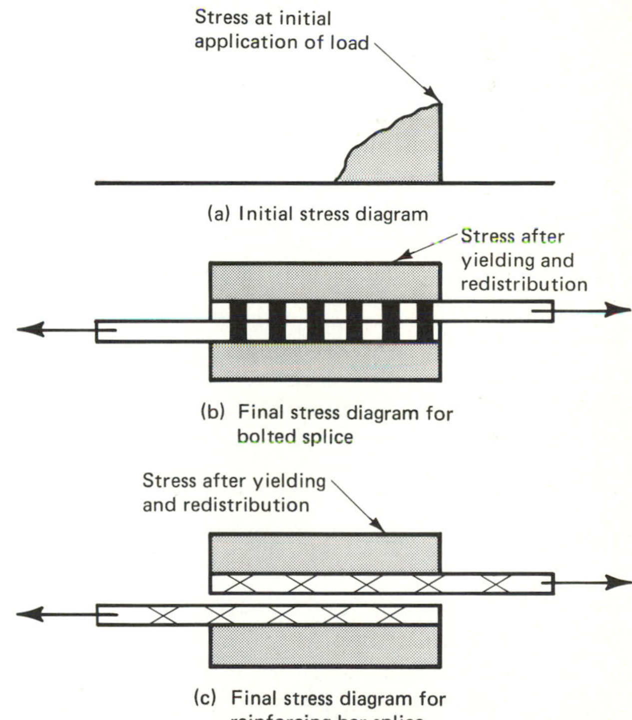

Figure 4–49. Tension Transfer Behavior

Compression Transfer. Compressive forces are transferred by direct bearing, lap splices (similar to tension lap splices), and by embedment with accompanying skin friction and bond stresses. Direct bearing is the only new transfer mechanism introduced (to the previous discussion). Bearing behavior also relies on local yielding of material immediately opposite concentrated loads in order for distribution to take place (Fig. 4–50).

Shear Transfer. Shear forces are transferred by friction (including shear friction), splice plates relying on shear and bearing on bolts, welded splice plates, and

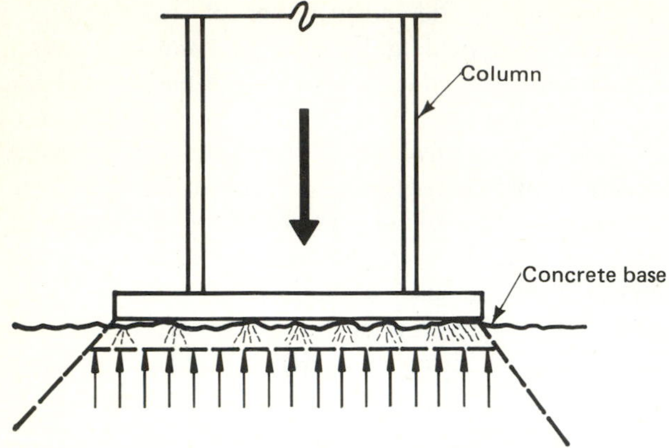

Figure 4–50. Compression Transfer Behavior

the shear strength of reinforcing steel and concrete at monolithic connections. The friction method is utilized for high-strength bolted connections of the friction type and for shear-friction connections in concrete. Figure 4–51 illustrates this behavior, in which the normal force is provided by tensioning of the bolts or by activation of the tensile strength of reinforcing steel. Bolted connections relying directly on the shear strength of the bolts are called *bearing connections* and their behavior is similar to the splice connections already discussed.

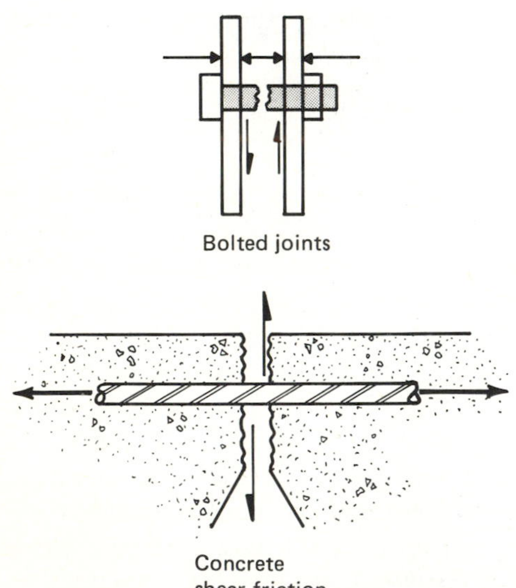

Figure 4–51. Behavior of Friction Connections

Shear transfer through monolithic connections can generally occur via compressive flow within a 45° cone (Fig. 4–52). If the material strength is not sufficient to get the load to this point, reinforcing must be provided to promote the transfer.

Moment Transfer. The transfer of moment forces is generally accomplished by combinations of tension, shear, and compressive mechanisms that simulate a "moment couple" (Fig. 4–53). This couple transfer introduces diagonal tension stress in the column core and stability considerations for thin-plate elements. Moment transfer from flat plate subsystems into columns also activates shear mechanisms in the surrounding slab to balance the applied moment (Fig. 4–54).

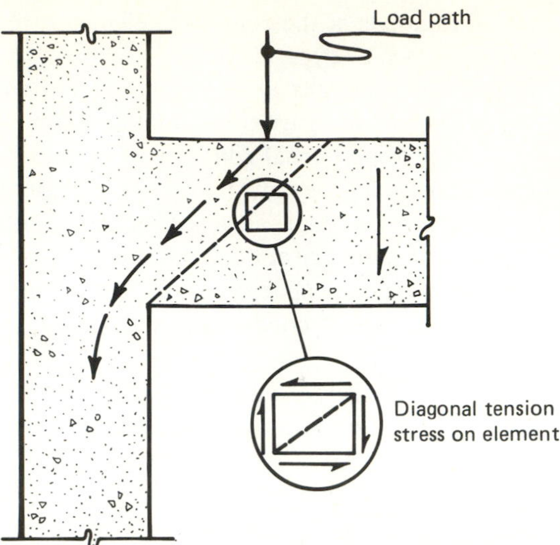

Figure 4–52. Shear Transfer at Monolithic Connections

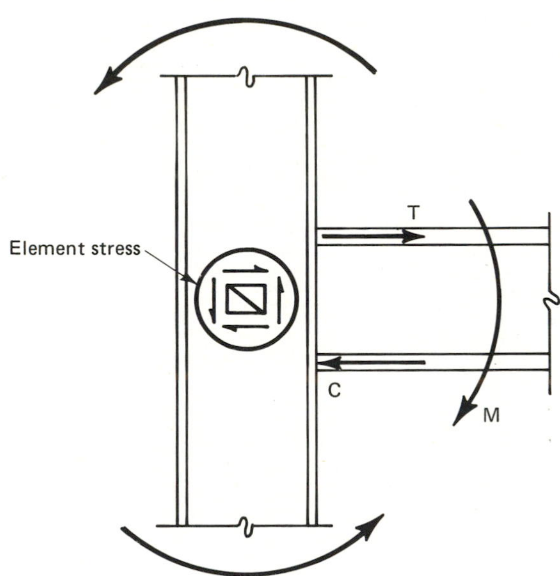

Figure 4–53. Behavior of Moment Connections

The *moment–rotation* behavior of moment connections is as important as the pure transfer of forces. Structural analysis methods generally assume either "fixed" (no rotation) or "free" (free rotation) conditions at connections. These two conditions are illustrated in Fig. 4–55. Actual connections fall somewhere in between these ideal assumptions as shown by the various curves for Type 1, Type 2, and Type 3 connections. Note that the "simple beam" connection actually carries a small amount

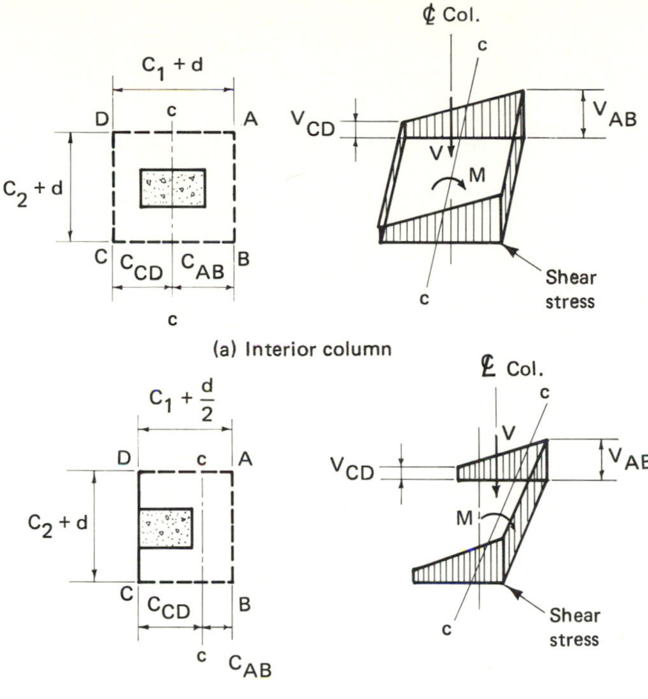

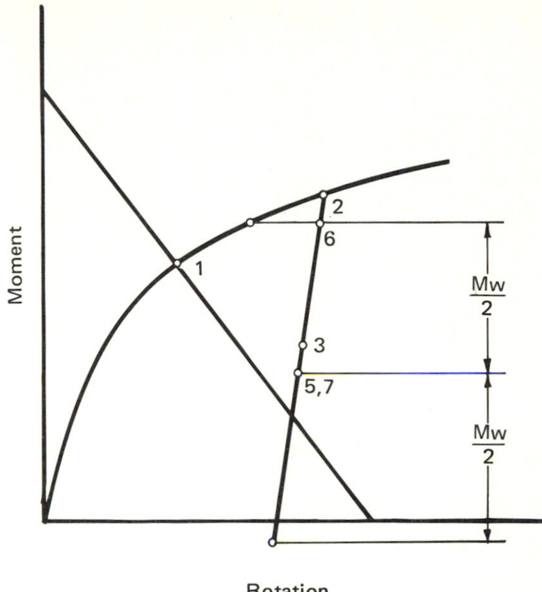

Figure 4–54. Flat Slab Moment Transfer

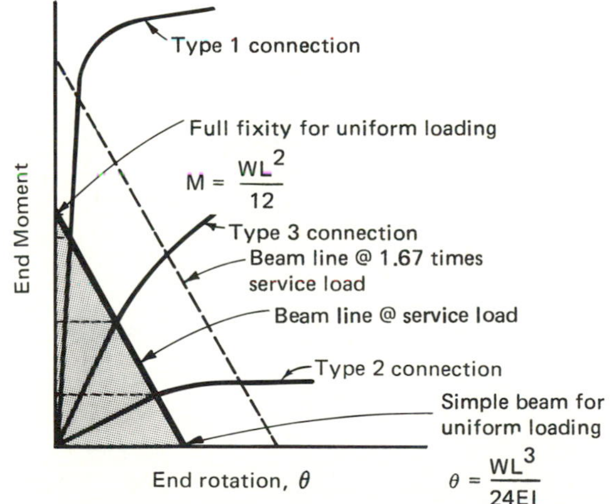

Figure 4–55. Moment–Rotation Curves for Steel Connections

of moment, approximately 20% of the fixed-end moment. It is conservative to neglect this moment restraint for beam design, but the uncomputed stress in the connection itself (primarily increased tension and shear on bolts) may deserve special attention for nonstandard connections.

The semirigid steel connection (Type 3) warrants special consideration. Under gravity load only, an end moment equal to point 1 in Fig. 4–56 will exist, although this degree of "fixity" will usually be ignored in calculating positive moments at beam midspan. When wind loads cause additional moments at the connection, the rotation

Figure 4–56. Semirigid Connection Behavior

will increase to point 2 on the curve. With subsequent unloading (no wind), the connection will retain some permanent deformation and unload along a path (with a slope parallel to the initial slope) to point 3. Wind loads from the opposite direction move the rotation to point 4 and unloading moves it back to point 5. Finally, wind from the original direction moves the rotation to point 6 and unloading back to point 7. At this point, the connection has gone through a complete "shakedown" cycle and operates thereafter between points 4 and 6. Point 4 represents a very small end moment and confirms the practice of designing the beam for no end restraining moment. These semirigid connections obviously must have sufficient ductility to accommodate the rotations without overstressing the connecting elements.

A limited amount of redistribution of moments at continuous connections is allowed by ACI, based on yielding in the negative moment region at supports. Limited experimental data are available for the moment-rotation characteristics of connections. Extrapolation of the tests that do exist and approximate analysis techniques are used to properly size moment connection elements.

Approximate Analysis

Shears, moments, and axial forces at connections come directly either from analysis of the individual connecting elements or (in the case of frames) from analysis of frame behavior. Simple hinged connections are assumed to transfer only shear and axial forces. Moment connections are either assumed fixed or semirigid. Semirigid moment connections in steel frames are assumed fixed for wind loads and free for gravity loads. Analysis of semirigid connections must include considerations of connection ductility and effect on member and frame stiffness.

Semirigid, Steel Beam-Column Connections. The moment–rotation characteristics of three types of semirigid connections can be analytically determined by the procedures of Lothers.[6] The moment–rotation behavior of a connection is defined by a factor Z, which depends on specific connection dimensions and thickness of connection elements. The factor Z is defined as follows:

$$Z = \frac{\phi}{M} \tag{4–80}$$

where $1/Z$ is the slope of the assumed moment–rotation curve tangent to the initial slope of experimentally determined curves (Fig. 4–57). The following equations are given by Lothers and the tabulated data by DeFalco and Marino.[7]

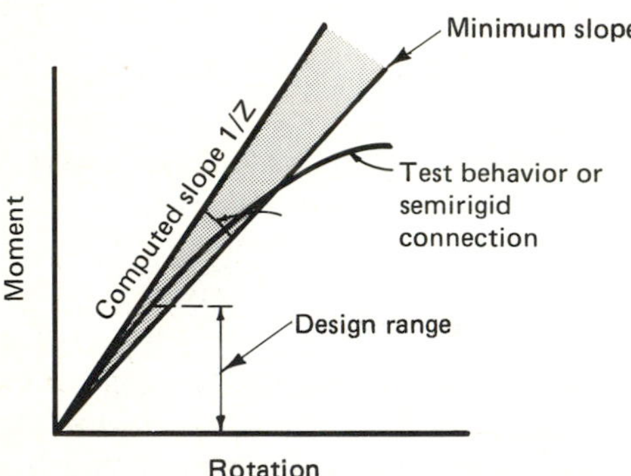

Figure 4–57. Semirigid Connection: Moment versus Rotation Curve

Type A connection (Fig. 4–58). For a Type A connection, the equation for Z is as follows:

$$Z = \frac{6g^3}{Eht^3y^2}\,\frac{g + g_1}{4g + g_1} \tag{4–81}$$

Type B connection (Fig. 4–59). For a Type B connection, the equation is

$$Z = \frac{4g'^3}{Et'^3b'(y' - g' - t')(y' + 2/3q)}\,\frac{g' + g_1'}{4g' + g_1'} \tag{4–82}$$

Type C connection (Fig. 4–60). For a Type C connection, the equation is

[6] John E. Lothers, *Advanced Design in Structural Steel,* copyright 1960, pp. 389, 397, 399. Reprinted by permission of Prentice-Hall, Inc., Englewood Cliffs, N.J. 07632.

[7] Fred DeFalco and Frank J. Marino, "Column Stability in Type 2 Construction," *AISC Engineering Journal* (April 1962), p. 67.

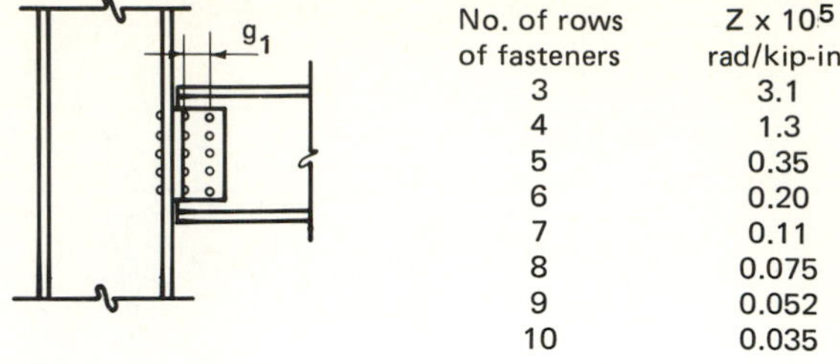

No. of rows of fasteners	Z x 10^5 rad/kip-in.
3	3.1
4	1.3
5	0.35
6	0.20
7	0.11
8	0.075
9	0.052
10	0.035

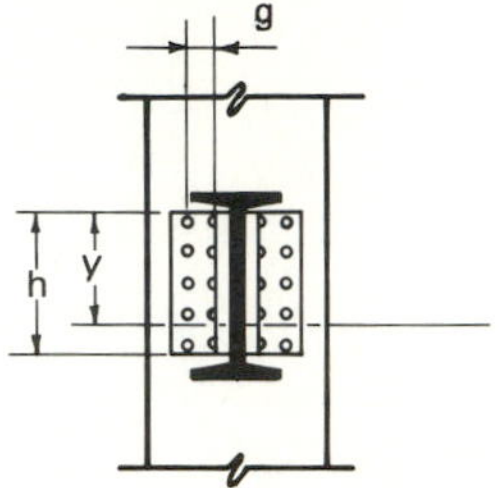

Figure 4–58. Type A Semirigid Connection

Table for 6 x 4 x ¾ connection angles

Depth of beam (in.)	Z x 10^5 rad/kip-in.
8	0.046
10	0.036
12	0.028
14	0.023
16	0.018
18	0.014
21	0.012
24	0.010
27	0.0078
30	0.0066
33	0.0055
36	0.0046

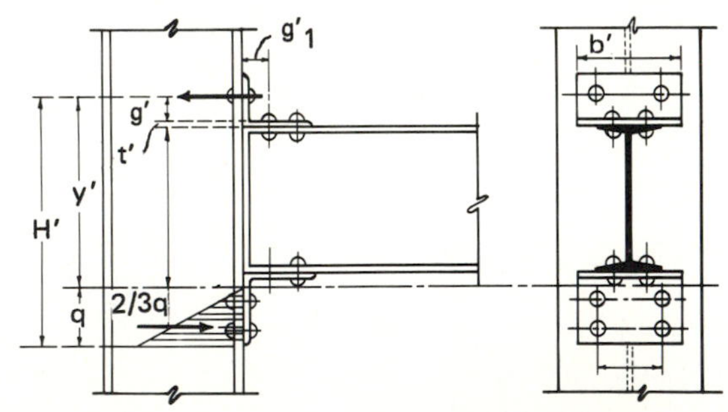

Figure 4–59. Type B Semirigid Connection

$$Z = \frac{12g'^3}{Eb't'^3(y+d)\{(y'+2H') + \left[\dfrac{2\alpha h}{y'+d}\right][y^2+(H-h)(2y-h)]\}} \times \frac{g'+g_1'}{4g'+g_1'} \tag{4–83}$$

where

$$\alpha = \frac{1}{b'}\left(\frac{g't}{gt'}\right)^3\left(\frac{4g+g_1}{4g'+g_1'}\right)\left(\frac{g'+g_1'}{g+g_1}\right)$$

Table for 6 x 4 x ¾ connection angles

Depth of beam (in.)	Z x 10^5 rad/kip-in.
8	0.046
10	0.036
12	0.028
14	0.024
16	0.020
18	0.015
21	0.013
24	0.011
27	0.0087
30	0.0076
33	0.0065
36	0.0054

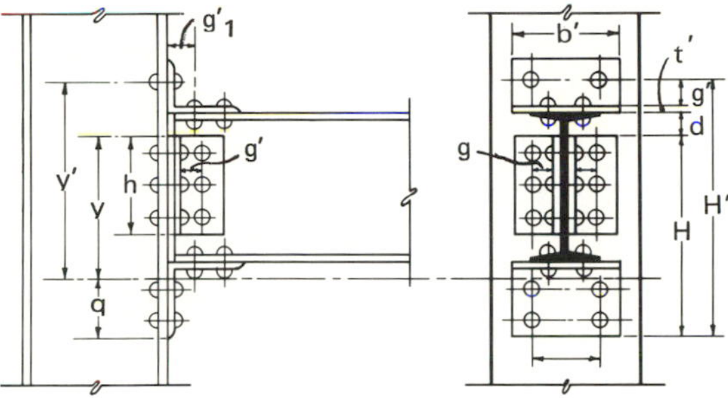

Figure 4–60. Type C Semirigid Connection

DeFalco and Marino extended the work of Lothers for a Type D connection and also derived an equation to account for a modified girder stiffness.

Type D connection (Fig. 4–61). DeFalco and Marino derive a value for this connection as follows:

$$Z = \frac{3}{t \times d^2 \times E} \tag{4–84}$$

Figure 4–61. Type D Semirigid Connection

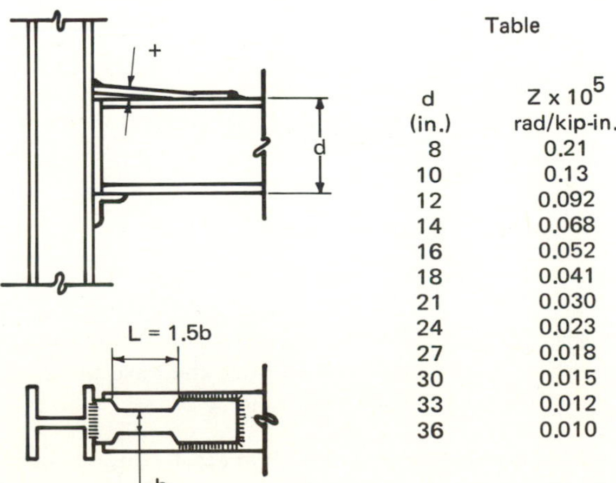

Table

d (in.)	Z x 10^5 rad/kip-in.
8	0.21
10	0.13
12	0.092
14	0.068
16	0.052
18	0.041
21	0.030
24	0.023
27	0.018
30	0.015
33	0.012
36	0.010

Reduced stiffness for frame analysis. Semirigid connections provide less frame stiffness than fully rigid connections. This effect will mean greater lateral drift, which will affect moment magnification ($P\Delta$). A modified girder stiffness can be used which incorporates the Z factor as follows:

$$k_r = \left(\frac{I_g}{L_g}\right)_m = \frac{3(I/L)}{4(L'/L) - (L/L')} \tag{4–85}$$

where k_r = relative stiffness of beam or girder with partially rigid connections
$L' = L + 3EIZ$
L = span of beam or girder
I = moment of inertia of beam or girder

This equation may be substituted for I_g/L_g when using the alignment charts, other approximate methods, or frame analysis by matrix methods.

The following example illustrates the approximate analysis of semirigid connections.

EXAMPLE PROBLEM 4–13: Approximate Analysis of Semirigid Beam-Column Connection

A Type *B* beam-column connection connects a W18 × 55 girder to a W10 × 49 column. The girder spans 25 ft (7.62 m) and supports a gravity load of 2.5 kips/ft (36,482 N/m). Bolts are $\frac{3}{4}$ in. (19 mm) (A325), number and arrangement as shown in Fig. 4–59. Determine the moment–rotation characteristics of the connection and check its capacity to carry a wind moment of 30 ft-kips (40,680 Nm).

Solution

1. Plot the moment rotation curve. From Fig. 4–59, $Z = 0.014 \times 10^5$ rad/kip-in.

   Beam line points:

$$\text{FEM} = \frac{wl^2}{12} = \frac{2.5(25)^2}{12} = 130.2 \text{ ft-kips (176,551 Nm)}$$

   Simple beam rotation:

$$\phi = \frac{wl^3}{24EI}$$

$$= \frac{2.5(25)^3 144}{24(30 \times 10^3)(890)} = 0.0088$$

   Figure 4–62 shows the plotted curve.

2. Check connection for wind moment.

$$M = 30 \text{ ft-kips}$$

$$T = C = \frac{M}{d} = \frac{30(12)}{18} = 20 \text{ kips (88,960 N)}$$

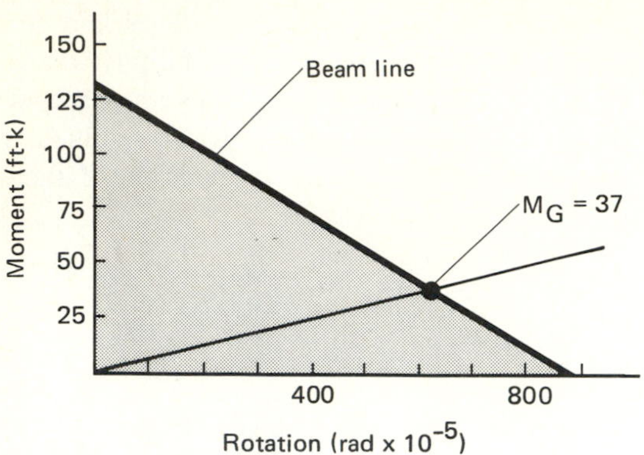

Figure 4–62. Example Problem 4–13

Bolt tension:

$$\frac{20 \text{ kips}}{2(0.4418)} = 22.6 \text{ ksi} < 44 \quad \text{(O.K.)}$$

$$= 155.8 \text{ MPa} < 303.4 \text{ MPa}$$

Bolt shear:

$$\frac{20 \text{ kips}}{4(0.4410)} = 11.3 \text{ ksi} < 17.5 \quad \text{(O.K.)}$$

$$= 77.9 \text{ MPa} < 120.7 \text{ MPa}$$

Angle tension:

$$\frac{20 \text{ kips}}{(8 - 2 \times 0.75)0.75} = 4.1 \text{ ksi} < 22 \quad \text{(O.K.)}$$

$$= 28.3 \text{ MPa} < 151.7 \text{ MPa}$$

3. Check connection for ductility. Total moment on connection at start of "shakedown" is equal to M_G and M_W. Limit stresses to 85% of material strength, F_u. From Fig. 4–62, M_G = 37 ft-kips (50,172 Nm).

$$M_G + M_w = 37 + 30 = 57 \text{ ft-kips (77,292 MPa)}$$

$$T = C = \frac{M}{d} = \frac{57(12)}{18} = 38 \text{ kips (169,024 N)}$$

Check bolts:

$$F_u = 120 \text{ ksi (827.4 MPa)}$$

Bolt tension:

$$\frac{38 \text{ kips}}{2(0.4418)} = 43.0 \text{ ksi} < 0.85F_u \quad \text{(O.K.)}$$

$$= 296.5 \text{ MPa}$$

Bolt shear:

$$\frac{38 \text{ kips}}{4(4418)} = 21.5 \text{ ksi} < 0.85F_u \quad \text{(O.K.)}$$

$$= 14,812 \text{ MPa}$$

Check steel material:

$$F_u = 58 \text{ ksi}$$

Angle tension:

$$\frac{38 \text{ kips}}{8 - 2 \times 0.75} = 7.79 \text{ ksi} < 0.85F_u \quad \text{(O.K.)}$$

$$= 53.7 \text{ MPa}$$

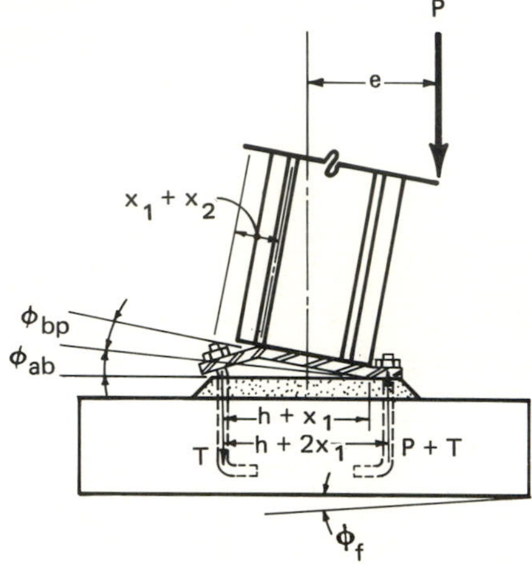

Figure 4–63. Column Base Rotation

Semirigid Column Bases. Column bases rotate due to rotation between the footing and soil, bending in the base plate, and elongation in the anchor bolts (Fig. 4–63). An approximate method for estimating the total rotation of the base is[8]

$$\phi_b = \phi_f + \phi_{bp} + \phi_{ab} \tag{4–86}$$

If the axial load is large enough so that there is no tension in the anchor bolts, ϕ_{bp} and ϕ_{ab} are zero, and $\phi_b = \phi_f$. Otherwise,

$$\phi_b = M(\gamma_f + \gamma_{bp} + \gamma_{ab}) = Pe(\gamma_f + \gamma_{bp} + \gamma_{ab})$$

$$\gamma_f = \frac{1}{k_s I_f} \tag{4–87}$$

$$\gamma_{bp} = \frac{(x_1 + x_2)^3 \,[2e/(h + 2x_1) - 1]}{6eE_s I_{bp}(h + x_1)} \geq 0 \tag{4–88}$$

$$\gamma_{ab} = \frac{g[2e/(h + 2x_1) - 1]}{2eA_b E_s(h + x_1)} \geq 0 \tag{4–89}$$

where γ_f, γ_{bp}, γ_{ab} = flexibility coefficients of the footing/soil interaction, the base plate, and the anchor bolts

[8] *PCI Design Handbook,* 2nd ed. (Chicago: Prestressed Concrete Institute, 1978), pp. 4–21.

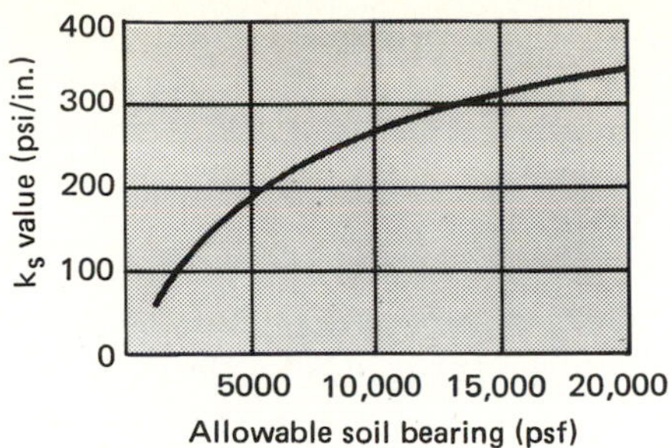

Figure 4–64. Coefficient of Subgrade Reaction

k_s = coefficient of subgrade reaction from Fig. 4–64
I_f = moment of inertia of the footing (plan dimensions)
E_s = modulus of elasticity of steel
I_{bp} = moment of inertia of the base plate (vertical cross-sectional dimensions)
A_b = total area of anchor bolts which are in tension
h = width of the column in the direction of bending
x_1 = distance from face of column to the center of the anchor bolts, positive when anchor bolts are outside the column, and negative when anchor bolts are inside the column
x_2 = distance from the face of the column to the base plate anchorage
g = assumed length over which elongation of the anchor bolt takes place = one-half of development length + projection for deformed anchor bolts or the length to the hook + projection for smooth anchor bolts

Approximate Design

Assuming equal effectiveness of all connectors, the required number of bolts or length of weld for connections in *shear* or *tension* can be determined from the tables in Appendix 4E. The required tension and compressive embedments and lap splices for reinforcing steel are also shown.

Approximate design of *rigid moment* connections can be accomplished by calculating an equivalent moment couple as shown in Fig. 4–65. Select a connection type with established moment–rotation characteristics for a Type 1 connection (such as shown in Fig. 4–66) and provide required tension and compression connection elements (flange area, bolts, welds, etc.).

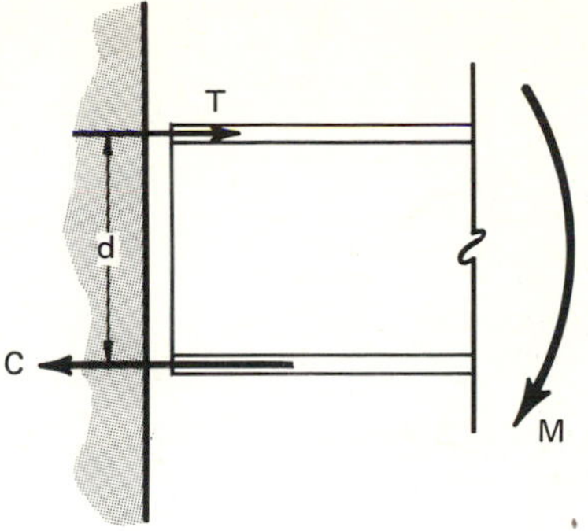

Figure 4–65. Model for Approximate Design of Rigid Moment Connections

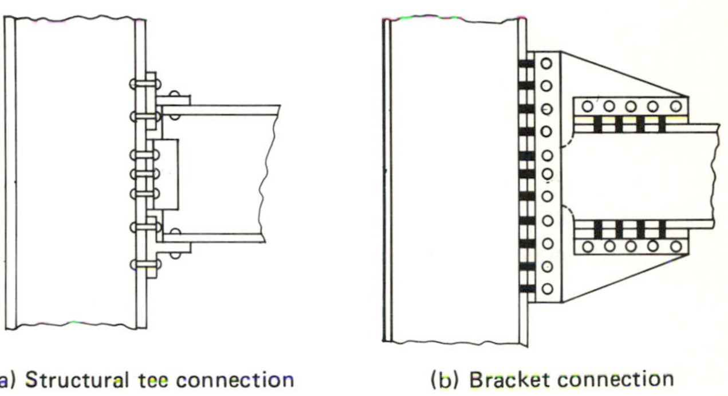

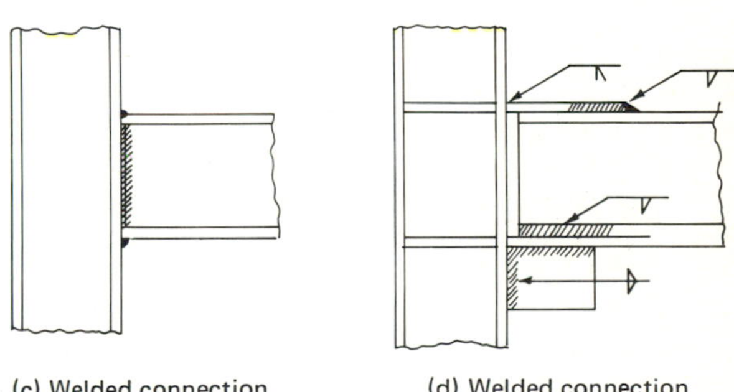

Figure 4–66. Rigid Connection Types

4–8 ARCHES

Arches are curved elements that use their geometrical shape to carry loads (across horizontal spans) primarily in compression. Much like the cable which functions in axial tension and achieves strength from its sag, the arch achieves its strength from the opposite curvature (convex upward). Unlike the cable, however, the arch must be rigid in order to hold it shape, and this rigidity introduces secondary forces (moments and shears) that are not present in cables.

The arch form is present in many types of structures, ranging from masonry arches over wall openings, to domelike roof systems, to bridge structures. A vertical

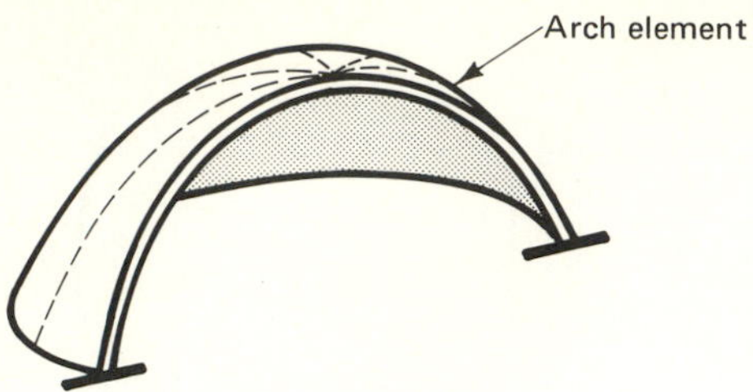

Figure 4–67. Arch–Dome Relationship

slice through a dome shell (Fig. 4–67) reveals the arch as the fundamental element of this system. It is also present in other shell forms.

History and Development

The first arches were constructed of stone materials, probably around the fourth or fifth century B.C. Stone materials were cut in wedge-shaped sections (called voussoirs) to form a curved arch form called the voussoir arch. The arch became a popular form for bridge structures, aqueducts, and vaulted roof systems. The behavior of the arch was not always understood, and many of these early structures experienced problems with the thrust produced at the base.

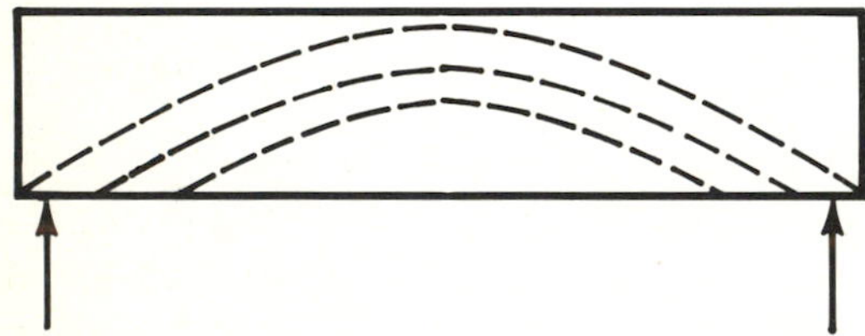

Figure 4–68. Beam Behavior Related to Arch Action

The arch form continued to develop and for a period of time was used to explain the behavior of the beam element (Fig. 4–68). Floor systems were even constructed with this concept in mind before the beam form fully developed (Fig. 4–69). Major structural applications were limited to bridges for a long time after the early developments in Europe, except for continuing applications in churches and other vaulted ceiling spaces, as developments in rigid frame construction provided more economical solutions for moderate spans. The arch has now reappeared in our vocabulary, however, as major arenas have been built in recent years using this element form.

Figure 4–69. Early Floor Systems

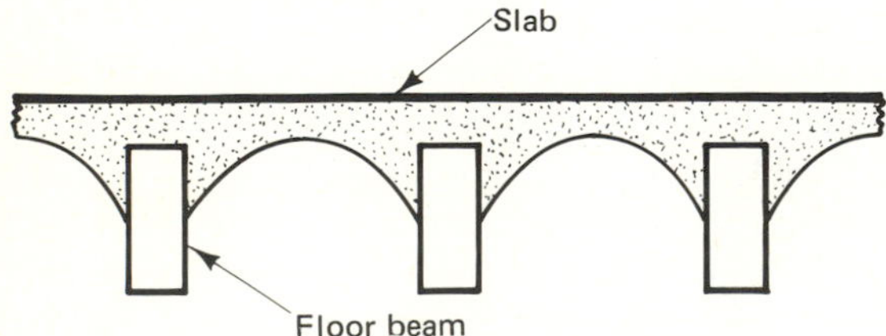

Types and Uses

The basic types of arch elements are shown in Fig. 4–70. These arches may be statically *determinate* (with three hinges) or *indeterminate* to the third degree with complete fixity at the base and at the crown, or any combination in between. The three-hinged determinate arch is most

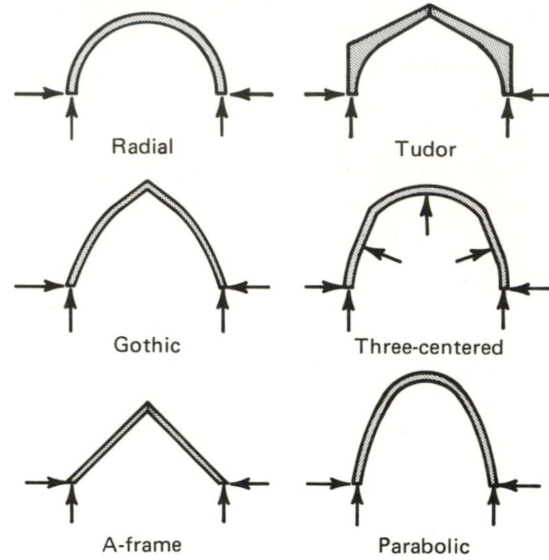

Figure 4–70. Types of Arches

commonly used due to its simplicity and freedom from temperature stresses and the influence of support settlement. It is often used in conjunction with tension or compression rings to achieve a complete framing system.

Major uses continue to be in bridge structures and large roof systems, where both vertical clearance and long

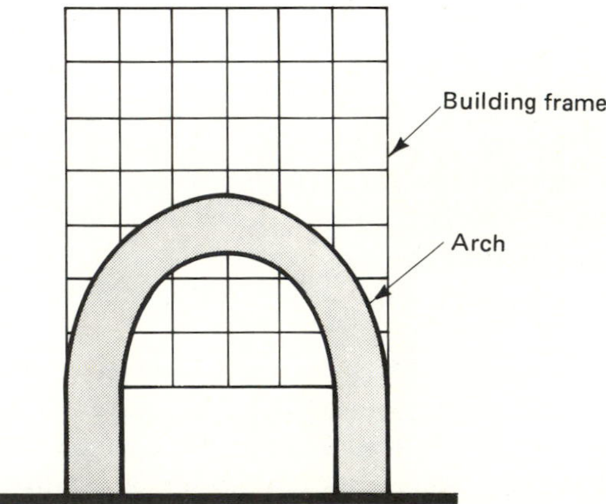

Figure 4–71. Arch Application in Conjunction with High-Rise Building Form

span strength are key requirements. The arch form is very competitive with other long-span elements for spans from 100 to 300 ft (30.5–91 m). The use of the arch form in conjunction with more traditional framing (Fig. 4–71) suggests that other applications await discovery by creative structural engineers.

Behavior

Pure axial compression in arches is achieved only under uniform loading and an arch shape that equals the funicular polygon for the load (Fig. 4–72). A parabolic curve represents the polygon shape for a uniform load per unit of horizontal length. Any other loads will introduce bending in the arch in addition to the axial thrust. This bending behavior was not a problem for the early voussoir arches due to the dead weight of the arch and supported fill material, and relatively smaller live load. Today's arches, however, are lighter and carry a greater live load, resulting in combined bending and compression in the arch for varying load patterns. These forces can be calculated at any section once the arch vertical and horizontal (thrust) reactions are known.

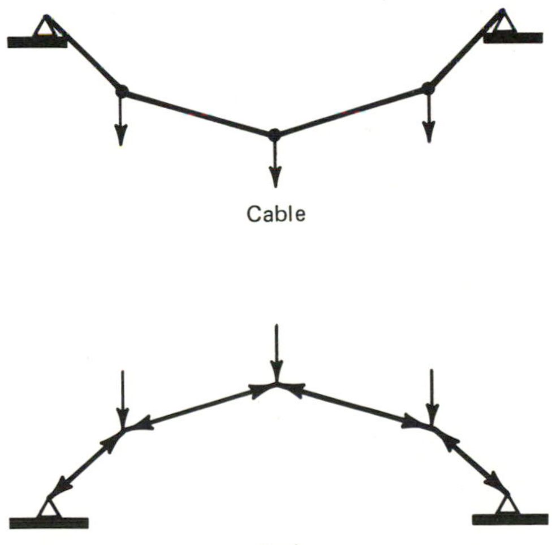

Figure 4–72. Funicular Polygon for Arch

Since the arch element is a beam-column type of element, it is subject to stability considerations (Fig. 4–73), including:

- "Snap-through" buckling
- Antisymmetric buckling
- Buckling out of the plane

Very shallow arches may exhibit the snap-through type of buckling behavior, but this is not a problem for most civil engineering structures. Normal arches will fail in the antisymmetric mode, and very high arches may buckle out of the plane if not braced in the lateral direction.

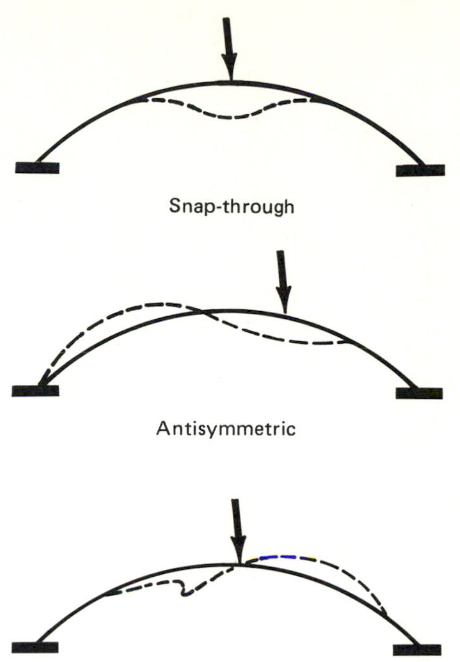

Figure 4–73. Types of Arch Instability

Approximate Analysis

Each type of arch requires a slightly different approximate analysis method depending on its geometry and degree of indeterminacy.

Three-Hinged Arch. The three-hinged arch is statically determinate; therefore, the reactions can be readily calculated, using $\Sigma M = 0$ about the crown hinge. Figure 4–74 gives the reactions for some common types of loading. Once the reactions have been determined, points of maximum moment and compressive thrust can be found by plotting moment and thrust diagrams. The following example illustrates the approximate analysis method.

EXAMPLE PROBLEM 4–14: Approximate Arch Analysis

Determine the points of maximum moment and axial load for the three-hinge arch shown in Fig. 4–75.

Solution

1. Determine arch reactions (Fig. 4–74).

$$R_v = \frac{wl}{2} = \frac{2000(150)}{2} = 150 \text{ kips } (667{,}200 \text{ N})$$

$$R_H = \frac{wl^2}{8h} = \frac{2.0(150)^2}{8(30)} = 187.5 \text{ kips } (834{,}000 \text{ N})$$

2. Plot moment diagrams.

$$M_x = R_v x - R_H(20 + y) - \frac{wx^2}{2}$$

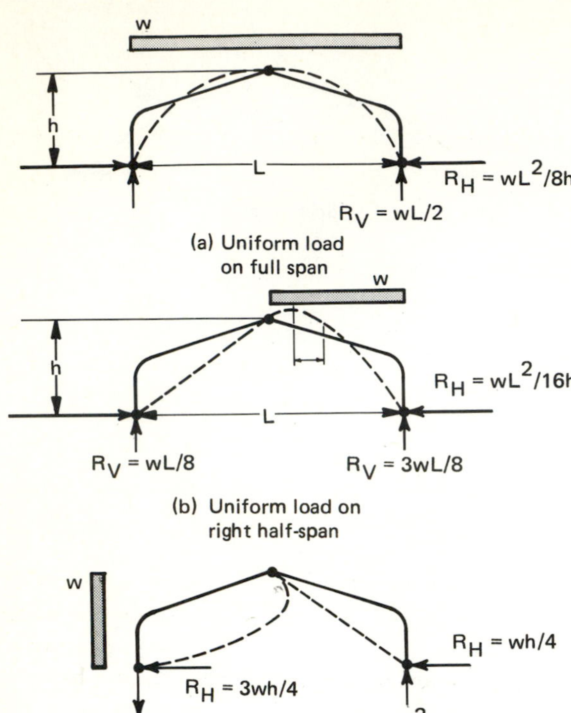

Figure 4–74. Three-Hinge Arch Reactions

Figure 4–75. Example Problem 4–14

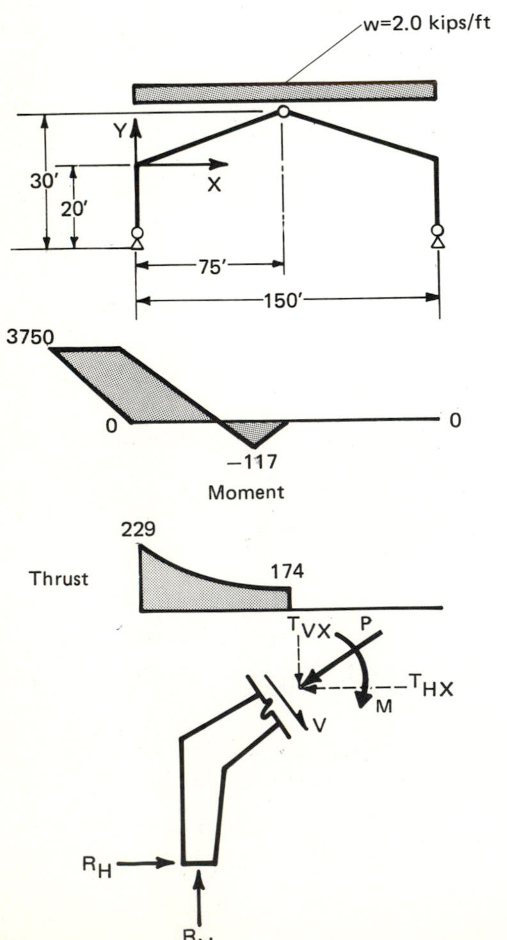

3. Plot axial load diagram.

$$T_{Hx} = 187.5 \text{ kips}$$

$$T_{vx} = V_L - wx$$

$$P_x = \frac{75}{80.8}\, T_{Hx} + \frac{30}{80.8}\, T_{vx}$$

4. Determine maximum moment and axial load.

Max. $M = 3750$ ft-kips (5,085,000 Nm) at knee
Max. $P = 229$ kips (1,018,592 N) at knee

Two-Hinged Arch. The two-hinged arch is indeterminate to one degree. The vertical reactions, however, can be determined directly from statics. With $I = I_c \sec \phi$ and $f/L < \frac{1}{8}$, the approximate influence line for the horizontal reaction is given by (Fig. 4–76)

$$H = \frac{3Lk(1-k)}{4f} \tag{4–90}$$

where I_c = moment of inertia about the member centroidal axis
I = moment of inertia translated to a horizontal axis
ϕ = angle between member centroidal axis and horizontal axis

with $H_{\max} = 3L/16f$. For full-span uniform load the thrust is

$$H = \frac{wL^2}{8f} \tag{4–91}$$

Figure 4–76. Influence Line for Two-Hinge Arch Reactions

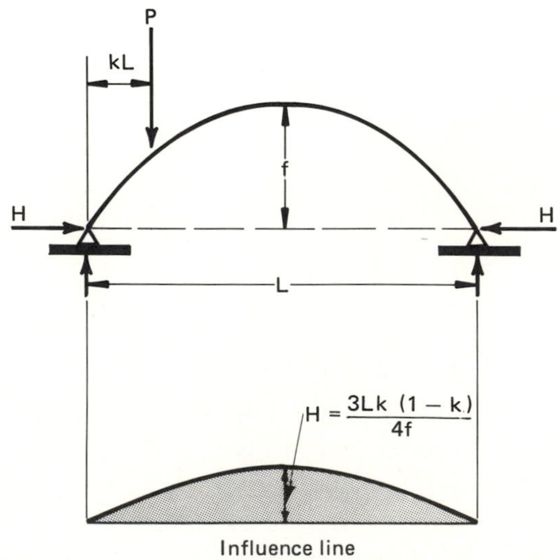

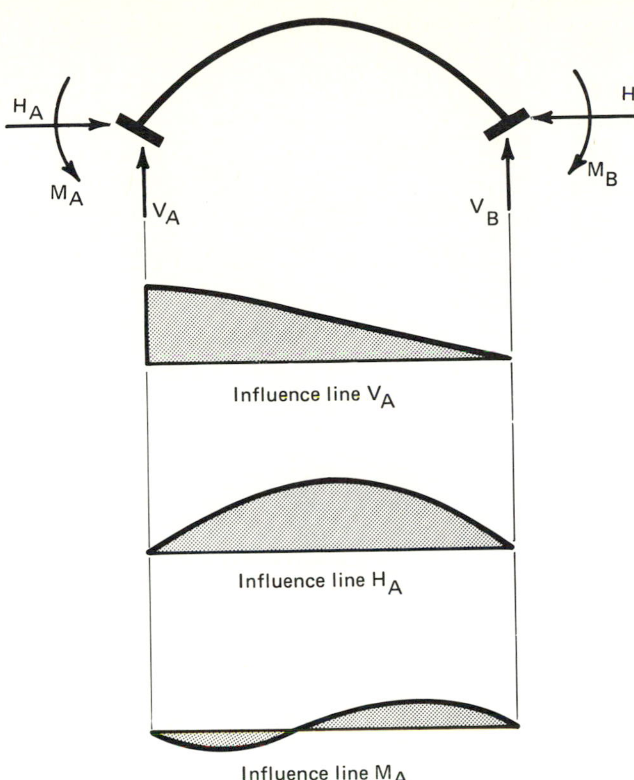

Figure 4-77. Influence Lines for Fixed Arch Reactions

Fixed Arch. With $I = I_c \sec \phi$ and $f/L < \frac{1}{8}$, the approximate formulas for a fixed arch are (Fig. 4-77):

$$V_A = (1 + 2k)(1 - k)^2$$

$$H_A = \frac{15}{4}\frac{L}{f}(k - k^2)^2 \tag{4-92}$$

$$M_A = \frac{L}{2}(1 - k)^2(5k^2 - 2k)$$

For full-span uniform vertical load per unit length of span:

$$H_A = \frac{wL^2}{8f}$$

$$V_A = \frac{wL}{2} \tag{4-93}$$

$$M_A = M_B = \tfrac{1}{12}\, wL^2$$

$$M_c = \tfrac{1}{24}\, wL^2$$

Approximate Design

The preliminary design of arches requires selection of the type of arch to be used and its proportions. Optimum rise/span ratio is 1/5, and the arch geometry should be selected to minimize bending moments under the anticipated loading.

Approximate member sizing for combined bending and axial load will follow the procedures already discussed in Section 4-6 for beam-columns. The distance between bracing points will establish the unbraced lengths for determining F_a and F_b. No moment magnification is required for arch design.

Buckling can be checked by the following equation:

$$H_{cr} = \frac{\beta EI}{L^2} \tag{4-94}$$

where β is given in Table 4-10. Approximate safety factors must be applied to Eq. 4-94.

Table 4-10 Arch Buckling Constants[a]

		Values of β for buckling of parabolic arches[b]						
f/L		*0*	*0.1*	*0.2*	*0.3*	*0.4*	*0.5*	*Source*
Three-hinged	$I = \text{const.}$	29.8	28.5	24.9	20.2 (19.8)	15.4 (13.6)	—	Stuessi
	$I = I_c \sec \phi$	29.7	29.4	27.7	25.3 (25.1)	22.6 (19.4)	19.8 (15.0)	Dischinger
Two-hinged	$I = \text{const.}$	39.4	35.6	28.4	19.4	13.7	9.6	Lockschin
	$I = I_c \sec \phi$	39.4	37.2	31.6	25.1	19.4	15.0	Dischinger
Fixed	$I = \text{const.}$	80.8	75.8	63.1	47.9	34.8	—	Stuessi
	$I = I_c \sec \phi$	80.8	78.4	70.8	61.1	51.1	41.8	Dischinger

[a] Edwin H. Gaylord, Jr. and Charles N. Gaylord, *Structural Engineering Handbook* (New York: McGraw-Hill Book Company, 1968), pp. 17-49.

[b] Values are for symmetrical buckling. Figures in parentheses are for unsymmetrical buckling.

4–9 TRUSSES

Trusses are assemblies of tension ties and column elements arranged in a pattern consisting of a top and bottom horizontal member (called chords) and vertical and diagonal members (called web members). This geometric arrangement is very efficient in converting bending loads into axial tension and compression forces in the truss members. Termed *triangulation of loads,* this technique is the most efficient way to use structural material (Fig. 4–78). Although trusses could technically be classified as a subsystem, we will treat them as elements because of their interchangeability with beams and their dominant length dimension.

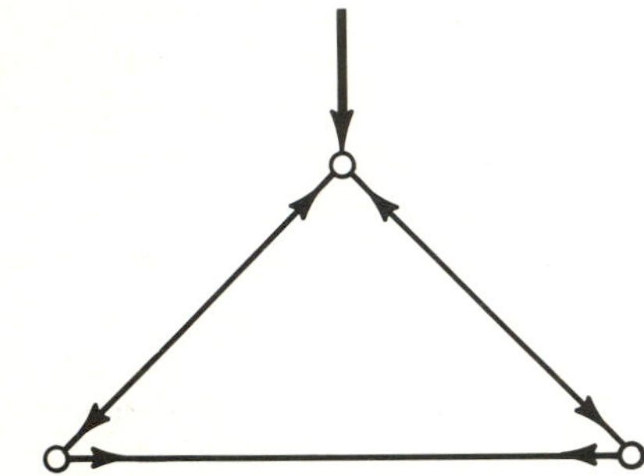

Figure 4–78. Triangulation of Loads

The pure truss form carries only axial loads. Actual trusses, however, are constructed with continuous chords and some degree of connection restraint or eccentricity, creating *secondary* bending stresses when loaded. Also, loads are often applied to the chords between the panel points, which transforms the chord members into beam-column elements.

History and Development

The truss form was probably developed during the Roman Empire, and the oldest surviving roof truss is at the sixth-century church of the Monastery of St. Catherine on Mt. Sinai. Continued development for roof construction fostered a number of different truss patterns, including the English innovation (in the fifteenth century) of the hammer beam truss used to span the width of Westminster Hall. The truss gradually found its way into bridge construction, with a publication by Palladio in 1570 describing the first detailed bridge design.

During the first half of the nineteenth century, light trusses (as we now know them) fully emerged as the demand for road and railway bridges increased. Use continued for roof construction, and development of prefabrication techniques led to bar joist elements and manufactured timber trusses. The truss also became a major element in tower structures and in space truss systems for large column-free spans. Trusses now are a major element in roof and floor systems, bridges, towers, platforms, and space structures.

Types and Uses

Various truss types are shown in Fig. 4–79. The most common types are the Warren and Pratt trusses. The truss finds a wide variety of uses, including the following:

- Bar joists
- Preengineered timber trusses
- Transmission towers
- Bridge spans
- Offshore platforms
- Bracing subsystems
- Plane truss roof systems
- Space truss roof systems

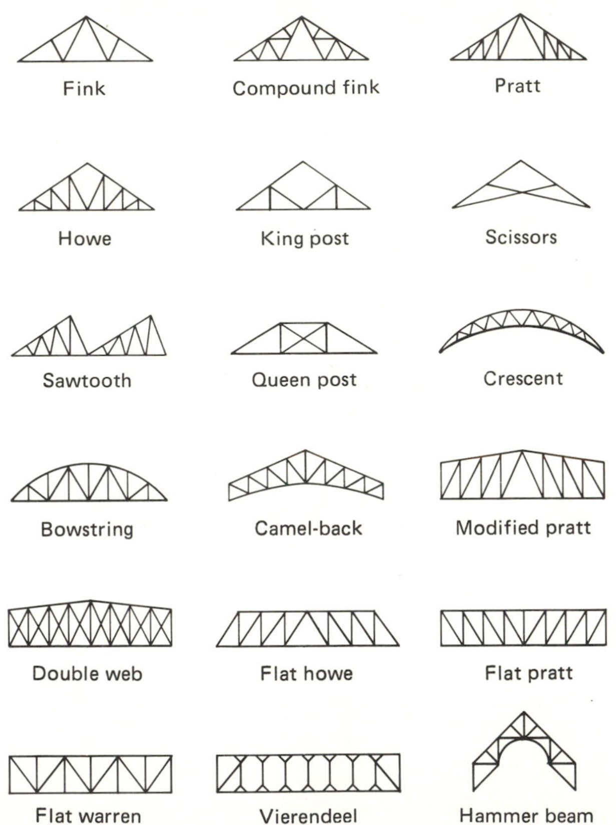

Figure 4–79. Truss Types

Bridge trusses are known as *deck trusses* if the roadway is placed on top, *through trusses* if the roadway is placed in the plane of the bottom chord, and *half-through* if an intermediate position is adopted. Bridge construction uses trusses in conjunction with cables for *suspension bridges,* as well as in cantilever spans.

Trusses are competitive elements anytime spans exceed 70 ft (21 m). Even longer spans may dictate the use of primary trusses in one direction with secondary trusses framing between them. The space truss fully incorporates primary trusses in two directions in a grid fashion.

Behavior

Pure truss behavior involves only tension and compression forces in the members since joints are considered hinged (Fig. 4–80). Centroidal axes that intersect at a common point at connections lead to only axial force transfer and the hinged assumption works very well. The individual members then act as either tension tie elements or column elements, with the usual stability considerations for the columns. The truss assembly acts as a pseudobeam, with moment resistance provided by the top and bottom chord couple and shear resistance by the diagonal web members (Fig. 4–81).

Just as beams can buckle laterally out of the plane of the loads, trusses also are susceptible to lateral stability problems. The compression chord essentially acts as a long column and requires lateral bracing at regular intervals. The tension chord is not as critical, but minimum kl/r values must be maintained to avoid extreme flexibility, which may lead to bowing and irregular curvature in and out of plane.

The ideal pin-connected truss (exhibiting "pure" truss behavior) is not a common element. Instead, chords are frequently made continuous, and joints are either welded or bolted with some degree of planned (or accidental) eccentricity. Loads are also frequently placed between panel points, putting the chord member in direct bending. The major contribution to secondary bending usually occurs due to deformation of the members: tension members lengthen and compression members shorten, while the angles between the members remain constant due to the connection fixity (Fig. 4–82). Secondary bending stresses are usually associated with truss behavior.

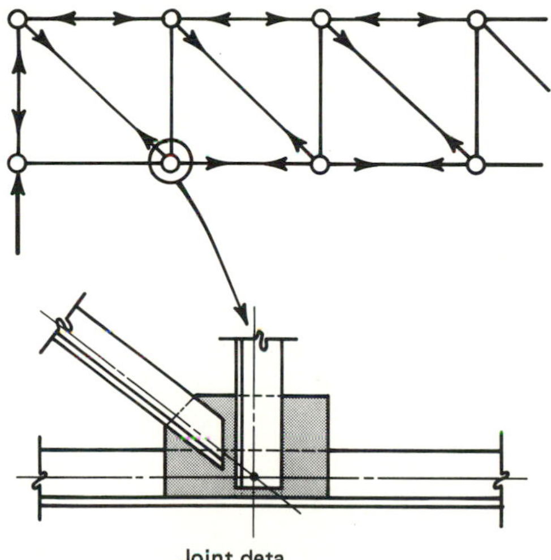

Figure 4–80. Pure Truss Action with Hinged Joints

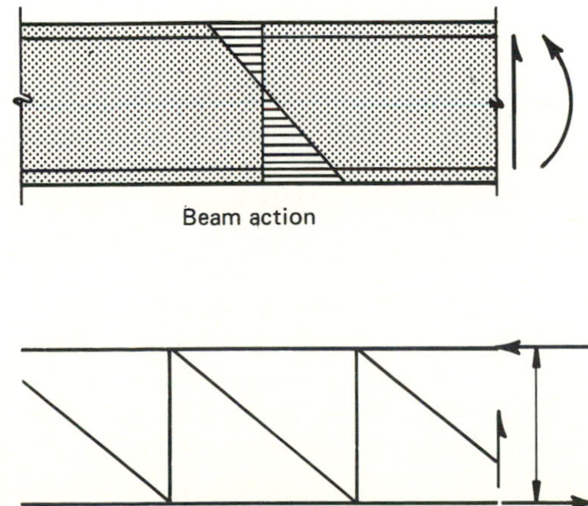

Figure 4–81. Beam Analogy to Truss Behavior

Figure 4–82. Secondary Bending in Trusses Due to Member Deformation

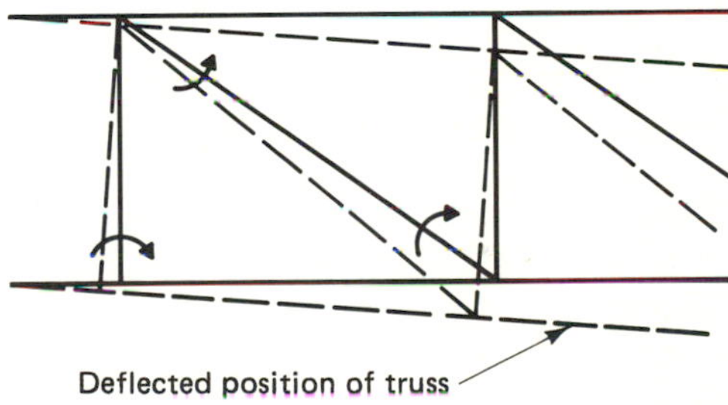

Approximate Analysis

Truss analysis requires that all loads be resolved to the panel points. This is normally done by assuming that the chord member acts as a simple beam spanning between panel points and the end reactions represent the panel loads (Fig. 4–83). The maximum moment in the chord for later use in member design is generally assumed to be $\omega l^2/8$ or $\omega l^2/10$ (if continuity exists). Once loads are established, several techniques are available for approximate truss analysis.

Determinate trusses can be analyzed readily, and the maximum moment and shear determined by the "method of sections" applied at midspan (for moment) and at the supports (for shear). Figure 4–84 shows this procedure.

Indeterminate trusses of the type shown in Fig. 4–85 can be analyzed by making an assumption regarding the proportionate panel shear to be carried by the tension diagonal and the compression diagonal. For wind bracing, it is common practice to assume that the compression member buckles and that all the load is carried by the

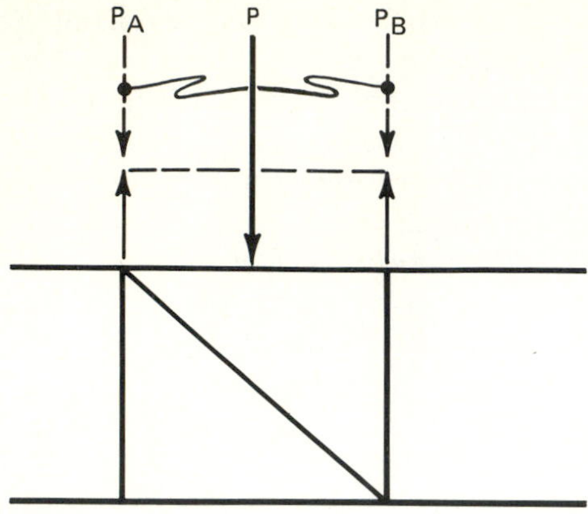

Figure 4–83. Transfer of Loads to Panel Points for Truss Analysis

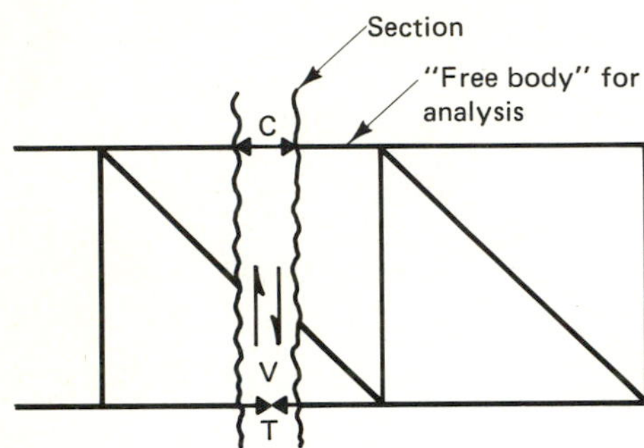

Figure 4–84. Determinate Truss Analysis by Method of Sections

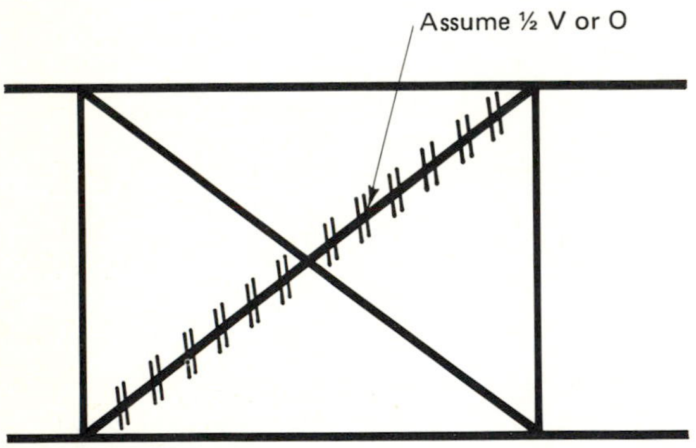

Figure 4–85. Indeterminate Truss Analysis by Assumptions for Redundant Member

tension member. For heavier trusses, it may be appropriate to assume that one-half of the shear load goes to each diagonal.

Continuous truss analysis is best handled by assuming that the truss acts as a continuous beam. Inflection points can thus be assumed or moment coefficients used to determine points of maximum moment.

Approximate Design

Optimum truss depths range from $0.08L$ to $0.12L$, with $0.10L$ customarily used for approximate design. The length between panel points should be approximately equal to the depth. Subverticals may be required if this panel spacing exceeds the economical span of the deck material. Once the depth and panel points are established, the approximate analysis techniques can be used to obtain the maximum tension and compression forces in the members and the bending moments due to location of the load.

Member design then proceeds for the tension tie elements, column elements, and beam-column elements. Secondary stresses are generally not considered during preliminary design.

Allowable stress for A36 steel trusses may be assumed as follows:

$$F_t = 22.0 \text{ ksi} \times 0.8 \text{ (reduction for net area)}$$
$$= 17.6 \text{ ksi (121 MPa)}$$

$$F_c = 15.0 \text{ ksi (103 MPa) (assumes that } kl/r = 80)$$

where F_t = allowable tension stress for preliminary design
F_c = allowable compression stress for preliminary design

The following example illustrates the approximate method.

EXAMPLE PROBLEM 4–15: Preliminary Truss Design

Select a simple span steel truss (A36 steel) to span 120 ft (36.6 m). Under a uniform load of 40 psf (1915 N/m²) with trusses spaced at 75 ft (7.6 m) on center. Provide an estimate of total truss weight.

Solution

1. Determine truss type. Assume a flat top chord is required and use a Pratt-type member arrangement.
2. Determine truss depth. Use depth $= \frac{1}{10} \times$ span $= \frac{1}{10} \times$ 120 ft = 12 ft (3.66 m).
3. Calculate maximum forces.
   Max. moment:

$$\frac{wl^2}{8} = \frac{(0.04 \times 25)(120)^2}{8} = 1800 \text{ ft-kips (2,440,800 Nm)}$$

Max. shear:

$$\frac{wl}{2} = \frac{(0.04 \times 25)120}{2} = 60 \text{ kips (266,880 N)}$$

4. Select chord members.
   Chord force (c or t):

$$\frac{M}{d} = \frac{1800}{12} = 150 \text{ kips (667,200 N)}$$

Area required for top chord:

$$\frac{150}{15} = 10 \text{ in.}^2 \text{ (6450 mm}^2\text{)}$$

Use two angles: 4 in. × 4 in. × $\frac{3}{4}$ in. (A = 10.9 in.^{2}).

$$\text{weight} = 37.0 \text{ lb/ft}$$

Area required for bottom chord:

$$\frac{150}{17.6} = 8.52 \text{ in.}^2 \text{ (5495 mm}^2\text{)}$$

Use two angles: 4 in. × 4 in. × $\frac{5}{8}$ in. (A = 9.22 in.^{2}).

$$\text{weight} = 31.4 \text{ lb/ft}$$

5. Select web members. Assume panel points at 12 ft (3.66 m); diagonals are all in tension for a Pratt truss.
   Maximum diagonal force:

$$[60 \text{ kips} - 0.04 \text{ ksf (25 ft)(6 ft)}]1.414 = 76.5 \text{ kips (339,382 N)}$$

Area required:

$$\frac{76.3}{17.6} = 4.33 \text{ in.}^2 \text{ (2793 mm}^2\text{)}$$

Use one angle: 4 in. × 4 in. × $\frac{5}{8}$ in. (A = 4.61 in.^{2}).

weight = 15.7 lb/ft

Minimum diagonal force:

$$(60 \text{ kips} - 54 \text{ kips})1.414 = 8.5 \text{ kips (37,808 N)}$$

Area required:

$$\frac{8.5}{17.6} = 0.48 \text{ in.}^2 \text{ (310 mm}^2\text{)}$$

$$\text{min. } \frac{kl}{r} = 240, \qquad \text{min. } r = \frac{12 \times 12 \times 1.414}{240} = 0.85$$

Use one angle: 3 in. × 3 in. × $\frac{3}{16}$ in. (A = 1.09 in.^{2}).

$$\text{weight} = 3.71 \text{ lb/ft}$$

Max vertical force at support: 60 kips (266,880 N).
Area required:

$$\frac{60}{15} = 4 \text{ in.}^2 \text{ (2580 mm}^2\text{)}$$

Use two angles: $3\frac{1}{2}$ in. × $3\frac{1}{2}$ in. × $\frac{5}{16}$ in. (A = 4.18 in.^{2}).

$$\text{weight} = 14.4 \text{ lb/ft}$$

Minimum vertical force at midspan: 12 kips (53,376 N).
Area required:

$$\frac{12}{15} = 0.8 \text{ in.}^2 \text{ (516 mm}^2\text{)}$$

$$\text{min. } \frac{kl}{r} = 200, \qquad \text{min. } r = \frac{12 \times 12}{200} = 0.72$$

Use two angles: $2\frac{1}{2}$ in. × $2\frac{1}{2}$ in. × $\frac{3}{16}$ in. (A = 1.80 in.^{2}).

$$\text{weight} = 6.14 \text{ lb/ft}$$

6. Estimate truss weight.
   Top chord:

$$120 \text{ ft} \times 37.0 \text{ lb/ft} = 4{,}440 \text{ lb}$$

Bottom chord:

$$120 \text{ ft} \times 31.4 \text{ lb/ft} = 3{,}768 \text{ lb}$$

Diagonals:

$$10 \times 12 \text{ ft} \times 1.414 \times \frac{15.7 + 3.71}{2} = 1{,}646 \text{ lb}$$

Verticals:

$$11 \times 12 \text{ ft} \times \frac{14.4 + 6.14}{2} = 1{,}355 \text{ lb}$$

Subtotal = 11,209 lb

Add 15% for connections = 1,120 lb

Total weight = 12,329 lb (54,839 N)

4–10 TENSION AND COMPRESSION RINGS

Tension and compression rings are closed circular elements that carry loads that are applied transverse to the member axis (Fig. 4–86). Pure tension and compression "hoop stresses" will result when the load is uniform in the radial direction; for other cases bending stresses will also be present. Rings are fundamental elements in pipes, tanks, aircraft, aerospace structures, and long-span roof systems (Fig. 4–87).

History and Development

Rings have been fundamental elements in the mechanical engineering field for a long time. The design of pipes, pressure vessels, and machine components has made extensive use of ring theory, and the aviation and shipbuild-

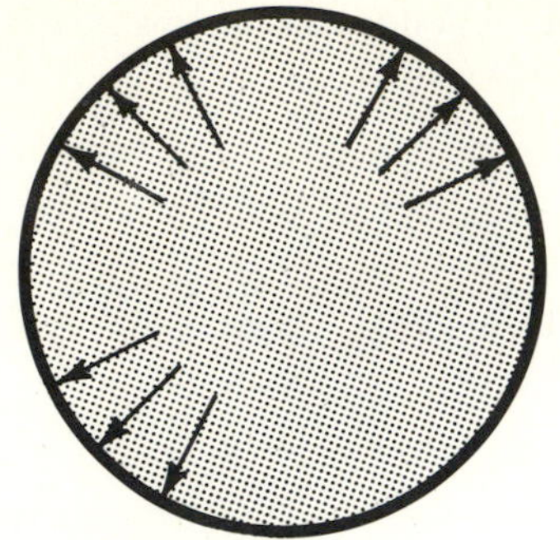

Figure 4–86. Ring Element

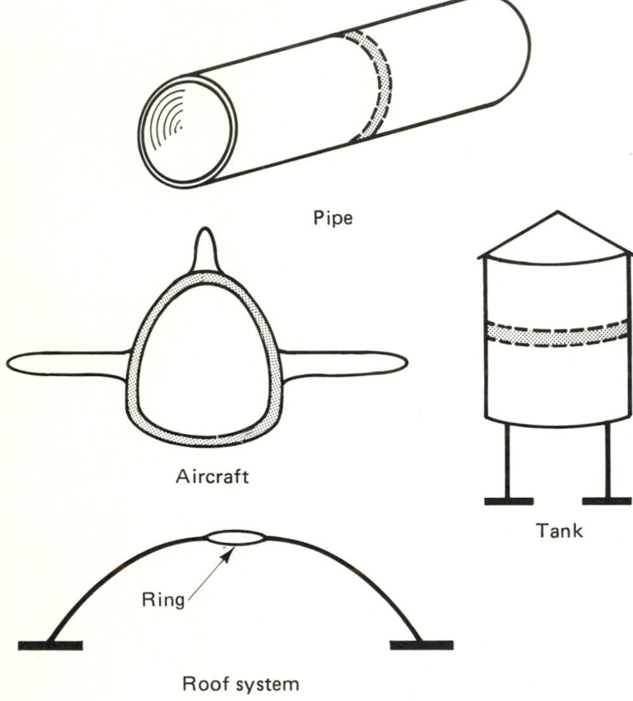

Figure 4–87. Ring Element Uses

ing industries have traditionally employed ring design for the construction of the main body and auxiliary components of air and sea vessels.

Civil engineers have used ring theory in designing culverts and pipe sections, and structural engineers have used the "hoop stress" principle in designing circular tanks for containment of water, or other materials. However, the ring has not been a standard part of our structural engineering vocabulary for very long. The introduction of major shell structures and other types of long-span roof systems during the midpart of the twentieth century has now established the ring as an important element, worthy of study and application to structural systems design.

Types and Uses

Circular rings and elliptical rings are the major types used in structural systems. The most common uses include concrete tanks (above and below ground), storage silos, water tanks, and dome-shaped roof systems. These uses involve loads that primarily induce axial tension and compression in the ring, making it an ideal element to resist the thrust created by loads on other intersecting elements.

Rings that receive loads directed inward toward the center are stressed in compression and called *compression rings.* Compression rings are commonly used at the crown of arch–dome systems and at the perimeter of cable systems (Fig. 4–88). Rings that receive loads directed outward away from the center are called *tension rings.* Tension rings are commonly used at the perimeter of arch–dome systems and at the midpoint of cable systems (Fig. 4–89). Rings in tank structures (Fig. 4–90) can act in either tension or compression depending on whether the greater load stems from containment of material or from external forces (e.g., soil, groundwater, etc.).

Figure 4–88. Compression Rings

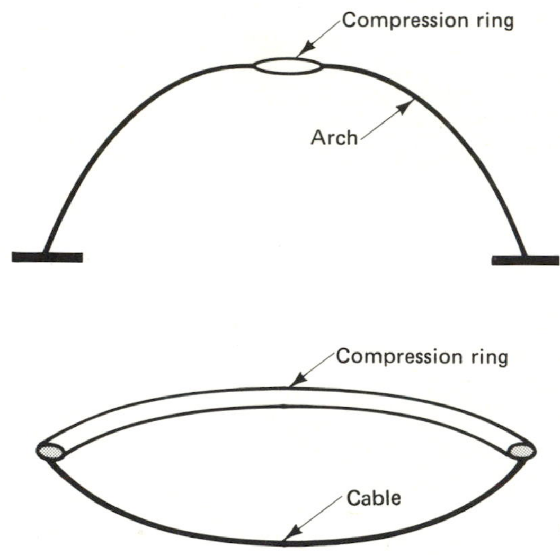

Figure 4–89. Tension Rings

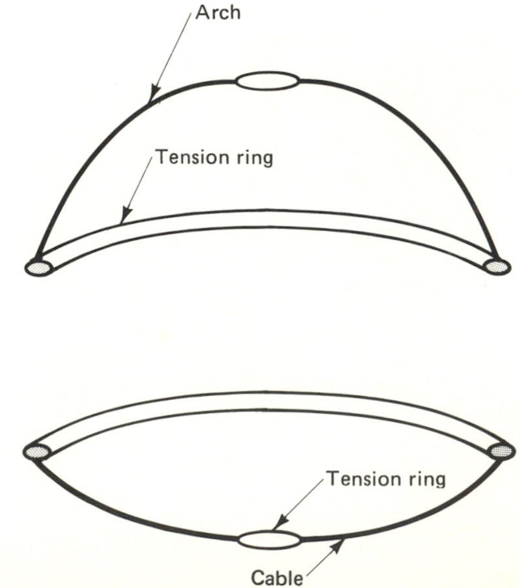

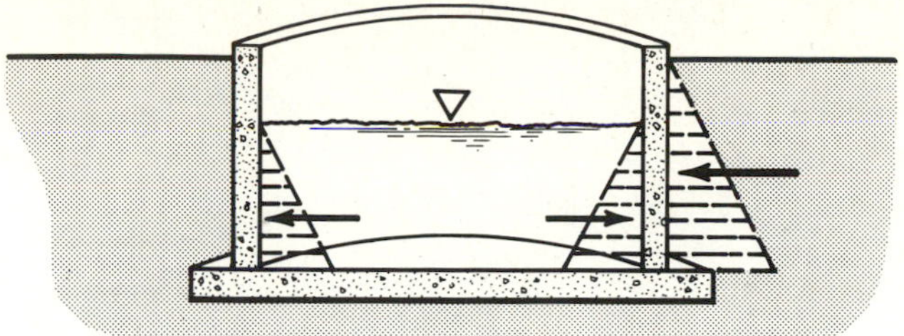

Figure 4–90. Tank Structure Rings

Behavior

The behavior of the ring element is similar to the arch. Uniform loads around the circumference of circular rings induce only axial tension and compression. Loads on elliptical rings and nonuniform loads on circular rings also induce bending stresses.

Rings are indeterminate structures under nonuniform loads, but the redundancy can be reduced from three redundants to one if the loading is symmetrical (Fig. 4–91). This reduces the thrust (H) to zero. Under these conditions, one-half of the ring can be treated as a circular arch with an unknown moment at A or B. The maximum possible moment at A is the fixed-end moment:

$$\text{FEM}_A = \frac{Pl}{8} = \frac{P\,(2R)}{8} = 0.25PR \qquad (4\text{–}95)$$

Some moment less than this value can be assumed and the subsequent analysis will be determinate. Thus the ring acts as a circular beam under symmetrical loading. Even many types of unsymmetrical loading will have little influence on stresses at the opposite side of the ring, and symmetrical conditions can be assumed for analysis.

Stability of rings is not often a problem, but buckling can take place for slender sections. Approximate buckling formulas are contained in the next section.

Approximate Analysis

Uniform loads on circular rings produce pure *hoop stresses* which can be determined by the following formula:

$$C \quad \text{or} \quad T = pR \qquad (4\text{–}96)$$

where C = compressive hoop force
T = tension hoop force
p = uniform pressure against ring
R = radius of ring

The buckling load for this case is expressed by

$$p_{cr} = \frac{3EI}{R^3} \qquad (4\text{–}97)$$

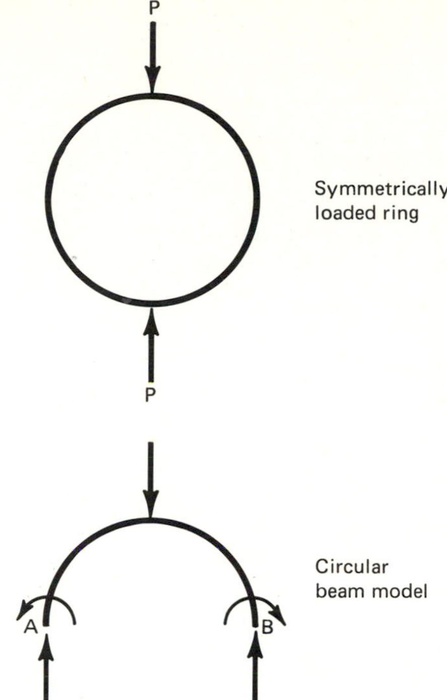

Figure 4–91. Curved Beam Analogy for Ring Elements

Concentrated loads on circular rings (if symmetrical) can be analyzed by the curved beam analogy discussed earlier or reference made to published formulas such as those shown in Fig. 4–92.

Figure 4–92. Ring Formulas

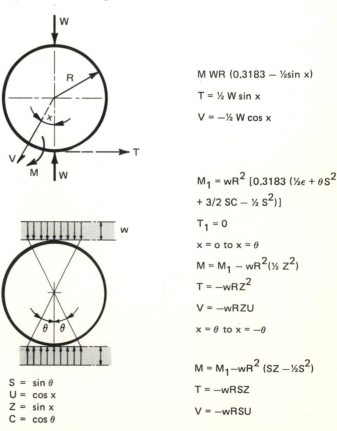

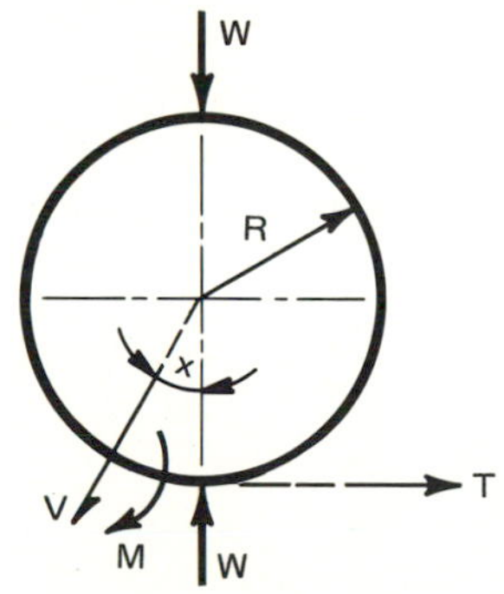

M WR (0.3183 – ½sin x)

T = ½ W sin x

V = –½ W cos x

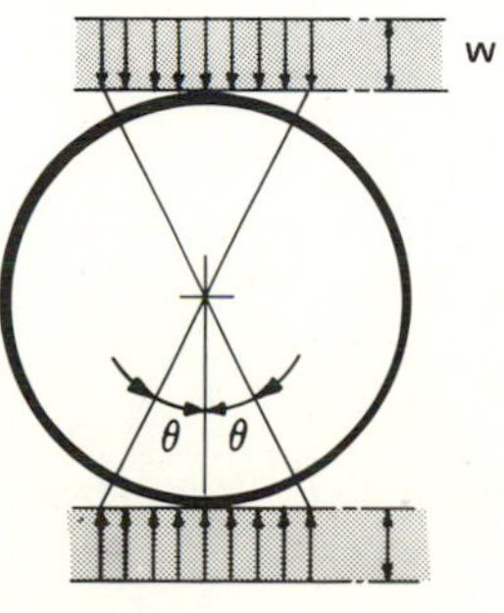

$M_1 = wR^2\,[0.3183\,(\tfrac{1}{2}\epsilon + \theta S^2 + 3/2\,SC - \tfrac{1}{2}S^2)]$

$T_1 = 0$

$x = o$ to $x = \theta$

$M = M_1 - wR^2(\tfrac{1}{2}Z^2)$

$T = -wRZ^2$

$V = -wRZU$

$x = \theta$ to $x = -\theta$

$M = M_1 - wR^2\,(SZ - \tfrac{1}{2}S^2)$

$T = -wRSZ$

$V = -wRSU$

S = sin θ
U = cos x
Z = sin x
C = cos θ

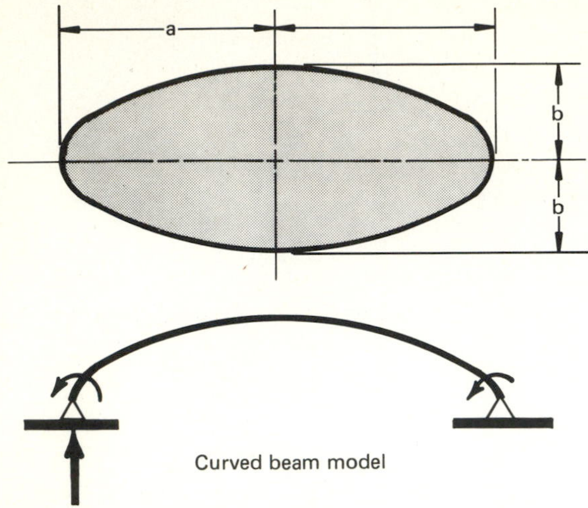

Figure 4–93. Elliptical Ring Analysis with Curved Beam Model

Elliptical rings experience both bending and axial stress under load. By working with the long side, symmetrical loads can be analyzed with the equivalent curved beam concept used earlier (Fig. 4–93). This procedure is less accurate for elliptical rings, however. Under uniform outward radial pressure, M_c will vary from $-0.172pa^2$ for b/a of 0.3 to zero for b/a of 1.0 (a circular ring). M_A varies from $0.282pa^2$ to zero. Interpolation for other values of b/a between 0.30 and zero will provide similar approximate values for the moments. The following example illustrates the approximate analysis of elliptical rings.

EXAMPLE PROBLEM 4–16: Approximate Elliptical Ring Analysis

For the elliptical ring shown in Fig. 4–94 calculate the maximum axial tension and the maximum positive and negative bending moments.

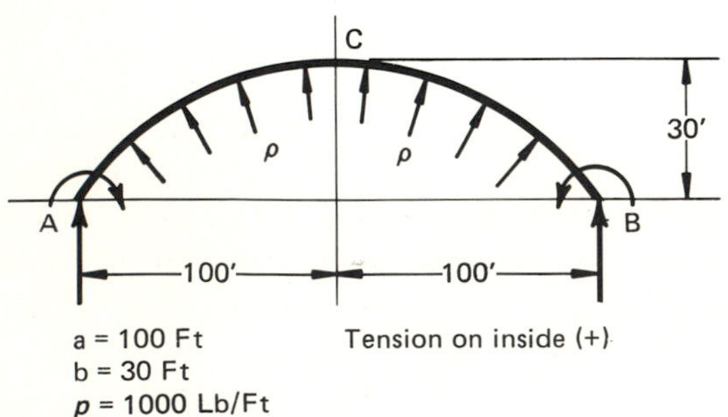

Figure 4–94. Example Problem 4–16

Solution

1. Calculate maximum possible moment at A for guidance.

$$\text{FEM}_A = \frac{pl^2}{12} = \frac{p(2a)^2}{12} = 0.333pa^2$$

2. Based on $b/a = 0.30$, assume values for M_A and M_c:

$$\begin{aligned} M_A &= 0.282pa^2 = 0.282(1000)(100)^2 \\ &= 2820 \text{ ft-kips } (3{,}823{,}920 \text{ Nm}) \end{aligned}$$

$$\begin{aligned} M_c &= \text{simple beam } M + M_A \\ &= \frac{-p(2a)^2}{8} + 0.282pa^2 \\ &= -0.22pa^2 \quad (\text{say} = -0.17pa^2) \\ &= -0.17(1000)(100)^2 \\ &= -1700 \text{ ft-kips } (2{,}305{,}200 \text{ Nm}) \end{aligned}$$

$$\text{Ring tension} = pa = 1000(100) = 100 \text{ kips } (444{,}800 \text{ N})$$

Approximate Design

Design of ring elements involves no new procedures. Once forces are determined by the approximate analysis techniques, the column, tension-tie, and beam column approximate design methods can be used to select a preliminary member size.

4–11 FINITE ELEMENTS

A *finite element* is a small subdivision of a planar or shell surface for which mathematical expressions can be developed to describe behavior (Fig. 4–95). The finite element technique is primarily a method of analysis rather than a construction element of the kind represented by beams and columns. In spite of the fact that we do not build with this element, a physical understanding of its behavior is of great help in applying the analysis technique and in understanding structures of irregular shape and geometry.

Figure 4–95. Finite Element

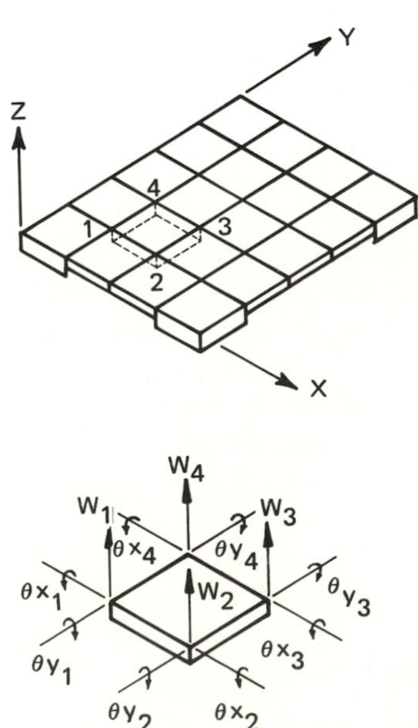

History and Development

The finite element was developed to describe the behavior of complex surfaces for which closed-form, exact mathematical expressions could not be developed. The idea of subdividing surfaces such as walls and slabs into strips for analysis has been used for a long time. A common technique was the use of a grid of beams (Fig. 4–96) to approximate the mathematical model of a slab. Each beam represented the properties of an equivalent width of slab and the analysis gave results at the intersection of the beams. Since the beams could be described exactly by mathematical expressions, the inaccuracies of this method stemmed from the neglect of the slab surface and its related behavior (stretching, for example).

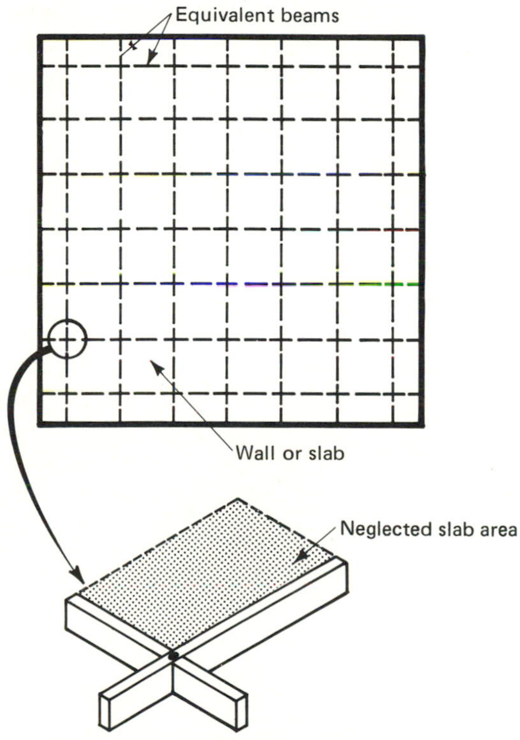

Figure 4–96. Grid Technique of Analysis for Continuous Surfaces

The finite-element technique was developed to describe better the slab surface behavior. Mathematical expressions describe the element surface and provide continuity between node points. Continuity of displacements across element boundaries, however, is not guaranteed (Fig. 4–97). The inaccuracies with this method are associated with this boundary effect.

Types and Uses

Common finite-element types are shown in Fig. 4–98. The rectangular elements are used for regular geometrical patterns; the triangular elements for boundary transitions and irregular patterns. If membrane (in-plane) forces are important in the analysis, the bending and stretching element can be used.

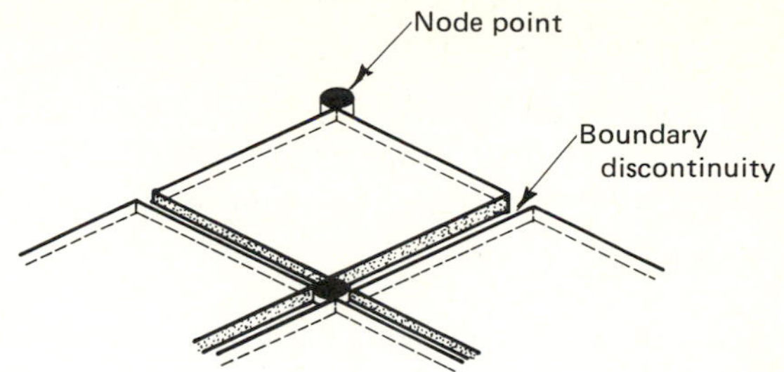

Figure 4–97. Finite-Element Boundary Displacement Behavior

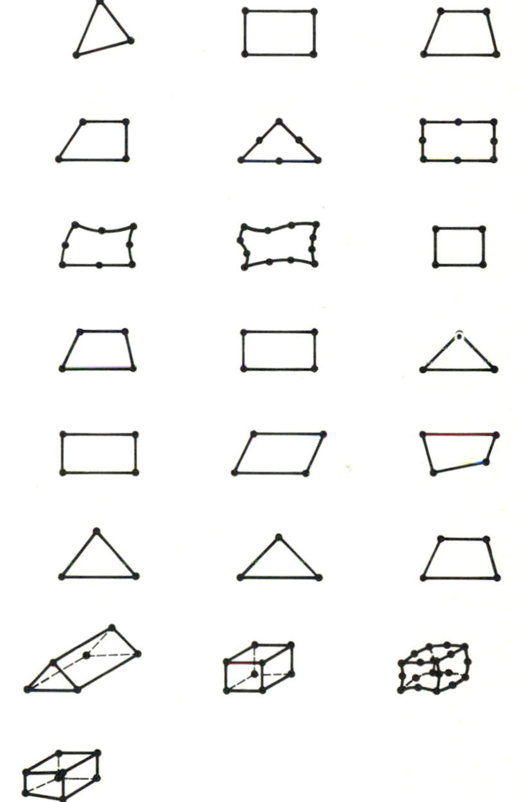

Figure 4–98. Finite-Element Types

Behavior

A finite element behaves as a miniature plate surface supported at the four corners. The joints at the corners are rigid and transfer shear, moment, tension, compression, and torsion forces. Adjacent elements are connected only at the nodes; the boundaries are not connected.

4–12 FOUNDATIONS

Foundations are elements that complete the load path, transferring forces directly into the ground. Although certain foundation elements are two-dimensional in nature (spread footings and mats), their classification as elements is consistent with the discrete size of most foundation types. The foundation element has many variations, ranging from the simple spread footing to a caisson, and distributes load to the soil either through direct bearing pressure, friction, or a combination of both. The element resembles either a beam, column, or slab in its structural action.

History and Development

The foundation element got its start when problems with deterioration of wood elements in contact with the ground prompted builders to lay grillages of stone to support column and wall loads. Such wall and column foundations had their beginnings in Egypt and later in the Greek and Roman Empires and consisted of natural rock, rafts, and timber piles. Excavation of soil and rock rarely went down more than a meter or two and, as a result, these footings were the first *shallow foundations.*

Near the beginning of the twentieth century, problems with space in the larger cities and improved construction materials made higher and heavier buildings possible. Many of these building sites, such as Chicago, Cleveland, and Detroit, were underlain by relatively thick deposits of soft or medium clay. Traditional foundations involving friction piles in the clay or shallow foundations near the surface would have settled excessively. So engineers were forced to extend the foundations to a deeper, more competent bearing material—and *deep foundations* were born. These early pier installations were constructed by excavating shafts through the soft soils to firm bearing strata below. Over the years, high-capacity drilled piers have been used with increasing frequency, especially for high-rise buildings and for industrial facilities with heavy column loads.

Piles and spread footings have also continued to develop, with concrete and steel developing as pile materials and uniform thickness concrete slabs replacing the early grillage systems. Today, there is a wide range of foundation types and materials to choose from.

Types and Uses

The major types of foundations are shown in Fig. 4–99. The *spread footing* has the widest usage and is generally the most economical system for bearing pressures ranging from 2000 to 10,000 psf (95,760 to 478,800 N/m^2). Heavier loads (which require excessive spread footing area) and settlement concerns dictate deeper foundations, such as *piles, drilled piers,* or *mats.* Mat foundations minimize differential settlement in softer soils and effectively spread the load over the entire building area. The mat may be designed as a "floating foundation," in which the total bearing pressure is equal to or less than the original soil overburden weight, thereby adding no additional load to the soil (Fig. 4–100).

Piles are slender elements that are driven into the ground with steam hammers (or drop hammers), except for the cast-in-place pile, which is installed by drilling.

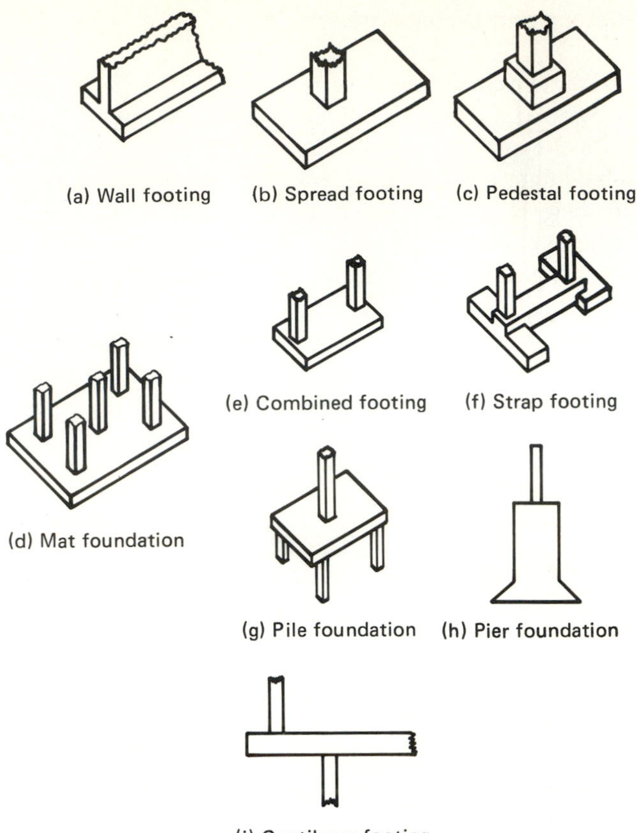

Figure 4–99. Foundation Types

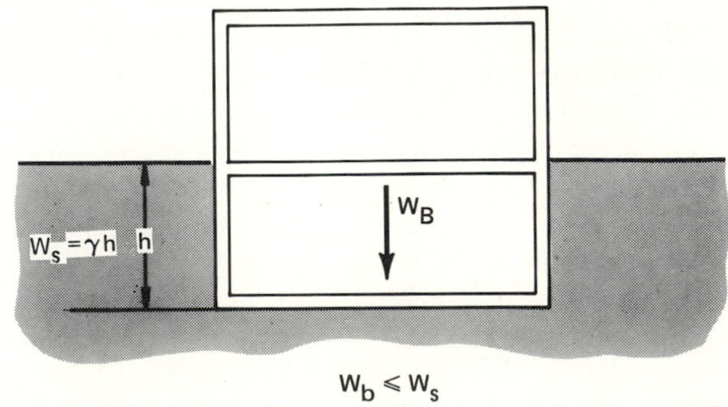

Figure 4–100. Floating Foundation

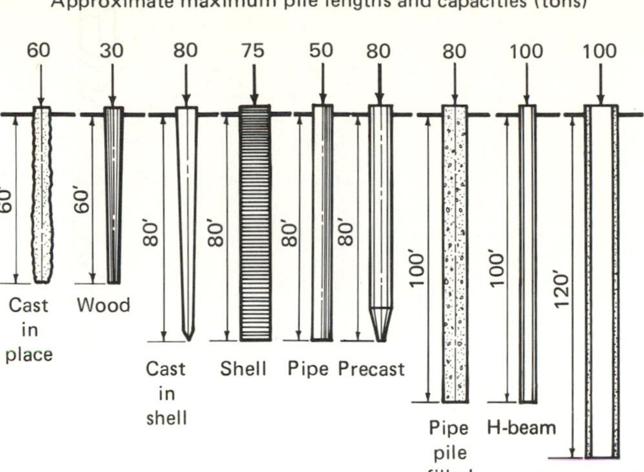

Figure 4–101. Pile Foundations

The major pile types are shown in Fig. 4–101. The usual spacing for piles is approximately 3 ft (0.91 m) center to center.

Drilled piers are larger shafts of concrete that are made by drilling large-diameter holes and filling with concrete. The drilled pier is used where piles cannot be driven due to obstructions and where heavier loads and firm bearing strata (at a reasonable depth) make piers more economical than piles. The term "caisson" is often used interchangeably with the term drilled pier, although caisson originally referred to the method of underwater construction in which pressurized-air chambers were sunk to the water bottom to provide a working environment for foundation construction. The major types of drilled piers are shown in Fig. 4–102.

Combined footings are used when either individual spread footings overlap or when columns would otherwise be located eccentrically on an individual footing (causing uneven soil pressure). *Strap footings* are used to structurally tie two or more spread footings together (with a strap beam) to minimize differential settlement and share in load transfer to the soil. *Cantilever footings* are used at property lines and at other locations where insufficient space is available to support a column symmetrically about its base.

Behavior

Foundation elements that transfer concentrated loads to the soil (by distributing them over a wider bearing area) undergo shear and bending stresses. This behavior is illustrated in Fig. 4–103, and the footing strength must be sufficient to resist the shear and bending forces. In an inverted position, the footing and column are similar to

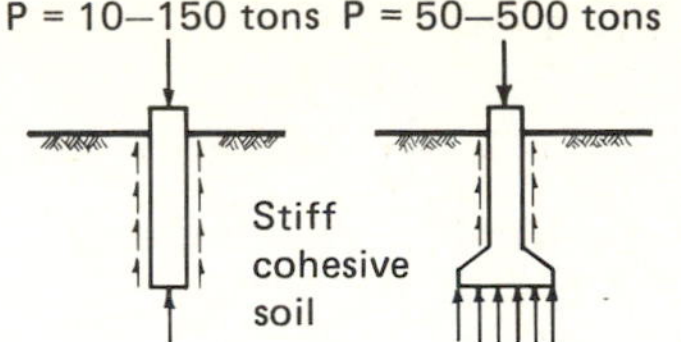

(a) Floating piers in homogeneous soil

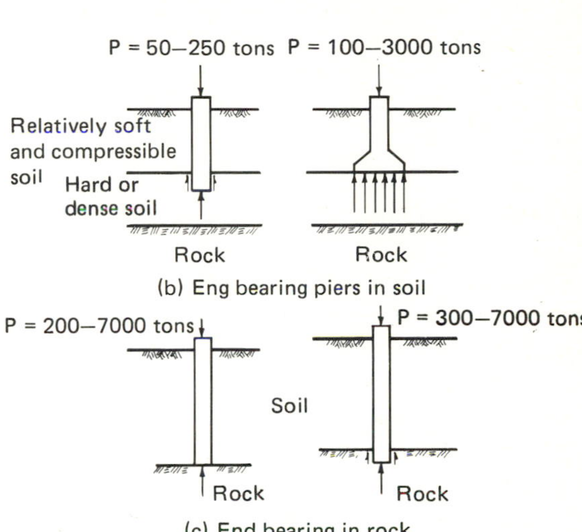

(b) Eng bearing piers in soil

(c) End bearing in rock

Figure 4–102. Drilled Pier Foundations

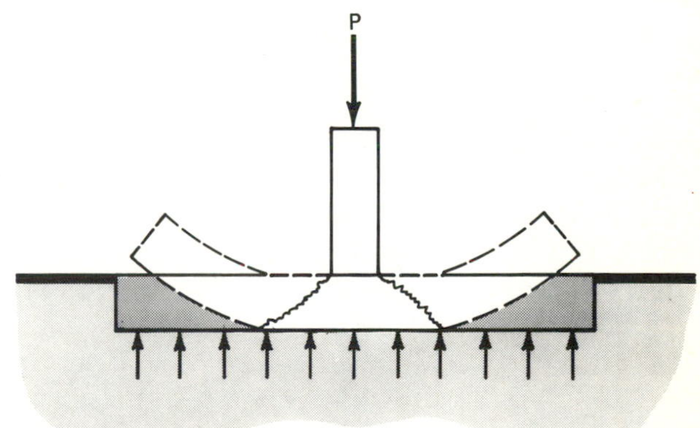

Figure 4–103. Spread Footing Behavior

a column supporting a floor slab, with floor load equal to the foundation bearing pressure. The distribution of soil pressure will vary with the relative stiffness of the soil and the footing, but is usually assumed uniform for most normal design.

Deep foundation elements transfer load to the soil via skin friction, adhesion, end bearing or a combination of these methods (Fig. 4–104). The elements act as columns under vertical load and as beam-columns under vertical and lateral loads. Additional vertical load increments may be imposed on the element by "negative skin friction" resulting from subsidence of the surrounding soil. This condition occurs when piles are driven through

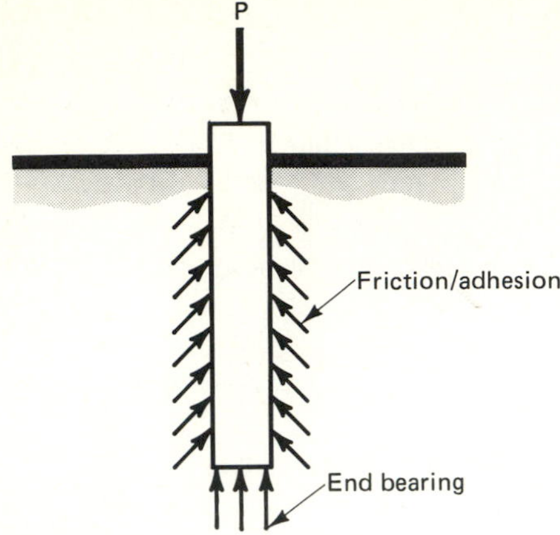

Figure 4–104. Deep Foundation Behavior

a layer of recently placed fill material, which may continue to consolidate the underlying original soil. Lateral loads can be resisted by a combination of passive soil pressure and pile bending strength, or batter piles can be provided. The usual assumption for lateral load analysis is to assume that the pile is fixed at a depth of 6 to 12 ft (1.83 to 3.66 m) below the surface, with the lower depths applicable for stiff piles and/or stiff soils and the greater depths applicable for more flexible piles and/or looser soils (Fig. 4–105).

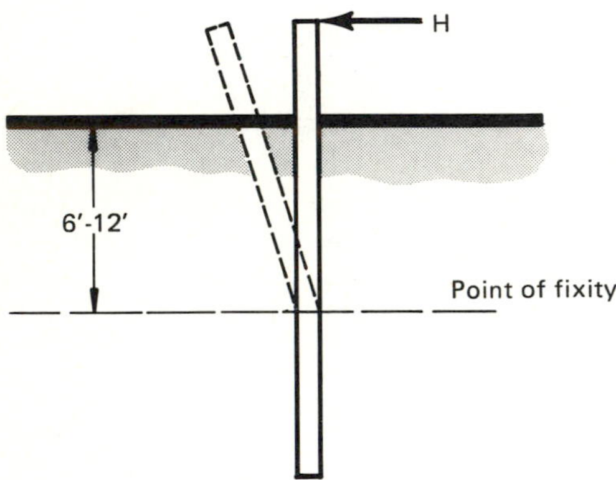

Figure 4–105. Lateral Load Analysis of Piles

Since pile elements act as columns, consideration must be given to possible stability problems due to off center loads, misalignment during pile driving, and shifting of the load center during earthquakes (Fig. 4–106). Many building codes require a minimum of three piles in an individual group to ensure stability, or other positive means of bracing single rows of piles. Earthquake regulations require pile and drilled pier foundations to be connected with ties capable of carrying tension or compression loads (in any one tie) of 10% of the larger column loading. This may be accomplished through the use of reinforced concrete slabs or tie beams.

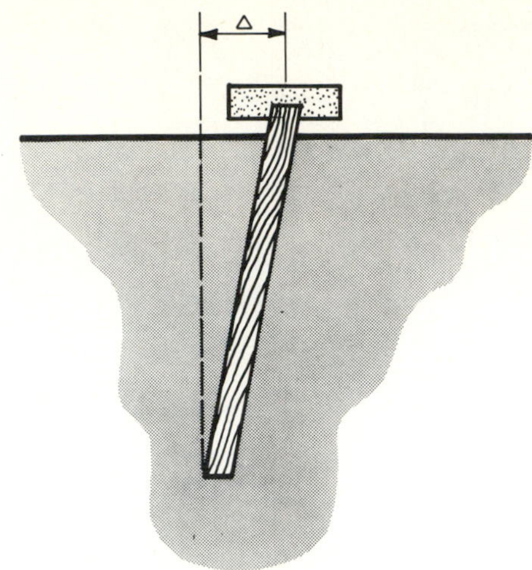

Figure 4–106. Stability of Piles

Approximate Analysis

Forces on foundation elements come directly from analysis of the column or beam-column elements being supported. The only analysis required is for lateral loads and moment forces (Fig. 4–107). Moments on spread footings are assumed to be resisted by a soil pressure diagram as shown in Fig. 4–108. The maximum pressure at the toe is given by the following expression:

$$f_s = \frac{P}{A} + \frac{Mc}{I} \tag{4–98}$$

where f_s = soil pressure
P = vertical column load
A = footing area

Figure 4–107. Moment Forces on Foundations

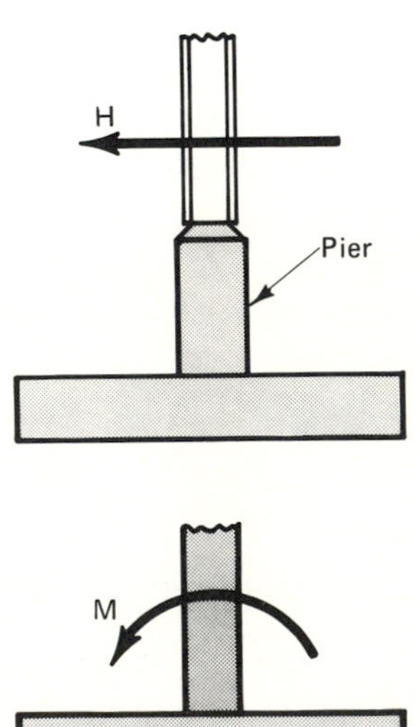

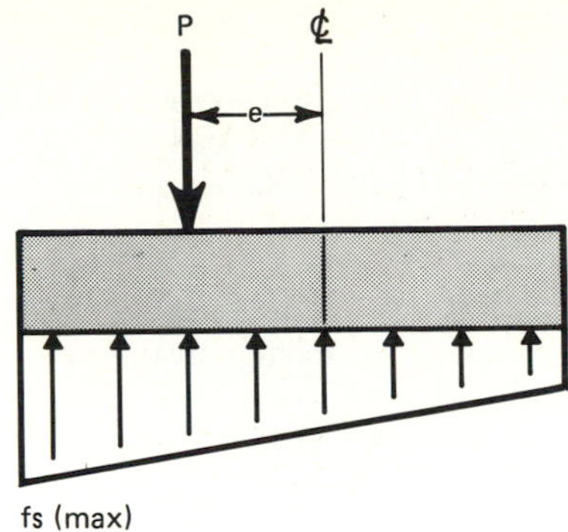

Figure 4–108. Moments on Spread Footings

M = applied moment = Pe
c = one-half footing length (in direction of bending)
I = moment of inertia in direction of bending

It is usual practice to allow the soil pressure (f_s) to exceed the allowable soil bearing pressure by 20 to 30% at the toe and to keep the resultant vertical force (Pe) within the middle one-third of the footing. Lateral loads on piles are analyzed by the assumptions discussed earlier and shown in Fig. 4–105.

Approximate Design

Spread Footings. Approximate design of foundation elements determines the element dimensions (area, length, thickness) and the number of elements required. For spread footings, the *area required* is given by the following equation:

$$A_f = \frac{P}{f_{sa}} \tag{4–99}$$

where P = column load (including any allowable live load reduction)
f_{sa} = allowable soil bearing pressure
A_f = footing area

In the absence of specific soil tests, f_{sa} can be assumed (for approximate design purposes) equal to a presumptive bearing pressure taken from Table 4–11.

Footing thickness (Fig. 4–109) is usually governed by shear considerations rather than by bending moment, with either punching shear or beam shear controlling. Table 4–12 provides coefficients for determining required

Table 4–11 Typical Presumptive Bearing Pressures[a] for Preliminary Design

Material	N (*standard penetration resistance*)	*kips/sq ft*	*kN/m²*
Loose sand, dry	5–10	1.5–3	70–140
Firm sand, dry	11–20	3–6	150–300
Dense sand, dry	31–50	8–12	400–600
Loose sand, inundated	5–10	0.8–1.6	40–80
Firm sand, inundated	11–20	1.6–3.5	80–170
Dense sand, inundated	31–50	5+	240+
Soft clay	2–4	0.6–1.2	30–60
Firm clay	5–8	1.5–2.5	70–120
Stiff clay	9–15	3–4.5	150–200
Hard clay	30+	8+	400+
Loose mica silty sand, damp	5–10	2.5–4.5	120–200
Firm mica silty sand, damp	11–20	4.5–7.5	200–350
Badly fractured or partially weathered rock	50+	10–25	500–1,200
Hard rock, occasional soft seams	RQD[b] = 50%	30–100	1,500–5,000
Massive hard rock	RQD[b] = 90%	200+	10,000+

[a] George F. Sowers, *Introductory Soil Mechanics and Foundations: Geotechnical Engineering*, 4th ed. (New York: Macmillan Publishing Co., Inc., 1979).

[b] Ratio of the length of intact rock in core sections longer than 4 in. (100 mm) to the distance drilled.

footing thickness. Reinforcing steel can be assumed equal to 0.85% for preliminary design. The following example illustrates the method.

Figure 4–109. Spread Footing Design

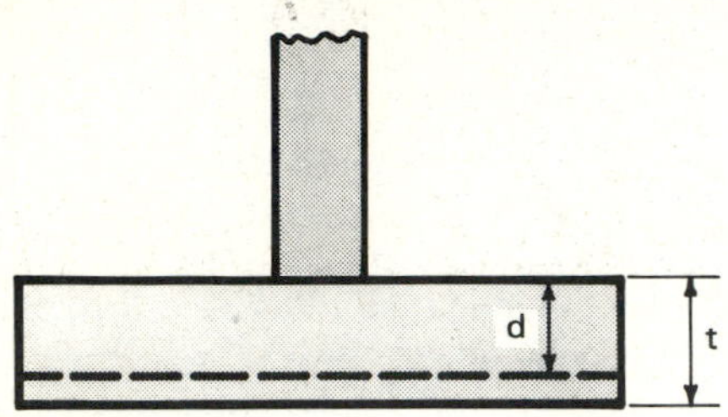

EXAMPLE PROBLEM 4–16: Preliminary Footing Design

Design a spread footing to carry a column dead load of 75 kips (333,600 N) and a live load of 50 kips (222,400 N). Use 3000 psi (20.7 MPa) concrete and allowable soil bearing of 3000 psf (143,640 N/m²).

Solution

1. Calculate required footing area, A_f, by Eq. 4–99.

$$A_f = \frac{P}{f_s} = \frac{125}{3} = 41.7 \text{ sq ft } (3.87 \text{ m}^2)$$

Use 6 ft 6 in. × 6 ft 6 in. (1.98 m × 1.98 m) footing (A_f = 42.75 ft²).

Table 4–12 Determination of Footing Depth[a]

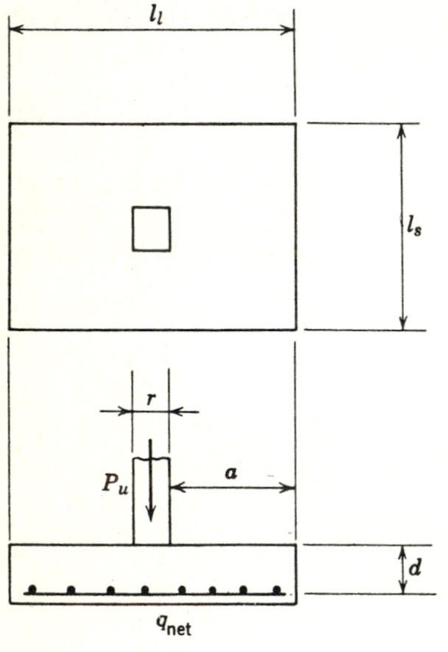

d = Depth of footing (in.)
a = Longer projection from face of col. (in.)
A_f = Area of footing = $l_l \times l_s$ (ft²)
A_c = Area of column (ft²)
r = Diameter of column with area $144A_c$ (in.) or length of side of square column (in.)
q_{net} = Net bearing pressure = P_u/A_f (psf)
f_c' = Design compression strength of standard cylinders (psi)
l_l = Length of longer side of footing (ft)
P_u = Ultimate column thrust (k)
l_s = Length of shorter side of footing (ft)

Values of A_f/A_c

$\sqrt{f_c'}$ / q_{net}	Punching Shear d/r															Beam Shear a/d
	0.4	0.5	0.6	0.7	0.8	0.9	1.0	1.1	1.2	1.3	1.4	1.5	1.6	1.7	1.8	
0.000	1	2	2	2	2	2	2	2	2	2	2	3	3	3	3	1.00
0.002	4	4	5	6	7	9	10	11	13	14	16	18	19	21	23	1.49
0.004	6	7	9	11	13	15	18	20	23	26	29	31	35	39	42	1.98
0.006	8	10	13	16	18	22	26	29	33	37	42	46	52	57	62	2.47
0.008	10	13	17	20	24	28	33	38	44	49	55	61	68	75	83	2.96
0.010	12	16	20	25	30	35	41	47	54	61	68	75	84	93	102	3.45
0.012	15	19	24	30	35	42	49	56	64	73	81	90	100	111	121	3.93
0.014	17	22	28	34	41	49	57	65	75	84	94	105	117	129	141	4.42
0.016	19	25	32	39	47	56	66	75	85	96	108	119	133	147	161	4.91
0.018	21	28	35	44	52	62	72	84	95	108	121	134	149	165	180	5.40
0.020	23	31	39	48	58	69	80	93	106	120	134	148	165	183	200	5.89
0.022	26	34	43	53	64	76	88	102	116	131	147	163	182	200	220	6.38
0.024	28	38	47	57	69	82	96	111	126	143	160	178	198	218	240	6.87
0.026	30	41	50	62	75	89	104	120	137	155	174	192	214	236	259	7.36
0.028	32	44	54	67	80	96	112	129	147	166	187	207	231	254	279	7.85
0.030	34	47	58	71	86	102	120	138	157	178	200	221	247	272	299	8.34

[a] Table by R. W. Furlong, in Phil M. Ferguson, *Reinforced Concrete Fundamentals*, 3rd ed. (New York: John Wiley & Sons, Inc., 1973), pp. 608–609.

2. Determine depth by Table 4–12.

$$q_{net} = \frac{P_u}{A_f} = \frac{1.4(75) + 1.7(50)}{42.25}$$

$$= 4.50 \text{ ksf } (215{,}460 \text{ N/m}^2)$$

$$\frac{\sqrt{f'_c}}{q_{net}} = \frac{\sqrt{3000}}{4500} = 0.012$$

$$\frac{A_f}{A_c} = \frac{42.25}{1.0} = 42.25, \; r = 12 \text{ in.}, \; a = 33 \text{ in.}$$

$$\frac{d}{r} = 0.9, \text{ required } d = 0.9 \text{ (12 in.)} = 10.8 \text{ in.}$$

$$\frac{a}{d} = 3.93, \text{ required } d = 33 \text{ in.}/3.93 = 8.4 \text{ in.}$$

Use $d = 11$ in., $t = 11 + 3 = 14$ in. (356 mm).

3. Determine reinforcing steel; assume that

$$A_s = 0.0085bd = 0.0085(6.5 \times 12)(11)$$

$$= 7.29 \text{ in.}^2 \text{ each way } (470 \text{ mm}^2)$$

Piles. Assumptions for pile capacity can be made by reference to Fig. 4–101. Estimates of pile length necessary to develop this capacity in friction require some knowledge of the soil properties. Preliminary soil investigation can often establish the soil as either primarily sand or clay and allow the following calculations:

For sand (Fig. 4–110):

$$Q_{ult} = (\tfrac{1}{2}\gamma l^2 k - \tfrac{1}{2}\gamma_\omega l_\omega^2)\pi d \tan\delta \qquad (4\text{–}100)$$

where Q_{ult} = ultimate load capacity neglecting point resistance (kips or N/9.807)
γ = saturated unit weight of soil, kips/ft³ or kg/m³)
l = length of pile, ft or m
γ_ω = unit weight of water, kips/ft² or kg/m³
l_ω = length of pile below water table, ft or m
d = pile diameter, ft or m
k = coefficient of lateral earth pressure
$\tan\delta$ = coefficient of friction between sand and pile

Figure 4–110. Pile Capacity in Sand

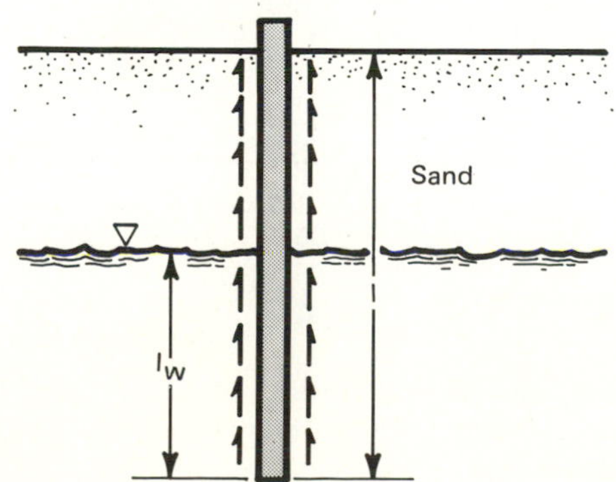

Values of k and $\tan\delta$ are difficult to evaluate. For driven or vibrated piles, k may be taken as 1.0, but should not exceed 0.4 if the piles are jetted into the sand. The coefficient of friction may be selected from Table 4–13.

Table 4–13 Coefficient of Friction, Cohesionless Soils against Piles and Similar Structures[a]

Material	*Coefficient of friction, tan δ*	δ
Wood	0.4	22°
Rough concrete, cast against soil	tan φ	φ
Smooth, formed concrete	0.3–0.4	17
Clean steel	0.2	11
Rusty steel	0.4	22
Corrugated metal	tan φ	φ

[a] George F. Sowers, *Introductory Soil Mechanics and Foundations: Geotechnical Engineering,* 4th ed. (New York: Macmillan Publishing Co., Inc., 1979).

For clay (Fig. 4–111):

$$Q_{ult} = c_a \pi d l \qquad (4\text{–}101)$$

where Q_{ult} = ultimate load capacity, kips
c_a = adhesion = $0.9c$ ($c < 1$ ksf) or $0.9 + 0.3(c - 1)$ for $c > 1$ ksf
c = undrained shear strength, ksf
d = pile diameter, ft
l = length of pile, ft

Values for friction and shear strength may be estimated based on the "Standard Penetration Resistance" (N) blow count and Fig. 4–112.

Figure 4–111. Pile Capacity in Clay

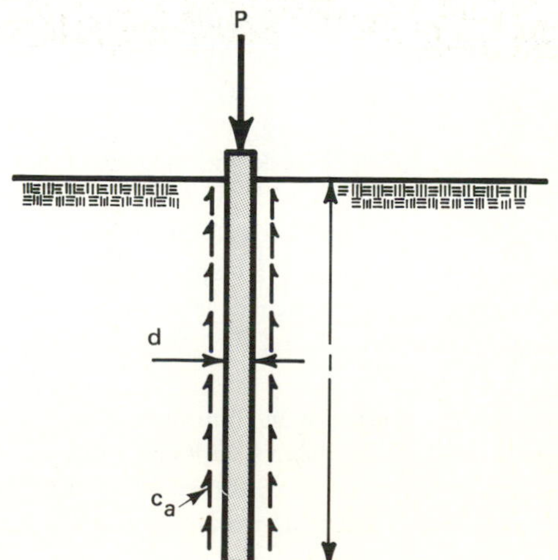

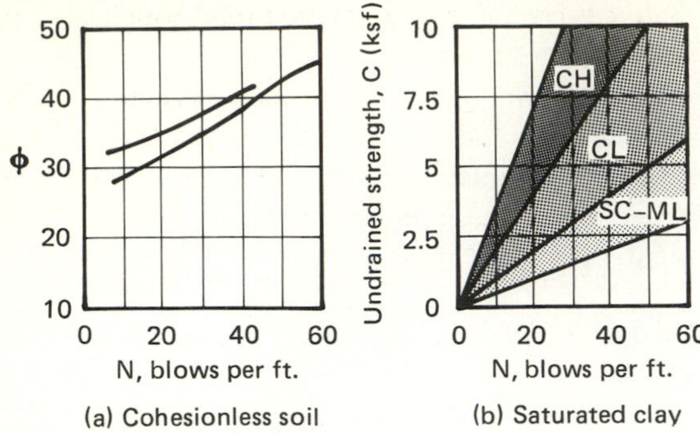

Figure 4–112. Approximate Relation of Standard Penetration Resistance to Soil Properties (From George F. Sowers, *Introductory Soil Mechanics and Foundations: Geotechnical Engineering,* 4th Ed., Macmillan Publishing Co., Inc., New York, 1979)

Equations 4–100 and 4–101 give approximate *ultimate* friction pile values. Safety factors of 2 to 3 should be applied to the working loads.

Drilled Piers. Drilled piers require a shaft area as follows:

$$A_p = \frac{P}{f_{sa}} \tag{4–102}$$

where A_p = area of pier, ft² or m²
f_{sa} = allowable end bearing, ksf or N/m²
P = column load (service load), kips or N

The depth of drilled piers depends on the soil and whether the pier will be end bearing (the usual case) or will utilize a combination of end bearing and friction. Table 4–11 can be consulted for approximate end bearing values. The skin friction procedures described earlier for piles may be used to estimate the added capacity due to friction.

4–13 SUMMARY

The *elements* described in this chapter are the basic *building blocks* of structural systems. We have concentrated on understanding the characteristic behavior of each element and on approximate analysis and design techniques, as a means of developing our ability to deal conceptually with problems.

With this background, we can now proceed to the next level of structural assembly, the *subsystem.* Chapter 5 describes the subsystem level.

PROBLEMS

4–1. Column Design. A reinforced concrete column supports three floors and a roof of an office building. Typical bays are 25 ft × 25 ft (7.62 m × 7.62 m) and the column is 12 ft (3.67 m) in height and is assumed braced at the top. The dead load of the floor system is 75 psf (3591 N/m²) and the floor live load is 50 psf (2394 N/m²) and roof live load is 20 psf (957.6 N/m²). Determine an approximate column size and reinforcing requirements.

4–2. Column Design. A steel column is 30 ft (9.14 m) tall and is braced at the top in the strong direction and in the weak direction. The column will carry 30 ft × 30 ft (9.14 m × 9.14 m) area of roof consisting of a five-ply built-up roof, 1 in. (25.4 mm) of rigid insulation, metal deck, and bar joist framing, for a total dead load of 20.0 psf (957.6 N/m²). The code-specified minimum live load for roof design is 20 psf (957.6 N/m²). Determine an approximate size for the steel column.

4–3. Cable Design. Determine an approximate cable size for the span and loads shown in Fig. 4–113. Consider the uniform load as case 1 and the concentrated load as case 2.

4–4. Beam Design. A three-span continuous girder (Fig. 4–114) supports several floor beams with end reactions as shown. Determine a preliminary reinforced-concrete beam size and reinforcement by using approximate methods of analysis and design.

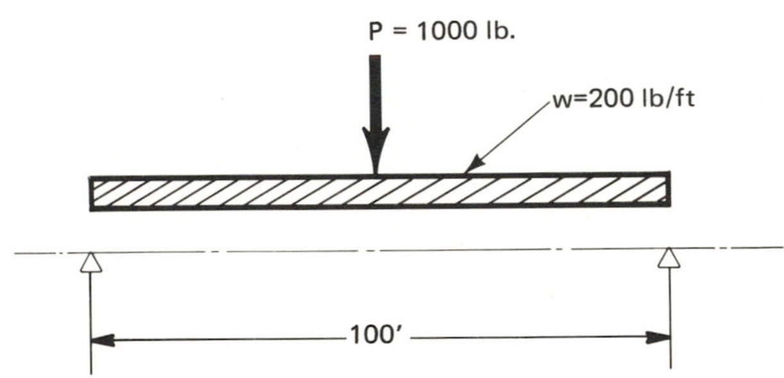

Figure 4–113. Problem 4–3

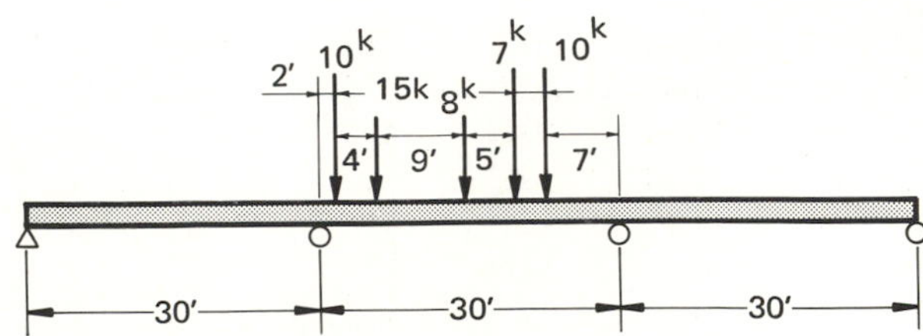

Figure 4–114. Problem 4–4

4–5. Beam-Column Design. Select a preliminary steel beam-column (A36) and concrete beam-column (3000 psi) to carry the following loads:

$$DL = 200 \text{ kips } (889{,}600 \text{ N})$$

$$LL = 100 \text{ kips } (44{,}800 \text{ N})$$

$$M_{DL} = 50 \text{ ft-kips } (67{,}800 \text{ Nm})$$

$$M_{LL} = 30 \text{ ft-kips } (40{,}680 \text{ Nm})$$

$$M_{WL} = 30 \text{ ft-kips } (40{,}680 \text{ Nm})$$

Story height is 12 ft (3.66 m) and the total lateral load (six columns per story) is 30 kips (133,400 N).

4–6. Connection Design. A steel beam-column connection must carry a moment of 60 ft-kips (81,360 Nm). The beam depth is 14 in. (356 mm). Provide a preliminary design based on a connection of your choice. Provide a sketch of your preliminary design.

4–7. Arch Design. Select an approximate member size for a three-hinged concrete arch to span 200 ft (60.9 m) under a uniform load (DL + LL) of 3000 psf (143,640 N/m²) and a concentrated live load at the crown of 600 lb (2668.8 N). [Assume a parabolic curve described by

$$Y = 4F\left(\frac{x}{l} - \frac{x^2}{l^2}\right)$$

where F is arch rise at midspan and X and Y are horizontal and vertical coordinates, respectively, with origin at the left support.]

4–8. Truss Design. Select an approximate steel truss (A36 steel) to span 100 ft (30.48 m) under a uniform roof load of 50 psf (2394 N/m²). Assume that trusses are spaced 30 ft (9.14 m) on center. Determine top and bottom chord sizes and maximum web member sizes only. Draw a sketch of your design and compare your estimated truss weight with the following approximate formula:

$$W = \sqrt{\frac{wa}{S}}\,(4L^2 + 60L) \text{ (ENR: 1919)}$$

where W = total roof truss weight
w = total load supported per horizontal square foot
S = average allowable stress used in design, psi
a = center to center spacing of trusses
L = truss span, ft

4–9. Footing Design. Provide an approximate design for a spread footing to support a column dead load of 100 kips (444,800 N) and a live load of 50 kips (222,400 N). Assume 3000-psi (20.7-MPa) concrete and 4000 psf (191,520 N/m²) of allowable soil bearing value. Column size is 12 in. × 12 in. (305 mm × 305 mm).

APPENDIX 4A: Steel Column Section Properties[a]

		Compression[b]		Bending[c]				Compression[b]		Bending[c]	
Shape	*Area (in.²)*	L_E (*ft*)	L_s (*ft*)	L_c (*ft*)	L_u (*ft*)	*Shape*	*Area (in.²)*	L_E (*ft*)	L_s (*ft*)	L_c (*ft*)	L_u (*ft*)
W6 x 15	4.43	15.2	24.1	6.3	12.0	W12 x 58	17.00	26.4	41.9	10.6	24.4
W6 x 20	5.87	15.8	25.1	6.4	16.4	W8 x 58	17.10	22.0	34.9	8.7	35.3
W8 x 24	7.08	16.9	26.8	6.9	15.2	W10 x 60	17.60	27.0	42.8	10.6	31.1
W8 x 28	8.25	17.0	27.0	6.9	17.5	W14 x 61	17.90	25.7	40.8	10.6	21.5
W8 x 31	9.13	21.2	33.6	8.4	20.1	W12 x 65	19.10	31.7	50.3	12.7	27.7
W10 x 33	9.71	20.4	32.4	8.4	16.5	W8 x 67	19.70	22.3	35.4	8.7	39.9
W8 x 35	10.30	21.3	33.8	8.5	22.6	W10 x 68	20.00	27.2	43.1	10.7	34.8
W10 x 39	11.50	20.8	33.0	8.4	19.8	W14 x 68	20.00	25.8	40.9	10.6	23.9
W8 x 40	11.70	21.4	33.9	8.5	25.3	W12 x 72	21.10	31.9	50.6	12.7	30.5
W12 x 40	11.80	20.3	32.2	8.4	16.0	W14 x 74	21.80	26.0	41.2	10.6	25.9
W14 x 43	12.60	19.8	31.4	8.4	14.4	W10 x 77	22.60	27.3	43.3	10.8	38.6
W12 x 45	13.20	20.4	32.4	8.5	17.7	W12 x 79	23.20	32.0	50.8	12.8	33.3
W10 x 45	13.30	21.1	33.5	8.5	22.8	W14 x 82	24.10	26.0	41.2	10.7	28.1
W14 x 48	14.10	20.0	31.7	8.5	16.0	W12 x 87	25.60	32.2	51.1	12.8	36.2
W8 x 48	14.10	21.8	34.6	8.6	30.3	W10 x 88	25.90	27.6	43.8	10.8	43.3
W10 x 49	14.40	26.7	42.3	10.6	26.0	W14 x 90	26.50	38.9	61.7	15.3	34.0
W12 x 50	14.70	20.6	32.7	8.5	19.6	W12 x 96	28.20	32.4	51.4	12.8	39.9
W12 x 53	15.60	26.0	41.2	10.6	22.0	W14 x 99	29.10	39.0	61.9	15.4	37.0
W14 x 53	15.60	20.2	32.0	8.5	17.7	W10 x 100	29.40	27.8	44.1	10.9	48.2
W10 x 54	15.80	26.9	42.7	10.6	28.2	W12 x 106	31.20	32.7	51.9	12.9	43.3

[a] Unbraced length values are for A36 steel.

[b] L_E = unbraced length $(kl/r) = C_c$ (126.1 for A36 steel); L_s = unbraced length $(kl/r) = 200$.

[c] L_c = unbraced length limit for $F_b = 0.66\ F_y$; L_u = unbraced length limit for $F_b = 0.60\ F_y$.

APPENDIX 4A: Steel Column Section Properties[a] (*Continued*)

Shape	Area ($in.^2$)	Compression[b] L_E (ft)	Compression[b] L_s (ft)	Bending[c] L_c (ft)	Bending[c] L_u (ft)
W14 x 109	32.00	39.2	62.2	15.4	40.6
W10 x 112	32.90	28.1	44.6	11.0	53.2
W12 x 120	35.30	32.9	52.2	13.0	48.2
W14 x 120	35.30	39.3	62.3	15.5	44.1
W14 x 132	38.80	39.5	62.6	15.5	47.7
W12 x 136	39.90	33.2	52.7	13.1	53.2
W14 x 145	42.70	41.4	65.7	16.4	52.6
W12 x 152	44.70	33.5	53.1	13.2	58.6
W14 x 159	46.70	42.0	66.6	16.4	57.2
W12 x 170	50.00	33.8	53.6	13.3	64.3
W14 x 176	51.80	42.2	66.9	16.5	62.6
W12 x 190	55.80	34.1	54.1	13.4	71.2
W14 x 193	56.80	42.5	67.4	16.6	68.1
W12 x 210	61.80	34.4	54.5	13.5	75.9

APPENDIX 4B: Steel Beam Section Properties

Shape	Section modulus, S ($in.^3$)	$F_y = 36$ ksi L_c (ft)	$F_y = 36$ ksi L_u (ft)
W8 x 10	7.8	4.2	4.7
W10 x 12[a]	10.9	3.9	4.3
W8 x 15[a]	11.8	4.2	7.2
M12 x 11.8	12.0	2.7	3.0
W10 x 15[a]	13.8	4.2	5.0
W12 x 14	14.9	3.5	4.2
W12 x 16	17.1	4.1	4.3
W10 x 19[a]	18.8	4.2	7.2
M14 x 18	21.1	3.6	4.0
W12 x 19	21.3	4.2	5.3
W10 x 22	23.2	6.1	9.4
W12 x 22	25.4	4.3	6.4
W10 x 26[a]	27.9	6.1	11.4
W14 x 22	29.0	5.3	5.6
W12 x 26	33.4	6.9	9.4
W10 x 33[a]	35.0	8.4	16.5
W14 x 26	35.3	5.3	7.0
W16 x 26	38.4	5.6	6.0
W12 x 30	38.6	6.9	10.8
W14 x 30	42.0	7.1	8.7
W10 x 39[a]	42.1	8.4	19.8
W12 x 35[a]	45.6	6.9	12.6
W16 x 31	47.2	5.8	7.1
W14 x 34	48.6	7.1	10.2
W10 x 45[a]	49.1	8.5	22.8
W12 x 40[a]	51.9	8.4	16.0
W10 x 49[a]	54.6	10.6	26.0
W14 x 38[a]	54.6	7.1	11.5
W16 x 36[a]	56.5	7.4	8.8
W18 x 35	57.6	6.3	6.7
W12 x 45[a]	58.1	8.5	17.7
W12 x 50[a]	64.7	8.5	19.6
W16 x 40	64.7	7.4	10.2
W18 x 40	68.4	6.3	8.2
W14 x 48[a]	70.3	8.5	16.0
W12 x 53[a]	70.6	10.6	22.0
W18 x 46[a]	78.8	6.4	9.4
W21 x 44	81.6	6.6	7.0
W18 x 50	88.9	7.9	11.0
W21 x 50	94.5	6.9	7.8
W18 x 55	98.3	7.9	12.1
W18 x 60[a]	108	8.0	13.3
W24 x 55	114	7.0	7.5
W18 x 65[a]	117	8.0	14.4
W21 x 62	127	8.7	11.2
W24 x 62	131	7.4	8.1
W21 x 68	140	8.7	12.4
W18 x 76[a]	146	11.6	19.1
W24 x 68	154	9.5	10.2
W21 x 83[a]	171	8.8	15.1
W24 x 76	176	9.5	11.8
W24 x 84	196	9.5	13.3
W27 x 84	213	10.5	11.0
W24 x 94	222	9.6	15.1
W27 x 94	243	10.5	12.8
W30 x 99	269	10.9	11.4
W30 x 108	299	11.1	12.3
W30 x 116	329	11.1	13.8
W33 x 118	359	12.0	12.6
W33 x 130	406	12.1	13.8
W36 x 135	439	12.3	13.0
W33 x 141	448	12.2	15.4
W36 x 150	504	12.6	14.6
W36 x 160	542	12.7	15.7
W36 x 170	580	12.7	17.0
W36 x 182	623	12.7	18.2

[a] Not the lightest section for given section modulus.

APPENDIX 4B: Steel Beam Section Properties (*Continued*)

Shape	Section modulus, S (*in.*³)	$F_y = 36$ *ksi* L_c (*ft*)	L_u (*ft*)	Shape	Section modulus, S (*in.*³)	$F_y = 36$ *ksi* L_c (*ft*)	L_u (*ft*)
W36 x 194	664	12.8	19.4	W36 x 245	895	17.4	28.6
W36 x 201	684	16.6	24.9	W36 x 260	953	17.5	30.5
W36 x 210	719	12.9	20.9	W36 x 280	1030	17.5	33.1
W36 x 221	757	16.7	27.6	W36 x 300	1110	17.6	35.3
W36 x 230	837	17.4	26.8				

APPENDIX 4C: Prestressed Beam Section Properties[a]

Size	A (*in.*²)	I (*in.*⁴)	Y_b (*in.*)	Y_t (*in.*)	S_b (*in.*³)	S_t (*in.*³)	Weight (*plf*) *N*	*LW*
		Double-Tee Sections						
8 ft–0 in. x 12 in.	287	2,872	9.13	2.87	315	1,001	299	229
		4,389	10.45	3.55	420	1,236	499	429
8 ft–0 in. x 14 in.	306	4,508	10.51	3.49	429	1,292	319	244
		6,539	11.97	4.03	546	1,623	519	444
8 ft–0 in. x 16 in.	325	6,634	11.93	4.07	556	1,630	339	260
		9,306	13.52	4.48	688	2,077	539	460
8 ft–0 in. x 18 in.	344	9,300	13.27	4.73	701	1,966	358	275
		12,749	15.00	5.00	850	2,550	558	475
8 ft–0 in. x 20 in.	363	12,551	14.59	5.41	860	2,320	378	290
		16,935	16.45	5.55	1,029	3,051	578	490
8 ft–0 in. x 24 in.	401	20,985	17.15	6.85	1,224	3,063	418	320
		27,720	19.27	6.73	1,438	4,119	618	520
8 ft–0 in. x 32 in.	567	55,464	21.21	10.79	2,615	5,140	591	453
		71,886	23.66	10.34	3,038	6,952	791	653
		Single-Tee Sections						
8 ft–0 in. x 36 in.	570	68,917	26.01	9.99	2,650	6,899	594	455
		83,212	28.28	9.72	2,942	8,561	794	655
10 ft–0 in. x 48 in.	782	168,968	36.64	11.36	4,612	14,873	815	—
		Hollow Core Sections						
4 ft–0 in. x 6 in.	187	763	3.00	3.00	254	254	195	149
		1,640	4.14	3.86	396	425	295	249
4 ft–0 in. x 8 in.	215	1,666	4.00	4.00	416	416	224	172
		3,071	5.29	4.71	580	652	323	271
4 ft–0 in. x 10 in.	259	3,223	5.00	5.00	645	645	270	207
		5,328	6.34	5.66	840	941	370	307
4 ft–0 in. x 12 in.	262	4,949	6.00	6.00	825	825	273	209
		7,811	7.55	6.45	1,035	1,211	373	309

[a] *A* = cross-sectional area
I = moment of inertia
Y_b = distance from neutral axis to extreme bottom fiber
Y_t = distance from neutral axis to extreme top fiber
S_b = section modulus at bottom of member = I/Y_b
S_t = section modulus at top of member = I/Y_t
N = normal weight
LW = lightweight

APPENDIX 4C: Prestressed Beam Section Properties[a] (*Continued*)

Size	A (*in.*^{2})	I (*in.*^{4})	Y_b (*in.*)	Y_t (*in.*)	S_b (*in.*^{3})	S_t (*in.*^{3})	Weight (*plf*) *N*	Weight (*plf*) *LW*
			L-Shaped Beams					
18 in. x 20 in.	288	9,696	9.00	—	1,077	882	300	—
18 in. x 24 in.	360	16,762	10.80	—	1,552	1,270	375	—
18 in. x 28 in.	408	26,611	12.59	—	2,114	1,727	425	—
18 in. x 32 in.	456	39,695	14.42	—	2,753	2,258	475	—
18 in. x 36 in.	504	56,407	16.29	—	3,463	2,862	525	—
18 in. x 40 in.	576	77,568	18.00	—	4,309	3,526	600	—
18 in. x 44 in.	624	103,153	19.85	—	5,197	4,271	650	—
18 in. x 48 in.	672	133,705	21.71	—	6,159	5,086	700	—
18 in. x 52 in.	720	169,613	23.60	—	7,187	5,972	750	—
18 in. x 56 in.	768	211,264	25.50	—	8,285	6,927	800	—
18 in. x 60 in.	816	259,046	27.41	—	9,451	7,949	850	—
			Inverted-Tee Beams					
24 in. x 20 in.	336	10,981	8.29	—	1,325	938	350	—
24 in. x 24 in.	432	19,008	10.00	—	1,901	1,358	450	—
24 in. x 28 in.	480	30,131	11.60	—	2,598	1,837	500	—
24 in. x 32 in.	528	44,969	13.27	—	3,388	2,401	550	—
24 in. x 36 in.	576	63,936	15.00	—	4,262	3,045	600	—
24 in. x 40 in.	672	87,845	16.57	—	5,301	3,749	700	—
24 in. x 44 in.	720	116,877	18.27	—	6,397	4,542	750	—
24 in. x 48 in.	768	151,552	20.00	—	7,578	5,413	800	—
24 in. x 52 in.	816	192,275	21.76	—	8,836	6,358	850	—
24 in. x 56 in.	864	239,445	23.56	—	10,163	7,381	900	—
24 in. x 60 in.	912	293,460	25.37	—	11,567	8,474	950	—
			AASHTO Girders					
Type II (36 in.)	369	50,979	15.83	—	3,220	2,527	384	—
Type III (45 in.)	560	125,390	20.27	—	6,186	5,070	583	—
Type IV (54 in.)	789	260,741	24.73	—	10,544	8,908	822	—

APPENDIX 4D: Timber Beam Section Properties

Nominal size, b × d (*in.*)	Dressed size, b × d (*in.*)	I (*in.*^{4})	S (*in.*^{3})	Board measure per linear foot
1 × 3	$\frac{3}{4} \times 2\frac{1}{2}$	1.0	0.8	$\frac{1}{4}$
1 × 4	$\frac{3}{4} \times 3\frac{1}{2}$	2.7	1.5	$\frac{1}{3}$
1 × 6	$\frac{3}{4} \times 5\frac{1}{2}$	10.4	3.8	$\frac{1}{2}$
1 × 8	$\frac{3}{4} \times 7\frac{1}{4}$	23.8	6.6	$\frac{2}{3}$
1 × 10	$\frac{3}{4} \times 9\frac{1}{4}$	49.5	10.7	$\frac{5}{6}$
1 × 12	$\frac{3}{4} \times 11\frac{1}{4}$	89.0	15.8	1
2 × 3	$1\frac{1}{2} \times 2\frac{1}{2}$	2.0	1.6	$\frac{1}{2}$
2 × 4	$1\frac{1}{2} \times 3\frac{1}{2}$	5.4	3.1	$\frac{2}{3}$
2 × 6	$1\frac{1}{2} \times 5\frac{1}{2}$	20.8	7.6	1
2 × 8	$1\frac{1}{2} \times 7\frac{1}{4}$	47.6	13.1	$1\frac{1}{3}$
2 × 10	$1\frac{1}{2} \times 9\frac{1}{4}$	98.9	21.4	$1\frac{2}{3}$
2 × 12	$1\frac{1}{2} \times 11\frac{1}{4}$	178.0	31.6	2
2 × 14	$1\frac{1}{2} \times 13\frac{1}{4}$	290.8	43.9	$2\frac{1}{3}$
3 × 4	$2\frac{1}{2} \times 3\frac{1}{2}$	8.9	5.1	1
3 × 6	$2\frac{1}{2} \times 5\frac{1}{2}$	34.7	12.6	$1\frac{1}{2}$
3 × 8	$2\frac{1}{2} \times 7\frac{1}{4}$	79.4	21.9	2
3 × 10	$2\frac{1}{2} \times 9\frac{1}{4}$	164.9	35.7	$2\frac{1}{2}$
3 × 12	$2\frac{1}{2} \times 11\frac{1}{4}$	296.6	52.7	3
3 × 14	$2\frac{1}{2} \times 13\frac{1}{4}$	484.6	73.2	$3\frac{1}{2}$
3 × 16	$2\frac{1}{2} \times 15\frac{1}{4}$	738.9	96.9	4
4 × 4	$3\frac{1}{2} \times 3\frac{1}{2}$	12.5	7.1	$1\frac{1}{3}$
4 × 6	$3\frac{1}{2} \times 5\frac{1}{2}$	48.5	17.6	2

APPENDIX 4D: Timber Beam Section Properties (*Continued*)

Nominal size, b × d *(in.)*	*Dressed size,* b × d *(in.)*	I *(in.⁴)*	S *(in.³)*	*Board measure per linear foot*
4 × 8	$3\frac{1}{2} \times 7\frac{1}{4}$	111.1	30.7	$2\frac{2}{3}$
4 × 10	$3\frac{1}{2} \times 9\frac{1}{4}$	230.8	49.9	$3\frac{1}{3}$
4 × 12	$3\frac{1}{2} \times 11\frac{1}{4}$	415.3	73.8	4
4 × 14	$3\frac{1}{2} \times 13\frac{1}{4}$	678.5	102.4	$4\frac{2}{3}$
4 × 16	$3\frac{1}{2} \times 15\frac{1}{4}$	1,034.4	135.7	$5\frac{1}{3}$
6 × 6	$5\frac{1}{2} \times 5\frac{1}{2}$	76.3	27.7	3
6 × 8	$5\frac{1}{2} \times 7\frac{1}{2}$	193.4	51.6	4
6 × 10	$5\frac{1}{2} \times 9\frac{1}{2}$	393.0	82.7	5
6 × 12	$5\frac{1}{2} \times 11\frac{1}{2}$	697.1	121.2	6
6 × 14	$5\frac{1}{2} \times 13\frac{1}{2}$	1,127.7	167.1	7
6 × 16	$5\frac{1}{2} \times 15\frac{1}{2}$	1,706.8	220.2	8
6 × 18	$5\frac{1}{2} \times 17\frac{1}{2}$	2,456.4	280.7	9
6 × 20	$5\frac{1}{2} \times 19\frac{1}{2}$	3,398.5	348.6	10
6 × 22	$5\frac{1}{2} \times 21\frac{1}{2}$	4,555.1	423.7	11
6 × 24	$5\frac{1}{2} \times 23\frac{1}{2}$	5,948.2	506.2	12
8 × 8	$7\frac{1}{2} \times 7\frac{1}{2}$	263.7	70.3	$5\frac{1}{3}$
8 × 10	$7\frac{1}{2} \times 9\frac{1}{2}$	535.9	112.8	$6\frac{2}{3}$
8 × 12	$7\frac{1}{2} \times 11\frac{1}{2}$	950.5	165.3	8
8 × 14	$7\frac{1}{2} \times 13\frac{1}{2}$	1,537.7	227.8	$9\frac{1}{3}$
8 × 16	$7\frac{1}{2} \times 15\frac{1}{2}$	2,327.4	300.3	$10\frac{2}{3}$
8 × 18	$7\frac{1}{2} \times 17\frac{1}{2}$	3,349.6	382.8	12
8 × 20	$7\frac{1}{2} \times 19\frac{1}{2}$	4,634.3	475.3	$13\frac{1}{3}$
8 × 22	$7\frac{1}{2} \times 21\frac{1}{2}$	6,211.5	577.8	$14\frac{2}{3}$
8 × 24	$7\frac{1}{2} \times 23\frac{1}{2}$	8,111.2	690.3	16
10 × 10	$9\frac{1}{2} \times 9\frac{1}{2}$	678.8	142.9	$8\frac{1}{3}$
10 × 12	$9\frac{1}{2} \times 11\frac{1}{2}$	1,204.0	209.4	10
10 × 14	$9\frac{1}{2} \times 13\frac{1}{2}$	1.947.8	288.6	$11\frac{2}{3}$
10 × 16	$9\frac{1}{2} \times 15\frac{1}{2}$	2,948.1	380.4	$13\frac{1}{3}$
10 × 18	$9\frac{1}{2} \times 17\frac{1}{2}$	4,242.8	484.9	15
10 × 20	$9\frac{1}{2} \times 19\frac{1}{2}$	5,870.1	602.1	$16\frac{2}{3}$
10 × 22	$9\frac{1}{2} \times 21\frac{1}{2}$	7,867.9	731.9	$18\frac{1}{3}$
10 × 24	$9\frac{1}{2} \times 23\frac{1}{2}$	10,274.1	874.4	20
12 × 12	$11\frac{1}{2} \times 11\frac{1}{2}$	1,457.5	253.5	12
12 × 14	$11\frac{1}{2} \times 13\frac{1}{2}$	2,357.9	349.3	14
12 × 16	$11\frac{1}{2} \times 15\frac{1}{2}$	3,568.7	460.5	16
12 × 18	$11\frac{1}{2} \times 17\frac{1}{2}$	5,136.1	587.0	18
12 × 20	$11\frac{1}{2} \times 19\frac{1}{2}$	7,105.9	728.8	20
12 × 22	$11\frac{1}{2} \times 21\frac{1}{2}$	9,524.3	886.0	22
12 × 24	$11\frac{1}{2} \times 23\frac{1}{2}$	10,274.1	1,058.5	24
14 × 16	$13\frac{1}{2} \times 15\frac{1}{2}$	4,189.4	540.6	$18\frac{2}{3}$
14 × 18	$13\frac{1}{2} \times 17\frac{1}{2}$	6,029.3	689.1	21
14 × 20	$13\frac{1}{2} \times 19\frac{1}{2}$	8,341.7	855.6	$23\frac{1}{3}$
14 × 22	$13\frac{1}{2} \times 21\frac{1}{2}$	11,180.7	1,040.1	$25\frac{2}{3}$
14 × 24	$13\frac{1}{2} \times 23\frac{1}{2}$	14,600.1	1,242.6	28

APPENDIX 4E: Allowable Connection Loads

Steel-Bolted Connections: Allowable Loads (kips)[a]

Material designation	*Loading case*[b]	*Nominal diameter,* d *(in.)*							
		$\frac{5}{8}$	$\frac{3}{4}$	$\frac{7}{8}$	1	$1\frac{1}{8}$	$1\frac{1}{4}$	$1\frac{3}{8}$	$1\frac{1}{2}$
A307 bolts	Tension	6.1	8.8	12.0	15.7	19.9	24.5	29.7	35.3
	Shear	3.1	4.4	6.0	7.9	9.9	12.3	14.8	17.7
A325 bolts	Tension	13.5	19.4	26.5	34.6	43.7	54.0	65.3	77.7
	Shear (F)	5.4	7.7	10.5	13.7	17.4	21.5	26.0	30.9
	Shear (N)	6.4	9.3	12.6	16.5	20.9	25.8	31.2	37.1
	Shear (X)	9.2	13.3	18.0	23.6	29.8	36.8	44.5	53.0
A490 bolts	Tension	16.6	23.9	32.5	42.4	53.7	66.3	80.2	95.4
	Shear (F)	6.7	9.7	13.2	17.3	21.9	27.0	32.7	38.9
	Shear (N)	8.6	12.4	16.8	22.0	27.8	34.4	41.6	49.5
	Shear (X)	12.3	17.7	24.1	31.4	39.8	49.1	59.4	70.7

[a] *Manual of Steel Construction,* 8th ed. (Chicago: American Institute of Steel Construction, 1980), pp. 4–3 to 4–6.

[b] F, friction-type connection; N, bearing-type connection with threads *included* in shear plane; X, bearing-type connection with threads *excluded* from shear plane.

[c] Bearing values based on fastener spacing of 3 in.

[d] F_u = 58 ksi

APPENDIX 4E: Allowable Connection Loads (*Continued*)

Material designation	*Loading case*[b]	Nominal diameter, d (in.) $\frac{5}{8}$	$\frac{3}{4}$	$\frac{7}{8}$	1	$1\frac{1}{8}$	$1\frac{1}{4}$	$1\frac{3}{8}$	$1\frac{1}{2}$
Plate thickness[d]	*Bearing*[c]								
$\frac{1}{8}$			8.2	9.3	9.1				
$\frac{3}{16}$			12.2	13.9	13.6				
$\frac{1}{4}$			16.3	18.6	18.1				
$\frac{5}{16}$			20.4	23.2	22.7				
$\frac{3}{8}$			24.5	27.9	27.2				
$\frac{7}{16}$			28.5	32.5	31.7				
$\frac{1}{2}$			32.6	37.2	36.3				
$\frac{9}{16}$			36.7	41.8	40.8				
$\frac{5}{8}$				46.4	45.3				
$\frac{11}{16}$				51.1	49.8				
$\frac{3}{4}$					54.4				
$\frac{13}{16}$					58.9				
$\frac{7}{8}$					63.4				
$\frac{15}{16}$									
1			65.3	74.3	72.5				

Steel Welded Connections: Allowable Loads (kips/in.)

Fillet weld size (in.)	*Grade 60* $F_v = 18.0$ *ksi*	*Grade 70* $F_v = 21.0$ *ksi*	*Grade 80* $F_v = 24.0$ *ksi*
$\frac{1}{8}$	1.59	1.86	2.12
$\frac{3}{16}$	2.39	2.78	3.18
$\frac{1}{4}$	3.18	3.71	4.24
$\frac{5}{16}$	3.98	4.64	5.30
$\frac{3}{8}$	4.78	5.57	6.37
$\frac{7}{16}$	5.57	6.50	7.43
$\frac{1}{2}$	6.37	7.42	8.49
$\frac{9}{16}$	7.16	8.35	9.55
$\frac{5}{8}$	7.96	9.28	10.61
$\frac{3}{4}$	9.55	11.14	12.73
1–0	12.74	14.85	16.98

Concrete Connections: Required Development Length (in.)

Bar size	f'_c	*Tension lap class*[a] *A*	*B*	*C*	*Compression lap*	*Tension embedment with standard hook*	*Compression embedment*
3	3000	12(12)	16(16)	20(21)	12	8(9)	8
	4000	12(12)	16(16)	20(21)		7(8)	8
	5000	12(12)	16(16)	20(21)		5(7)	8
4	3000	12(17)	16(22)	20(29)	15	8(11)	11
	4000	12(17)	16(22)	20(29)		7(10)	10
	5000	12(17)	16(22)	20(29)		5(9)	9
5	3000	15(21)	20(27)	26(36)	19	8(14)	14
	4000	15(21)	20(27)	26(26)		7(13)	12
	5000	15(21)	20(27)	26(26)		5(11)	11

[a] Values in parentheses are for top bar conditions.

Concrete Connections (*continued*)

Bar size	f'_c	*Tension lap class*[a] *A*	*B*	*C*	*Compression lap*	*Tension embedment with standard hook*	*Compression embedment*
6	3000	19(27)	25(35)	33(46)	23	10(20)	17
	4000	18(25)	24(33)	31(43)		8(17)	14
	5000	18(25)	24(33)	31(43)		6(16)	14
7	3000	26(38)	34(48)	45(63)	26	13(29)	19
	4000	23(32)	30(42)	39(54)		10(24)	17
	5000	21(29)	27(38)	36(50)		7(22)	16
8	3000	35(48)	45(63)	59(82)	30	17(38)	22
	4000	30(42)	39(55)	51(71)		12(31)	19
	5000	27(38)	35(49)	46(64)		9(27)	18
9	3000	44(61)	57(80)	74(104)	34	22(48)	25
	4000	38(53)	49(69)	65(90)		16(40)	22
	5000	34(48)	44(62)	58(81)		12(35)	20
10	3000	56(78)	72(101)	95(132)	38	31(60)	28
	4000	48(67)	63(88)	82(115)		23(50)	24
	5000	43(60)	56(78)	73(103)		18(43)	23
11	3000	68(96)	89(124)	116(163)	42	41(74)	31
	4000	59(83)	77(108)	101(141)		32(61)	27
	5000	53(74)	69(96)	90(126)		26(52)	25

Timber Connections[a]

The allowable bolt load acting at an angle θ with the grain of the main member is given by the following equation:

$$P\text{ allow} = \frac{PQ}{P\sin^2 + Q\cos^2\theta}$$

where P = allowable load parallel to the grain

Q = allowable load perpendicular to the grain

P and Q are given for one bolt loaded in double shear:

				Douglas fir-larch (dense), southern pine (dense)		*California redwood (close grain) Douglas fir-larch, southern pine, southern cypress*		*Douglas fir south*	
Length of bolt in main member, l (*in.*)	*Diameter of bolt,* D (*in.*)	l/D	*Projected area of bolt,* Al × D	*Parallel to grain,* P	*Perpendicular to grain,* Q	*Parallel to grain,* P	*Perpendicular to grain,* Q	*Parallel to grain,* P	*Perpendicular to grain,* Q
$1\frac{1}{2}$	$\frac{1}{2}$	3.00	0.750	1100	500	940	430	870	370
	$\frac{5}{8}$	2.40	0.938	1380	570	1180	490	1090	420
	$\frac{3}{4}$	2.00	1.125	1660	630	1420	540	1310	470
	$\frac{7}{8}$	1.71	1.313	1940	700	1660	600	1530	520
	1	1.50	1.500	2220	760	1890	650	1750	570

[a] Extract from *National Design Specification for Wood Construction, 1977 Edition* (Washington, D.C.: National Forest Products Association, 1977).

APPENDIX 4E: Allowable Connection Loads (*Continued*)

Timber Connections[a] (*Continued*)

Length of bolt in main member, l (*in.*)	*Diameter of bolt,* D (*in.*)	l/D	*Projected area of bolt,* Al × D	*Douglas fir-larch (dense), southern pine (dense)* — *Parallel to grain,* P	*Perpendicular to grain,* Q	*California redwood (close grain) Douglas fir-larch, southern pine, southern cypress* — *Parallel to grain,* P	*Perpendicular to grain,* Q	*Douglas fir south* — *Parallel to grain,* P	*Perpendicular to grain,* Q
2	$\frac{1}{2}$	4.00	1.000	1370	670	1170	570	1080	500
	$\frac{5}{8}$	3.20	1.250	1820	760	1550	650	1440	560
	$\frac{3}{4}$	2.67	1.500	2210	840	1890	720	1740	630
	$\frac{7}{8}$	2.29	1.750	2580	930	2200	790	2040	690
	1	2.00	2.000	2960	1010	2520	870	2330	750
$2\frac{1}{2}$	$\frac{1}{2}$	5.00	1.250	1480	840	1260	720	1170	620
	$\frac{5}{8}$	4.00	1.563	2140	950	1820	810	1690	710
	$\frac{3}{4}$	3.33	1.875	2710	1060	2310	900	2140	790
	$\frac{7}{8}$	2.86	2.188	3210	1160	2740	990	2530	860
	1	2.50	2.500	3680	1270	3150	1080	2910	940
3	$\frac{1}{2}$	6.00	1.500	1490	1010	1270	860	1180	750
	$\frac{5}{8}$	4.80	1.875	2290	1140	1960	970	1810	850
	$\frac{3}{4}$	4.00	2.250	3080	1270	2630	1080	2430	940
	$\frac{7}{8}$	3.43	2.625	3770	1390	3220	1190	2980	1040
	1	3.00	3.000	4390	1520	3750	1300	3460	1130
$3\frac{1}{2}$	$\frac{1}{2}$	7.00	1.750	1490	1140	1270	980	1180	870
	$\frac{5}{8}$	5.60	2.188	2320	1330	1980	1130	1830	990
	$\frac{3}{4}$	4.67	2.625	3280	1480	2800	1260	2590	1100
	$\frac{7}{8}$	4.00	3.063	4190	1630	3580	1390	3300	1210
	1	3.50	3.500	5000	1770	4270	1520	3950	1320
4	$\frac{1}{2}$	8.00	2.000	1490	1180	1270	1010	1180	960
	$\frac{5}{8}$	6.40	2.500	2330	1510	1990	1290	1840	1130
	$\frac{3}{4}$	5.33	3.000	3340	1690	2850	1440	2630	1260
	$\frac{7}{8}$	4.57	3.500	4450	1860	3800	1590	3510	1380
	1	4.00	4.000	5470	2030	4670	1730	4320	1510
$4\frac{1}{2}$	$\frac{5}{8}$	7.20	2.813	2330	1640	1990	1400	1840	1270
	$\frac{3}{4}$	6.00	3.375	3350	1900	2860	1620	2650	1410
	$\frac{7}{8}$	5.14	3.938	4530	2090	3870	1790	3580	1560
	1	4.50	4.500	5770	2280	4930	1950	4560	1700
	$1\frac{1}{4}$	3.60	5.625	7980	2670	6820	2280	6300	1990
$5\frac{1}{2}$	$\frac{5}{8}$	8.80	3.438	2330	1650	1990	1410	1840	1380
	$\frac{3}{4}$	7.33	4.125	3350	2200	2860	1880	2640	1720
	$\frac{7}{8}$	6.29	4.813	4570	2550	3900	2180	3600	1900
	1	5.50	5.500	5930	2790	5070	2380	4680	2080
	$1\frac{1}{4}$	4.40	6.875	8940	3260	7640	2790	7060	2430

APPENDIX 4F: Open Web Steel Joist Properties

Section	Wt (lb/ft)	Resisting moment (in.-kips)	Maximum end reaction (lb)
8 H 3	5.0	91	2,400
10 H 3	5.0	116	2,500
10 H 4	6.1	148	2,800
12 H 3	5.2	140	2,800
12 H 4	6.2	180	3,200
12 H 5	7.1	222	3,600
12 H 6	8.2	260	3,900
14 H 3	5.5	165	3,200
14 H 4	6.5	212	3,500
14 H 5	7.4	259	3,800
14 H 6	8.6	307	4,200
14 H 7	10.0	369	4,600
16 H 4	6.6	221	3,800
16 H 5	7.8	289	4,300
16 H 6	8.6	344	4,600
16 H 7	10.3	413	4,900
16 H 8	11.4	478	5,200
18 H 5	8.0	325	4,500
18 H 6	9.2	383	4,800
18 H 7	10.4	466	5,200
18 H 8	11.6	540	5,400
20 H 5	8.4	365	4,800
20 H 6	9.6	406	5,100
20 H 7	10.7	499	5,400
20 H 8	12.2	602	5,600
22 H 6	9.7	422	5,400
22 H 7	10.7	526	5,600
22 H 8	12.0	653	5,800
22 H 9	13.8	776	6,700
22 H 10	15.2	873	7,200
22 H 11	16.9	1,009	8,100
24 H 6	10.3	462	5,600
24 H 7	11.5	576	5,800
24 H 8	12.7	716	6,000
24 H 9	14.0	851	7,000
24 H 10	15.5	957	7,500
24 H 11	17.5	1,106	8,200
26 H 8	12.8	784	6,700
26 H 9	14.8	925	7,200
26 H 10	16.2	1,040	7,600
26 H 11	17.9	1,203	8,300
28 H 8	13.5	846	6,700
28 H 9	15.2	1,000	7,200
28 H 10	16.8	1,124	7,700
28 H 11	18.3	1,300	8,400

Section	Wt (lb/ft)	Resisting moment (in.-kips)	Maximum end reaction (lb)
30 H 8	14.2	909	6,800
30 H 9	15.4	1,075	7,500
30 H 10	17.3	1,207	8,100
30 H 11	18.8	1,397	8,700
18 LH 02	13	476	
18 LH 03	14	519	
18 LH 04	16	599	
18 LH 05	17	690	
18 LH 06	19	770	
18 LH 07	21	863	
18 LH 08	22	995	
18 LH 09	24	1,100	
20 LH 02	13	516	
20 LH 03	14	583	
20 LH 04	16	660	
20 LH 05	17	770	
20 LH 06	19	866	
20 LH 07	21	969	
20 LH 08	22	1,097	
20 LH 09	24	1,241	
20 LH 10	27	1,380	
24 LH 03	14	641	
24 LH 04	16	715	
24 LH 05	17	833	
24 LH 06	19	1,049	
24 LH 07	21	1,178	
24 LH 08	22	1,248	
24 LH 09	24	1,517	
24 LH 10	27	1,685	
24 LH 11	29	1,843	
28 LH 05	16	850	
28 LH 06	19	1,130	
28 LH 07	21	1,273	
28 LH 08	21	1,362	
28 LH 09	24	1,682	
28 LH 10	27	1,838	
28 LH 11	29	1,967	
28 LH 12	33	2,161	
28 LH 13	36	2,257	
32 LH 06	18	1,223	
32 LH 07	20	1,373	
32 LH 08	21	1,489	
32 LH 09	24	1,868	
32 LH 10	26	2,063	
32 LH 11	28	2,258	
32 LH 12	33	2,670	
32 LH 13	36	3,004	
32 LH 14	37	3,098	

APPENDIX 4F: Open Web Steel Joist Properties (*Continued*)

Section	Wt (lb/ft)	Resisting moment (in.-kips)	Maximum end reaction (lb)	Section	Wt (lb/ft)	Resisting moment (in.-kips)	Maximum end reaction (lb)
32 LH 15	41	3,199		52 DLH 13	36	5,264	
36 LH 07	20	1,436		52 DLH 14	40	6,024	
36 LH 08	20	1,582		52 DLH 15	45	6,761	
36 LH 09	23	2,020		52 DLH 16	50	7,295	
36 LH 10	26	2,230		52 DLH 17	55	8,388	
36 LH 11	28	2,435		56 DLH 11	29	4,065	
36 LH 12	32	2,916		56 DLH 12	31	4,672	
36 LH 13	36	3,424		56 DLH 13	36	5,659	
36 LH 14	37	3,812		56 DLH 14	40	6,393	
36 LH 15	41	4,153		56 DLH 15	45	7,311	
40 LH 08	20	1,670		56 DLH 16	50	7,889	
40 LH 09	23	2,189		56 DLH 17	55	9,075	
40 LH 10	25	2,390		60 DLH 12	31	4,925	
40 LH 11	27	2,621		60 DLH 13	36	5,983	
40 LH 12	32	3,168		60 DLH 14	39	6,588	
40 LH 13	36	3,744		60 DLH 15	45	7,798	
40 LH 14	37	4,262		60 DLH 16	50	8,489	
40 LH 15	41	4,666		60 DLH 17	55	9,785	
40 LH 16	47	5,530		60 DLH 18	62	11,275	
44 LH 09	22	2,218		64 DLH 12	31	5,112	
44 LH 10	25	2,438		64 DLH 13	36	6,218	
44 LH 11	27	2,650		64 DLH 14	39	7,053	
44 LH 12	31	3,254		64 DLH 15	45	8,135	
44 LH 13	35	3,878		64 DLH 16	50	9,093	
44 LH 14	36	4,387		64 DLH 17	55	10,469	
44 LH 15	41	5,155		64 DLH 18	62	12,067	
44 LH 16	47	5,991		68 DLH 13	36	6,409	
44 LH 17	54	6,710		68 DLH 14	39	7,380	
48 LH 10	25	2,502		68 DLH 15	43	8,157	
48 LH 11	27	2,710		68 DLH 16	50	9,683	
48 LH 12	31	3,401		68 DLH 17	55	11,014	
48 LH 13	35	4,064		68 DLH 18	62	12,734	
48 LH 14	36	4,783		68 DLH 19	70	14,566	
48 LH 15	41	5,488		72 DLH 14	39	7,620	
48 LH 16	47	6,345		72 DLH 15	43	8,678	
48 LH 17	54	7,119		72 DLH 16	50	10,109	
52 DLH 10	27	3,541		72 DLH 17	55	11,384	
52 DLH 11	29	3,885		72 DLH 18	62	13,250	
52 DLH 12	31	4,337		72 DLH 19	70	15,459	

REFERENCES

1. ACI Committee 318, *Building Code Requirements for Reinforced Concrete,* ACI–318–77. Detroit, Mich.: American Concrete Institute, 1977.
2. BEEDLE, LYNN S., et al., *Structural Steel Design.* New York: The Ronald Press Company, 1964.
3. Bethlehem Steel, *Cable Roof Structures.* Bethlehem, Pa., 1968.
4. CARPENTER, SAMUEL T., *Structural Mechanics.* New York: John Wiley & Sons, Inc., 1960.
5. FERGUSON, PHIL M., *Reinforced Concrete Fundamentals.* New York: John Wiley & Sons, Inc., 1973.
6. FINTEL, MARK, *Handbook of Concrete Engineering.* New York: Van Nostrand Reinhold Company, 1974.
7. FISHER, JOHN W., and JOHN H. A. STRUIK, *Guide to Design Criteria for Bolted and Riveted Joints.* New York: John Wiley & Sons, Inc., 1974.
8. FORD, J. S., D. C. CHAY, and J. E. BREEN, "Behavior of 'Unbraced Multipanel Concrete' Frames," *Journal of the American Concrete Institute,* 1978 (March–April, 1981, No. 2).
9. FURLONG, RICHARD W., and CARLOS REZENDE, "Alternate to ACI Analysis Coefficients," *Journal of the Structural Division, ASCE,* Vol. 105, No. ST-11 (November 1979).
10. GAYLORD, EDWIN H., JR., and CHARLES N. GAYLORD, *Design of Steel Structures.* New York: McGraw-Hill Book Company, 1972.
11. GAYLORD, EDWARD H., JR., and CHARLES N. GAYLORD, *Structural Engineering Handbook.* New York: McGraw-Hill Book Company, 1968.
12. HORNBOSTEL, CALEB, *Construction Materials.* New York: John Wiley & Sons, Inc., 1978.
13. JOHNSTON, BRUCE G., ed., *Guide to Stability Design Criteria for Metal Structures.* New York: John Wiley & Sons, Inc. 1976.
14. LOTHERS, JOHN E., *Advanced Design in Structural Steel.* Englewood Cliffs, N.J.: Prentice-Hall, Inc., 1960.
15. MAINSTONE, ROLAND S., *Developments in Structural Form.* Cambridge, Mass.: The MIT Press, 1975.
16. *Manual of Steel Construction,* 8th ed. Chicago: American Institute of Steel Construction, 1980.
17. MERRITT, FREDERICK S., *Building Construction Handbook.* New York: McGraw-Hill Book Company, 1975.
18. MERRITT, FREDERICK S., *Standard Handbook for Civil Engineers.* New York: McGraw-Hill Book Company, 1976.
19. MERRITT, FREDERICK S., *Structural Steel Designers Handbook.* New York: McGraw-Hill Book Company, 1972.
20. NILSON, ARTHUR H., *Design of Prestressed Concrete.* New York: John Wiley & Sons, Inc., 1978.
21. NORRIS, CHARLES H., and JOHN BENSON WILBUR, *Elementary Structural Analysis.* New York: McGraw-Hill Book Company, 1960.
22. OTTO, FREI, ed., *Tensile Structures.* Cambridge, Mass.: The MIT Press, 1969.
23. *PCI Design Handbook,* 2nd ed., Chicago: Prestressed Concrete Institute, 1978.
24. RAMSEY, C. G., and H. R. SLEEPER, *Architectural Graphic Standards.* New York: The American Institute of Architects/John Wiley & Sons, Inc. 1970.
25. ROARK, RAYMOND J., *Formulas for Stress and Strain.* New York: McGraw-Hill Book Company, 1954.
26. SALMON, CHARLES G., and JOHN E. JOHNSON, *Steel Structures: Design and Behavior.* Scranton, Pa.: Intext Educational Publishers, 1977.
27. SALVADORI, MARIO, and MATTHYS LEVY, *Structural Design in Architecture.* Englewood Cliffs, N.J.: Prentice-Hall, Inc., 1967.
28. SCHUELLER, WOLFGANG, *High-Rise Building Structures.* New York: John Wiley & Sons, Inc., 1977.
29. United States Steel, *Suspended Structures Concepts.* Pittsburgh, Pa., October 1966.
30. WHITE, RICHARD N., PETER GERGELY, and ROBERT G. SEXSMITH, *Structural Engineering,* Vol. 3: *Behavior of Members and Systems.* New York: John Wiley & Sons, Inc., 1974.
31. WILSON, FORREST, *Emerging Form in Architecture: Conversations with Lev Zetlin.* Boston: Cahners Books, 1975.
32. WINTER, GEORGE, and ARTHUR H. WILSON, *Design of Concrete Structures.* New York: McGraw-Hill Book Company, 1972.
33. WINTERKORN, HANS F., and HSAI-YANGFANG, *Foundation Engineering Handbook.* New York: Van Nostrand Reinhold Company, 1975.
34. ZIENKIEWICZ, O. C., and Y. K. CHEUNG, *The Finite Element Method in Structural and Continuum Mechanics.* New York: McGraw-Hill Book Company, 1967.

5

SUBSYSTEMS

5-1 INTRODUCTION

The structural elements of Chapter 4 serve no function by themselves but must be grouped or interconnected with other elements into "functional subsystems" in order to contribute to structural system stability. With the *subsystem* level of assembly, we build on our knowledge of element behavior and add another part of the systems design model (Fig. 5–1). Structural subsystems perform one (or more) of the following functional roles (Fig. 5–2):

- Vertical support
- Horizontal distribution
- Lateral distribution
- Lateral support
- Foundation support

The *horizontal distribution* subsystem operates in either parallel (to the ground) or oblique planes and distributes gravity loads horizontally to vertical support subsystems. The *vertical support* subsystem operates in a vertical plane and carries gravity loads from the various horizontal distribution levels to the ground. The *lateral distribution* subsystem operates in a horizontal plane, re-

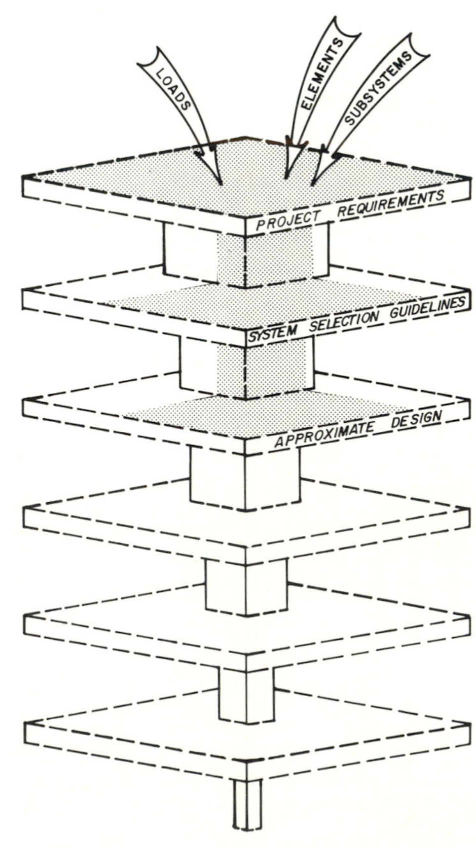

Figure 5–1. Structural Systems Design Model

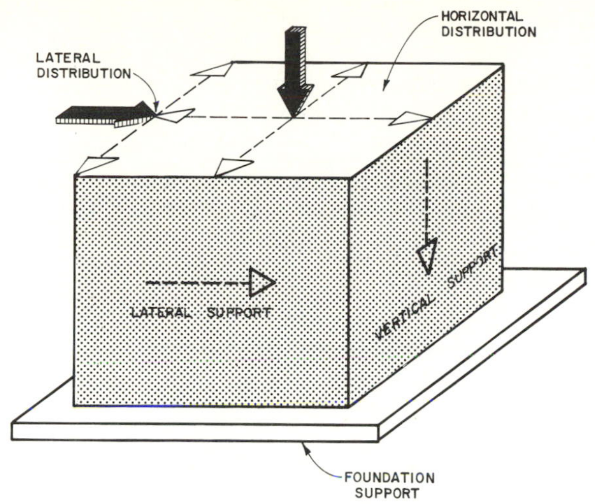

Figure 5–2. Functional Subsystems

ceiving lateral loads and distributing them to the lateral support subsystem. The *lateral support* subsystem receives lateral loads and carries them to the ground. The *foundation support* subsystem is defined by both vertical and horizontal planes and functions as the final linkage of all loads to the ground.

Subsystems, then, by definition are two-dimensional (planar) assemblies of elements, grouped together to serve a specific functional purpose. The elements may be physically connected (as in a frame) or may be simply grouped (as in spread footings, or bar joists spanning directly between vertical support subsystems).

Subsystem Design

Design at the element level involves choices of material and material arrangement. Design at the subsystem level requires choices to be made regarding the number and type of each element to be used, spacing between elements and connection concepts. The fact that choices are available for subsystem design implies that comparisons have to be made in order to select the optimum one. This sets up an iterative process of trial selection, approximate analysis, approximate element design, and comparison with other trial selections, an iterative process at the subsystem level which previews the expanded version which operates at the system level (Fig. 1–1). Project requirements (fire rating, durability, availability, etc.) usually narrow the range of choices initially and a *data base* of subsystem characteristics and selection guidelines will further narrow the choice to a manageable number for comparison.

The design choices at the subsystem level often affect the economy of the other subsystems in the structure. However, we will ignore this relationship at present, concentrating instead on optimizing the particular subsystem, with optimization of the total structure reserved for the Chapter 6 discussion of systems.

This chapter describes the major classes of subsystems and the various types within each classification. Each description includes a discussion of history and development, behavior, approximate analysis, and approximate design. The emphasis on behavior continues, expanded to cover two-dimensional space. Appendix 5A contains additional reference information on subsystem properties.

5–2 POST-AND-BEAM FRAMES

Post-and-beam frames are vertical support subsystems constructed of beam and column elements linked together with simple hinged or bearing connections (Fig. 5–3). The name given to this subsystem originated from the early construction process that simply stacked beams on top of column elements to create a vertical assembly. This subsystem picks up loads from intersecting horizontal distribution subsystems, and distributes them to its column elements, which in turn transmit the loads to the foundations. The post-and-beam frame is unstable without lateral support and must depend on some type of lateral support subsystem for stability. The girder elements of post-and-beam frames may act either as simple beams or may act as continuous beams if continued across several column elements.

Figure 5–3. Post-and-Beam Frame

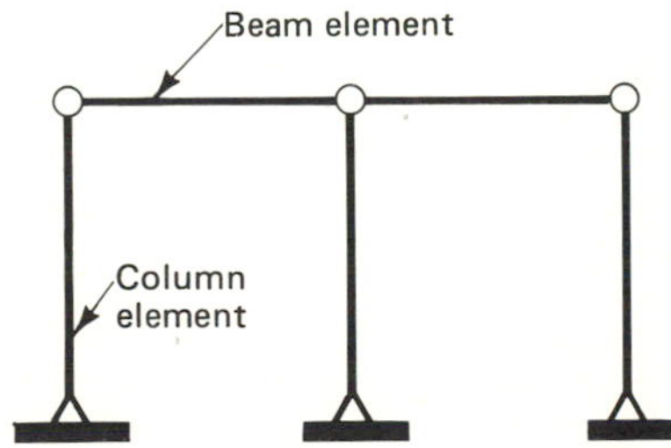

History and Development

The post-and-beam frame is one of the oldest subsystem assemblages, with early builders intuitively erecting stone and timber columns and beams to provide shelter. Stability was not generally a problem for these structures, for the timber columns relied on embedment in the ground for stability (actually serving as a beam-column), and the weight of the massive stone columns provided inherent stability. Other structures were of such modest proportions that lateral loads presented no significant problem; the structures were simply rebuilt after destruction by storms.

The subsystem continued to develop (along with the wall) as the primary form of vertical support. Adjoining massive stone and masonry walls generally provided adequate stability, and the simplicity of this form of construction made it very popular. Other forms of vertical support subsystems have emerged in the last century (rigid frame, flat slab/column, etc.), but the post-and-beam frame continues to be the most common.

Types and Uses

The major types of post-and-beam frames are shown in Fig. 5–4. The beam element is usually called a girder when it "doubles" as part of the horizontal distribution subsystem. It may also be a truss element, a composite beam, or even a cable element. The simple bearing connection is commonly used for residential and light commercial construction, and the hinged connection is typically used for multistory structures or for single-story structures where no girder continuity is desired. The rigid frame connected with semirigid connections is actually a post-and-beam frame under gravity load. Any structural material may be used to construct this subsystem; the most common are structural steel, reinforced concrete, prestressed concrete, and timber.

Figure 5–4. Types of Post-and-Beam Frames

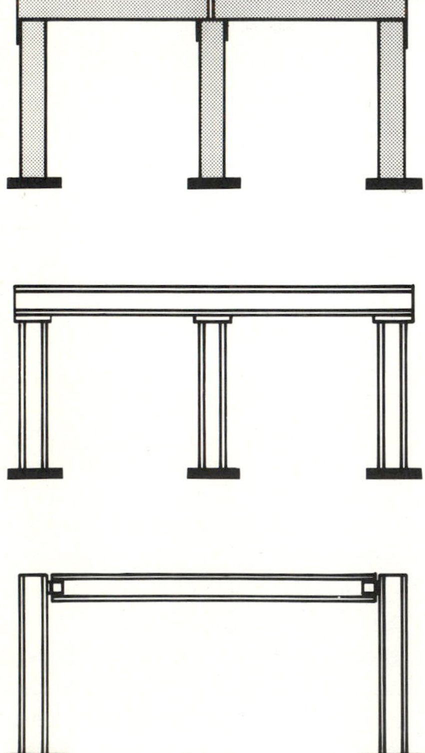

Behavior

The post-and-beam frame is unstable under both gravity and lateral loads and must depend on another subsystem for lateral support (Fig. 5–5). With lateral support provided, gravity loads are distributed to the columns by the girder acting as a simple beam (Fig. 5–6). The columns in turn act as elements pinned at both ends ($k = 1.0$) and transmit the loads in compression to the foundations.

Figure 5–5. Stability of Post-and-Beam Frames

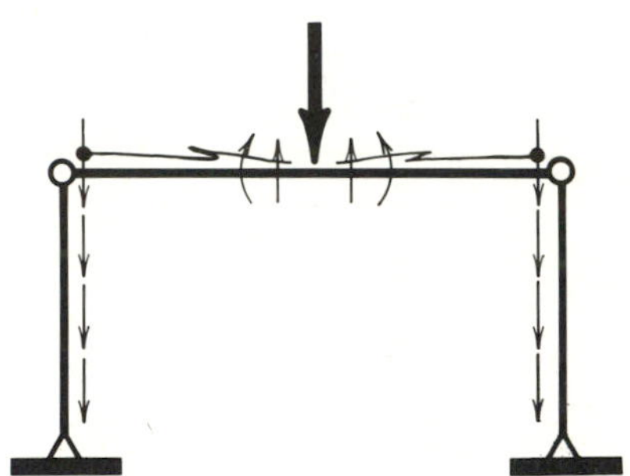

Figure 5–6. Load Path Behavior of Post-and-Beam Frames

The load path behavior of the post-and-beam frame is easily visualized and involves only bending behavior of the simple beam element and compressive behavior of the hinged-end column element. The beam-to-column connection must allow simple beam rotation of the girder end without the transfer of appreciable bending moment. This moment–rotation behavior can be verified with the procedures developed in Section 4–7. The column-footing connection is usually considered hinged, since only a slight amount of footing rotation is necessary to achieve hinged conditions.

Approximate Analysis

Loads on the post-and-beam frame come from intersecting horizontal distribution subsystems. These loads may come in the form of end reactions from floor beams, uniform loads from slab systems, or in any combination of uniform and concentrated loads (Fig. 5–7). We can use a convenient approximate method of analysis by replacing the exact loads with an "equivalent uniform load" based on the tributary width of horizontal distribution area sup-

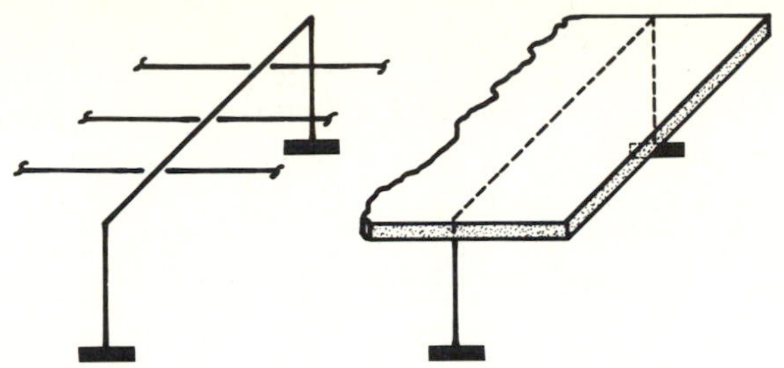

Figure 5-7. Loads on Post-and-Beam Frames

ported, using a markup factor (to account for the effect of concentrated loads on the moment) as described in Section 4-5. The following problem illustrates this concept.

EXAMPLE PROBLEM 5-1: Approximate Analysis of Post-and-Beam Frame

Determine the maximum moment and shear in the girder and the column axial load for the post-and-beam frame shown in Fig. 5-8.

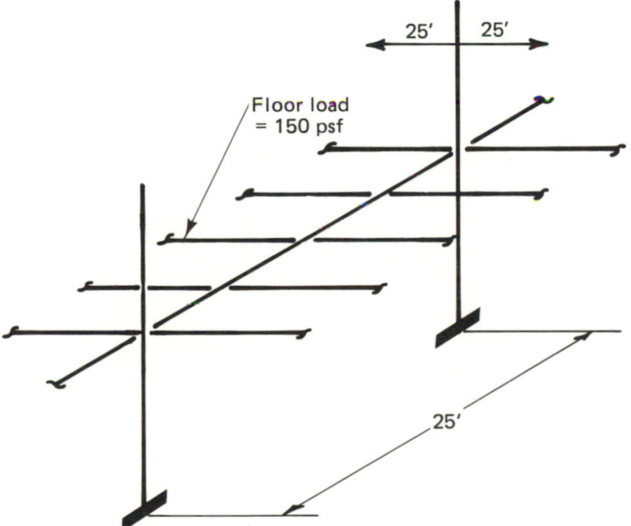

Figure 5-8. Example Problem 5-1: Post-and-Beam Frame

Solution

1. Calculate equivalent uniform load.

$$\text{Equivalent girder load} = 25 \text{ ft } (150 \text{ psf}) = 3750 \text{ lb/ft } (54{,}724 \text{ N/m})$$

2. Calculate girder forces. Apply 1.33 markup factor for concentrated load effect.

Maximum girder moment:

$$\frac{wl^2}{8} = 1.33\frac{(3.75)(25)^2}{8} = 389 \text{ ft-kips } (527{,}484 \text{ Nm})$$

Maximum girder shear:

$$\frac{wl}{2} = \frac{3.75(25)}{2} = 4.68 \text{ kips } (20{,}816 \text{ N})$$

3. Calculate column load.

$$\begin{aligned}\text{Maximum column load} &= wlb \\ &= 0.15(25)(25) \\ &= 93.8 \text{ kips } (417{,}222 \text{ N})\end{aligned}$$

or

$$2 \times \text{girder shear} = 2(468) = 93.6 \text{ kips}$$

Approximate Design

Approximate design of post-and-beam frames requires the determination of the following:

- Material selection
- Frame height
- Column spacing
- Element proportions

Frame height will generally be dictated by spatial requirements. Column spacing usually has a practical minimum distance based on functional considerations (equipment layout, storage patterns, etc.) and a practical maximum distance based on girder depth and cost-effective use of the girder element. The choice of the element type and material will be influenced by the project requirements and the load/span relationships.

Appendix 6A contains a data base of selection guidelines. Equipped with this resource, the project requirements, and our own imagination, we can select a preliminary scheme as illustrated in Example Problem 5-2.

EXAMPLE PROBLEM 5-2: Approximate Design of Post and Beam Frame

Select a preliminary post-and-beam frame for the north-south direction of the single-story warehouse shown in Fig. 5-9. Space requirements dictate a clear height of 20 ft (6.09 m) from the floor slab to the bottom of the girder element and a minimum spacing between columns of 20 ft (6.09 m). Roof loads are 20 psf DL (957.6 N/m^2), including allowance for the weight of the structure, and 20 psf LL (957.6 N/m^2). Preliminary soil information indicates that spread footings can be used with an allowable soil bearing pressure of 3000 psf (143,640 N/m^2). Assume the only fire requirements are that noncombustible material must be used.

Solution

1. *Material selection.* Steel generally has the best cost/strength ratio in flexure. The girder element will experience bending forces and more of the required material will be in the girder than in the columns. Assuming

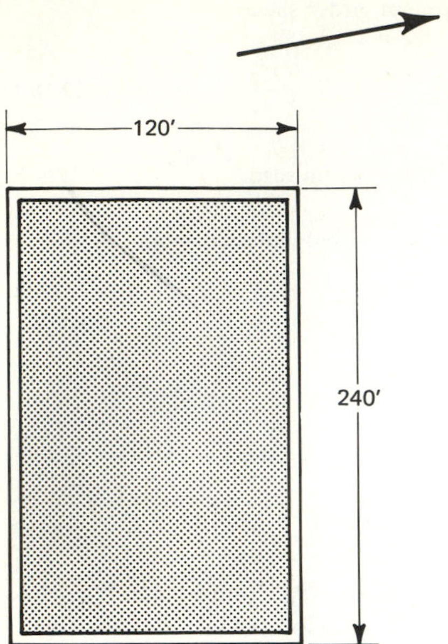

Figure 5–9. Example Problem 5–2: Plan View of Warehouse

no problems with availability or fire protection, select structural steel as the preliminary material.

2. *Column spacing/girder element selection.* Practical column spacings are 20 ft (6.09 m), 30 ft (9.14 m), 40 ft (12.19 m), 60 ft (18.29 m), and 120 ft (36.58 m). Reference to Appendix 6A eliminates 120 ft for consideration for economic reasons. The remaining spans are below the usual competitive range for trusses, but consider (for illustration purposes) a truss girder for the 60-ft (18.29-m) alternative.

3. *Design alternate 1.* 20-ft (6.09-m) bay, steel beams, and columns.

   a. Beams: Assume continuous-beam or "double-cantilever" construction and maximum moment of $wl^2/10$.

   $$\text{Min. } d = \tfrac{1}{2} \text{ span} = \tfrac{1}{2}(20) = 10 \text{ in. (254 mm)}$$

   $$w = 40 \text{ psf (30 ft)} = 1200 \text{ lb/ft (17,511 N/m)}$$

   $$M = \frac{wl^2}{10} = \frac{1.2(20)^2}{10} = 48 \text{ ft-kips (65,088 N/m)}$$

   Since lateral bracing is provided by another subsystem which is not a part of this example, assume that $F_b = 20.0$ ksi (137.9 MPa) as average allowable stress.

   $$S_{\text{req'd}} = \frac{M}{F_b} = \frac{48 \times 12}{20} = 28.8 \text{ in.}^3 \text{ (472,320 mm}^3\text{)}$$

   By Appendix 4B, use W14 × 22.

   $$\text{Total weight} = 22 \times 120 = (2640 \text{ lb})(11,742 \text{ N})$$

   b. Columns:

   $$\text{Axial load} = 20 \text{ ft}(30 \text{ ft})(40 \text{ psf}) = 24 \text{ kips (106,752 N)}$$

   $$kl = 1.0(21 \text{ ft}) = 21 \text{ ft (6.4 m)}$$

   $$\text{Max. allowable stress} = 22 \text{ ksi (151.7 MPa) at } kl = 0$$

   $$\text{Int. allowable stress} = 9.37 \text{ ksi (64.6 MPa) at } 22.6\sqrt{P}$$

   $$22.6\sqrt{24} = 110.7 \text{ in} = 9.2 \text{ ft (2.8 m)}$$

   $$\text{Min. allowable stress} = 3.72 \text{ ksi (25.6 MPa) at } 57\sqrt{P}$$

   $$57\sqrt{24} = 279 \text{ in.} = 23.3 \text{ ft (7.1 m)}$$

   Select $F_a = 4.64$ ksi (31.9 MPa).

   $$A_c = \frac{24}{4.64} = 5.17 \text{ in.}^2 \text{ (3334 mm}^2\text{)}$$

   By Appendix 4A, use W6 × 20 (5.87 in.²).

   $$\text{Total weight} = 24 \text{ ft}(20 \text{ plf})7 = (3360 \text{ lb})(14,945 \text{ N})$$

   c. Footings:

   $$\text{Column load} = 24 \text{ kips (106,752 N)}$$

   $$A_f = \frac{P}{f_{sa}} = \frac{24}{3} = 8 \text{ sq ft (0.74 m}^2\text{)}$$

   Use 3 ft 0 in. × 3 ft 0 in. (0.91 m × 0.91 m) footing. From Table 4–12,

   $$q_{\text{net}} = \frac{1.4(12) + 1.7(12)}{9.0} = 4.13 \text{ ksf (197,744 N/m}^2\text{)}$$

   $$\frac{\sqrt{f'_c}}{q_{\text{net}}} = \frac{\sqrt{3000}}{4.30} = 0.013, \qquad \frac{A_f}{A_c} = \frac{9.0}{1.0} = 9.0$$

   $$r = 12 \text{ in.}, \ \frac{d}{r} = 0.4, \ d = 0.4(12) = 4.8 \text{ in.}$$

   $$a = 12 \text{ in.}, \ \frac{a}{d} = 4.2, \ d = \frac{12}{4.2} = 2.86 \text{ in.}$$

   $$t = d + 3 = 5 + 3 = 8 \text{ in. (203 mm)}$$

   Material quantity: 7 footings × 9 ft² × 8 in./12 × 1/27 = 1.55 cu yd (1.19 m³)

4. Alternates 2, 3, and 4 are designed in a similar manner and results are shown tabulated in Table 5–1.

5. *Design alternate 5.* 60-ft (18.29-m) bay, steel truss, and columns.

Table 5–1 Example Problem 5–2

Alternate	Beam weight	Column weight	Total steel	Steel[a] cost	Footing volume	Footing[b] cost	Total cost
1. 20-ft bay	2,640	3,360	6,000	$3,060	1.55	$217	$3,277
2. 30-ft bay	4,200	2,400	6,600	$3,366	1.70	$238	$3,604
3. 40-ft bay	6,600	2,304	8,904	$4,541	2.37	$331	$4,872
4. 60-ft bay	11,280	2,232	13,512	$6,891	3.47	$485	$7,376
5. 60-ft bay with truss	3,752	2,232	5,984	$3,277	3.47	$485	$3,762

[a] Steel cost (Appendix 7A):
Rolled steel = $0.51/lb
Trusses = $0.57/lb
[b] Footing cost (Appendix 7A):
Footings = $140/cy

a. Truss. Assume that d = 1/10 span = 0.1(60 ft) = 6 ft (1.83 m).

$$M = \frac{wl^2}{8} = \frac{1.2(60)^2}{8} = 540 \text{ ft-kips } (732{,}240 \text{ Nm})$$

$$C = T = \frac{M}{d} = \frac{540}{6} = 90 \text{ kips } (400{,}320 \text{ N})$$

Assume allowable stresses in accordance with Section 4–9:

$$\text{Top chord} = \frac{90}{15.0} = 6.0 \text{ in.}^2 \ (3870 \text{ mm}^2)$$

$$\text{Bottom chord} = \frac{90}{17.6} = 5.11 \text{ in.}^2 \ (3295 \text{ mm}^2)$$

$$\text{Max. shear} = \frac{wl}{2} = \frac{1.2(60)}{2}$$

$$= 36 \text{ kips } (160{,}128 \text{ N})$$

$$\text{Diagonal tension} = 2(36) = 50.9 \text{ kips } (226{,}403 \text{ N})$$

$$\text{Diagonal area} = \frac{50.9}{17.6} = 2.89 \text{ in.}^2 \ (1864 \text{ mm}^2)$$

Total area:

$$\begin{aligned} \text{Top chord: } (60 \text{ ft} \times 12)(6.0) &= 4{,}320 \text{ in.}^3 \\ \text{Bottom chord: } (60 \text{ ft} \times 12)(5.11) &= 3{,}679 \\ \text{Diagonal: } (8.5 \text{ ft} \times 12)(10)(2.89) &= 2{,}947 \\ \text{Vertical: } (6 \text{ ft} \times 12)(11)(2.89) &= \underline{2{,}288} \\ & 13{,}234 \text{ in.}^3 \end{aligned}$$

$$\text{Weight} = \frac{13{,}234 \text{ in.}^3 (490 \text{ lb/ft}^3)}{1728}$$

$$= 3752 \text{ lb } (16{,}688 \text{ N})$$

6. *Results.* The results (Table 5–1) show the 20-ft bay to be the most economical.

5–3 FLOOR AND ROOF ASSEMBLIES

Floor and roof assemblies are primarily horizontal distribution subsystems consisting of several "levels" of horizontal framing. (They may also function as the lateral distribution subsystem if they are provided with enough stiffness to act as diaphragms.) This subsystem is constructed with interconnected beams, trusses, and slab elements that provide a floor or roof surface (deck), and one or more levels of supporting beam or truss elements.

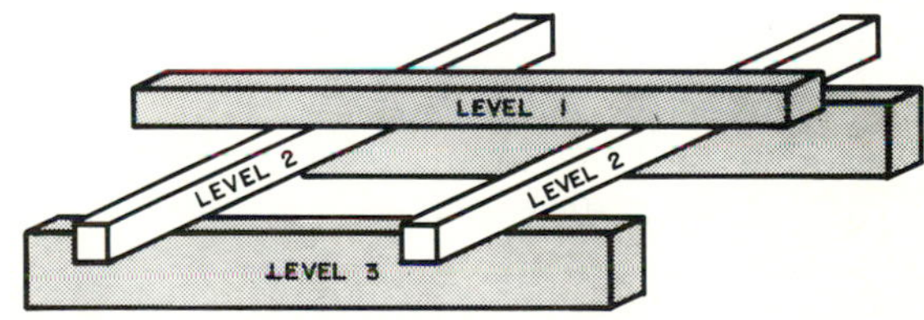

Figure 5–10. Framing Levels

The concept of framing "level" is illustrated in Fig. 5–10, which shows floor beams (level 2) picking up the load from the slab (level 1) and transferring it to other beams (level 3) framing directly to the vertical support subsystem. Normally, no more than two or three levels of framing will be included in a floor or roof assembly, and two or more of these levels may be incorporated in a single element. For example, the prestressed double-tee element provides both level 1 (deck) and level 2 (floor beam) framing.

History and Development

Early roof assemblies provided enclosure and protection from the environment. These surfaces were built with timber piles, thatched straw, mud, and other readily available natural materials. Later, stone became popular as a more permanent building material, leading to roofs (and later floors) constructed with stone beams and slabs.

Wood construction further developed with the use of heavy timber beams, trusses, and individual sawn lumber sections for decking. Stone and wood provided the major elements for floor and roof construction until the development of cast iron and concrete.

Wrought iron and later structural steel rapidly developed as key materials for bridge floor systems and multistory building structures. Concrete developed initially as a decking and surfacing material, and it was not until the late 1800s and early 1900s that reinforced concrete emerged as a material for structural elements. Concern for fire safety spawned widespread use of concrete floor and roof construction, including joist systems formed with clay tile units and various combinations of concrete with metal lath and steel bar reinforcement. Concrete was initially the major material used to "fireproof" steel beams and columns.

The development of a wide variety of fire protection methods has now qualified all major structural materials for floor and roof construction. Even wood can now be chemically treated to a noncombustible status. The choice between materials and types of elements is now chiefly based on economics, serviceability, and construction requirements.

Types and Uses

The major elements used to construct the floor and roof assembly include the following:

- Beams
- Trusses (bar joists)
- Deck assemblies
- Connections

Beams can be made from any of the major structural materials. Structural steel and reinforced concrete are popular for the great majority of structures, with wood a popular material for residential floor and roof subsystems. The steel bar-joist is the major truss form used in floors and roofs. Plywood joists and combination steel–wood composite joists are used for light commercial and residential construction.

Deck assemblies include a great variety of materials and forms of construction. They are two-dimensional assemblies which provide a continuous surface and may provide some of their own level 2 support. Common deck assemblies include the following:

- Plywood
- Timber planks
- Concrete pan joist
- Corrugated metal decks with concrete topping
- Precast concrete floor plank
- Precast, prestressed double tees
- One-way concrete slab
- Two-way concrete slab
- Precast concrete form plank with concrete topping
- Cementitious plank

The need for level 2 or level 3 support framing for these deck assemblies is determined primarily by their optimum span ranges and the spacing between the vertical support subsystems. Appendix 5A gives additional information regarding typical deck assemblies.

Connections for floor and roof framing are either bearing type, hinged, or moment resisting. Bearing connections are used for many types of deck assemblies, steel bar joists, and occasionally for beam elements. Hinged connections are used typically for structural steel framing, and the moment-resisting connection is inherently a part of concrete framing systems and is also used for continuous steel beam framing.

Behavior

Gravity Loads. Distribution of gravity loads horizontally involves bending and shear forces for all elements except arches and cables (which are not a part of our floor and roof assembly classification). The load path (Fig. 5–11) travels from level 1 deck elements—to level 2 floor beams—to level 3 beams—to the vertical support subsystems. The deflection of any individual element is dependent on its own stiffness and the stiffness of the members that support it (Fig. 5–12).

Elements connected by bearing or hinged connections transmit only shear forces along the load

Figure 5–11. Floor and Roof Assembly Load Path

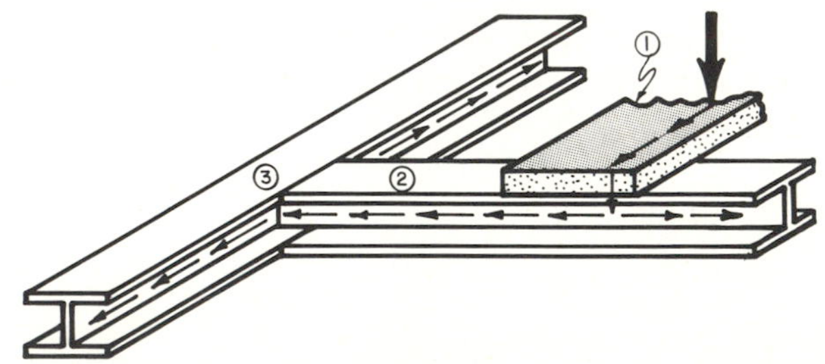

Figure 5–12. Floor and Roof Deflection Behavior

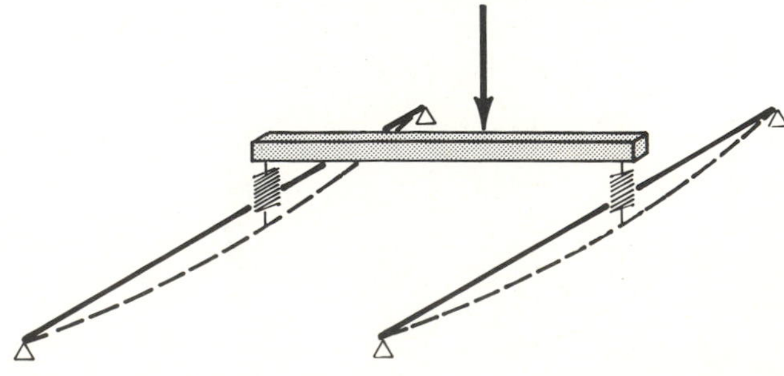

path. Moment-resisting connections, however, also transmit rotational forces which include torsion in transverse supporting beams. Torsion is especially present at perimeter spandrel beams of concrete construction (Fig. 5–13).

Figure 5–13. Torsion Forces

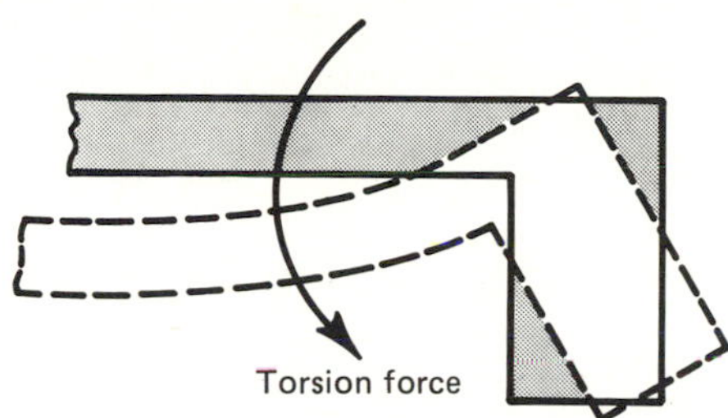

Lateral Loads. Floor and roof assemblies may be constructed with the necessary stiffness and load path continuity to distribute lateral loads (wind and earthquake) to lateral support subsystems. In this role, the floor or roof surface acts as a horizontal beam (also called a diaphragm) spanning between lateral support points (Fig. 5–14) and may effectively engage beam elements at the perimeter (transverse to the direction of the load) as top and bottom "chords" or "flanges," in which case the bending moment can be resolved into a tension and compression couple and considered resisted by the beam elements (with shear resisted by the diaphragm surface). In the absence of such flange elements to take the moment couple, the floor or roof must act as a deep plate taking both bending and shear forces. Either type of diaphragm behavior requires effective transfer of bending and shear forces in the plane of the floor or roof, necessitating careful detailing of connections between elements and between the diaphragm and the lateral support subsystem.

Figure 5–14. Diaphragm Action of Roof or Floor Subsystem

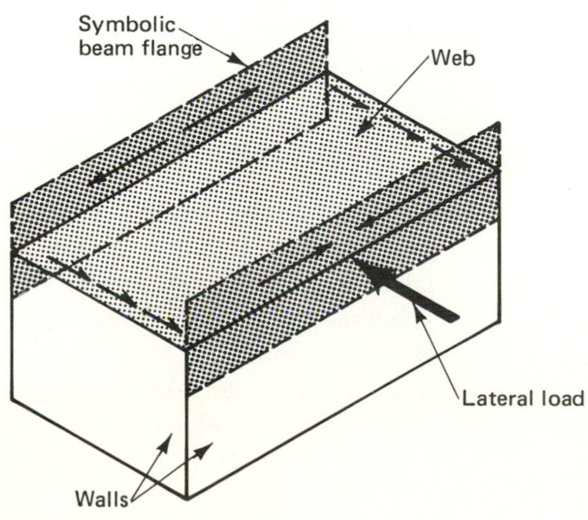

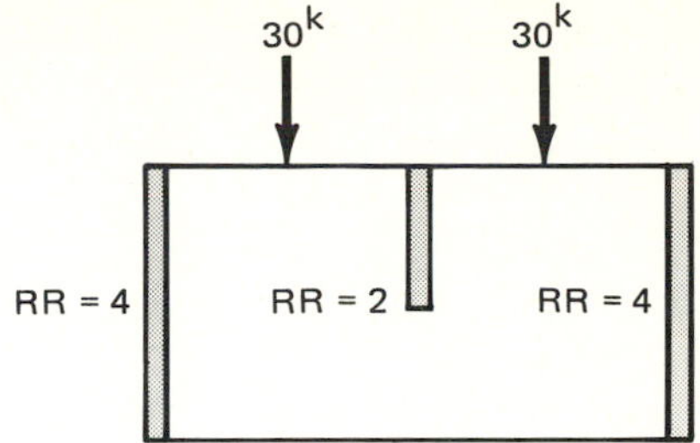

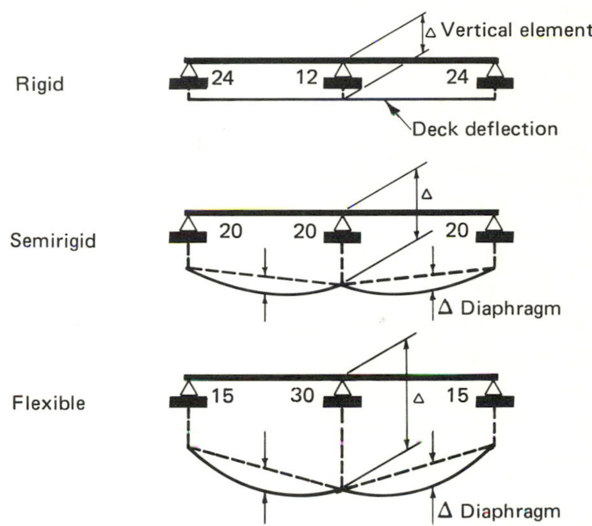

Figure 5–15. Relative Effects of Diaphragm Stiffness

The stiffness of the diaphragm has an important effect on the proportionate distribution of lateral loads to the various components of the lateral support system. This effect is shown in Fig. 5–15, which shows how diaphragm and lateral support element relative rigidities influence load distribution. At one extreme is the *rigid diaphragm,* which distributes the lateral loads to the lateral support elements in proportion to just *their* rigidities. At the other extreme is the *flexible diaphragm,* which transmits lateral loads based on tributary area. The *semirigid diaphragm* falls somewhere in between these two extremes and involves both aspects of behavior.

The effect of diaphragm stiffness on vertical wall deflection is illustrated in Fig. 5–16. To avoid cracking wall materials, the allowable deflection of horizontal diaphragms in buildings is expressed by the following formula:

$$\Delta = \frac{h^2 f}{1.44Et} \qquad (5\text{–}1)$$

where Δ = allowable deflection between adjacent supports of wall, in. (mm)

h = height of the wall between adjacent horizontal supports, in. (mm)

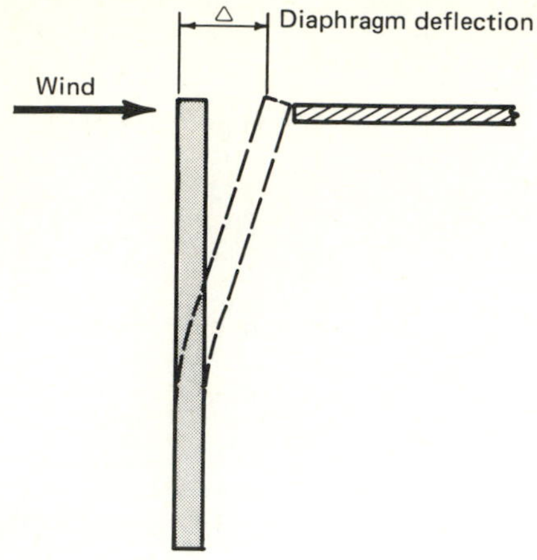

Figure 5–16. Wall Deflection Due to Diaphragm Deflection

t = thickness of wall, in. (mm)
f = allowable flexural compressive stress of wall material, psi (MPa)
E = modulus of elasticity of wall material, psi (MPa)

Approximate Analysis

Gravity Loads. No new concepts are involved in the preliminary analysis of floor and roof assemblies for gravity loads. The slab, beam, or truss elements distribute loads according to the assumptions discussed in Chapter 4 (Fig. 4–27) and tributary loads to each element can be readily determined. Girder elements typically pick up concentrated loads from floor beam end reactions, and these loads can be converted into equivalent uniform loads by the procedures of Chapter 4.

Lateral Loads. When the floor and roof assembly must also act as a diaphragm, bending moments and shears due to the lateral loads must be determined. The diaphragm simply acts as a simple span beam on its side and the following formulas apply:

For diaphragm shear:

$$S = \frac{wl}{2b} \qquad (5\text{–}2)$$

where S = maximum web shear, lb/ft (N/m)
w = uniform lateral load, lb/ft (N/m)
l = span of diaphragm between lateral support elements, ft (m)
b = width of diaphragm, parallel to direction of lateral load, ft (m)

For diaphragm bending stress:

$$F_b = \frac{Mc}{I} \qquad (5\text{–}3)$$

where F_b = maximum bending stress, ksi (N/m²)
$M = wl^2/8$ (ft-kips) (Nm)
$c = b/2$ (ft) (m)
I = moment of inertia in the plane of bending, ft^4 (m^4)

If the diaphragm has flanges, the moment can be resisted by a tension and compression couple computed as follows:

$$C = T = \frac{M}{b} \qquad (5\text{–}4)$$

The *deflection* of diaphragms can be evaluated with the stiffness factor (G') and the following expressions:

$$\Delta d = \Delta f + \Delta \omega \qquad (5\text{–}5)$$

$$\Delta \omega = \frac{q_{ave} L_1}{G'}, \qquad \Delta f = \frac{5wl^4}{384EI} \qquad (5\text{–}6)$$

where q_{ave} = average web shear over length L_1, kips/ft (N/m)
L_1 = distance between vertical resisting element and the point to which the deflection is to be determined, ft (m)
$\Delta\omega$ = web component of total deflection, Δd
Δf = flexural component of total deflection, Δd
G' = diaphragm stiffness, kips/in. (N/mm)
w = uniform lateral load
l = span of diaphragm
E = modulus of elasticity, diaphragm material
I = moment of inertia of diaphragm

Approximate Design

Preliminary design of floor and roof subsystems involves the following choices:

- Material selection
- Deck assembly selection
- Beam type, spacing, and direction
- Bay spacing
- Element proportions

All of these choices are influenced by project requirements and practical considerations such as use of compatible materials. The preliminary design process usually starts with selection of alternate deck assemblies. Each type of deck has an optimum span range which helps establish requirements for floor beam spacing (Appendix 5A). Bay spacing (each direction) becomes a balancing act between floor beam span and direction and girder span and direction. This sequence of choices can become quickly unmanageable, especially if the floor and roof assembly is only one of several alternate horizontal distribution subsystems being considered. Our experience helps narrow the range of options and Appendix 5A offers us additional help. The following example problem illustrates the preliminary design process.

EXAMPLE PROBLEM 5–3: Preliminary Design of Floor and Roof Subsystem

Select a preliminary roof system for the warehouse described in Example Problem 5–2.

Solution

1. *Material selection.* With no special fire protection or availability problems, select structural steel as the preliminary material.
2. *Deck assembly.* Reference to Appendix 5A suggests that an open rib metal deck roof assembly is generally the most economical for light roof loads.
3. *Floor beam spacing and type.* Optimum span for metal deck is approximately 5 ft (1.52 m) to 8 ft (2.44 m). Select 5 or 6 ft and 22-gage deck. Bar joists are recommended by Appendix 6A for light roof construction (with no fire protection requirements).
4. *Bay spacing and element direction.* Several options now exist for framing the roof:
   a. 20 ft × 20 ft (6.1 m × 6.1 m) bays with joists at 5 ft 0 in. (1.52 m) on center. Girders running either direction.
   b. 20 ft × 30 ft (6.1 m × 9.1 m) bays with joists at 5 ft 0 in. (1.52 m) on center. Girders spanning 20 ft (6.1 m).
   c. 20 ft × 30 ft (6.1 m × 9.1 m) bays with joists at 6 ft 0 in. (1.83 m) on center. Girders spanning 30 ft (9.1 m).
   d. 30 ft × 30 ft (9.1 m × 9.1 m) bays with joists at 6 ft 0 in. (1.83 m) on center. Girders spanning either direction.
5. Alternate A:
   a. Deck assembly: 22-gage metal deck at 20 ft × 20 ft = 400 sq ft (37.2 m²).
   b. Bar joists:

$$\text{Uniform load} = 40 \text{ psf } (5 \text{ ft}) = 200 \text{ plf } (2918 \text{ N/m})$$

$$\text{Span} = 20 \text{ ft}$$

$$M = \frac{wl^2}{8} = \frac{200(20)^2}{8} = 10{,}000 \text{ ft-lb} = 120 \text{ in.-kips } (13{,}560 \text{ Nm})$$

By Appendix 4F, use 12 H 3 at 5.2 plf (resisting M = 140 in.-kips).

$$5.2 \text{ lb/ft} \div 5 \text{ ft} = 1.04 \text{ psf } (49.8 \text{ N/m}^2)$$

   c. Girder:

$$\text{Uniform load} = 40 \text{ psf } (20 \text{ ft}) = 800 \text{ plf } (11{,}674 \text{ N/m})$$

$$\text{Span} = 20 \text{ ft}$$

$$M = \frac{wl^2}{8} = \frac{0.80(20)^2}{8} = 40 \text{ ft-kips } (54{,}240 \text{ Nm})$$

$$\text{Min. } d = \tfrac{1}{2} \times \text{span} = \tfrac{1}{2}(20 \text{ ft}) = 10 \text{ in.}$$

Assume 12 in. depth (304.8 mm).

$$F_b = \frac{\sqrt{10{,}320M}}{ld^2} = \frac{\sqrt{10{,}320(40)}}{5(12)^2} = 23.9 > 22$$

Select 22.0 ksi (151.7 MPa) and calculate S:

$$S_{\text{req'd}} = \frac{M}{F_b} = \frac{40 \times 12}{22} = 21.8 \text{ in.}^3 \text{ } (357{,}302 \text{ mm}^3)$$

By Appendix 4B, use W12 × 22.

$$22 \text{ lb/ft} \div 20 \text{ ft} = 1.10 \text{ psf } (52.7 \text{ N/m}^2)$$

6. *Tabulation of alternates.* The other alternates are devel-

Table 5–2 Example Problem 5–3

Alternate	*Deck quantity [sq ft (m²)]*	*Deck cost ($)*	*Joist weight [psf (N/m²)]*	*Joist cost ($)*	*Girder weight [psf (N/m²)]*	*Girder cost ($)*	*Total cost ($)*
A. 20 ft x 20 ft (6.1 m x 6.1 m)	400 (37.2)	0.66	1.04 (49.8)	0.42	1.10 (52.7)	0.56	1.64
B. 20 ft x 30 ft (6.1 m x 9.1 m)	400 (37.2)	0.66	1.56 (74.7)	0.62	0.87 (41.7)	0.44	1.78
C. 20 ft x 30 ft[a] (6.1 m x 9.1 m)	400 (37.2)	0.66	1.04 (49.8)	0.42	1.75 (83.8)	0.89	1.97
D. 30 ft x 30 ft (9.1 m x 9.1 m)	400 (37.2)	0.66	1.43 (68.5)	0.57	1.47 (70.4)	0.75	1.98

[a] Girder direction.

oped in a similar manner and the results are shown in Table 5–2. Alternate A is shown to be the most economical for this example.

Diaphragm Design. Use of a floor or roof assembly as a diaphragm requires both strength and stiffness properties and development of connections to transfer the diaphragm forces. Moment resistance is normally provided by the diaphragm "flange" elements acting in direct axial tension or compression. Shear strength of the diaphragm web depends on the type of deck assembly and supporting system used.

Concrete. Shear strength for concrete slabs is given by the following expression:

$$q_{ud} = \frac{6.0 f'_c t}{1 + 3L_v^2/t^2} \quad \text{(lb/ft)}$$

$$q_{ud} = \frac{500 f'_c t}{1 + 21{,}000 L_v^2/t^2} \quad \text{(N/m)} \qquad (5\text{–}7)$$

where q_{ud} = ultimate shear capacity, lb/ft (N/m)
f'_c = compressive strength of concrete but limited to a maximum of 3000 psi (20.7 MPa) for determination of q_{ud}
L_v = clear span between framing members, ft (m)
t = thickness of slab, in. (mm)

If the slab itself is supporting vertical loads, the span (L_v) must not be greater than $3t$. The shear capacity is determined by the requirements of ACI 318 but limited to the value determined by the formula. When poured slabs are not monolithic with the supporting framing members, the slab must be anchored by mechanical means at intervals not exceeding 4 ft on center.

The web stiffness factor G' is determined by the following formula:

$$G' = 0.0085 t \omega^{1.5} f'_c \quad \text{kips/in.}$$

$$G' = \frac{361 t \omega^{1.5} f'_c}{10^6} \quad \text{N/mm} \qquad (5\text{–}8)$$

where G' = web stiffness, kips/in. (N/mm)
t = thickness of the slab, in. (mm)
ω = weight of the concrete in lb/cu ft (N/m³); minimum value of ω will be 90 lb/cu ft
f'_c = compressive strength of the concrete at 28 days, psi (MPa)

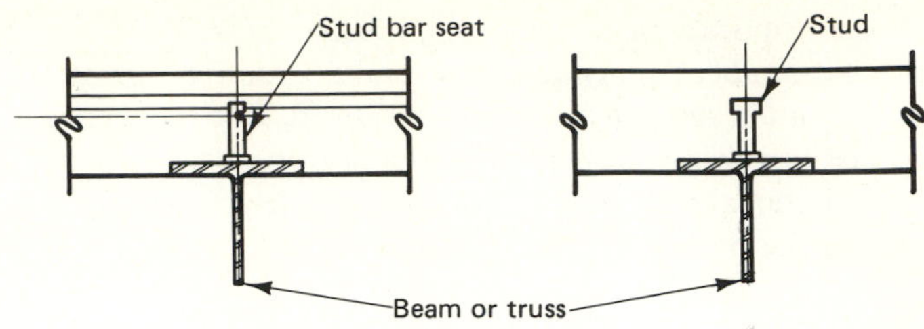

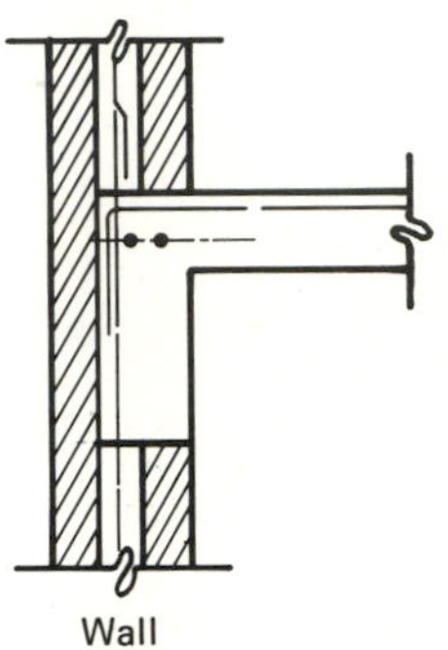

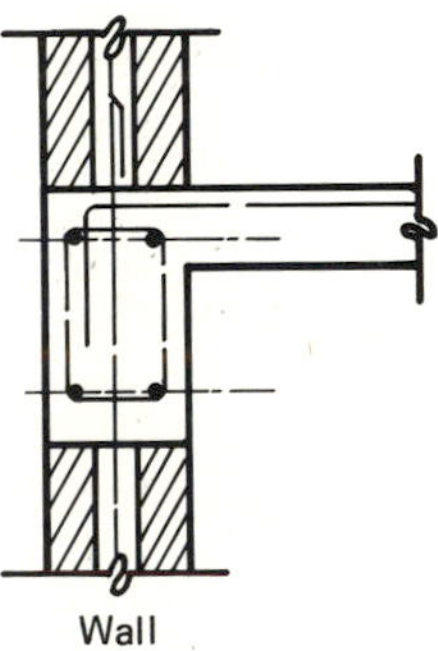

Figure 5–17. Reinforced Concrete Diaphragm Connections

Diaphragms of this type are in the rigid category of stiffness and are usually limited only by the appropriate deflection limitations. Typical connection details are shown in Fig. 5–17.

Metal deck. Stiffness and strength equations for metal deck diaphragms are given in Tables 5–3 and 5–4 for various types of metal deck diaphragms. These tables are taken from Ref. 16 and are based on a load factor of 2.75 applied to measured failure loads and standard weld patterns, consisting of welds at designated spacings across 30-in. (609.6-mm)-wide panels. Typical connections for metal deck diaphragms are shown in Fig. 5–18.

Table 5–3 Shear Diaphragm Design[a,b] Allowable Loads (Lb/Ft)

30/6 Pattern, thickness = 0.0295 in.[c]

Number of fasteners[d]	*Span (ft)* 3.0	3.5	4.0	4.5	5.0	5.5	6.0	6.5	7.0	K_1
0	356	309	269	236	210	189	172	156	143	0.05640
1	480	421	374	336	301	272	247	227	209	0.03244
2	589	521	465	421	383	352	321	296	272	0.02277
3	680	609	550	500	458	421	390	363	338	0.01754
4	758	685	625	572	525	487	450	421	394	0.01426

30/6 Pattern, thickness = 0.0358 in.

Number of fasteners	*Span (ft)* 4.0	4.5	5.0	5.5	6.0	6.5	7.0	7.5	8.0	K_1
0	320	281	250	225	203	187	170	158	147	0.05818
1	443	398	358	321	292	269	247	229	212	0.03364
2	552	500	454	418	381	350	323	300	280	0.01825
3	652	592	543	500	463	430	400	372	347	0.01825
4	741	678	623	576	536	500	469	440	414	0.01485

30/6 Pattern, thickness = 0.0474 in.

Number of fasteners	*Span (ft)* 5.0	5.5	6.0	6.5	7.0	7.5	8.0	8.5	9.0	K_1
0	318	285	260	236	216	200	185	172	161	0.05826
1	452	409	372	341	314	290	270	252	236	0.03412
2	578	529	485	445	410	381	354	332	312	0.02412
3	689	634	587	545	507	472	440	412	387	0.01865
4	790	732	680	634	594	558	525	492	463	0.01521

30/4 Pattern, thickness = 0.0295 in.

Number of fasteners	*Span (ft)* 3.0	3.5	4.0	4.5	5.0	5.5	6.0	6.5	7.0	K_1
0	323	285	254	229	205	183	167	152	140	0.06345
1	418	376	340	309	283	261	241	221	203	0.03466
2	489	447	409	376	349	323	301	281	265	0.02384
3	540	501	465	434	403	378	354	332	314	0.01817
4	578	543	510	480	450	425	400	378	358	0.01467

30/4 Pattern, thickness = 0.0358 in.

Number of fasteners	*Span (ft)* 4.0	4.5	5.0	5.5	6.0	6.5	7.0	7.5	8.0	K_1
0	301	272	243	218	198	180	165	152	141	0.06545
1	403	367	336	309	287	263	241	223	209	0.03595
2	485	447	414	383	358	334	314	296	274	0.02478
3	552	514	480	449	420	394	372	352	332	0.01891
4	605	569	536	503	476	449	425	403	383	0.01528

[a] Larry D. Luttrell, *Steel Deck Institute Diaphragm Design Manual* (St. Louis, Mo.: Steel Deck Institute, 1981).

[b] SDI standard deck with welded connections ($\frac{5}{8}$ in. diameter); number of spans = 3 or more.

[c] Pattern designation refers to width of sheet (e.g., 30 in.) and number of welds per sheet at both interior supports and sheet ends. Thickness designation is thickness of deck material.

[d] Number of fasteners refers to number of sidelap connections along the deck span.

Table 5–3 (*Continued*)

30/4 Pattern, thickness = 0.0474 in.

Number of fasteners	*Span (ft)* 5.0	5.5	6.0	6.5	7.0	7.5	8.0	8.5	9.0	K_1
0	309	276	250	229	210	194	180	167	156	0.06554
1	427	392	363	332	307	283	265	247	230	0.03649
2	525	487	454	423	398	374	349	327	307	0.02528
3	609	569	532	501	472	447	423	401	381	0.01934
4	680	640	603	570	540	512	487	463	441	0.01566

30/3 Pattern, thickness = 0.0295 in.

Number of fasteners	*Span (ft)* 3.0	3.5	4.0	4.5	5.0	5.5	6.0	6.5	7.0	K_1
0	236	210	187	169	150	134	121	110	100	0.09229
1	318	289	265	243	223	207	192	180	165	0.04179
2	369	343	320	300	280	261	245	230	218	0.02701
3	400	380	360	340	321	305	289	274	260	0.01995
4	421	405	387	370	354	338	323	309	294	0.01582

30/3 Pattern, thickness = 0.0358 in.

Number of fasteners	*Span (ft)* 4.0	4.5	5.0	5.5	6.0	6.5	7.0	7.5	8.0	K_1
0	223	201	178	160	143	130	120	109	101	0.09520
1	314	289	265	247	229	212	196	181	167	0.04341
2	380	354	332	310	292	274	260	245	232	0.02811
3	427	403	381	361	343	325	309	294	280	0.02079
4	460	440	420	401	383	367	350	334	321	0.01649

30/3 Pattern, thickness = 0.0474 in.

Number of fasteners	*Span (ft)* 5.0	5.5	6.0	6.5	7.0	7.5	8.0	8.5	9.0	K_1
0	225	201	183	165	150	140	129	118	110	0.09533
1	338	312	290	270	249	229	212	198	185	0.04418
2	421	394	370	349	329	310	294	278	261	0.02875
3	485	458	434	412	392	372	356	340	323	0.02131
4	532	509	487	465	445	425	407	390	374	0.01693

Table 5–4 Shear Stiffness of Metal Deck Diaphragm[a]

For typical Steel Deck Institute deck profiles of wide rib, intermediate rib, and narrow rib shapes, the following stiffness formulas apply:

$$\text{For } t = 0.0295 \text{ in.}, \quad G'_{22} = \frac{870}{3.78 + D_n\lambda + C} \text{ (kips/in.)}$$

$$t = 0.0358 \text{ in.}, \quad G'_{20} = \frac{1056}{3.78 + D_n\lambda + C} \text{ (kips/in.)}$$

$$t = 0.0474 \text{ in.}, \quad G'_{18} = \frac{1398}{3.78 + D_n\lambda + C} \text{ (kips/in.)}$$

[a] Larry D. Luttrell, *Steel Deck Institute Diaphragm Design Manual* (St. Louis, Mo.: Steel Deck Institute, 1981).

Table 5–4 (*Continued*)

where λ depends on the number of interior purlins or joists, n_p, as follows:

n_p	0	1	2	3	4	5	6
λ	1.00	1.00	0.90	0.80	0.71	0.64	0.58

The warping constant D_n depends on panel profile and end-of-panel fastener positions.[b]

$$D_n = \frac{D}{12L}$$

D is from the following table
L is the total panel (sheet) length (ft)

		D *values for warping* *End valley spaces (6 in.) between welds*		
Type	*Gage*	*One*	*Two*	*Three*
WR	22	1549	12864	26504
	20	1159	9623	19825
	18	761	6316	13013
IR	22	2712	14589	29131
	20	2028	10913	21790
	18	1331	7163	14303
NR	22	4271	15388	29303
	20	3195	11511	21919
	18	2097	7555	14387

C, the slip relaxation constant, is given by the following equation:

$$C = 12K_1L$$

where K_1 = constant from Table 5–3
L = total panel (sheet) length (ft)

[b] For nonuniform fastener spacing, use appropriate table value for percentage of sheet width at that spacing. One space = 6 in.

Figure 5–18. Metal Deck Diaphragm Connections

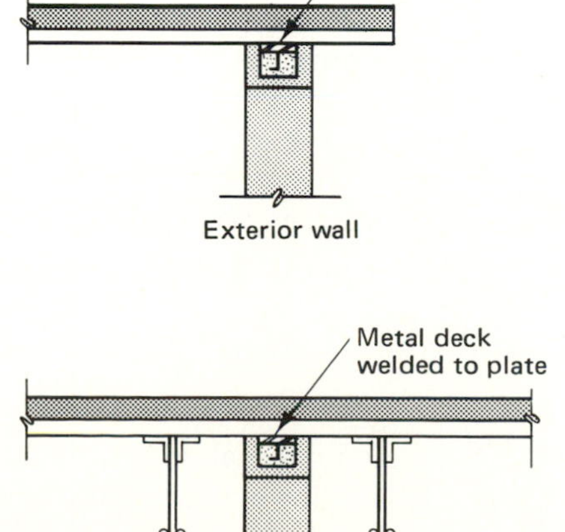

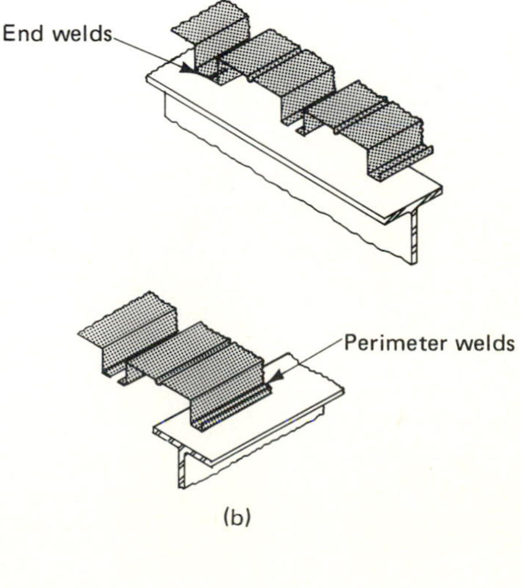

Recommended Shear (pounds per foot) for Horizontal APA Panel Diaphragms with Framing of Douglas Fir, Larch or Southern Pine[(a)] for Wind or Seismic Loading

Panel grade	Common nail size	Minimum nail penetration in framing (inches)	Minimum nominal panel thickness (inch): Veneer-faced panels	Minimum nominal panel thickness (inch): Other panels	Minimum nominal width of framing member (inches)	Blocked diaphragms — Nail spacing (in.) at diaphragm boundaries (all cases), at continuous panel edges parallel to load (Cases 3 & 4), and at all panel edges (Cases 5 & 6)[(b)]: 6 / Nail spacing (in.) at other panel edges (Cases 1, 2, 3 & 4): 6	4 / 6	$2\frac{1}{2}$[(c)] / 4	2[(c)] / 3	Unblocked diaphragms — Nails spaced 6" max. at supported edges[(b)]: Case 1 (No unblocked edges or continuous joints parallel to load)	All other configurations (Cases 2, 3, 4, 5 & 6)
APA Structural rated sheathing EXP 1 or EXT	6d	$1\frac{1}{4}$	$\frac{5}{16}$	—	2	185	250	375	420	165	125
					3	210	280	420	475	185	140
	8d	$1\frac{1}{2}$	$\frac{3}{8}$	—	2	270	360	530	600	240	180
					3	300	400	600	675	265	200
	10d	$1\frac{5}{8}$	$\frac{1}{2}$	—	2	320	425	640	730	285	215
					3	360	480	720	820	320	240
APA rated sheathing EXP 1, EXP 2 or EXT, APA Structural II rated sheathing EXP 1 or EXT, and other APA grades except species Group 5	6d	$1\frac{1}{4}$	$\frac{5}{16}$	$\frac{7}{16}$ or $\frac{3}{8}$	2	170	225	335	380	150	110
					3	190	250	380	430	170	125
			$\frac{3}{8}$	$\frac{1}{2}$	2	185	250	375	420	165	125
					3	210	280	420	475	185	140
	8d	$1\frac{1}{2}$	$\frac{7}{16}$ or $\frac{3}{8}$	$\frac{1}{2}$	2	240	320	480	545	215	160
					3	270	360	540	610	240	180
			$\frac{1}{2}$	$\frac{5}{8}$	2	270	360	530	600	240	180
					3	300	400	600	675	265	200
	10d	$1\frac{5}{8}$	$\frac{1}{2}$	$\frac{5}{8}$	2	290	385	575	655	255	190
					3	325	430	650	735	290	215
			$\frac{5}{8}$	$\frac{3}{4}$	2	320	425	640	730	285	215
					3	360	480	720	820	320	240

(a) For framing of other species: (1) Find species group of lumber in Table 8.1A, NFPA 1977 National Design Spec. (2) Find shear value from table for nail size, and for Structural 1 panels (regardless of actual grade). (3) Multiply value by 0.82 for Lumber Group III or 0.65 for Lumber Group IV.

(b) Space nails 12 in. oc along intermediate framing members for roofs, and 10 in. oc for floors.

(c) Where nails are spaced 2 in. or $2\frac{1}{2}$ in. oc, framing shall be 3-in. nominal or wider, and nails shall be staggered.

Notes: Design for diaphragm stresses depends on direction of continuous panel joints with reference to load, not on direction of long dimension of sheet. Continuous framing may be in either direction for blocked diaphragms.

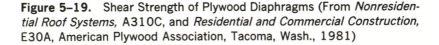

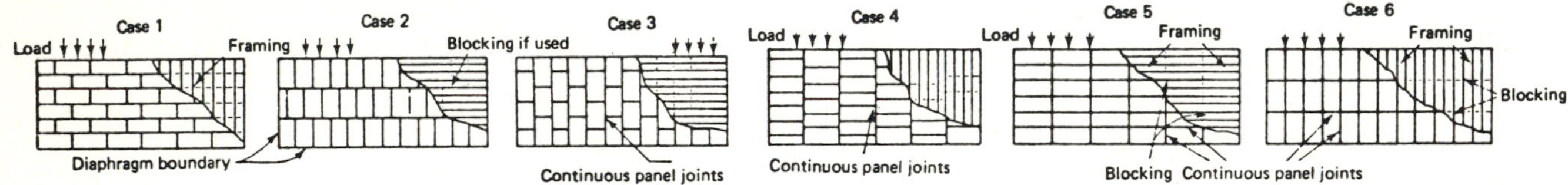

Figure 5–19. Shear Strength of Plywood Diaphragms (From *Nonresidential Roof Systems,* A310C, and *Residential and Commercial Construction,* E30A, American Plywood Association, Tacoma, Wash., 1981)

Plywood deck. Allowable shear for plywood diaphragm is given in Fig. 5–19. The G' value for these diaphragms is:

$$G' = \frac{(q_d)^2}{33q_{ave}} \quad \text{kips/in.} \qquad (5\text{–}9)$$

$$G' = \frac{(q_d)^2}{2.75q_{ave}} \quad \text{N/mm}$$

where G' = web stiffness, kips/in. (N/mm)
q_d = allowable shear specified in Fig. 5–19, lb/ft (N/m)
q_{ave} = average shear over span L_1, lb/ft (N/m)

Typical connections for plywood diaphragms are shown Fig. 5–20. The following example illustrates preliminary diaphragm design.

Recommended Shear (pounds per foot) for APA Panel Shear Walls with Framing of Douglas Fir, Larch, or Southern Pine[(a)] for Wind or Seismic Loading[(b)]

Panel grade	Minimum nominal panel thickness (in.) Veneer-faced panels	Other panels	Minimum nail penetration in framing (in.)	Panels applied direct to framing: Nail size (common or galvanized box)	Nail spacing at panel edges (in.) 6	4	3	2[(e)]	Panels applied over ½" gypsum sheathing: Nail size (common or galvanized box)	Nail spacing at panel edges (in.) 6	4	3	2[(e)]
APA Structural I rated sheathing EXP 1 or EXT	5/16	—	1¼	6d	200	300	390	510	8d	200	300	390	510
	3/8	—	1½	8d	230[(d)]	360[(d)]	460[(d)]	610[(d)]	10d	280	430	550	730
	1/2	—	1⅝	10d	340	510	665	870	—	—	—	—	—
APA rated sheathing EXP 1, EXP 2 or EXT; APA Structural II rated sheathing EXP 1 or EXT; APA panel siding (f) and other APA grades except species Group 5	5/16 or 1/4[(c)]	7/16 or 3/8	1¼	6d	180	270	350	450	8d	180	270	350	450
	7/16 or 3/8	1/2	1½	8d	220[(d)]	320[(d)]	410[(d)]	530[(d)]	10d	260	380	490	640
	1/2	5/8	1⅝	10d	310	460	600	770	—	—	—	—	
APA panel siding (f) and other APA grades except species Group 5	5/16[(c)]	—	1¼	Nail size (galvanized casing) 6d	140	210	275	360	Nail size (galvanized casing) 8d	140	210	275	360
	3/8	—	1½	8d	130[(d)]	200[(d)]	260[(d)]	340[(d)]	10d	160	240	310	410

[(a)] For framing of other species: (1) Find species group of lumber in the NFPA National Design Spec. (2)(a) For common or galvanized box nails, find shear value from table for nail size, and for Structural 1 panels (regardless of actual grade). (b) For galvanized casing nails, take shear value directly from table. (3) Multiply this value by 0.82 for Lumber Group III or 0.65 for Lumber Group IV.

[(b)] All panel edges backed with 2-inch nominal or wider framing. Install panels either horizontally or vertically. Space nails 6 inches oc along intermediate framing members for 3/8-inch and 7/16-inch panels installed on studs spaced 24 inches oc. For other conditions and panel thicknesses, space nails 12 inches oc on intermediate supports.

[(c)] 3/8-inch or 303-16 oc is minimum recommended when applied direct to framing as exterior siding.

[(d)] Shears may be increased 20 percent provided (1) studs are spaced a maximum of 16 inches oc, or (2) panels are 1/2-inch or greater in thickness, or (3) if panels are plywood and are applied with face grain across studs.

[(e)] Framing shall be 3-inch nominal or wider, and nails shall be staggered.

[(f)] 303-16 oc plywood may be 11/32-inch, 3/8-inch or thicker. Thickness at point of nailing on panel edges governs shear values.

Figure 5–19 (*Continued*)

EXAMPLE PROBLEM 5–4: Preliminary Diaphragm Design

Check the warehouse roof system selected in Example Problem 5–3 as a potential diaphragm to distribute wind loads 240 ft (73.2 m) in the east–west direction to lateral support elements at the end walls. Use a wind load of 25 psf (1197 N/m²) and an exterior wall of 12-in. (304.8-mm) concrete block, $F'_m = 1350$ psi (9.31 MPa), $f'_m = 445$ psi (3.07 MPa).

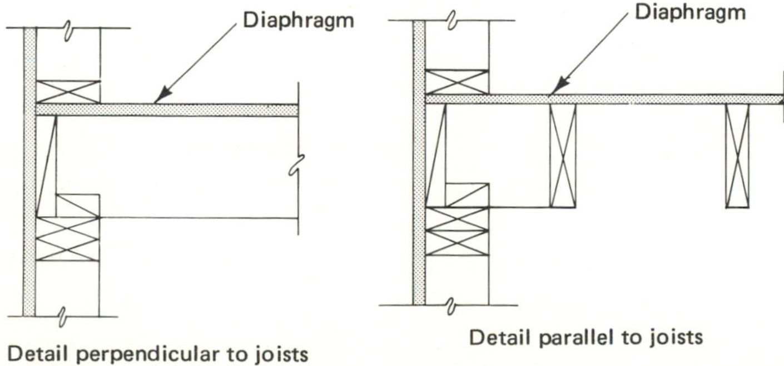

Figure 5–20. Plywood Diaphragm Connections

Solution

1. *Moment concept.* Since beam and girder framing is available at the perimeter to take diaphragm flange forces, consider the moment to be resisted by a tension and compression couple.

2. *Moment and shear calculation:*

$w = 25$ psf ($\frac{1}{2}$ wall height) $= 25(11) = 275$ lb/ft (4013 N/m)

$$M = \frac{wl^2}{8} = \frac{0.275(240)^2}{8} = 1980 \text{ ft-kips } (2{,}684{,}880 \text{ Nm})$$

$$C = T = \frac{M}{b} = \frac{1980}{120} = 16.5 \text{ kips } (73{,}392 \text{ N})$$

Design perimeter beams as "beam-columns" to carry 16.5 kips (73,392 N) of axial load.

$$q = \frac{wl}{2b} = \frac{275(240)}{2(120)} = 275 \text{ lb/ft } (4013 \text{ N/m})$$

3. *Check allowable shear:*

Joist span = 5 ft (1.52 m)

Deck thickness (t) for 22 gage = 0.0295 in. (0.749 mm)

From Table 5–3:

- A 30/3 pattern with two sidelap connections will carry 280 lb/ft.
- A 30/4 pattern with one sidelap connection will carry 283 lb/ft.
- A 30/6 pattern with one sidelap connection will carry 301 lb/ft.

Therefore, the shear capacity is O.K.

4. *Deflection calculation:*

a. Web deflection (Table 5–4):

Assume panel length (L) = 25 ft

Interior joists per panel = 4

Assume 30/4 pattern

$$G'_{22} = \frac{870}{3.78 + D_n\lambda + C}$$

$\lambda = 0.71$ for four interior joists

From Table 5–4: Since 30/4 pattern must have two-fifths of the sheet width at single spacing and three-fifths at three spaces:

$D = 2712$ for one space between end welds

$$D_1 = \frac{2712}{12(25)} = 9.04$$

$D = 29{,}131$ for three spaces between end welds

$$D_2 = \frac{2913}{12(25)} = 97.1$$

$$D_N = \tfrac{2}{5}(9.04) + \tfrac{3}{5}(97.1) = 61.9$$

$$C = 12K_1L, \qquad K_1 = 0.03466 \text{ (from Table 5–3)}$$

$$= 12(0.03466)(25) = 10.4$$

$$G' = \frac{870}{3.78 + (61.9)(6.71) + 10.4} = 14.9 \text{ kips/in.}$$

By Eq. 5–6,

$$\Delta\omega = \frac{q_{ave}L_1}{G'} = \frac{(0.275/2)(120)}{14.9} = 1.11(28.2 \text{ mm})$$

b. Flexural deflection:

$$\Delta f = \frac{5wl^4}{384EI}$$

Assume that $I = 2Ad^2$, where A = flange steel area, and assume a nominal allowable compressive stress in the flange column (since it also carries bending stresses due to gravity load).

$$A = \frac{16.5 \text{ kips}}{3 \text{ ksi}} = 5.5 \text{ in.}^2 \text{ (3548 mm}^2\text{)}$$

$$I = 2(5.5)(60 \times 12)^2$$
$$= 5.7 \times 10^6 \text{ in.}^4 \text{ (}2.37 \times 10^{12} \text{ mm}^4\text{)}$$

$$\Delta f = \frac{5(0.275)(240)^4(1728)}{384(30 \times 10^3)(5.7 \times 10^6)}$$

$$= 0.12 \text{ in. (3.05 mm)}$$

c. Total deflection:

$$\text{Total } \Delta = 1.11 + 0.12 = 1.23 \text{ in. (31.2 mm)}$$

5. *Check wall deflection.* From Eq. 5–1,

$$\text{Allowable } \Delta = \frac{h^2f}{0.01Et}$$

$$= \frac{(22)^2(445)}{0.01(1.35 \times 10^6)12} = 1.32 \text{ in. (33.5 mm)}$$

$$= 1.32 > 1.23 \qquad \text{O.K.}$$

5–4 RIGID FRAMES

A *Rigid frame* is a vertical support subsystem and, also, quite often serves as a lateral support subsystem. It is constructed with beam and beam-column elements rigidly joined together with moment-resisting connections (Fig. 5–21). The rigid connections make this subsystem "indeterminate" from an analysis standpoint, creating an interdependence between member forces and member deformation not found in determinate structures.

The rigid frame derives its unique strength to resist both gravity and lateral loads from the moment interaction between the beams and beam-columns. The moment restraint at the ends of the elements leads to reduced

Figure 5–21. Rigid Frame Subsystem

positive bending moments for beams and reduced effective lengths (k factor) for the beam-columns under gravity load. Moment magnification effects, however, lead to increased stresses for slender beam-column elements, especially in the unbraced (sway) mode. Net increase in beam-column size due to lateral loads can often be kept to a minimum for moderate height structures due to reduced load factors for wind loading combinations. The efficiency gained in the beam elements must be weighed against the increased column sizes required for the beam-column action in determining the overall economy of the rigid frame.

History and Development

The rigid frame subsystem developed in the late 1800s as multistory structures of iron and reinforced concrete were constructed. Features of the bearing wall concept remained, however, as these early structures generally depended on masonry walls to carry lateral loads. As buildings became taller and lighter, and as engineers developed better understanding of this new method of assembly, lateral loads were increasingly assumed to be resisted by the frame itself. Engineers such as Hardy Cross provided the basis for our early understanding of rigid frame behavior.

The development of improved materials and connection methods (welding, bolting, etc.) paralleled analytical development, and rigid frame construction grew rapidly from the early 1900s to the present day. The trend toward lighter construction and more slender elements was clearly established and continues today. A large part of structural engineering research is still devoted to studying various aspects of the rigid frame and its elements: beam-column behavior, moment–rotation characteristics of connections, lateral stability, and so on.

Types and Uses

Typical types of rigid frames are shown in Fig. 5–22. These range from the single story *portal frame,* which is used extensively for warehouses, light manufacturing and general-purpose structures, to the high-rise building frame used for offices and housing. Bridge piers are generally rigid frames in a direction transverse to the roadway section, and rigid frame bents may also be used for the major longitudinal bridge structure. Another type of rigid frame is shown in Fig. 5–23: instead of rigid beam-to-column connections, the column-to-footing connection is made rigid, and stability is achieved from a series of column cantilevers, or "flagpoles." This concept is frequently used with precast construction.

Reinforced concrete frames almost always include the rigid frame subsystem. Flat slab, flat plate, beam-

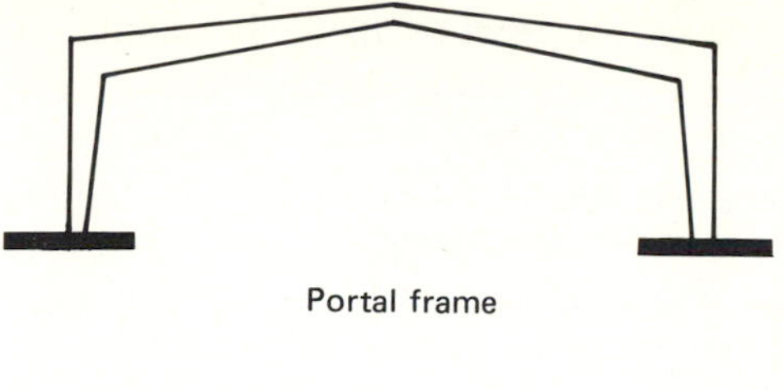

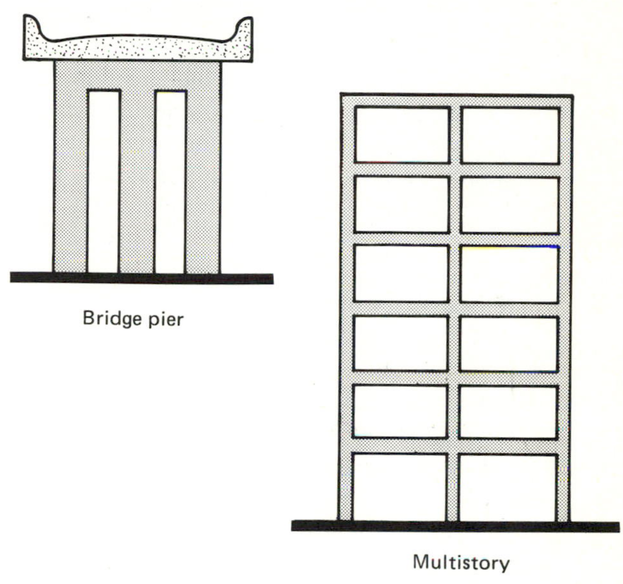

Figure 5–22. Rigid Frame Types

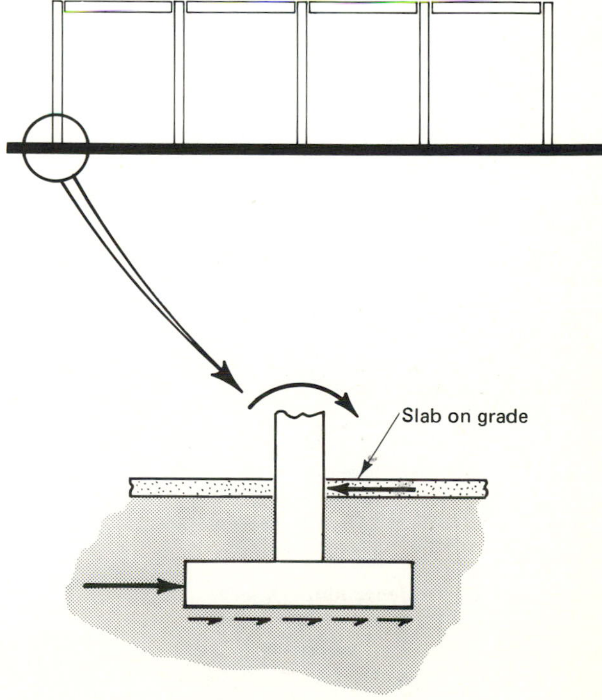

Figure 5–23. Cantilever Column Rigid Frame

and-slab, and pan joist construction all involve rigid frame action. In the flat slab and flat plate systems, an equivalent width of slab serves as the beam element. These structures may rely on the frame for lateral stability or may utilize shear walls, core areas, or bracing for lateral support.

The connection elements used to join beams and beam-columns include the moment-resisting steel connection (AISC Type 1), the semirigid steel connection (AISC Type 3), and various types of monolithic steel and concrete connections. Monolithic steel connections are primarily used for "bents" in which an entire section of the bent (including the connection) is fabricated in a continuous piece. Monolithic concrete connections require reinforcing details to develop the joint moment and to provide ductility, especially in earthquake areas.

Behavior

Rigid frame behavior can be studied by isolating a segment of a total frame, called a *subassemblage,* and observing the effects of axial load and bending moment on the individual members and their interaction with each other. Such a subassemblage is shown in Fig. 5–24(a).

If the subassemblage is restrained against sway, the effects of column axial load and bending moment applied at the joint are shown in Fig. 5–24(b). The bending moment is resisted by the four intersecting members in proportion to their relative stiffness (I/L) and creates the curvature shown. The axial load is resisted directly by the beam-column element but causes an additional column moment (moment magnification) due to the $P\Delta$ effect at midheight of the column.

Figure 5–24. Rigid Frame Subassemblage

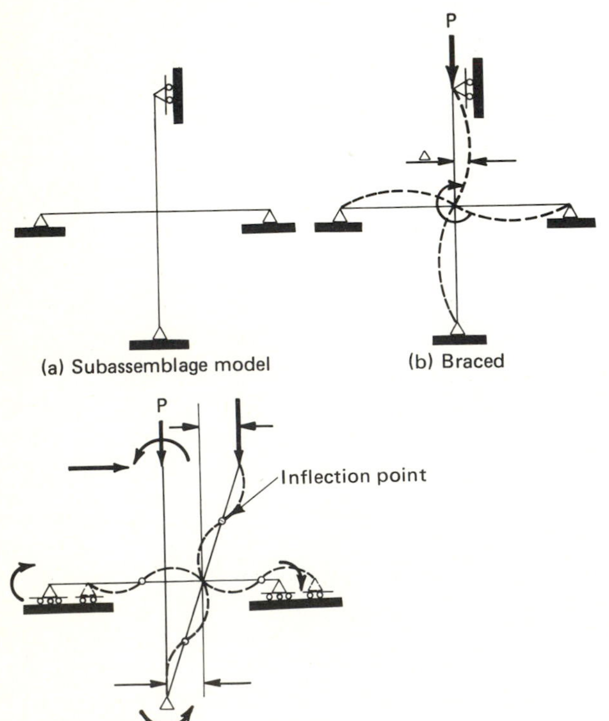

When sway is not restrained, the effects of lateral load, column load, and joint moment are shown in Fig. 5–24(c). Moment magnification effects due to $P\Delta$ are now greatest at the ends of the column and create additional bending moments in the beams equal to $P\Delta$. The connection moment restraint provided at joints A and B creates reverse curvature (S-curve) in the column, with the inflection point approximately at midheight. Assuming the inflection point at this location gives a column moment as follows (Fig. 5–25):

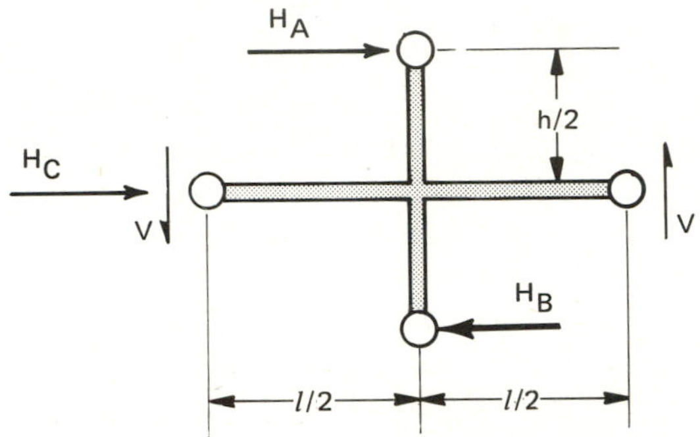

Figure 5–25. Subassemblage Free Body

$$M_c = \frac{Hh}{2} + \frac{P\Delta}{2} + M_G \qquad (5\text{–}10)$$

where M_c = column moment
H = lateral load (H_A or H_B)
h = story height
P = column axial load
Δ = sway
M_G = moment transferred to column by beam gravity moment

$P\Delta/2$ is obviously the moment magnification for the sway (unbraced) condition. This moment magnification is slightly less than the approximate value for δ_s given by Eq. 4–72.

Similar points of inflection occur in the beams, which gives a beam moment as follows:

$$M_b = \frac{Vl}{2} + \frac{P\Delta}{2} + M_G \qquad (5\text{–}11)$$

where l = beam span length
V = beam shear created by lateral load

$$= \frac{H_A(h/2) + H_B(h/2)^1}{l}$$

[1] Substitute h for $h/2$ for first-story columns with hinged bases.

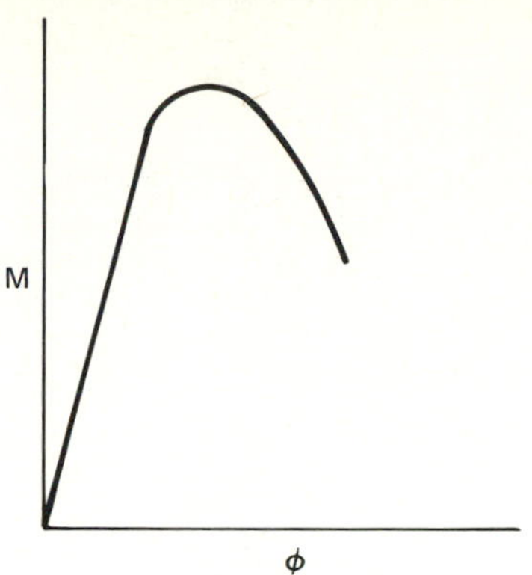

Figure 5–26. Ultimate Strength of Frame Subassemblage

The ultimate strength of the subassemblage is shown by the curve in Fig. 5–26, in which failure may be due to material failure, joint instability, or frame instability. Material failure occurs when the combined stress from moment and axial load causes crushing or rupture of the material. Joint instability may occur when the beam-columns reach a point on their moment–rotation curves (Fig. 5–27) at which further rotation leads to a decreasing ability to carry moment. *Rotation capacity* is thus very important in evaluating the ultimate strength of rigid frames.

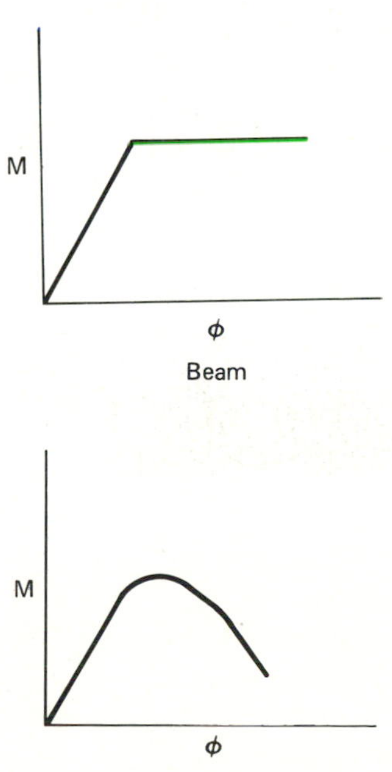

Figure 5–27. Ultimate Strength of Beam-Columns

Frame instability occurs when the frame is not restrained from sway and Δ becomes very large with the addition of small increments of lateral load, similar to the column buckling phenomenon. The stiffness properties of the frame determine the loads that can be carried without frame buckling occurring.

Approximate Analysis

Approximate analysis of rigid frames requires calculation of the following loads, forces, and stiffness factors:

- Moments due to gravity loads
- Moments due to lateral loads
- Column axial loads
- Lateral drift (Δ)
- Stability check

Since rigid frames are indeterminate, all of these factors (except column axial loads) depend on the member properties, which are not known initially. We must, therefore, make assumptions for analysis and later check their accuracy when individual members are sized using the approximate design methods.

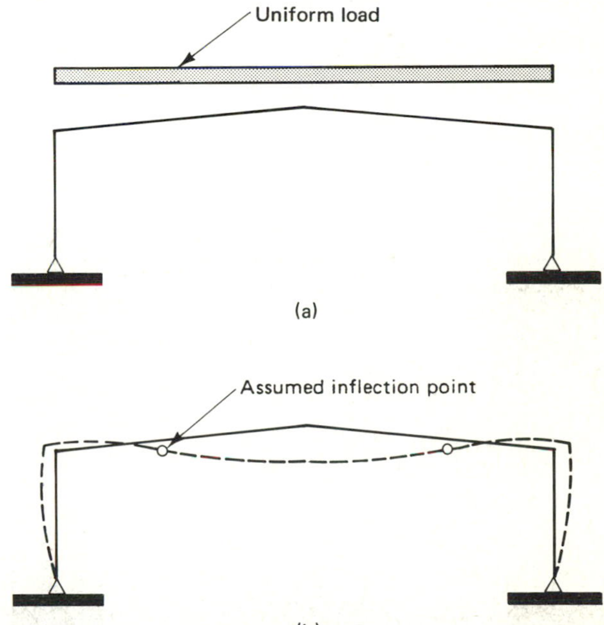

Figure 5–28. Portal Rigid Frames

Portal Frames. Portal frames such as shown in Fig. 5–28 can be analyzed by assuming inflection points based on a sketch of the loads and the deformed shape (Fig. 5–28b). Formulas for calculating portal frame reactions for several common loading cases are given in Fig. 5–29.

Multistory Frames. Beam moments due to *gravity loads* can be approximately determined by assuming inflection points corresponding to a fixed-end beam condition (Fig. 5–30). Column moments due to gravity load are the greatest under a "checkerboard" loading scheme (Fig. 5–31), and the unbalanced joint moment is equal

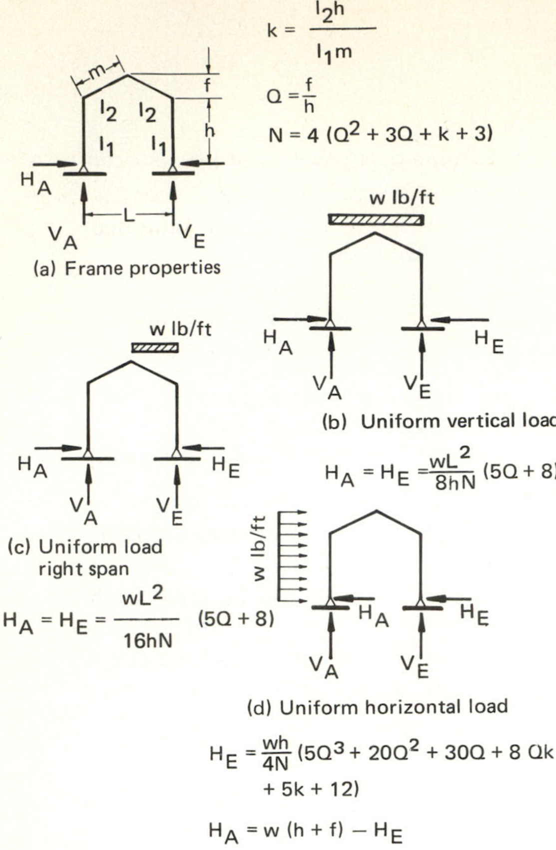

Figure 5–29. Portal Frame Reactions

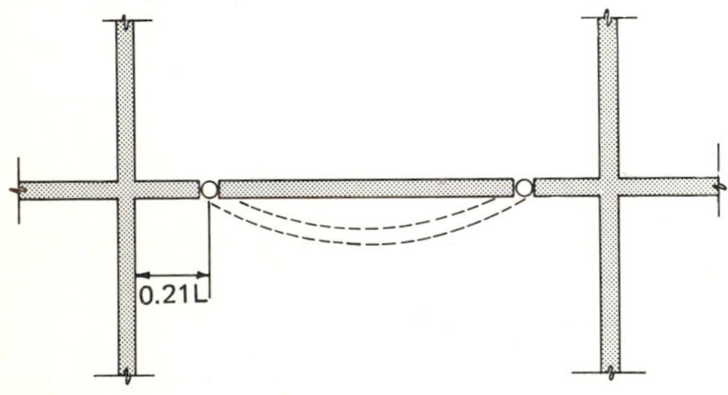

Figure 5–30. Location of Inflection Points for Rigid Frame Girders

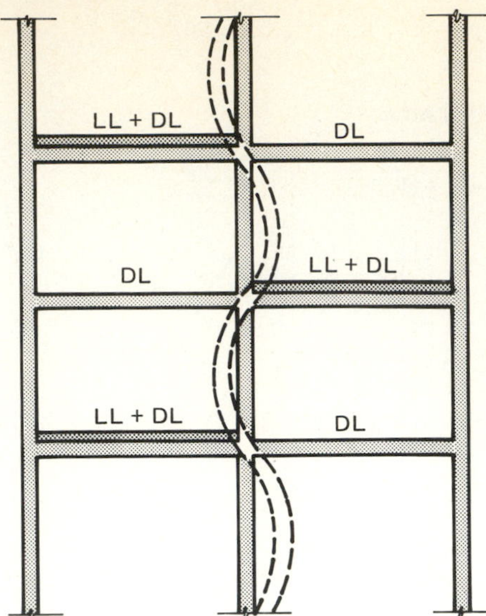

Figure 5–31. Placement of Live Load for Maximum Column Moment

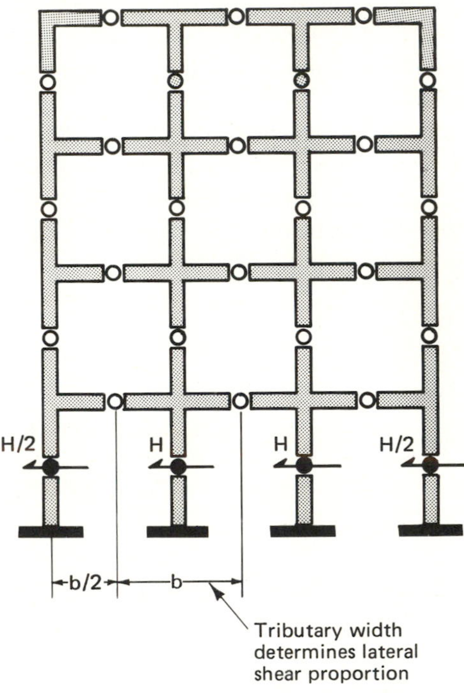

Figure 5–32. Portal Method of Lateral Load Analysis

to the difference in the fixed-end moments of the intersecting beams. In the absence of specific member properties to determine distribution of this unbalanced joint moment, one-fourth to one-half can be assumed taken by each column.

Moments due to *lateral loads* can be calculated by assuming inflection points at midheight of the columns and midspan of the beams as discussed earlier (Fig. 5–32). This is known as the "portal method" and is widely used for preliminary analysis. The lateral load carried by each column is proportional to its tributary width. Lower-story columns with hinged bases will obviously not have inflection points at midcolumn height but, instead, will carry a moment equal to the total column height times the lateral shear.

Column axial loads are approximately equal to the tributary area supported by each column plus the additional load produced by the overturning effects of the lateral load. The additional load in a column due to overturning can be assumed to be proportional to its distance from the centroidal axis of the column group (Fig. 5–33).

Determination of *lateral drift* (Δ) can be calculated for each story by using the concept of story stiffness developed in Ref. 9 and shown below:

$$S_T = \frac{12E}{h^3} \sum_{1}^{m} \left(\frac{I_c}{1 + U\psi} \right) - \frac{\Sigma P}{h} \qquad (5\text{–}12)$$

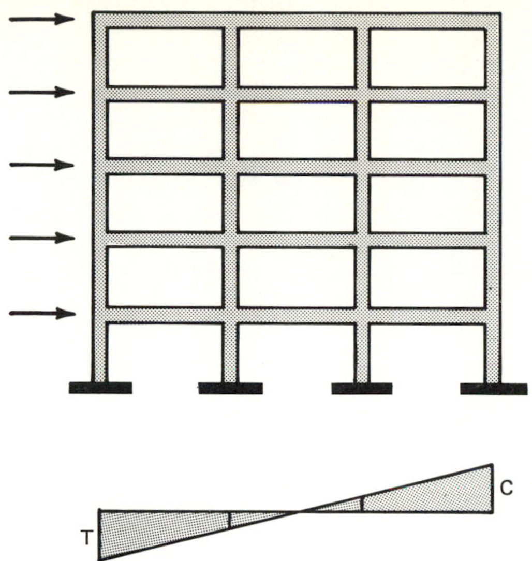

Figure 5–33. Axial Column Load Created by Lateral Overturning Forces

where S_T = story stiffness
E = modulus of elasticity
I_c = column moment of inertia
h = story height

$$\psi = \frac{I_c/h}{\Sigma\, Ib/Lb} \text{ at a joint}$$

$\Sigma\, P$ = total gravity loads on the story
U = ratio of beam's restraining moment to column moment; equal to 2 for intermediate stories, 1 for top story, and $\frac{3}{2}$ for bottom story with pinned base

By neglecting the P/h term as being of minor importance, the equation can be divided into two components as follows:

$$\frac{1}{S_T} = \frac{1}{S_c} + \frac{1}{S_g} \quad (5\text{–}13)$$

$$S_c = \frac{12E\,\Sigma\, I_c}{h^3},$$

$$S_g = \frac{12E\,\Sigma\, I_g/l}{h^2} \text{ (top-story column)} \quad (5\text{–}14)$$

$$S_g = \frac{6E\,\Sigma\, I_g/l}{h^2} \text{ (intermediate-story column)}$$

$$S_g = \frac{8E\,\Sigma\, I_g/l}{h^2} \text{ (bottom-story column)}$$

where S_c = column stiffness contribution
S_g = girder stiffness contribution
l = overall dimension of bay
$\Sigma\, I_g/l$ = summation of girder stiffness at each joint (two girders at interior joint; one at exterior)

Lateral drift is then given as follows:

$$\Delta = \frac{H}{S_T} = \frac{H}{S_g} + \frac{H}{S_c} \quad (5\text{–}15)$$

where H is the total lateral load on the story.

When $S_c/S_g > 0.5$, increasing the beam size will influence Δ greater than increasing column size. When $S_c/S_g < 0.5$, just the opposite is true.

Equation 5–12 can also be used to check overall *frame stability* by the following rearrangement:

$$\Sigma\, P = \frac{12E}{h^2} \sum_{1}^{m} \frac{I_c}{1 + 2\psi} \quad (5\text{–}16)$$

These approximate analysis techniques are illustrated in Example Problem 5–5.

Approximate Design

Design of rigid frames involves the following choices:

- Material selection
- Story height
- Frame height
- Column spacing
- Relative stiffness between columns and girders
- Functional role: vertical support only, or combined vertical and lateral support
- Element proportions

Reinforced concrete and structural steel are the most common materials used for rigid frame construction. Precast, prestressed concrete is also used, although connections are more difficult. Story height is usually dictated by spatial requirements, while overall frame height depends on such factors as optimum use of land, code restrictions, economic and functional comparisons of various combinations of floor area, and number of stories. Many of these factors involve total system optimization (as opposed to subsystem optimization), which will be addressed in Chapter 6. Column spacing has both a functional and structural acceptable range, usually between 20 and 40 ft (6.09 and 12.2 m) for multistory buildings and 40 to 150 ft (12.2 to 45.7 m) for single-story portal frames.

Relative stiffness between columns and girders affects moment distribution and lateral drift. In most conventional building frames, increasing the girder size is more effective in reducing lateral drift. The decision as to whether the rigid frame should function as the lateral support subsystem (in addition to its vertical support role) is usually based on the increment of cost necessary to carry lateral loads in the frame versus carrying them with another type of lateral support subsystem (i.e., shear walls or bracing). Moderate height buildings with sufficient number of columns per story can often carry lateral loads with very little premium, due mostly to the allowable stress increase for lateral load combinations. Figure 5–34 shows a comparison of premium cost versus number of stories.

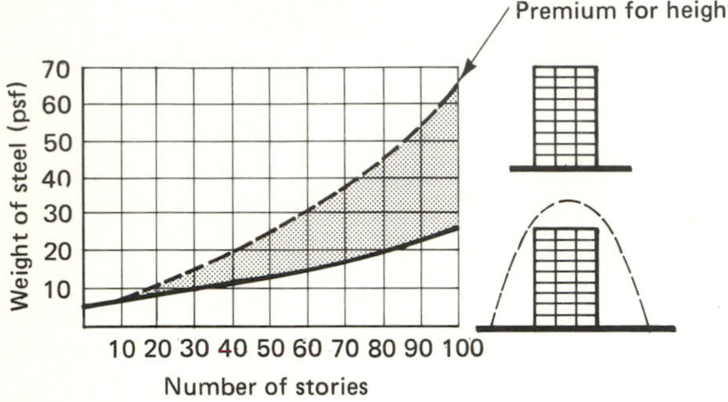

Figure 5–34. Lateral Load Premium

The following example problem illustrates the consideration of many of the choices involved in rigid frame design.

EXAMPLE PROBLEM 5–5: Preliminary Rigid Frame Design

Develop an approximate design for a six-story rigid-frame subsystem which is to be part of an office building structural system (Fig. 5–35). Minimum story height is 12 ft (3.66 m) and the building width (in direction of the frame) is 100 ft (30.5 m). The building location is Dallas, Texas. Determine the premium cost (if any) if the frame also carries the wind load. Spacing between frames is 20 ft (6.09 m).

Figure 5–35. Example Problem 5–5

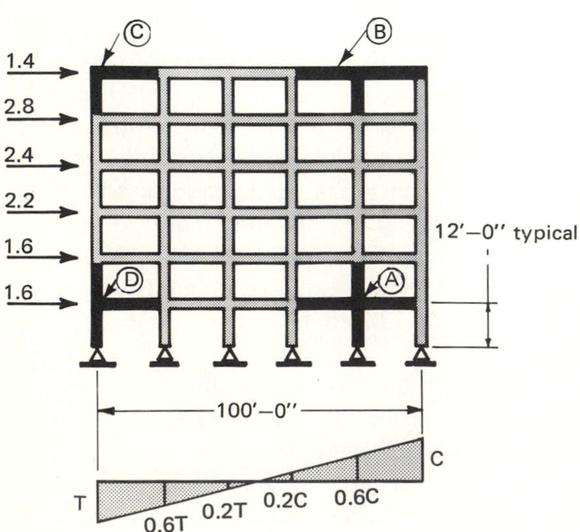

Solution

1. *Project requirements:*
   - Function: Office building
     - Story height = 12 ft (3.66 m)
     - Building height = 72 ft (21.9 m)
     - Building width = 100 ft (30.5 m)
     - Provide vertical support and consider lateral support role.
   - Esthetics: Assume no expression of structural frame.
   - Serviceability: Normal deflection limits.
     - Fire rating: Type II.
     - No special durability or other requirements.
   - Construction: Assume that time is extremely critical and that economy is paramount.
     - Both steel and reinforced concrete have acceptable levels of quality control in the geographical area.

2. *Loads:*
   a. *Live load:* 50 psf (offices)
      20 psf (minimum roof live load)
   b. *Dead load:* 80 psf (pan joist floor and roof)
      2 psf (mechanical and electrical)
      2 psf (ceiling system)
      10 psf (assumed girder weight)
      94 psf for floor (4,500 N/m²)
      8 psf (built-up roof and insulation)
      102 psf for roof (4883 N/m²)
   c. *Wind load:* Determine design pressure for walls.
      From Fig. 3–50, basic wind speed for Dallas:

$$V = 70 \text{ mph}$$

From Table 3–7, importance coefficient:

$$I = 1.07 \text{ for Category II}$$

From Table 3–8, gust response factor:

$$G_h = 1.67 \text{ for Exposure } A \text{ and } Z = 72 \text{ ft}$$

From Table 3–9, external pressure coefficient:

$$C_p = 0.8$$

$$p_z = q_z G_h C_p = 0.00256 k_z (IV)^2 G_h C_p$$

Height	k_z	q_z	p_z	p_h
0–15	0.12	1.7	2.3	3.9
20	0.15	2.2	2.9	3.9
25	0.17	2.4	3.2	3.9
30	0.19	2.7	3.6	3.9
40	0.23	3.3	4.4	3.9
50	0.27	3.9	5.2	3.9
60	0.30	4.3	5.7	3.9
70	0.33	4.7	6.3	3.9

$$p_z = 0.00256k_z[1.07(70)]^2(1.67)(0.8)$$

$$p_h = q_h G_h C_p = 4.7(1.67)(0.5)$$

3. *Material selection.* Material selection involves several serviceability and construction considerations. Normally, steel systems would be compared with reinforced-concrete systems and value engineering comparisons made. A reinforced-concrete rigid frame will be assumed for this example, with the following material properties:

$$\text{Concrete: } f'_c = 3000 \text{ psi (20.7 MPa)}$$

$$\text{Reinforcing steel: } f_y = 60{,}000 \text{ psi (413.7 MPa)}$$

4. *Column spacing.* Practical choices are 6 at 16.7 ft (5.1 m); 5 at 20 ft (6.1 m); 4 at 25 ft (7.6 m). Select 5 at 20 ft (6.1 m) assuming no interference with office or core layout.
5. *Gravity load analysis/design:*
   a. *Floor girder:* Assume inflection points for fixed-end beam.

$$w_u = 1.7(0.050 \text{ ksf})20 \text{ ft} + 1.4(0.094 \text{ ksf})20 \text{ ft} = 4.3 \text{ kips/ft (62,750 N/m)}$$

$$-M = \frac{w_u l^2}{12} = \frac{4.3(20)^2}{12} = 143 \text{ ft-kips (193,908 N/m)}$$

$$+M = \frac{w_u l^2}{8} = \frac{4.3(11.6)^2}{8} = 72.3 \text{ ft-kips (98,038 N/m)}$$

$$\text{Min. depth } (h) = \frac{20 \times 12}{21} = 11.4 \text{ in. (289.6 mm) (ACI coeff.)}$$

For F_b = 1.26 ksi/0.85% reinforcement (8.69 MPa):

$$S_{\text{req'd}} = \frac{M}{F_b} = \frac{143 \times 12}{1.26} = 1362 \text{ in.}^3 \text{ (22,336,800 mm}^3)$$

$$h^2 = \frac{1362(6)}{(12)}; \quad h = 26.09 \text{ in. (662.7 mm)}$$

Assume that b = 12 in. (304.8 mm). For F_b = 2.15 ksi/1.6% reinforcement (14.8 MPA):

$$S_{\text{req'd}} = 798 \text{ in.}^3 \text{ (13,079,220 mm}^3)$$

$$h^2 = \frac{798(6)}{12}; \qquad h = 19.9 \text{ in. (505.5 mm)}$$

Final design will allow some moment redistribution. Keep girder size minimum for effect on building height and better balance with positive moment.

$$\text{Shear } (V_u) = \frac{w_u l}{2} = \frac{4.3(20)}{2} = 43 \text{ kips (191,264 N)}$$

$$\text{Shear stress } v = \frac{43}{12(20)} = 0.179 \text{ ksi (1.23 MPa)}$$

Try 12 in. × 20 in. (304.8 mm × 508 mm) girder (1.6% steel).

   b. *Roof girder:*

$$w_u = 1.4(0.102 \text{ ksf})20 \text{ ft} + 1.7(0.020 \text{ ksf})20 = 3.54 \text{ kips/ft (51,659 N/m)}$$

$$-M_u = \frac{w_u l^2}{12} = \frac{3.54(20)^2}{12} = 118 \text{ ft-kips (160,008 N/m)}$$

$$+M_u = \frac{w_u l^2}{8} = \frac{3.54(11.5)^2}{8} = 59.5 \text{ ft-kips (80,682 N/m)}$$

$$S_{\text{req'd}} = \frac{M}{F_b} = \frac{118 \times 12}{1.26} = 1123 \text{ in.}^3 \text{ (18,405,970 mm}^3)$$

$$h^2 = \frac{1123(6)}{12}; \qquad h = 23.69 \text{ in. for } F_b = 1.26 \text{ ksi}$$

$$h = 18.14 \text{ in. for } F_b = 2.15 \text{ ksi}$$

Try 12 in. × 20 in. (304.8 mm × 508 mm) girder (1.6% steel).

   c. *Interior column at first story:*
   (1) Axial load:

$$\text{Roof DL} = 0.102 \text{ ksf (20)(20)} = 40.8 \text{ kips (181,478 N)}$$

$$\text{Floor DL} = 0.094 \text{ ksf (20)(20)5} = 188 \text{ kips (836,224 N)}$$

$$\text{Roof LL} = 0.020 \text{ ksf (20)(20)} = 8 \text{ kips (35,584 N)}$$

LL reduction:

$$L = L_o\left(0.25 + \frac{15}{\sqrt{A_I}}\right)$$

$$\text{Area} = 5 \text{ floors} \times 20 \text{ ft} \times 20 \text{ ft} = 2000 \text{ sf}$$

$$A_I = 4 \times 2000 \text{ sf} = 8000 \text{ sf}$$

By Eq. 3–1:

$$L = 50 \text{ psf}\left(0.25 + \frac{15}{\sqrt{8000}}\right) = 20.9 \text{ psf}$$

$$\text{Maximum reduction} = 0.40(50) = 20.0 \text{ psf}$$

$$\text{Uniform floor LL} = 20.9 \text{ psf (assume 20.0 psf)}$$

$$\text{Total floor LL} = (0.02 \text{ ksf})(20)(20)5$$
$$= 40 \text{ kips } (177{,}920 \text{ N})$$

$$P_u = 1.4(40.8 + 188) + 1.7(8 + 4)$$
$$= 401 \text{ kips } (1{,}783{,}648 \text{ N})$$

(2) Moment: Unbalanced moment at joint:

$$M_o = 0.083(w_u l^2 - w_D l^2)$$

$$w_u = 4.3 \text{ kips/ft } (62{,}750 \text{ N/m})$$

$$w_D = 1.4(20)(0.094) = 2.6 \text{ kips/ft } (37{,}942 \text{ N/m})$$

$$M_o = 0.083[4.3(20)^2 - 2.6(20)^2]$$

$$= 56.4 \text{ ft-kips } (76{,}478 \text{ Nm})$$

Assume that one-half goes to each column.

$$M_u = \frac{56.4}{2} = 28.2 \text{ ft-kips } (38{,}239 \text{ Nm})$$

(3) Calculate required area.

$$A_{\text{req'd}} = \frac{P_u'}{F_a'} + \frac{\delta(6)}{h}\frac{M_u}{F_b'} \qquad (4\text{–}79)$$

Try $h = 12$ in. (304.8 mm).

$$\frac{kl}{r} = \frac{1.0(12 \times 12)}{0.3(12)} = 40$$

Since this is close to 34, neglect moment magnification and set $\delta_b = 1.0$.

$$P_u' = P_u - 0.030bdf_c'$$
$$= 401 - .30(12)(0.8 \times 12)3$$
$$= 297 \text{ kips } (1{,}321{,}056 \text{ N}) \qquad (4\text{–}77)$$

$$A = \frac{297}{2.77} + \frac{1.0(6)}{12}\frac{28.2 \times 12}{(3.13)}$$
$$= 107.2 + 54 = 161.2 \text{ in.}^2 \ (103{,}974 \text{ mm}^2)$$

Try a 12 in. × 12 in. (304.8 mm × 304.8 mm) column (5% steel).

(4) Check moment distribution.

$$\text{Girder stiffness} = k_g = \frac{I_g}{L}$$

$$I_g = \frac{12(20)^3}{12}$$
$$= 8000 \text{ in.}^4 \ (3.32 \times 10^9 \text{ mm}^4)$$

$$k_g = \frac{8000}{20} = 400$$

$$\text{Column stiffness} = k_c = \frac{I_c}{L}$$

$$I_c = \frac{12(12)^3}{12}$$
$$= 1728 \text{ in.}^4 \ (7.18 \times 10^8 \text{ mm}^4)$$

$$k_c = \frac{1728}{12} = 144 \text{ kips } (10{,}115 \text{ Nm})$$

$$M_c = \frac{k_c}{\Sigma k} M_o = \frac{144}{1088}(56.4)$$
$$= 7.46 \text{ ft-kips } (10{,}115 \text{ Nm})$$
$$< 28.2 \text{ ft-kips } (38{,}329 \text{ Nm})$$

Initial distribution is o.k.

d. *Exterior column at first story:*

(1) Axial load:

$$P_u = 200 \text{ kips } (889{,}600 \text{ N})$$
$$(\tfrac{1}{2} \times \text{int. column load})$$

(2) Moment:

$$M_o = 0.083(4.3)(20)^2 = 142 \text{ ft-kips } (192{,}552 \text{ Nm})$$

Assume that one-half goes to each column.

$$M_u = \frac{142}{2} = 71 \text{ ft-kips } (96{,}276 \text{ Nm})$$

(3) Calculate required area.

$$P_u' = 200 - 0.30(12)(0.8 \times 12)3 = 96$$

$$A = \frac{96}{2.77} + \frac{1.0(6)}{12}\frac{71 \times 12}{3.13}$$
$$= 34.7 + 136 = 171 \text{ in.}^2 \ (110{,}295 \text{ mm}^2)$$

Try a 12 in. × 12 in. (304.8 mm × 304.8 mm) column (6% steel).

e. *Interior column at top story:*

(1) Axial load:

$$P_u = 1.4(40.8) + 1.7(8) = 70.7 \text{ kips } (314{,}473 \text{ N})$$

(2) Moment: Unbalanced moment at joint:

$$M_o = 0.083[3.54(20)^2 - 2.85(20)^2]$$

$$w_D = 1.4(102)20 = 2.85 \text{ kips/ft}$$

$$M_o = 22.9 \text{ ft-kips}$$

$$M_o = \frac{144}{288}(22.9) = 11.5 \text{ ft-kips } (15{,}594 \text{ Nm})$$

(3) By inspection, use a 12 in. × 12 in. (304.8 m × 304.8 m) column (2% steel).

f. *Exterior column at top story:*

(1) Axial load:

$$P_u = \frac{70.7}{2} = 35.4 \text{ kips } (157{,}459 \text{ N})$$

(2) Moment:

$$M_o = 0.083(3.54)(20)^2 = 117 \text{ ft-kips}$$

$$k_c = \frac{I}{L} = \frac{1728}{12} = 144$$

$$k_g = \frac{I}{L} = \frac{8000}{20} = 400$$

$$M_o = \frac{144}{544}(117) = 30.9 \text{ ft-kips } (41{,}900 \text{ Nm})$$

(3) By inspection, use a 12 in. × 12 in. (304.8 m × 304.8 m) column (2% steel).

g. *Cost tabulation:*

(1) Floor girders: Assume that $-A_s$ extends to the one-third point.

$$-A_s \text{ at } 1.6\% = 0.016 \text{ (12 in.} \times \text{20 in.)} (0.67 \times 20 \times 12) = 617 \text{ in.}^3$$

$$+A_s \text{ at } 0.85\% = 0.0085(12 \text{ in.} \times 20 \text{ in.})(20 \times 12) = 489 \text{ in.}^3$$

$$\text{No. of stirrups} = \frac{L}{d/2} = \frac{20(12)}{9} = 26.6 \quad \text{(assume 27)}$$

Assume No. 3 stirrup: A_s = 0.11 in.^{2}.

$$\text{Stirrup steel} = 27(18 \text{ in.} + 18 \text{ in.} + 10 \text{ in.})(0.11 \text{ in.}^2) = 136 \text{ in.}^3$$

$$\text{Total steel} = 617 + 489 + 136 = 1242 \text{ in.}^3 \ (2.03 \times 10^7 \text{ mm}^3)$$

$$\text{Weight of steel} = 1242 \text{ in.}^3(0.284 \text{ lb/in.}^3) = 352 \text{ lb (1565 N)}$$

$$\text{Concrete quantity} = \frac{12(20)(20)}{144(27)} = 1.23 \text{ cu yd } (0.93 \text{ m}^3)$$

Check: 352 lb/1.23 cu yd = 286 lb/cu yd (1682 N/m^3). (Compare with parameter values for typical construction. In this case, 286 lb/cu yd seems a little high.)

(2) Roof girders: Based on previous calculation for floor girder, assume that steel = 200 lb/cu yd (1176 N/m^3).

(3) Interior columns:

First-story steel at 5%:

$$0.05(12 \text{ in.} \times 12 \text{ in.})(12 \text{ ft} \times 12 \text{ in.}) = 1036 \text{ in.}^3$$

Top story steel at 2%:

$$0.02(12 \text{ in.} \times 12 \text{ in.})(12 \text{ ft} \times 12 \text{ in.}) = 414 \text{ in.}^3$$

Assume average steel = (1036 + 414)/2 = 725 in.^{3}. Increase one-third for splice = 725 × 1.3 = 942 in.^{3}

$$\text{Column ties} = \frac{12 \text{ ft} \times 12 \text{ in.}}{12 \text{ in.}} = 12 \text{ spaces}$$

(Assume No. 3 ties 12 in. o.c.)

$$\text{Tie steel} = 12(10 + 10 + 10 + 10)(0.11) = 52.8 \text{ in.}^3$$

$$\text{Total steel} = (942 \text{ in.}^3 + 52.8 \text{ in.}^3)0.284 \text{ lb/in.}^3 = 282 \text{ lb (1254 N)}$$

$$\text{Concrete} = \left(\frac{12 \text{ in.} \times 12 \text{ in.}}{144}\right)\frac{12 \text{ ft}}{27} = 0.44 \text{ cu yd } (0.33 \text{ m}^3)$$

Check:

$$\frac{282 \text{ lb}}{0.44 \text{ cu yd}} = 640 \text{ lb/cu yd } (3800 \text{ N/m}^3)$$

(4) Exterior columns: By inspection, use same quantities as for interior columns.

Table 5–5 shows the complete cost tabulation.

Table 5–5 Estimated Costs for Example Problem 5–5

	Gravity load			*Gravity + wind*		
	Columns	*Girders*	*Cost*	*Columns*	*Girders*	*Cost*
Concrete	15.8 cy (11.9 m^3)	29.5 cy (22.3 m^3)	$ 2,862	16.3 cy (13.3 m^3)	29.5 cy (22.3 m^3)	$ 2,894
Reinforcing steel	10,152 lb (45,156 N)	10,057 lb (44,733 N)	$ 7,477	10,859 lb (48,300 N)	10,057 lb (44,733 N)	$ 7,739
Forms	1,728 sf (160.7 m^2)	1,000 sf (93 m^2)	$ 8,081	1,751 sf (162.8 m^2)	1,000 sf (93 m^2)	$ 8,148
			$18,420			$18,781

$$\text{Unit cost} = \frac{\$18{,}420}{45.3 \text{ cy}} = \$406/\text{cy}$$

6. *Wind load analysis/design*
   a. *Interior column at first story:*

(1) Calculate forces. Using subassembly A (Fig. 5–35), assume uniform load = 9.0 psf for preliminary computations.

Total wind load at second story:

$$9.0(54\text{ ft})(20\text{ ft}) = 9.7\text{ kips }(43{,}146\text{ N})$$

Total wind load at first story:

$$9.0(56\text{ ft})(20\text{ ft}) = 10.1\text{ kips }(44{,}925\text{ N})$$

Column shear:

$$\frac{9.7}{5} = 1.9\text{ kips};\qquad \frac{10.1}{5} = 2.0\text{ kips }(8896\text{ N})$$

(2) Based on member properties computed for gravity load design, calculate approximate drift. By Eq. 5–15,

$$\Delta = \frac{H}{S_T}$$

Use $I = 0.4I_g$ for beams and $I = 0.8I_g$ for columns, since this is a concrete system.

$$\begin{aligned} S_c &= \frac{12E\,\Sigma\,Ic}{h^3} \\ &= \frac{12(3\times 10^3)(0.8)(1728\times 6)}{(12\times 12)^3} \\ &= 100\text{ kips/in. }(17{,}512\text{ N/mm}) \end{aligned}$$

$$\begin{aligned} S_g &= \frac{8E\,\Sigma\,Ig/l}{h^2} \\ &= \frac{8(3\times 10^3)(0.4)(8000\times 10)(20\times 12)}{(12\times 12)^2} \\ &= 154\text{ kips/in. }(26{,}968\text{ N/mm}) \end{aligned}$$

$$\begin{aligned} \Delta_u &= \frac{H_u}{S_T} = \frac{0.75(1.7)(10.1)}{154} + \frac{0.75(1.7)(10.1)}{100} \\ &= 0.21\text{ in. }(5.4\text{ mm}) \end{aligned}$$

(3) Calculate moment magnifier due to wind. By Eq. 4–72,

$$\begin{aligned} Q &= \frac{\Sigma\,P_u\Delta_u}{H_u h_s} = \frac{(0.75)(401\times 5)0.21}{(0.75)(1.7\times 10.1)(12\times 12)} \\ &= 0.17 > 0.04 \end{aligned}$$

$$\delta_s = \frac{1}{1-Q_u} = \frac{1}{1-0.17} = 1.20$$

(4) Calculate the column moment (assume hinged footing).

$$\begin{aligned} M_c &= H(h) + M_G = 2.0\text{ kips }(12\text{ ft})(1.7) + 28.2 \\ &= 40.8\text{ ft-kips (WL)} + 28.2\text{ ft-kips (LL)} \\ &= 69.0\text{ ft-kips }(93{,}564\text{ Nm}) \end{aligned}$$

(5) Calculate axial load due to overturning.

$$M_{OT} = \Sigma\,Hh$$

$$\begin{aligned} \Sigma\,Hh &= 0.009\text{ ksf}(72\text{ ft})(35\text{ ft})(20\text{ ft}) \\ &= 466.6\text{ ft-kips }(632{,}710\text{ Nm}) \end{aligned}$$

Resisting moment:

$$2[1.0C(50) + 0.6C(30) + 0.2C(10)] = 140C$$

$$C = \frac{466.6}{140} = 3.33\text{ kips }(14{,}812\text{ N})$$

(6) Calculate the total axial load.

$$\begin{aligned} P_u &= 401 + 0.6(3.33)1.7 \\ &= 404\text{ kips }(1{,}796{,}992\text{ N}) \end{aligned}$$

(7) Calculate required column size by Eq. 4–79.

$$A = \frac{P'_u}{F'_a} + \frac{\delta(6)}{h}\frac{M}{F'_b}$$

$$\begin{aligned} P'_u &= 404 - 0.30(12)(0.8\times 12)3 \\ &= 300\text{ kips} \end{aligned}$$

$$\begin{aligned} A &= \frac{0.75(300)}{2.77} + \frac{1.0(6)(28.2\times 12)(0.75)}{12(3.13)} \\ &\quad + \frac{1.20(6)(40.8\times 12)(0.75)}{12(3.13)} \\ &= 81.2 + 40.5 + 70.4 \\ &= 192.1\text{ in.}^2\ (123{,}905\text{ mm}^2) \end{aligned}$$

Use a 12 in. × 14 in. (304.8 mm × 355.6 mm) column (6% steel).

b. *Interior column at second story:*

(1) Calculate the column moment.

$$\begin{aligned} M_c &= H\left(\frac{h}{2}\right) + M_G = 1.9\left(\frac{12}{2}\right) + 28.2 \\ &= 11.4 + 28.2 = 39.6\text{ ft-kips }(53{,}698\text{ Nm}) \end{aligned}$$

(2) Calculate the axial load.

$$\begin{aligned} \text{Dead load} &= 1.4[(0.094)(20)(20)4 + 40.8] \\ &= 267\text{ kips} \\ \text{Live load} &= 1.7[(0.40)(0.050)(20)(20)4 + 8] \\ &= \underline{68}\text{ kips} \\ &\quad 335\text{ kips }(1{,}490{,}080\text{ N}) \end{aligned}$$

(3) Calculate the required column size. Assume that $\delta_b = 1.0$ and $\delta_s = 1.0$.

$$P'_u = 335 - 0.30(12)(0.8\times 12)3 = 231\text{ kips}$$

$$A = \frac{0.75(231)}{2.77} + \frac{0.75(1.0)(6)(39.6 \times 12)}{12(3.13)}$$
$$= 119.5 \text{ in.}^2 \text{ (77,078 mm}^2\text{)}$$

This is less than required for gravity load design. No change needed.

c. *Exterior columns at first and second stories:* Assume 12 in. × 14 in. (304.8 mm × 355.6 mm) column at first story and no change at second story.

d. *Exterior and interior columns at top story:* By inspection, no change.

e. *Floor girder at second floor:*

(1) Calculate wind shear.

$$V = \frac{1.9(6 \text{ ft}) + 2.0(12 \text{ ft})}{20 \text{ ft}} = 1.77 \text{ kips (7873 N)}$$

(2) Calculate wind moment.

$$M_w = 1.77(10 \text{ ft}) = 17.7 \text{ ft-kips (24,001 Nm)}$$

(3) Calculate M_u.

$$M_u = M_w + M_G = 1.7(17.7) + 143$$
$$= 173 \text{ ft-kips (234,588 Nm)}$$

$$S_{\text{req'd}} = \frac{M}{F_b} = \frac{0.75(173)(12)}{2.15}$$
$$= 724 \text{ in.}^3 \text{ (11,873,600 mm}^3\text{)}$$

This is less than required for gravity load design, therefore no change.

f. *Floor girder at roof:* By inspection, no change.

7. *Premium for wind load:* Table 5–5 indicates that an added cost of only $361 is required to carry the wind.

5–5 GRIDS

Grids are horizontal distribution subsystems constructed of intersecting parallel beam elements joined by rigid or continuous connections (Fig. 5–36). Loads are distributed in two orthogonal directions by this subsystem in proportion to the relative stiffness of the intersecting members. The grid derives its strength from the interaction between the intersecting members, similar to the rigid frame but with loads applied perpendicular to the plane of the subsystem.

This subsystem is indeterminate and thus requires preliminary member properties for analysis. Its two-way load distribution has prompted its use as a mathematical model for other more complex two-way systems such as slabs, walls, and space frames. In addition to bending stiffness, many grid members also possess significant torsional stiffness, which assists in the load sharing between the individual members.

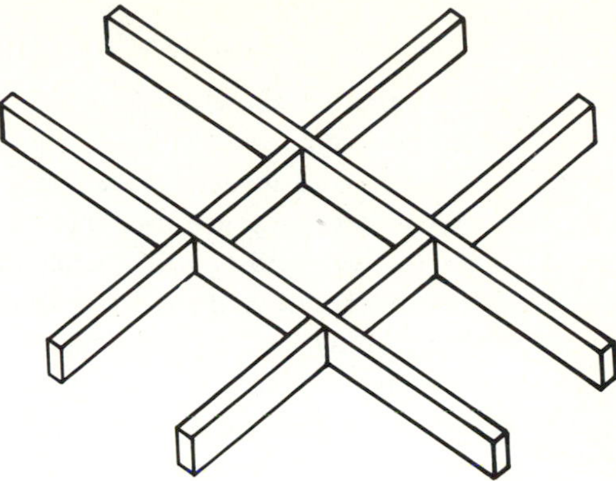

Figure 5–36. Grid Subsystem

History and Development

Probably the earliest grid structure was constructed quite accidentally in an effort to span distances longer than the length of available timber members (Fig. 5–37). This method of floor framing was conceived for just that purpose, and without any analytical understanding, just practical knowledge and experience. Other practical grid structures developed from systems of overlapping two-way timbers. This intuitive use of the grid form has continued for residential and other light frame construction.

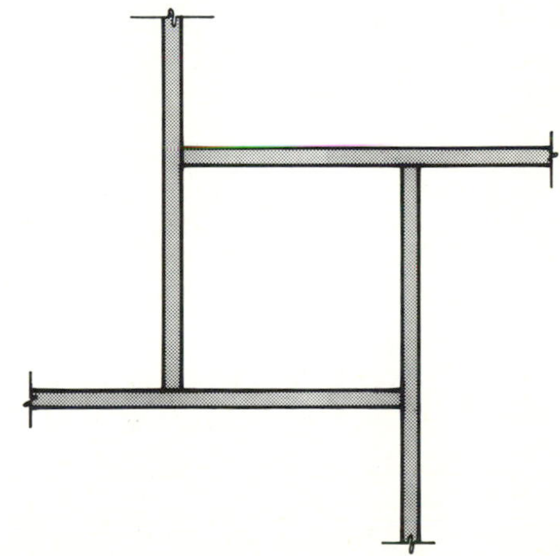

Figure 5–37. Early Grid Structure

True grid frameworks, with intersecting members in the same plane, developed with the advent of steel and reinforced concrete materials and connection methods during the late nineteenth and early twentieth centuries. Grids of major structural significance first appeared during the 1930s in response to the need for longer-span roof systems. Various concrete floor systems also incorporated many of the properties of the grid—the waffle slab, for example. Current use of the grid subsystem continues to be primarily for roof systems and concrete floor systems.

Types and Uses

The major types of grids are shown in Fig. 5–38. These types include closely spaced girders, widely spaced girders with beams in between, and girders placed diagonally. As single-layer assemblies, they are used primarily for roof systems. Double-layer grids (with connecting web members) are a basic component of the space frame subsystem which we will study later in this chapter.

Grid members may consist of rectangular beams, wide-flange rolled sections, plate girders, or trusses. Structural steel is the most common material used for grid construction because of its efficient cost-flexural strength ratio. Connections between steel members are normally made by welding or high-strength bolting.

Figure 5–38. Types of Grids

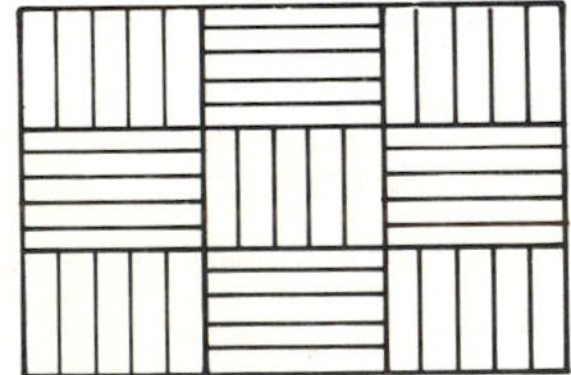

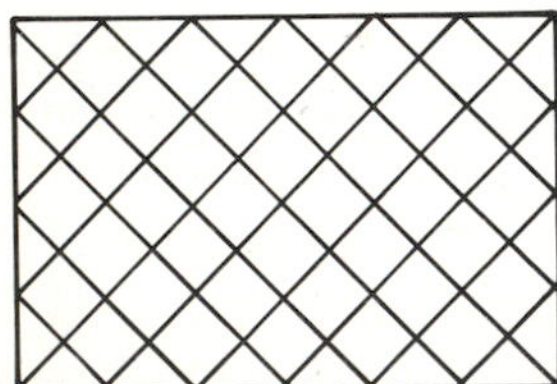

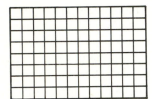

Behavior

We can study the behavior of the grid using a simple two-beam model as shown in Fig. 5–39. A load applied at the intersection will be partially supported by each beam, with P_1 carried by beam 1 and P_2 carried by beam 2. The deflection at the intersecting point must be the same for both beams, giving the following relationship:

$$\delta_1 = \delta_2, \qquad \frac{P_1 l_1^3}{48EI_1} = \frac{P_2 l_2^3}{48EI_2} \qquad (5\text{–}17)$$

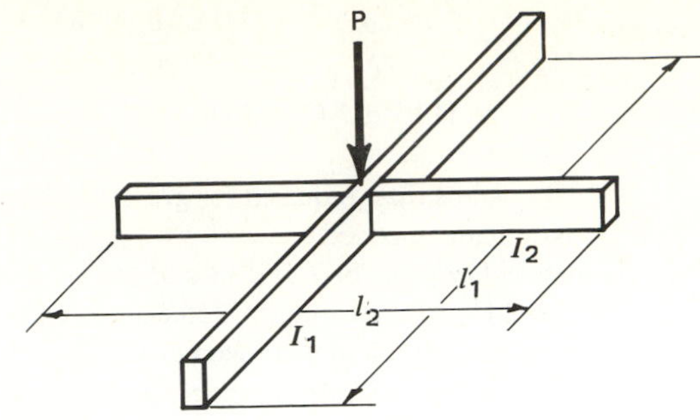

Figure 5–39. Grid Study Model

$$\frac{P_1}{P_2} = \frac{(l_2)^3 I_1}{(l_1)^3 I_2}$$

From this expression, one-half of the load goes each direction for beams with equal I's and equal spans. We can further understand the nature of the grid by examining the effects of varying the moments of inertia and varying the relative spans. At l_2/l_1 equals 1.33, P_1/P_2 equals 2.37, indicating that approximately 70% of the load is carried by beam 1. At l_2/l_1 equals 2.0, P_1/P_2 equals 8.0, indicating that 89% of the load is carried by beam 1. It is apparent from this study that very little two-way action is involved when the ratio of the two spans exceeds 2.

Load transfer in multimember grids is further aided by the torsional interaction between the members (Fig. 5–40). The torsional forces increase the stiffness of the system and its ability to transfer load. This effect is illustrated by the following example.

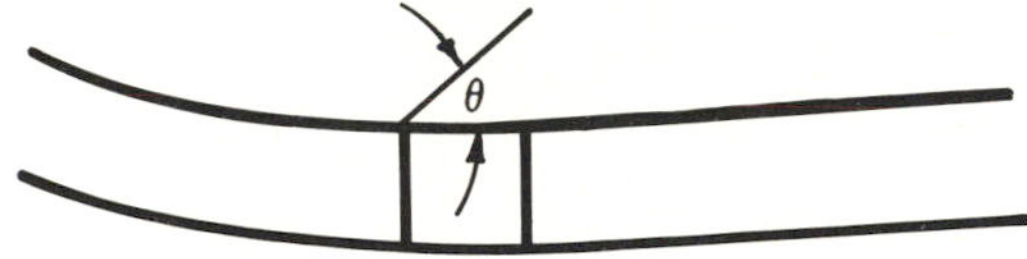

Figure 5–40. Torsional Action in Grids

EXAMPLE PROBLEM 5–6: Influence of Torsion on Grid Behavior

Determine the deflection for the two-member concrete grid shown in Fig. 5–41, neglecting torsion and then considering torsion. Assume that the grid members are 12 in. wide × 24 in. deep (304.8 mm × 609.6 mm).

Figure 5–41. Example Problem 5–6

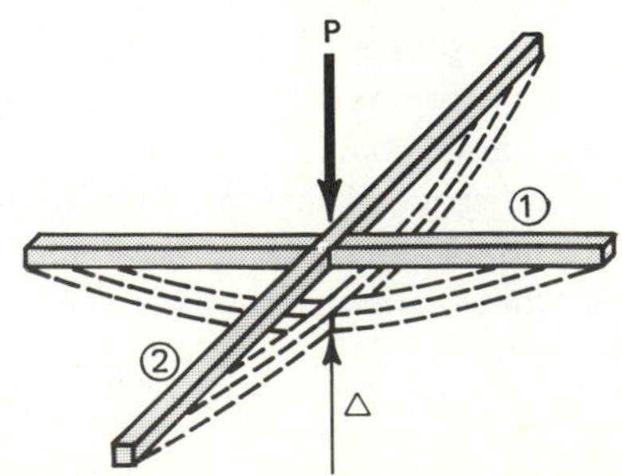

Solution

1. Neglecting torsion effects:

$$\Delta = \frac{Pa^2b^2}{3EIl}; \qquad a = 20 \text{ ft}, \quad b = 40 \text{ ft}$$

$P_1 = P_2$ since spans and member properties are equal

$$I = \frac{12(24)3}{12} = 13{,}824 \text{ in.}^4$$

$$\Delta = \frac{5(20)^2(40)^2(1728)}{3(3 \times 10^3)(13{,}824)60} = 0.74 \text{ in. (18.8 mm)}$$

2. Considering torsion effects:
   a. Calculate fixed end moments.

$$\text{FEM}_a = \frac{6EI\Delta}{la^2} = \frac{6(3 \times 10^3)(13{,}824)\Delta}{(20 \times 12)^2} = 4320\Delta$$

$$\text{FEM}_b = \frac{6EI\Delta}{lb^2} = \frac{6(3 \times 10^3)(13{,}284)\Delta}{(40 \times 12)^2} = 1080\Delta$$

   b. Calculate moment distribution factors.

$$S_b = \text{bending stiffness} = \frac{4EI}{l}$$

$$S_t = \text{torsional stiffness} = \frac{Gk}{l}$$

Assume that $G = 0.4E$; k = torsional constant $\simeq \Sigma\, bt^3/3$; $I = 13{,}824$ in.^{4}; $k = 55{,}296$ in.^{4}

Stiffness factors:

$$S_{ba} = \frac{4(3 \times 10^3)(13{,}824)}{(20 \times 12)} = 6.9 \times 10^5$$

$$S_{bb} = \frac{4(30 \times 10^3)(13{,}824)}{(40 \times 12)} = 3.5 \times 10^5$$

$$S_{ta} = \frac{0.4(3 \times 10^3)(55{,}296)}{(20 \times 12)} = 2.8 \times 10^5$$

$$S_{tb} = \frac{0.4(3 \times 10^3)(55{,}296)}{(40 \times 12)} = 1.4 \times 10^5$$

$$\Sigma\, S = 14.6$$

   c. Calculate moment distribution.

$$M_o = \text{unbalanced joint moment} = (4320\Delta - 1080\Delta) = 3240\Delta$$

$$T = \text{torsion force} = M_o \left(\frac{S_{ta} + S_{ta}}{\Sigma S}\right)$$

$$= (3240\Delta)\frac{2.8 + 1.4}{14.6} = 932\Delta$$

   d. Calculate deflection.

$$\text{Deflection due to torsion} = \frac{T}{6EI}(0.147L^2)$$

(Ref. 21, p. 104)

$$\Delta = 0.74 \text{ in.} - \frac{T}{6EI}(0.147L^2)$$

$$= 0.74 \text{ in.} - \frac{932\Delta(0.147)}{6(3 \times 10^3)(13{,}824)}(60 \times 12)^2$$

$$= 0.74 \text{ in.} - 0.285\Delta$$

$$= 0.57 \text{ in. (14.5 mm)}$$

Neglect of the torsion force is conservative as far as deflections are concerned. Torsional stress can also be neglected if sufficient ductility is available to prohibit cracking and premature joint failure.

Approximate Analysis

Grid subsystems are highly indeterminate but can be approximately analyzed for symmetrical framing and symmetrical loads by the method of "consistent deflections" which we used earlier in discussing behavior. For a simply supported grid, the unknowns generally can be selected so that there is one unknown and one equation for each interior node. The number of equations required can be reduced if the framing is made symmetrical about perpendicular axes and the loading is symmetrical. For symmetrical grids subjected to unsymmetrical loading, the amount of work involved in analysis often can be decreased by resolving loads into components which are symmetrically placed, but varying in direction. Figure 5–42 illustrates this process for a single unsymmetrical load on a grid. The analysis requires the solution of four sets of simultaneous equations, but there are fewer equations in each set than for an unsymmetrical loading formulation. The number of unknowns may be further de-

Figure 5–42. Unsymmetrical Grid Loading

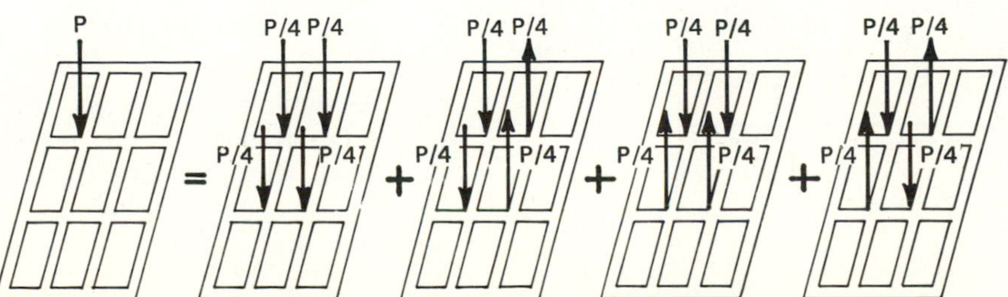

creased when the proportion of a load to be assigned to a girder at a node can be determined by inspection or simple computation. For example, for a square orthogonal grid under symmetrical or unsymmetrical loading, each girder at the central node carries half the load applied at that location.

For analysis of simply supported grid girders, influence coefficients for deflection induced by a unit load are useful. The following sections provide coefficients for several types of loading.

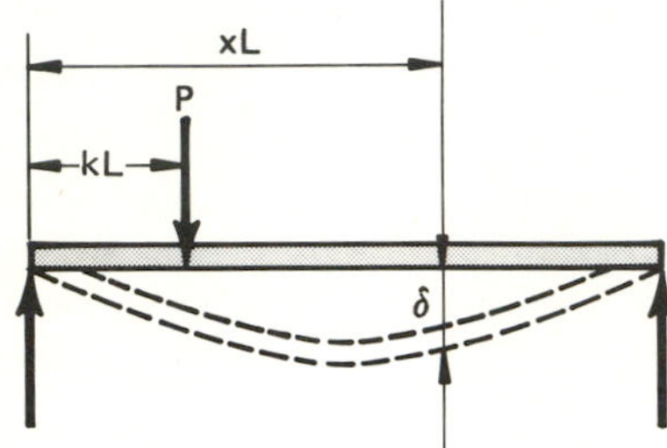

Figure 5–43. Single Load on Beam

Single Load. The deflection at a distance xL (from the support) produced by a concentrated load P at a distance kL (Fig. 5–43) is given by

$$\delta = \frac{PL^3}{6EI} x(1-k)(2k-k^2-x^2) \quad \text{for } 0 \leqslant x \leqslant k \quad (5\text{–}18)$$

$$\delta = \frac{PL^3}{6EI} k(1-x)(2x-x^2-k^2) \quad \text{for } k \leqslant x \leqslant 1 \quad (5\text{–}19)$$

where L = span of simply supported girder
E = modulus of elasticity of the member material
I = moment of inertia of girder cross section

Two Equal Downward Loads. The deflection at a distance xL from one support of the girder produced by concentrated loads P at distances kL and $(1-k)L$ from that support (Fig. 5–44) is given by

$$\delta = \frac{PL^3}{EI}\frac{x}{6}(3k-3k^2-x^2) \quad \text{for } 0 \leqslant x \leqslant k \quad (5\text{–}20)$$

$$\delta = \frac{PL^3}{EI}\frac{k}{6}(3x-3x^2-k^2) \quad \text{for } k \leqslant x \leqslant \tfrac{1}{2} \quad (5\text{–}21)$$

Figure 5–44. Two Equal Downward Loads on a Beam

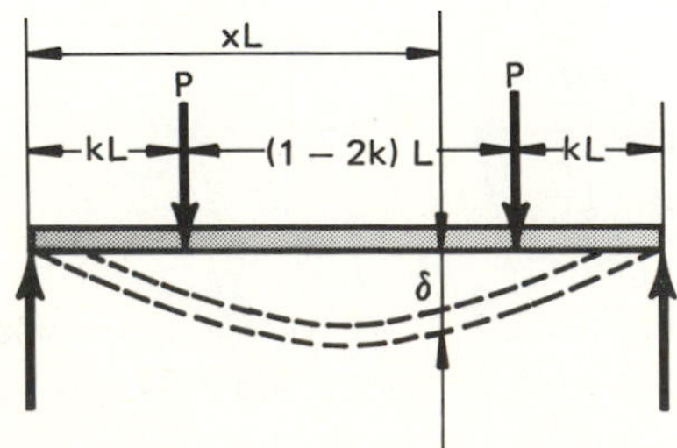

Equal Upward and Downward Loads. The deflection at a distance xL from one support of the girder produced by a downward concentrated load P at distance kL from the support and an upward concentrated load P at a distance $(1-k)L$ from the support (antisymmetric loading) is given by (Fig. 5–45)

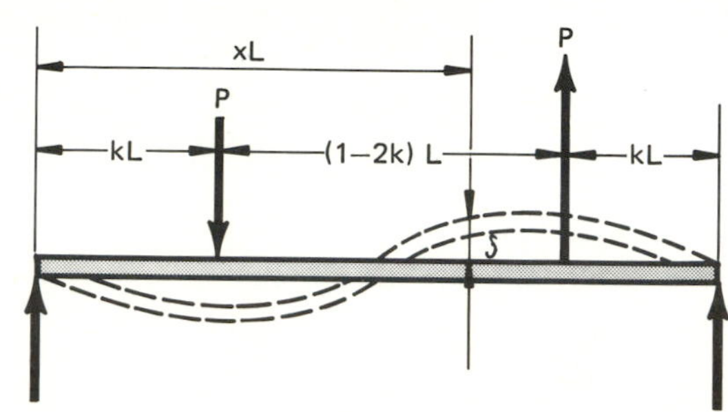

Figure 5–45. Equal Upward and Downward Loads on a Beam

$$\delta = \frac{PL^3}{EI}\frac{x}{6}(1-2k)(k-k^2-x^2) \quad \text{for } 0 \leqslant x \leqslant k \quad (5\text{–}22)$$

$$\delta = \frac{PL^3}{EI}\frac{k}{6}(1-2x)(x-x^2-k^2) \quad \text{for } k \leqslant x \leqslant \tfrac{1}{2} \quad (5\text{–}23)$$

Analysis Procedure. The loading carried by grid framing can be converted into concentrated loads at the nodes. It is convenient to assume that one girder at the node is subjected to an unknown force X_j, and that the other girder carries a force $P - X_j$. With one set of girders uncoupled from the other set, the deflection produced by these forces can be determined by Eqs. 5–18 to 5–23. The following example illustrates the application of the method of consistent deflections.

EXAMPLE PROBLEM 5–7: Analysis of Grid Subsystem

Determine the load distribution for an orthogonal grid with a square boundary (Fig. 5–46). All girders are simply supported at their ends and continuous at interior nodes. Assume that all girders have equal moments of inertia.

Solution

1. Establish load distribution for symmetrical loads. Because of symmetry, only five different nodes must be considered—numbered from 1 to 5 in Fig. 5–46(a). By symmetry: $X_1 = P/2$ for girders 4–4 [Fig. 6–46(b)]; $X_3 = P/2$ for girders 5–5 [Fig. 5–46(c)].
2. Establish unknown loads. Let X_2 = load on girder 4–4 (x direction); $P - X_2$ = load on girder 5–5 (y direction). These loads are shown in Fig. 5–46(b) and (c). Reactions on the boundary girder BC are shown in Fig. 5–46(d).

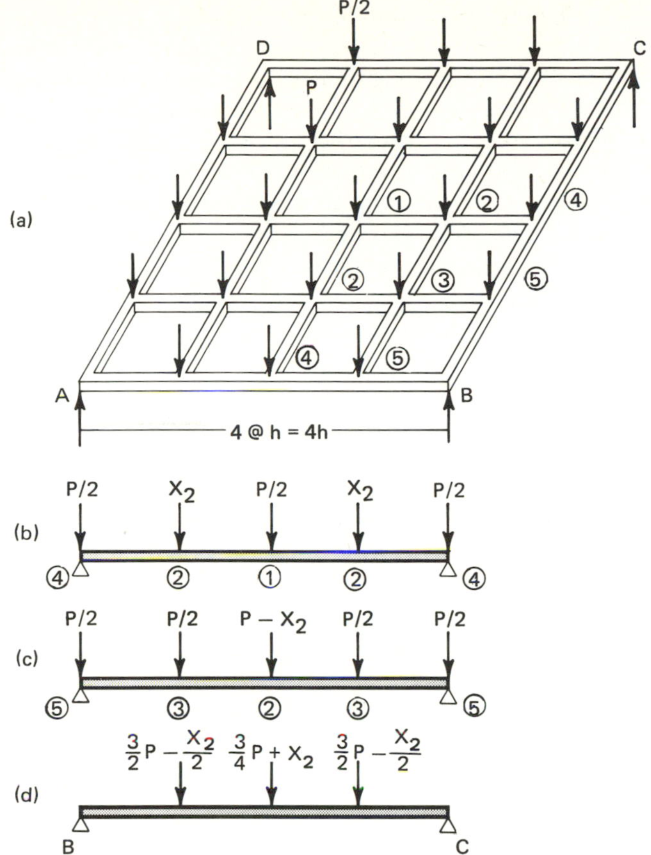

Figure 5–46. Example Problem 5–7

3. Calculate deflection of girder 4–4 at node 2.

$$\delta_2' = \delta_2 + \delta_4$$

a. Deflection at 2 due to $P/2$ by Eq. 5–18 ($X = \frac{1}{4}$, $k = \frac{1}{2}$):

$$\delta = \frac{(P/2)L^3}{6EI} (\tfrac{1}{4})(1 - \tfrac{1}{2})[2(\tfrac{1}{2}) - (\tfrac{1}{2})^2 - (\tfrac{1}{4})^2]$$

$$= \frac{L^3}{48EI}\frac{11P}{32}$$

b. Deflection at 2 due to X_2 by Eq. 5–20 ($X = \frac{1}{4}$, $k = \frac{1}{4}$):

$$\delta = \frac{X_2L^3}{6EI} (\tfrac{1}{4})[3(\tfrac{1}{4}) - 3(\tfrac{1}{4})^2 - (\tfrac{1}{4})^2] = \frac{L^3}{48EI}(X_2)$$

$$\delta_2 = \frac{L^3}{48EI}\left(\frac{11P}{32} + X_2\right)$$

c. Deflection at 4 due to $(\frac{3}{4}P + X_2)$ by Eq. 5–18 [Fig. 6–46(d)] ($X = \frac{1}{2}$, , $= \frac{1}{2}$):

$$\delta = \frac{(\frac{3}{4}P + X_2)L^3}{6EI} (\tfrac{1}{2})(1 - \tfrac{1}{2})[2(\tfrac{1}{2}) - (\tfrac{1}{2})^2 - (\tfrac{1}{2})^2]$$

$$= \frac{L^3}{48EI}\left(\frac{3P}{4} + X_2\right)$$

d. Deflection at 4 due to $(3P/2 - X_2/2)$ by Eq. 5–21 ($X = \frac{1}{2}$, $k = \frac{1}{4}$):

$$\delta = \frac{(3P/2 - X_2/2)L^3}{6EI} (\tfrac{1}{2})[3(\tfrac{1}{4}) - 3(\tfrac{1}{4})^2 - (\tfrac{1}{2})^2]$$

$$= \frac{L^3}{48EI}\left(\frac{33P}{16} - \frac{11X_2}{16}\right)$$

$$\delta_4 = \frac{L^3}{48EI}\left(\frac{45P}{16} + \frac{5X_2}{16}\right)$$

$$\delta_2' = \frac{L^3}{48EI}\left(\frac{11P}{32} + \frac{45P}{16} + X_2 + \frac{5X_2}{16}\right)$$

4. Calculate deflection of girder 5–5 at node 2 (similar to step 3).

$$\delta_2' = \frac{L^3}{48EI}\left(\frac{27P}{16} + \frac{129P}{64} - X_2 + \frac{3X_2}{16}\right)$$

5. Equate deflections at node 2:

$$\frac{11P}{32} + X_2 + \frac{45P}{16} + \frac{5X_2}{16} = \frac{27P}{16} - X_2 + \frac{129P}{64} + \frac{3X_2}{16}$$

Solution of the equation yields

$$X_2 = \frac{35P}{136} = 0.257P \quad \text{and} \quad P - X_2 = \frac{101P}{136} = 0.743P$$

With these forces known, the bending moments, shears, and deflections of the girders can be computed by conventional methods.

Approximate Design

Design of grid subsystems involves the following choices:

- Material selection
- Boundary dimensions and aspect ratio
- Type of grid
- Girder spacing
- Element proportions

The depth of grids will normally be between $\frac{1}{20}$ and $\frac{1}{40}$ of the span. Preliminary estimates of the load sharing between the two beam systems can be made by assuming that they act as a continuous plate. The approximate methods of Section 5–7 can then be used for preliminary design.

5–6 SPACE FRAMES

Space frames are horizontal distribution subsystems consisting of two parallel plane grids interconnected by verti-

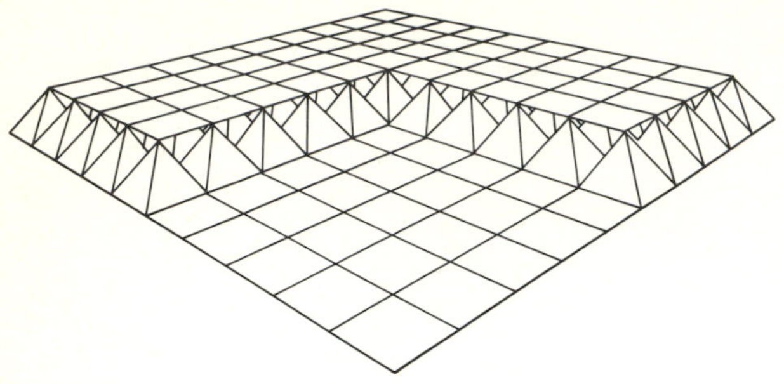

Figure 5–47. Space Frame Subsystem

cal and inclined web members (Fig. 5–47). The space frame can be visualized as a three-dimensional development of the grid subsystem and the truss element, with external loads distributed in three or more directions in space. Other common names given to this structural form include *space truss* and *space grid.* "Space truss" is often used in reference to pin-connected networks and "space frame" in reference to rigid or continuous joints. We will use the term "space frame" to denote any type of three-dimensional truss assembly.

The space frame is the most complex of the structural subsystems, involving a high degree of indeterminacy and requiring special analysis techniques. The structural concept is basically a three-dimensional application of the triangulation principle common to truss elements. The addition of the third dimension allows the distribution of loads in all directions and provides numerous alternative load paths throughout the system.

History and Development

The space frame is actually a planar version of the more general triangulated skeletal frames that were first developed around 1860. Schwedler constructed the first fully triangulated framed dome in Berlin in 1863. Perhaps the most significant breakthrough was the work done by Alexander Graham Bell, who worked extensively with tetrahedral space cells in experiments for kites in the years around 1900 (Fig. 5–48).

Space frame construction enjoyed only limited use until the mid-twentieth century and the development of computer analysis techniques—a 3000-joint space frame simply could not be analyzed by manual methods. Since the advent of the computer, large-span space frames have been erected for numerous buildings around the world and several patented systems and standardized components for moderate span systems now exist.

Types and Uses

The major types of space frames are shown in Fig. 5–49, with the major identifying characteristic being the relationship of the top and bottom grids. There are four types of top grid-to-bottom grid relationships:

- *Direct grids:* two parallel grids similar in design with one layer directly over the top of the other.
- *Offset grids:* two parallel grids similar in design with one grid offset from the other in plan but directionally the same.

Figure 5–48. Alexander Graham Bell Space Frames (Courtesy of Library of Congress)

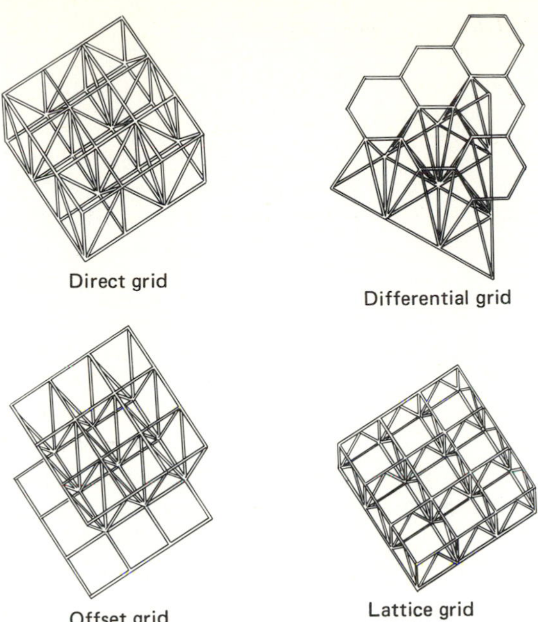

Figure 5–49. Types of Space Frames

- *Differential grid:* two parallel grids that may be of different design and therefore directionally different but are chosen to coordinate and form a regular pattern.
- Lattice grid: the upper and lower members are braced to form a girder prior to erection.

The direct grid may have diagonal members in the plane of the top and bottom grid or may simply act as a two-way space truss if these members are omitted. The offset grid, which can be visualized as a series of repeating pyramids, is generally the most efficient form. Together with the direct grid, it is used quite commonly for major roof systems, arenas, hangers, and so on.

Behavior

The behavior of the two-way space truss is similar to the grid, so we will not discuss this form. We can study the true three-dimensional space frames by isolating their basic cellular structure. The pyramid is the basic unit for the offset grid and is shown in Fig. 5–50. A concentrated load applied at the apex will create direct axial compression forces in the four diagonal web members. Secondary bending forces will only be present as a result of connection eccentricity and deformation restraint provided by the connection. Minimizing secondary bending forces at the connection is difficult, and accurate mathematical modeling of the true connection condition is very critical.

If member A in the pyramid buckles under load, the remaining members still form a stable unit and will each pick up a share of the load previously carried by

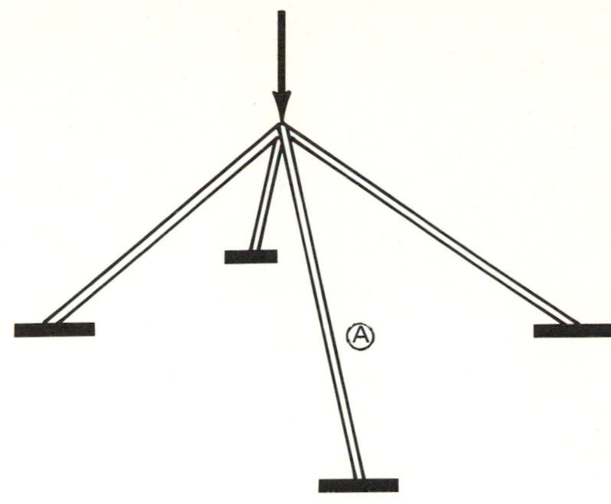

Figure 5–50. Basic Space Frame Unit

the buckled member. This characteristic of space frames—providing for alternative load paths—has prompted a method of ultimate strength analysis termed *limit analysis.* Using this analysis procedure, the true ultimate strength of a space frame is determined by successively "failing" individual members until the overall collapse of the subsystem occurs. The ultimate strength by this method is generally far greater than the strength predicted by linear elastic analysis methods.

The behavior of the complete subsystem can be compared to that of a reinforced concrete slab. The variation in load transfer with aspect ratio is similar, and the three-dimensional geometry of the space frame simulates the torsional restraint provided by the slab.

Approximate Analysis

Approximate analysis of space frames has not been well developed beyond the use of analogous plate bending formulas. Table 5–6 gives moment coefficients for a simply supported plate which can be used for a space frame supported on all sides. Many space frames are supported on piers, however, and we must use other methods of approximate analysis.

For the space frame shown in Fig. 5–51, a "strip method" can be used which divides the space frame into bands spanning between the column support points. The width of the band is a matter of judgment but usually should not exceed $l/4$ in each direction from the column line. The remaining areas of the frame outside the band can be analyzed using strips that carry load relative to assumed yield line patterns. Example Problem 5–8 illustrates this method.

Approximate Design

Design of space frames involves the following choices:

- Material selection

Table 5–6 Slab Moment Coefficients[a]

Moments	Short span Span ratio, short/long 1.0	0.9	0.8	0.7	0.6	0.5 and less	Long span/ all span ratios
Case 1: Interior panels							
Negative moment at:							
Continuous edge	0.033	0.040	0.048	0.055	0.063	0.083	0.033
Discontinuous edge	—	—	—	—	—	—	—
Positive moment at midspan	0.025	0.030	0.036	0.041	0.047	0.062	0.025
Case 2: One edge discontinous							
Negative moment at:							
Continuous edge	0.041	0.048	0.055	0.062	0.069	0.085	0.041
Discontinuous edge	0.021	0.024	0.027	0.031	0.035	0.042	0.021
Positive moment at midspan	0.031	0.036	0.041	0.047	0.052	0.064	0.031
Case 3: Two edges discontinuous							
Negative moment at:							
Continuous edge	0.049	0.057	0.064	0.071	0.078	0.090	0.049
Discontinuous edge	0.025	0.028	0.032	0.036	0.039	0.045	0.025
Positive moment at midspan	0.037	0.043	0.048	0.054	0.059	0.068	0.037
Case 4: Three edges discontinuous							
Negative moment at:							
Continuous edge	0.058	0.066	0.074	0.082	0.090	0.098	0.058
Discontinous edge	0.029	0.033	0.037	0.041	0.045	0.049	0.029
Positive moment at midspan	0.044	0.050	0.056	0.062	0.068	0.074	0.044
Case 5: Four edges discontinuous							
Negative moment at:							
Continuous edge	—	—	—	—	—	—	—
Discontinuous edge	0.033	0.038	0.043	0.047	0.053	0.055	0.033
Positive moment at midspan	0.050	0.057	0.064	0.072	0.080	0.083	0.050

l_s = length of short span for two-way slabs. The span shall be considered as the center-to-center distance between supports or the clear span plus twice the thickness of slab, whichever value is the smaller

w = total uniform load per sq ft.

$M = (\text{coeff.})wl_s^2$

[a] ACI Committee 318, *Building Code Requirements for Reinforced Concrete,* ACI 318–63 (Detroit, Mich.: American Concrete Institute, 1963), p. 130.

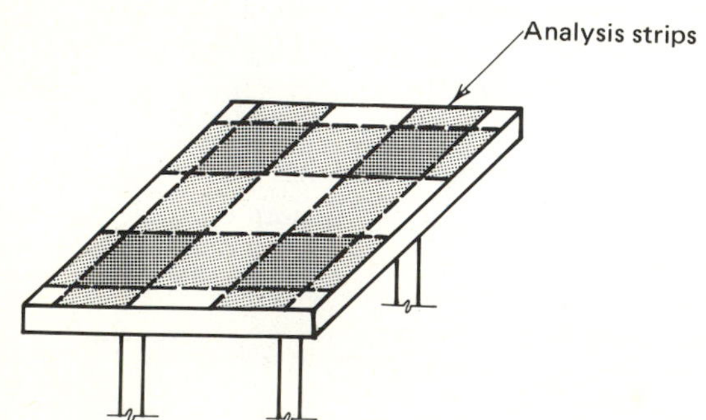

Figure 5–51. Strip Method of Space Frame Analysis

- Boundary dimensions and aspect ratio
- Type of grid pattern
- Depth
- Support location
- Element type and proportions

Structural steel is the most common material used for space frame construction. Aluminum and composite materials are also used for moderate span systems. The most efficient space frames are approximately square in shape, with corner or wall support located at a distance in from the corners equal to approximately 0.3 of the span. This support location creates cantilever segments that counteract positive bending moment at the midpoint

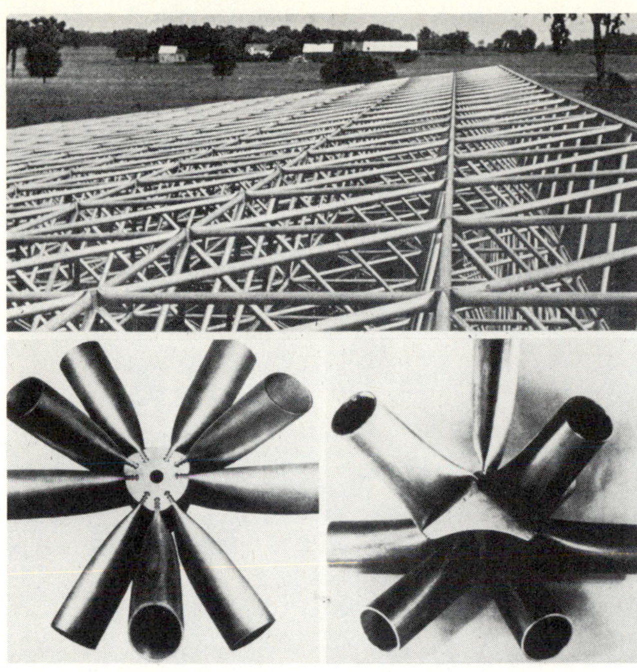

Figure 5–52. Space Frame Connections (Reprinted from *Space Grid Structures* by John Borrego by permission of The MIT Press, Cambridge, Massachusetts, 1968.)

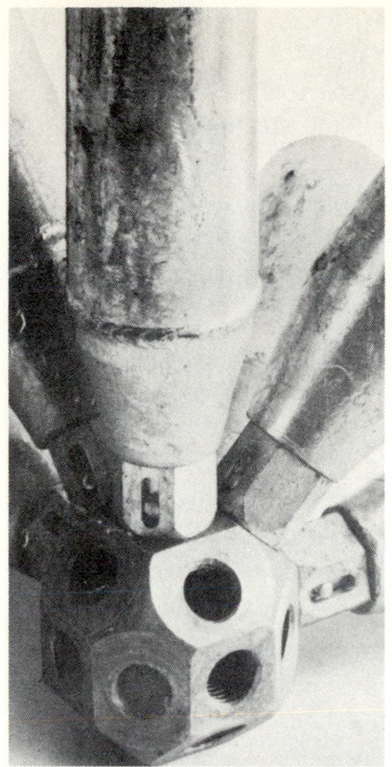

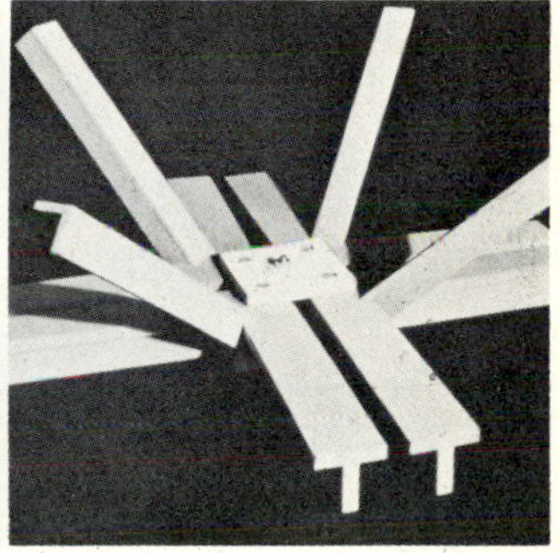

Figure 5–53. Example Problem 5–8

of the interior span. The optimum depth/span ratio for space frames ranges from $\frac{1}{15}$ to $\frac{1}{25}$ of the span.

Many types of tension and compression elements have been used for space frames. These include double angles, pipe, tubing, and crucifix shapes. The major design difficulty is the detailing and fabrication of cost-effective connections. Figure 5–52 shows typical details of element types and connection details.

The following example illustrates the approximate analysis and design of space frames.

EXAMPLE PROBLEM 5–8: Preliminary Space Frame Design

Develop an approximate design and cost estimate for a space frame roof system 300 ft × 300 ft (91.4 m × 91.4 m) in size (Fig. 5–53). The owner desires supports only near the corners. Roof live load is 20 psf (958 N/m²).

Solution

1. *Project requirements:*

Function:	Arena roof
	Covered area = 300 ft × 300 ft (91.4 m × 91.4 m)
	Provide horizontal distribution
Esthetics:	Assume that roof system will be exposed to the interior.
Serviceability:	Camber roof for drainage.
	Fire rating: Type IV.
Construction:	Assume that time is critical and that economy is paramount. Project is signif-

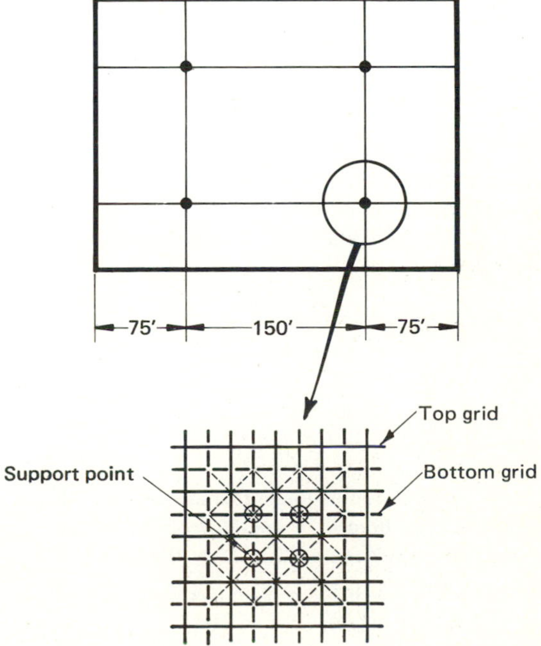

icant in size and will attract high-quality fabricators and erectors.

2. *Loads:*
   a. Live load: 20 psf (958 N/m²).
   b. Dead load: Assume 40 psf (1916 N/m²).
3. *Material Selection:* Choose structural steel A572.
4. *Establish Geometry:*
   a. Type of grid pattern: Assume an offset grid, 10 ft × 10 ft (3.05 m × 3.05 m) with bottom grid offset 5 ft (1.52 m). Top grid is 295 ft × 295 ft (89.9 m × 89.9 m).
   b. Depth: Assume 10 ft (3.05 m) for this example.
   c. Support location: Locate corner supports as shown in Fig. 5–53 to balance positive moment of interior span.
5. *Approximate analysis:* Use the strip method shown in Fig. 5–51.
   a. Calculate the moment.

$$w = 0.06 \text{ ksf } (2873 \text{ N/m}^2)$$

l_c (cantilever length) = 75 ft (22.9 m)

$$\text{Max. } M = \frac{wl_c^2}{2} = \frac{150(0.06)(75)^2}{2}$$
$$= 25{,}312 \text{ ft-kips } (34{,}323{,}072 \text{ Nm})$$

Assume column band width = 75 ft (22.9 m).

$$\text{Moment per foot of width} = \frac{25{,}312}{75}$$
$$= 337 \text{ ft-kips/ft } (1{,}498{,}824 \text{ Nm/m})$$

   b. Calculate the shear.

$$\text{Support reaction} = \frac{(0.06)(300)(300)}{4}$$
$$= 1350 \text{ kips } (6{,}004{,}800 \text{ N})$$

6. *Approximate design:*
   a. Determine roof covering. Select a 2-in. (50.8-mm) metal deck for a 10-ft (3.05-m) span.
   b. Determine the size of the top chord.

$$C = \frac{M_{max}}{d}$$
$$= \frac{337 \text{ ft-kips/ft(10 ft)}}{10 \text{ ft}} = 337 \text{ kips } (1{,}498{,}976 \text{ N})$$

Assume a double-angle member and top flange continuously braced by metal deck for buckling about the y axis. Assume that kl = 0.8 (10 ft) = 8 ft.

Try 6 in. × 6 in. × $\frac{5}{8}$ in., (152 mm × 152 mm × 15.9 mm) angles (48.4 lb/ft).

Minimum size of top chord:

$$C = \frac{M_{min}}{d}, \qquad M_{min} = \frac{10(0.06)(75)^2}{2}$$
$$= 1687.5 \text{ ft-kips } (2{,}288{,}250 \text{ Nm})$$
$$= \frac{1687.5}{10} = 168 \text{ kips } (747{,}264 \text{ N})$$

Try 4 in. × 4 in. × $\frac{7}{16}$ in. (102 mm × 102 mm × 11.1 mm) (22.6 lb/ft).

   c. Determine the size of the bottom chord.

Maximum size of bottom chord:

$$T = 337 \frac{\text{ft-kips/ft(10 ft)}}{10 \text{ ft}} = 337 \text{ kips } (1{,}498{,}976 \text{ N})$$

$$\text{Net area required} = \frac{337^k}{30 \text{ ksi}} = 11.2 \text{ in.}^2 \ (7224 \text{ mm}^2)$$

Try 6 in. × 6 in. × $\frac{9}{16}$ in. (152 mm × 152 mm × 14.3 mm) angles (48.8 lb/ft).

Minimum size of bottom chord:

$$T = \frac{M_{min}}{d}$$
$$= \frac{1687.5}{10} = 168 \text{ kips } (747{,}264 \text{ N})$$

$$\text{Net area required} = \frac{168}{30} = 516 \text{ in.}^2 \ (332{,}820 \text{ mm}^2)$$

Try 4 in. × 4 in. $\frac{7}{16}$ in. (102 mm × 102 mm × 11.1 mm) (22.6 lb/ft).

   d. Determine the size of the diagonals.

Size of diagonals at support: Assume support spreads load to four cells (Fig. 5–53), with 16 members sharing in the load.

$$\text{Vert. force in one diagonal} = \frac{1350}{16} \text{ kips}$$
$$= 84 \text{ kips } (373{,}632 \text{ N})$$

$$\text{Length of diagonal} = \sqrt{(5)^2 \times (50^2 + (10)^2}$$
$$= 12.2 \text{ ft } (3.72 \text{ m})$$

$$\text{Resultant force in diagonal} = \frac{12.2(84)}{10}$$
$$= 102 \text{ kips } (453{,}696 \text{ N})$$

Assume that k = 1.0. Try two angles: 4 in. × 4 in. × $\frac{3}{4}$ in. (102 mm × 102 mm × 19.1 mm) (37.0 lb/ft).

Minimum size of diagonals:

$$\text{Shear at band} = \frac{wl}{2} = \frac{0.06(150)}{2}$$
$$= 4.5 \text{ kips/ft } (65{,}669 \text{ N/m})$$

$$\text{Shear in one diagonal} = \frac{4.5 \text{ kips/ft}(5 \text{ ft})}{2}\,\frac{12.2}{10}$$

$$= 13.7 \text{ kips } (60{,}938 \text{ N})$$

Try 3 in. × 3 in. × $\frac{1}{4}$ in. (76.2 mm × 76.2 mm × 6.4 mm) (9.8 lb/ft).

7. *Cost estimate:*
   a. Estimate the quantities.
      (1) Top chord:

$$\text{Average weight} = \frac{48.4 + 22.6}{2} = 35.5 \text{ lb/ft}$$

$$290 \text{ ft}(30)(2)(35.5 \text{ lb/ft}) = 617{,}700 \text{ lb}$$

      (2) Bottom chord:

$$\text{Average weight} = \frac{43.8 + 22.6}{2} = 33.2 \text{ lb/ft}$$

$$300 \text{ ft}(31)(2)(33.2 \text{ lb/ft}) = 617{,}520 \text{ lb}$$

      (3) Diagonals:

$$\text{Average weight} = \frac{37.0 + 9.8}{2} = 23.4 \text{ lb/ft}$$

$$4(12.2 \text{ ft})(29.5)(29.5)(23.4 \text{ lb/ft}) = 993{,}755 \text{ lb}$$

      (4) Connections: Assume 100 lb per connection.

$$(60)(60)(100 \text{ lb}) = 360{,}000 \text{ lb}$$

$$\text{Total weight} = 2{,}588{,}975 \text{ lb } (11{,}515{,}761 \text{ N})$$

$$\text{Weight per sq ft} = \frac{2{,}588{,}975}{300 \text{ ft} \times 300 \text{ ft}}$$

$$= 28.7 \text{ psf } (1374 \text{ N/m}^2)$$

   b. Cost estimate:

$$2{,}588{,}975 \text{ lb at } \$1.00/\text{lb} = \$2{,}588{,}975$$

5–7 SLABS

Slabs are horizontal distribution subsystems constructed of reinforced concrete (Fig. 5–54). The slab subsystem can be considered the final development of the two-way structural form initiated by the grid and advanced by the space frame. The basic elements in the subsystem are "finite elements" which collectively form a continuous plane surface with flexural and torsional rigidity. Other elements, such as beams, drop panels, and column capitals, may also be included to assist the slab in transferring moments and shears to the columns. Slabs that function as one-way systems spanning between supporting beams are not included in our definition, as these slabs act very similarly to beam elements.

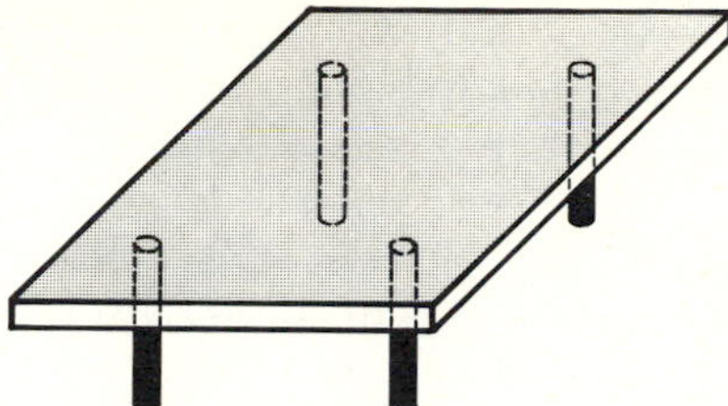

Figure 5–54. Slab Subsystem

Slabs are complex structural forms and are indeterminate, like grids and space frames. However, the slab can be defined by continuous mathematical expressions that make it easier to analyze than the other two-way forms. As we have already seen, the analogous slab is often used to assist in the analysis of both grids and space frames.

History and Development

The first slabs were supported on all four sides by beams or walls. Experiments in doing away with beams completely and supporting the floors on regular spaced freestanding columns began in the United States and Switzerland at the start of the twentieth century (Fig. 5–55). Experimental verification of slab capacity through load tests was used for design of these slabs, as rational analysis methods had not been developed. The Soo Line Terminal in Chicago, one of the early reinforced-concrete slab structures (1913), was load tested with 10 railway cars at 200.8 kips (893,158 N) each, a locomotive, and track work. In the same era, one panel of the Schwinn Bicycle Factory in Chicago was loaded to 450 psf (21,546 N/m²) for an entire year. Rational analysis of slabs has always lagged

Figure 5–55. Early Slab Forms

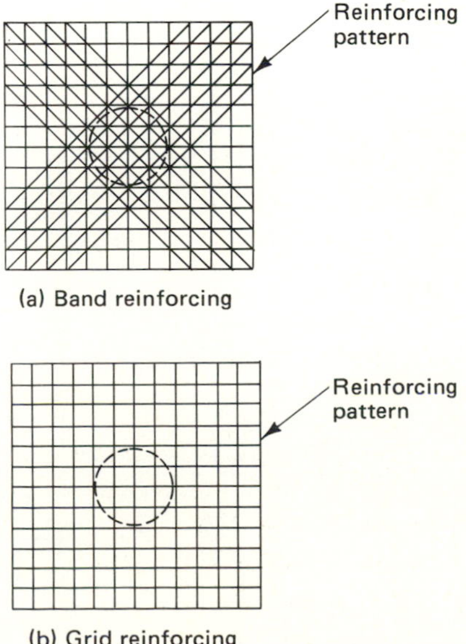

behind design and construction practices. In fact, more than 1000 flat slab buildings had already been built when the first rational analysis was published by Nichols in 1914.

Slab design took on a split personality at this point, a division that continued until adoption of the 1971 ACI Code. Slabs supported on beams or walls were designed using analytical techniques based on the theory of elasticity, while beamless slabs relied on the results of continuing load tests and semiempirical procedures. The slab tables (Table 5–6) of earlier codes are typical of the approach to supported slabs (called two-way slabs). Beamless slabs (called flat slabs) were treated with approximate methods, after 1941 by the equivalent frame concept.

No methods were available for handling slabs that were intermediate between these two cases (e.g., flexible beam supports), and factors of safety varied considerably between the two methods. The 1971 ACI Code combined the two approaches to slab design by treating all slabs as a general case with beams varying from zero stiffness to infinite stiffness.

Types and Uses

The major types of slabs are shown in Fig. 5–56. The slab shown in Fig. 5–56(a) functions more as a deck assembly component of a floor or roof subsystem. The other types of slabs span from column to column with a variety of patterns and shapes. The choice of slab type depends largely on the loads and span. The flat plate system is suitable for lighter loads and moderate spans, characteristics that make it a popular system for residential construction (apartments, hotels, etc.). Slabs with beams or drop panels are more suitable for heavier loads, including lateral loads from wind or earthquake. Office buildings, institutional buildings, and parking garages make use of the heavier systems.

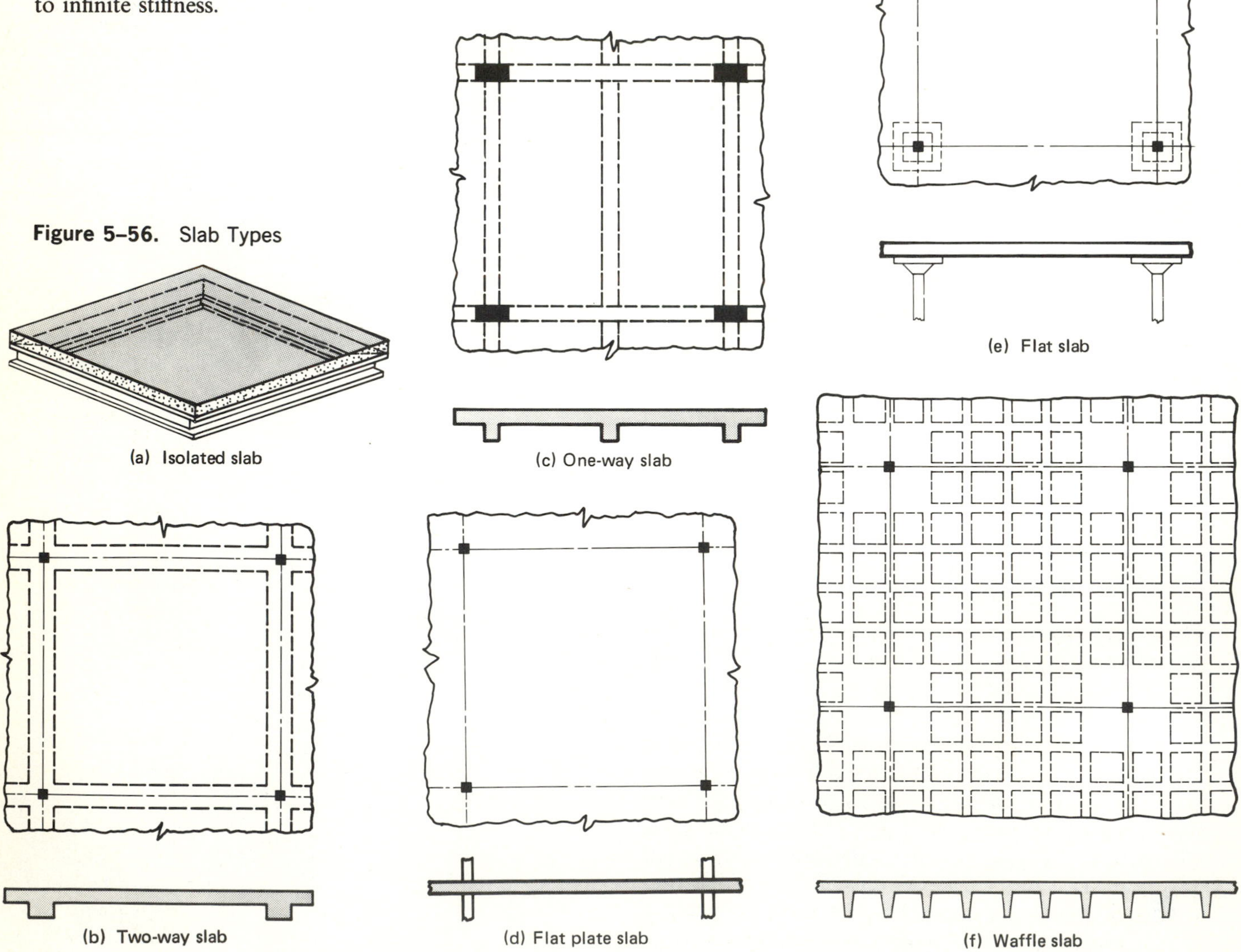

Figure 5–56. Slab Types

Behavior

Isolated Two-Way Slabs. Slabs that are deck assemblies [Fig. 5–56(a)] and do not form a part of an overall monolithic slab/column subsystem can be considered as supported on nondeflecting perimeter beams, walls, or columns, as shown in Fig. 5–57. The behavior of these slabs is similar to the grid and space frame, involving two-way distribution of loads and flexural and torsional forces. Because of the torsional forces, the maximum moment in a square simply supported slab is only $0.048wA^2$ instead of $0.0675wA^2$ for pure bending action.

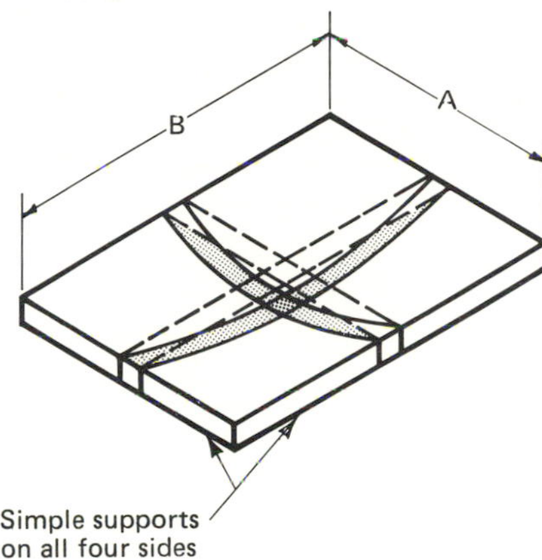

Figure 5–57. Isolated Two-Way Slabs

The ultimate strength of these slabs can be predicted by locating *yield lines* that are hinges which form as stresses reach yield and plastic deformation begins. Yield lines may be located with the help of the following guidelines (Fig. 5–58):

- Yield lines are generally straight.
- Axes of rotation generally lie along lines of support (the support line may be a real hinge, or it may establish the location of a yield line that acts as a plastic hinge).
- Axes of rotation pass over any columns.
- A yield line passes through the intersection of the axes of rotation of adjacent slab segments.

Once yield lines are located, the structure is determinate and equilibrium equations can be written to solve for the slab capacity.

Slab/Column Systems. Slabs that are integral with column elements and form a complete horizontal distribution and vertical support subsystem can be studied as analogous rigid frames, with the slab (and any integral beams) functioning as the girder element. Using the subassembly concept developed for rigid frames, the equivalent slab rigid frame is shown in Fig. 5–59. The width of the slab extends to the centerline of the slab on either side of the column.

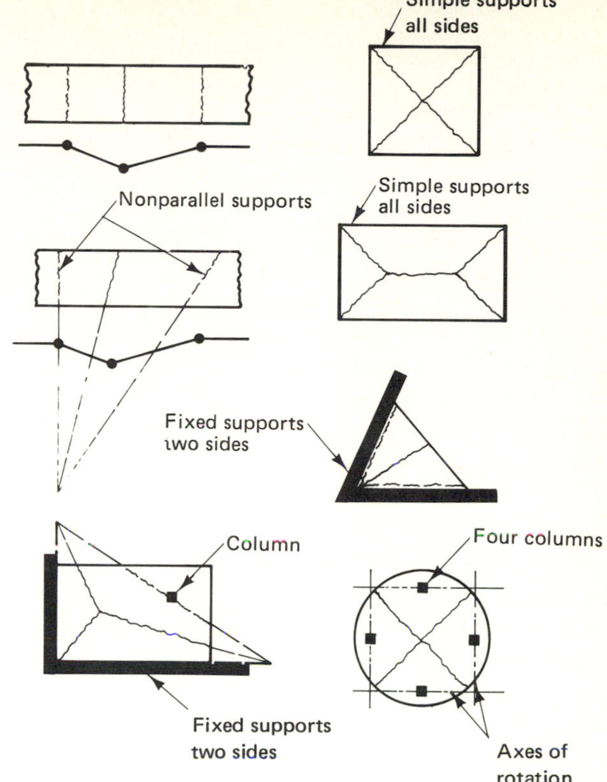

Figure 5–58. Yield Lines in Slabs

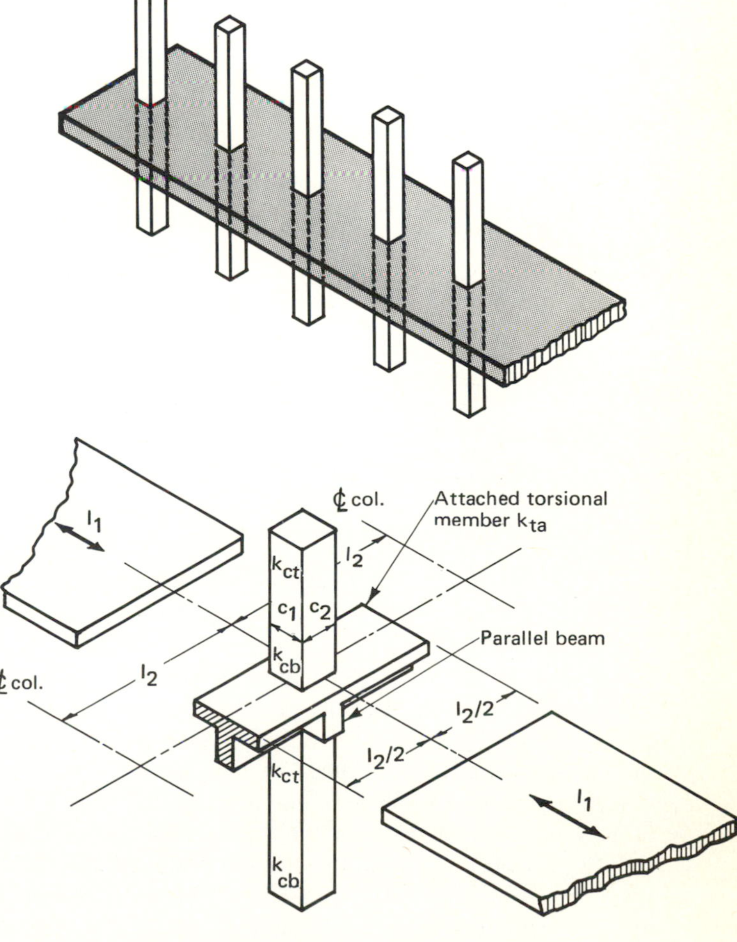

Figure 5–59. Slab/Rigid Frame Subassemblage

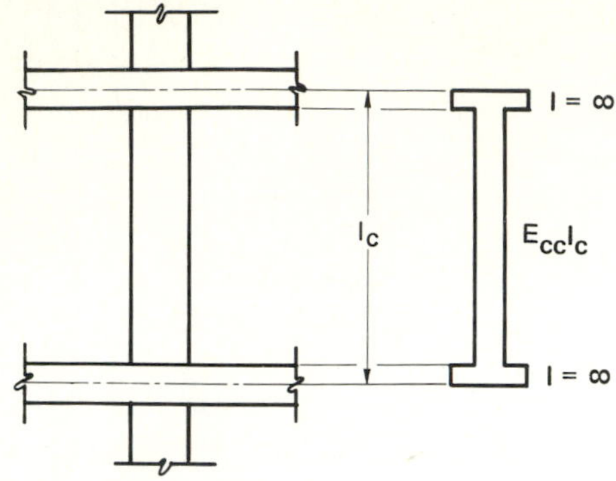

(a) Column stiffness modification

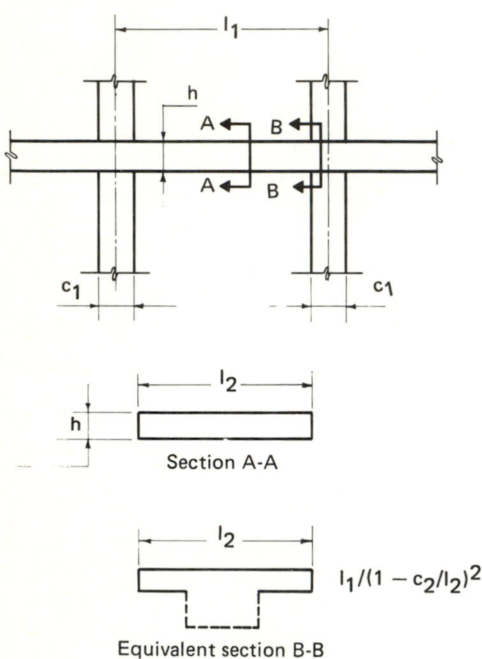

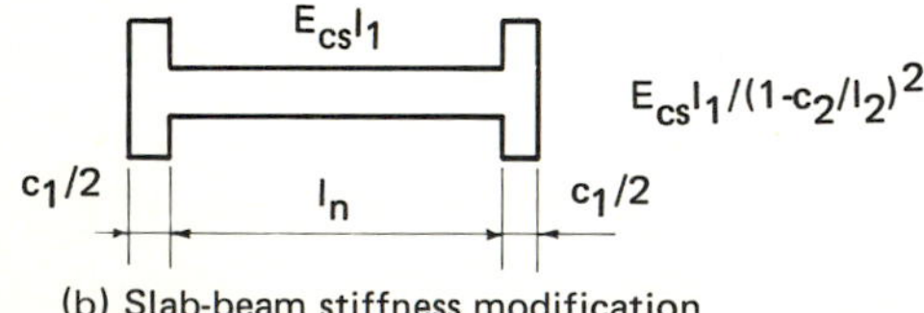

(b) Slab-beam stiffness modification

Figure 5–60. Slab and Column Stiffness Modifications

Some modifications to column and girder stiffness are required to better simulate actual slab behavior. The stiffness of the "girder" element is increased by the effect of the large change in stiffness at the column [Fig. 5–60(a)]. The stiffness of the beam-column is increased to infinity in the area between the top and bottom of the slab [Fig. 5–60(b)]. The stiffness of the beam-column element, however, must also be decreased because the column stiffness does not extend uniformly across the width of the entire slab. Instead, the equivalent column stiffness outside the limits of the actual column is provided by the torsional stiffness of a portion of the transverse slab or transverse beam framing into the column (Fig. 5–61).

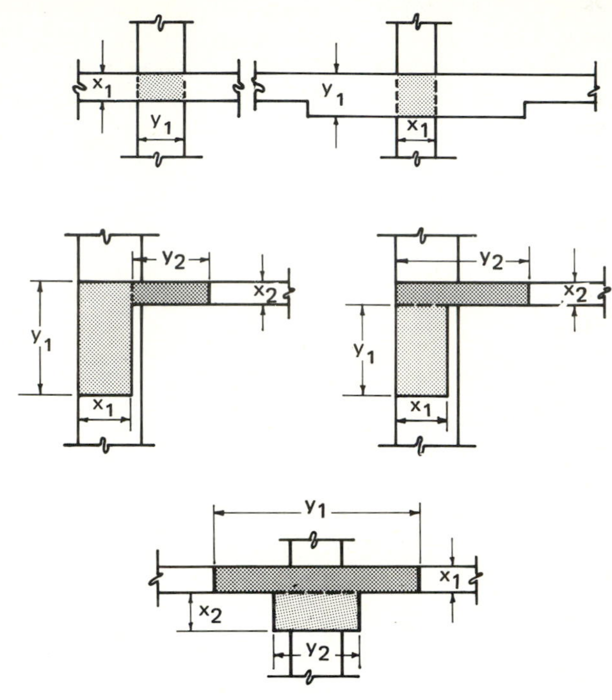

Figure 5–61. Torsional Elements

This torsional stiffness is given by the following equation:

$$K_T = \sum \frac{9E_{cs}C}{l_2(1 - c_2/l_2)^3} \qquad (5\text{–}24)$$

where E_{cs} = modulus of elasticity of slab concrete

l_2, c_2 = dimensions shown in Fig. 5–59

$$C = \Sigma(1 - 0.63x/y)x^3y/3 \qquad (5\text{–}25)$$

x, y = dimension of rectangular sections of the torsional member shown in Fig. 5–59; x is short dimension, y is larger dimension

The equivalent column stiffness is then

$$K_{ec} = \frac{\Sigma\, k_c}{1 + \Sigma\, k_c/K_T} \qquad (5\text{–}26)$$

where

k_c = stiffness of column, which is approximately

$$= \frac{4I}{l_c - 2h}$$

With these stiffness modifications, the slab subsystem behaves very similarly to the rigid frame. Concepts of moment distribution, lateral load analysis (portal method), and inflection points are applicable.

Approximate Analysis

Isolated Slabs. These slabs can be analyzed by the use of coefficients (Table 5–6) or by the yield-line method discussed earlier. The following example illustrates both methods.

EXAMPLE PROBLEM 5–9: Preliminary Isolated Slab Design

Determine the maximum moment in the slab shown in Fig. 5–62.

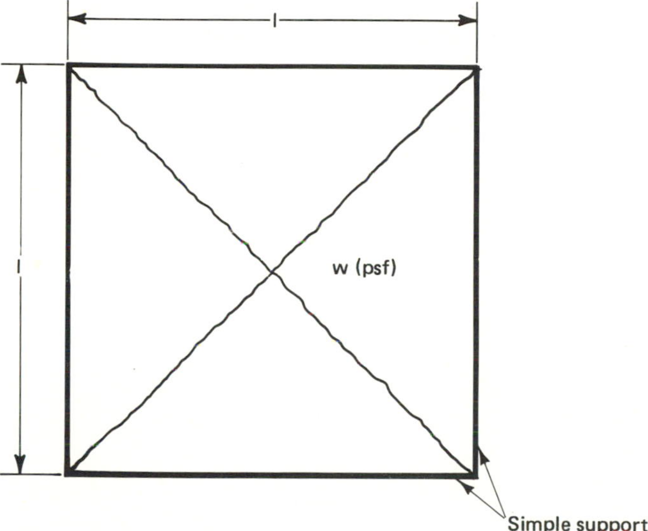

Figure 5–62. Example Problem 5–9

Solution

1. Calculate the slab moment by Table 5–6:

$$M = 0.05\ wl^2$$

2. Calculate the slab moment by the yield-line theory:

$\Sigma\ M$ about support $= 0$

$$\frac{wL^2}{4}\frac{L}{6} - \frac{2ML}{2}\frac{1}{2} = 0$$

$$M = 0.042wl^2 \text{ (ultimate moment)}$$

Slab/Column Systems. The ACI Code provides for both the *direct design method* and the *equivalent frame method* for slab analysis. Both make use of the rigid frame subassembly model. Since the direct design method is simpler, we will use it for our approximate analysis technique for gravity loads.

Slab moments. The absolute sum of the gravity load moments in the slab is (by statics) equal to the following:

$$M_o = \frac{w_u l_2 l_n^2}{8} \qquad (5\text{–}27)$$

where w_u = factored design load
l_2 = transverse span of the panel
l_n = clear span of slab between columns

For an interior span and an exterior span with exterior edge fully restrained, the distribution of the total moment between positive and negative approximates the fixed-end beam distribution as follows:

Negative design moment: 0.65
Positive design moment: 0.35

For an end span, the reduced restraint provided by the exterior column requires the following approximate distribution:

Interior negative design moment: 0.70
Positive design moment: 0.52
Exterior negative design moment: 0.26

Distribution of the positive and negative moments to the various sections of the slab and beam is approximately as follows:

Interior span negative moment:	
Column strip	75%
Middle strip	25%
Exterior span negative moment:	
Column strip	100%
Middle strip	0%
Positive moment:	
Column strip	75%
Middle strip	25%
Beam moments:	
85% of column strip moments	

Column moments. Column moments are the result of distribution of the unbalanced slab moments as follows:

$$M_c = 0.07[(w_d + 0.5w_l)l_2 l_n^2 - w_d' l_2'(l_n')^2] \times \frac{\Sigma k_c}{\Sigma k_c + \Sigma k_s}$$

where w_d', l_2', and l_n' refer to the shorter span

w_d = factored dead load
w_l = factored live load
k_c = column stiffness
k_s = slab stiffness

Approximate lateral load analyis of unbraced slab/column systems may be performed using the portal method developed earlier, or the equivalent frame method of the ACI Code. For lateral drift calculations, such as Eq. 5–15, requiring the use of member properties, the moment of inertia of slab–beams must be reduced to one-third that based on the gross area of concrete to approximate the cracked–section properties. The moment of inertia of the columns may be based on the gross section properties.

Irregular Column Layouts. The "spatter column" layout of Fig. 5–63(a) is difficult to analyze due to the irregular nature of the column spacing. The *strip method* is an approximate technique for handling slabs of this nature, including slabs with openings. In this method [Fig. 5–63(b)], arbitrary lines of zero shear are drawn on a sketch of the slab plan to indicate directions of load transfer. Columns are connected with slab bands that function as beams to carry the load from the slab strips to the columns. The width of the bands is a matter of judgment but usually is assumed equal to the face width of the column plus one-fourth of the transverse span on either side. Individual slab strips support only the load designated to go in the direction of the strip span.

Moments in the slab bands spanning between columns will depend on whether continuity exists across the column and on the restraining moment of the column. Inflection points can be assumed for continuous slabs, and distribution to columns can be assumed in proportion to the relative stiffness of the members.

Figure 5–63. Irregular Column Layout

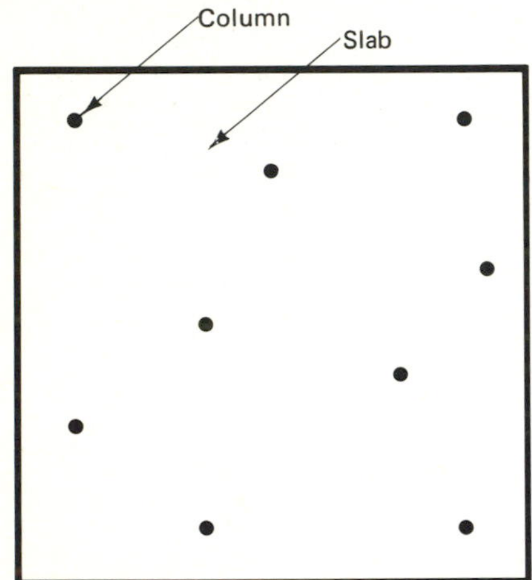

(a) Irregular column layout

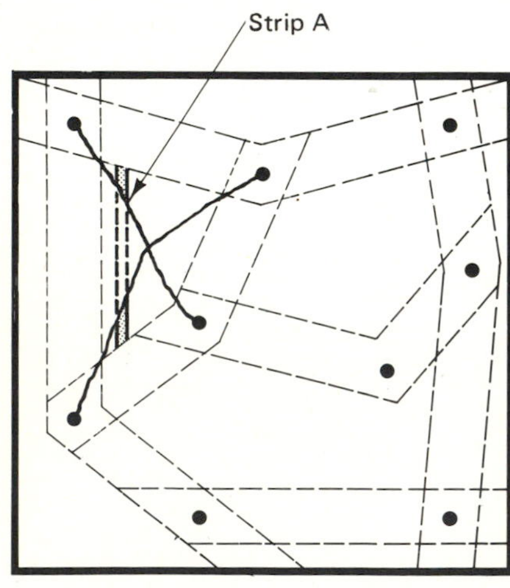

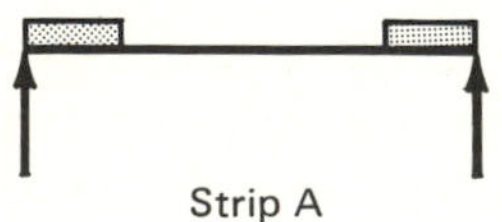

Strip A

(b) Strip method of analysis

EXAMPLE PROBLEM 5–10: Preliminary Slab Design With Irregular Column Layout

For the 9-in. (22,816-mm) concrete slab and column layout shown in Fig. 5–64, calculate approximate slab moments. Live load is 50 psf (2394 N/m²).

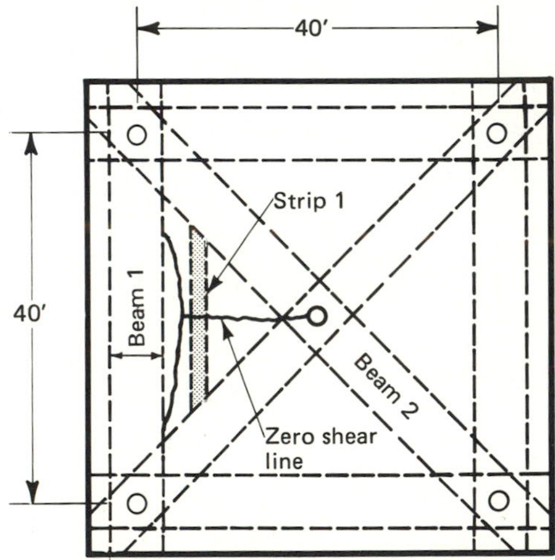

Figure 5–64. Example Problem 5–10

Solution:

1. *Establish model for analysis.* Assume the load distribution and beam layout shown in Fig. 5–64. Beam 1 is a simple span member and has the largest span. The load distribution pattern is selected to minimize the load carried by this beam. Beam 2 is a two-span continuous member for which standard inflection points or moment coefficients can be used.
2. *Calculate the slab moments.* Maximum span for strip $1 \simeq 15$ ft (4.6 m) (allowing for beam width).

$$w_u = 1.4(113) + 1.7(50) = 243 \text{ psf } (11{,}635 \text{ N/m}^2)$$

$$M_u = \frac{w_u l^2}{8} = \frac{0.243(15)^2}{8} = 6.83 \text{ ft-kips } (9{,}261 \text{ Nm})$$

3. *Calculate the beam moments.*

   a. Beam 1: Assume a beam width of 9 ft (2.74 m).

$$M_u = \frac{w_u l^2}{8} - \frac{w_u l_c^2}{2}$$

$$M_u = \frac{9(0.243)(40)^2}{8} - \frac{9(0.243)(4)^2}{2}$$

$$= 420 \text{ ft-kips } (569{,}520 \text{ Nm})$$

   b. Beam 2: Assume a beam width of 10 ft (3.05 m). Assume an equivalent uniform load equal to two-thirds of the triangular peak load.

$$w_u = \frac{2}{3}(0.243)(14.1 \text{ ft})$$

$$= 2.22 \text{ kips/ft } (32{,}396 \text{ N/m})$$

$$+M = \frac{w_u l^2}{11} = \frac{2.22(28.2)^2}{11}$$

$$= 160.5 \text{ ft-kips } (217{,}638 \text{ Nm})$$

$$-M = \frac{w_u l^2}{9} = \frac{2.22(28.2)^2}{9}$$

$$= 196 \text{ ft/kips } (265{,}776 \text{ Nm})$$

Approximate Design

Design of slab subsystems involves the following choices:

- Material selection (concrete strength, etc.)
- Column spacing
- Slab type
- Slab depth
- Column height
- Element proportions

Slab/Column System. Figure 5–65 shows preliminary design charts for several slab systems. Slab *depth* is governed by the larger of the two following equations:

$$h = \frac{l_n(800 + 0.005 f_y)}{36{,}000 + 5000\beta\{\alpha_m - 0.5(1 - \beta_s)[1 + (1/\beta)]\}} \tag{5–28}$$

$$h = \frac{l_n(800 + 0.005 f_y)}{36{,}000 + 5000\beta(1 + \beta_s)} \tag{5–29}$$

But the slab thickness need not exceed the value given by

$$h = \frac{l_n(800 + 0.005 f_y)}{36{,}000} \tag{5–30}$$

which equals $l_n/32$ for 60,000-psi (413.7-MPa) steel,

where l_n = clear span in the long direction measured face to face of columns for slabs without beams and face to face of beams for slabs with beams

β = ratio of the long to the short clear span

β_s = ratio of the sum of the lengths of the sides of the panel that are continuous with other panels to the total perimeter of the panel

α_m = average value of the ratios of beam-to-slab stiffness on all sides of a panel:

$$\alpha = \frac{E_{cb} I_b}{E_{cs} I_s}$$

E_{cb} = modulus of elasticity of the beam concrete

E_{cs} = modulus of elasticity of the column concrete

I_b = the gross moment of inertia about the centroidal axis of a section made up of the beam and the slab on each side of the beam above or below the slab (whichever is greater) but not exceeding four times the slab thickness

I_s = moment of inertia of the gross section of the slab taken about the centroidal axis and equal to $h^3/12$ times the slab width, where the width is the same as for α

Equations 5–28 and 5–29 account for the influence of beams on the slab stiffness. Slabs without beams are governed directly by Eq. 5–30.

Moment design. Once the slab thickness is determined, the design loads can be established and the approximate analysis method used to obtain slab and column moments. If beams are present on the column lines, 85% of the column strip moment should be allotted to the beam. Moment design of the slab section is similar to beam design and is illustrated in Example Problem 5–11.

Shear design. The maximum allowable beam shear at a distance d from the face of walls or beam without special reinforcement is

$$\phi V_c = \phi 2 \sqrt{f'_c}\, b_\omega d \tag{5–31}$$

The maximum allowable punching shear at a distance $d/2$ from the face of the column, capital, or drop panel is

$$\phi V_c = \phi 4 \sqrt{f'_c}\, b_o d \tag{5–32}$$

Beam shear does not usually control. If slab shear exceeds these allowable values, provisions must be made for increased thickness, drop panels, capitals, or special reinforcement.

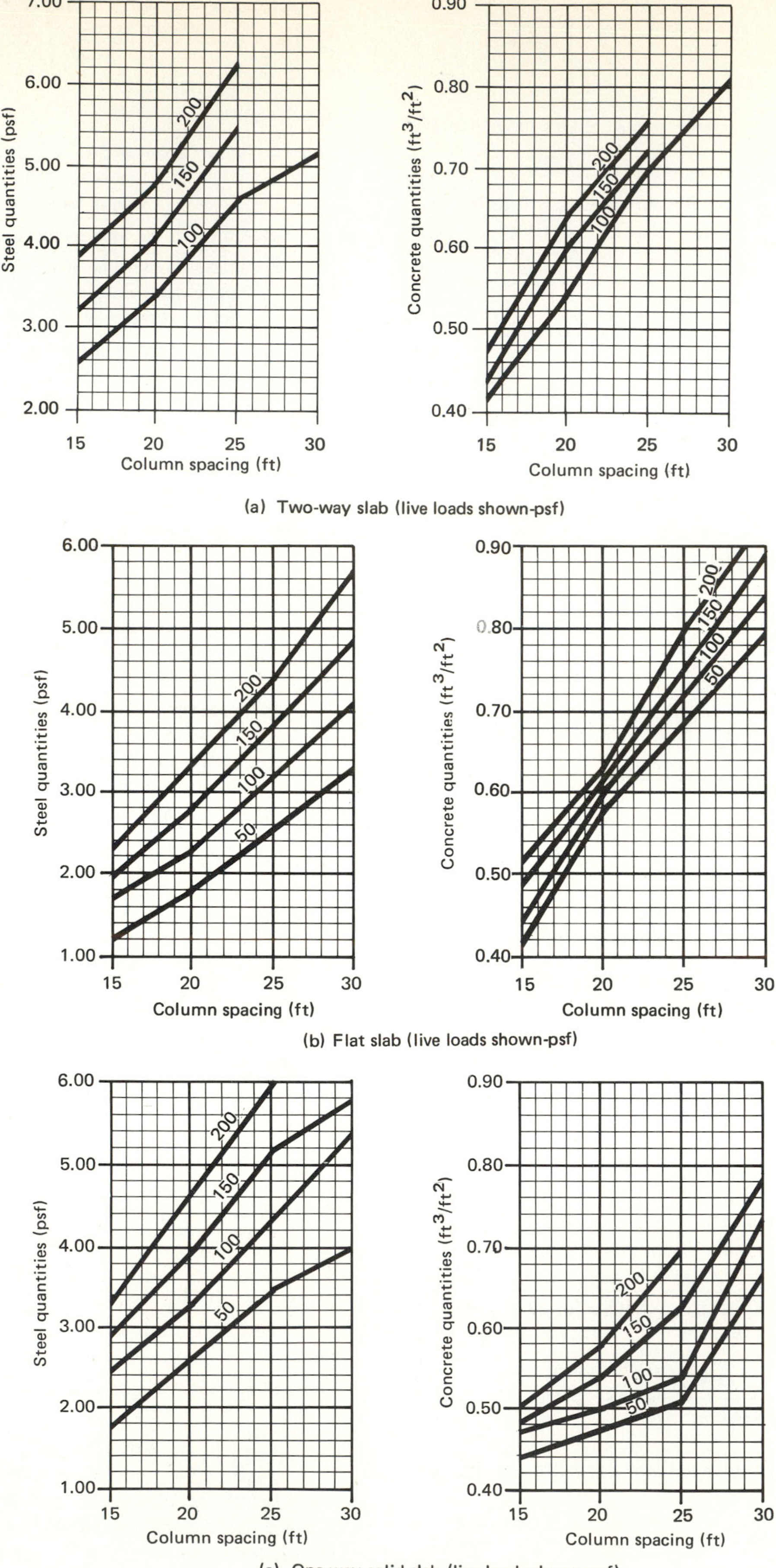

Figure 5–65. Slab Design Charts

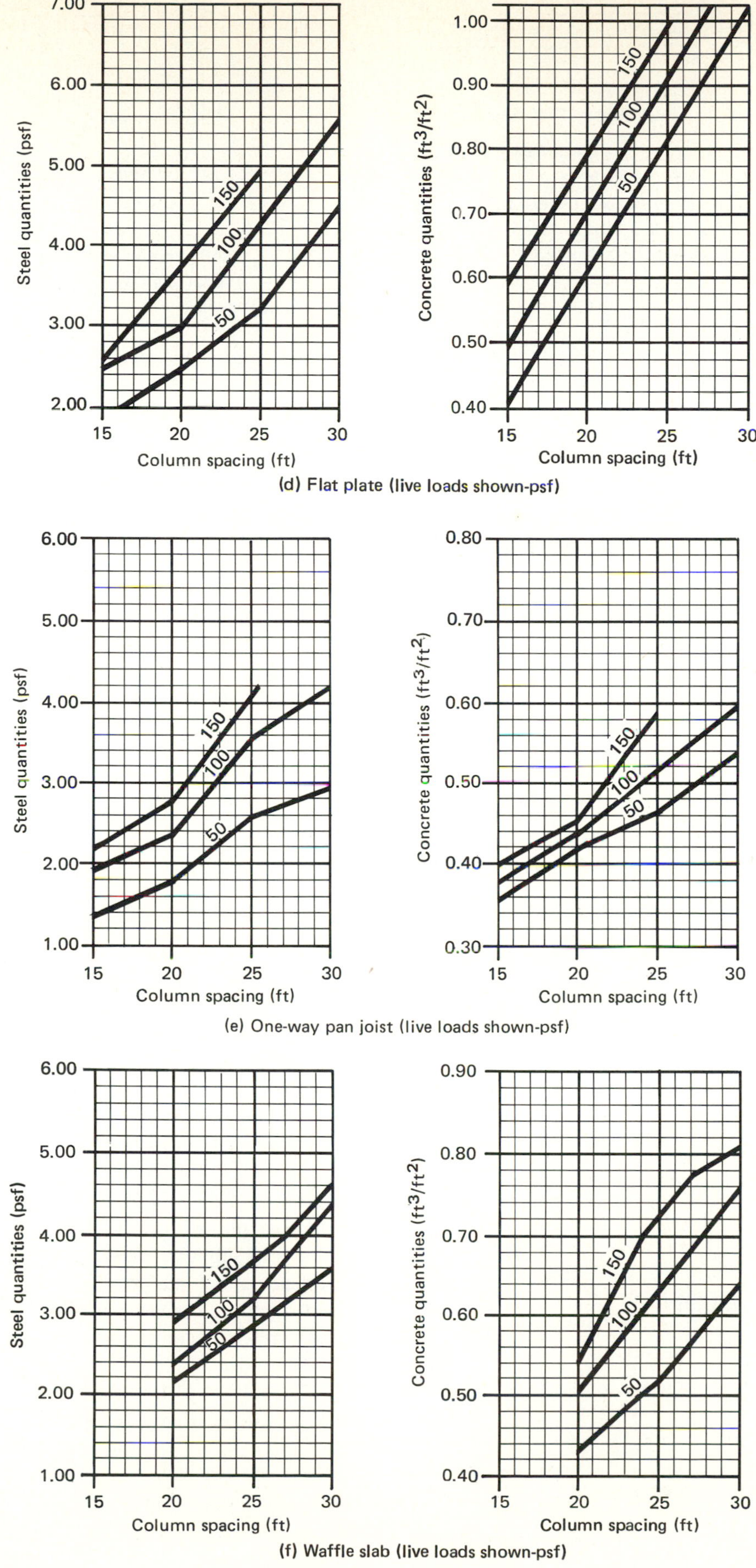

(d) Flat plate (live loads shown-psf)

(e) One-way pan joist (live loads shown-psf)

(f) Waffle slab (live loads shown-psf)

Figure 5–65. *(Continued)*

EXAMPLE PROBLEM 5–11: Preliminary Flat Slab Design

Design a flat slab with no beams to span between columns with bay dimensions of 20 ft × 16 ft (6.1 m × 4.9 m). Assume that the live load equals 50 psf. Determine the slab depth and steel requirements for a typical interior panel in the 20-ft (6.1-m) direction. Assume 3000-psi (20.7-MPa) concrete.

Solution

1. *Determine the slab thickness.* By Eq. 5–30 (assume a 12-in. square column):

$$h = \frac{l_n}{32} = \frac{(20 \times 12) - 12}{32} = 7.13 \text{ in.}$$

Try a $7\frac{1}{2}$-in. (190-mm) thickness.

2. *Determine loads.* Live load = 50 psf (given).
   Dead load:

$7\frac{1}{2}$-in. slab	=	94 psf
Ceiling	=	1 psf
Suspension system	=	2 psf
Mech., elect.	=	5 psf
		102 psf

$$w_u = 1.7(50) + 1.4(102) = 227.8 \text{ psf} \quad (10{,}907 \text{ N/m}^2)$$

3. *Check shear.* Assume that d = 6.25 in. (159 mm).

$$b_o = 4(12 + 6.25) = 73.0 \text{ in. (1854 mm)}$$

$$V_u = 0.228[20 \times 16 - \frac{(12 + 6.25)^2}{12}] = 72.4 \text{ kips (322,035 N)}$$

$$\phi V_c = (0.85)(4\sqrt{3000})(73)(6.25) = 84{,}965 \text{ lb} = 84.9 \text{ kips}$$

Since $\phi V_c > V_u$, the shear is O.K.

4. *Calculate total moment.* By Eq. 5–27,

$$M_o = \frac{w_u l_2 l_n^2}{8}$$

$$= \frac{0.228(16)(19)^2}{8} = 164.6 \text{ ft-kips (223,198 Nm)}$$

5. *Slab design for negative moment.*

$$-M = 0.65(164.6) = 106.9 \text{ ft-kips (144,956 Nm)}$$

Distribute 75% to column strip:

$$M_{cs} = 0.75(106.9) = 80.2 \text{ ft-kips (108,751 Nm)}$$

Column strip width = 8 ft

$$S = \frac{96(7.5)^2}{6} = 900 \text{ in.}^3$$

$$f_b = \frac{M}{S} = \frac{80{,}200 \times 12}{900} = 1069 \text{ psi (7.4 MPa)}$$

Since 0.85% reinforcement = 1260 psi allowable, use A_s = 0.0085(12)(7.5) = 0.76 in.²/ft. Distribute 25% to middle strip:

$$M_{ms} = 0.25(106.9) = 26.7 \text{ ft-kips (36,204 Nm)}$$

$$f_b = \frac{26{,}700 \times 12}{900} = 356 \text{ psi (2.5 MPa)}$$

Use A_s minimum = 0.0018(12)(7.5) = 0.16 in.²/ft.

6. *Slab design for positive moment* (similar procedure to step 5).

5–8 WALLS

Walls are continuous plane surface subsystems which can be visualized as consisting of adjacent beam-column strips in either a one-way or grid assembly, or as an assembly of finite elements. Walls have several potential functional purposes, including vertical support, lateral support, and lateral distribution. As a vertical support subsystem, the wall is similar to a series of beam-column elements. As a lateral distribution subsystem, the wall is similar to either a one-way slab or two-way slab. As a lateral support subsystem, the wall acts as a vertical cantilever beam extending from the foundations.

Nonstructural purposes, such as weather protection, insulation, and light emittance, make the wall subsystem interdisciplinary by nature. System cost comparisons inevitably involve trade-offs among the structural system, the mechanical system, and the architectural system.

History and Development

The history of the wall dates back to the earliest shelters built by people, with walls providing weather protection and vertical support. As structures became larger, walls were made more massive in order to achieve self-contained stability. The *bearing wall* was the major vertical support subsystem until frame construction started during the late 1800s.

The advent of frame construction meant that walls could now depend on other parts of the structural system for stability. Wall construction became more slender and the term *curtain wall* was coined to describe walls that provided enclosure but served no structural purpose.

The development of engineered masonry concepts and the evolution of precast and cast-in-place systems have reintroduced the wall as a major vertical support and lateral support subsystem. Walls are now used extensively in bearing-wall systems and as *shear walls* to provide lateral support and stability to framed systems.

Types and Uses

Several types of walls are shown in Fig. 5–66. These include many of the following:

- Masonry walls
- Precast concrete walls
- Cast-in-place concrete walls
- Stud walls
- Metal siding
- Prefabricated panel systems

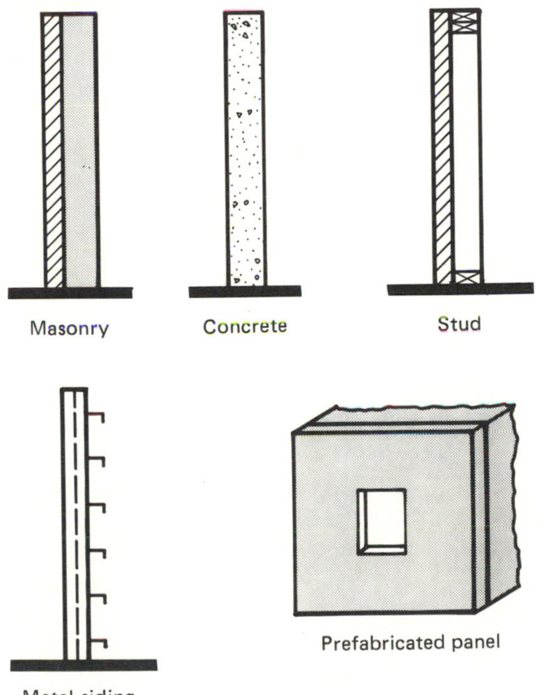

Figure 5–66. Types of Walls

Bearing walls are normally constructed of masonry or concrete. Residential construction utilizes wood-stud bearing walls extensively. Shear walls (which may also function as bearing walls) can be either masonry, concrete, wood, or metal. Shear walls made of individual components, such as plywood and metal siding, must function as vertical diaphragms in order to provide lateral support.

Exterior walls, regardless of their structural purpose, must be capable of transmitting lateral loads (wind and earthquake) to the main structural system. This means that even curtain walls really have a structural mission. Curtain walls are generally supported (vertically) by the structural frame at each floor level, although some masonry curtain walls may be supported only at the foundation due to their weight. Lateral support in this case must still be provided at each floor level.

Behavior

Wall behavior can be studied in relation to the three types of functional purposes that walls provide. As a *vertical support* subsystem, walls serve the same function as columns. In-plane compressive forces are transmitted through the wall to the foundation (Fig. 5–67). If these compressive forces are uniform along the length of the wall, the wall behaves very similar to a column element, with the usual concerns for compressive strength and stability. Eccentric loads on bearing walls may induce end moments that force the wall into beam-column type of behavior (Fig. 5–68).

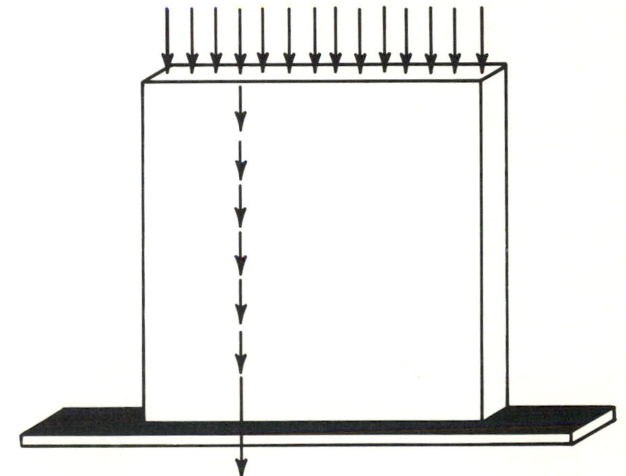

Figure 5–67. Wall Behavior

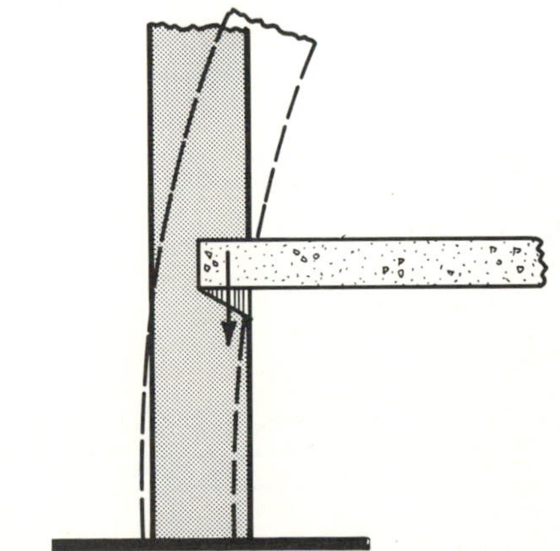

Figure 5–68. Eccentric Load Behavior

Concentrated loads on bearing walls spread laterally as they follow a load path from their point of application to the ground (Fig. 5–69). This lateral distribution is a result of the natural stiffness of the wall materials in compression and the tendency for "arching action" to be the

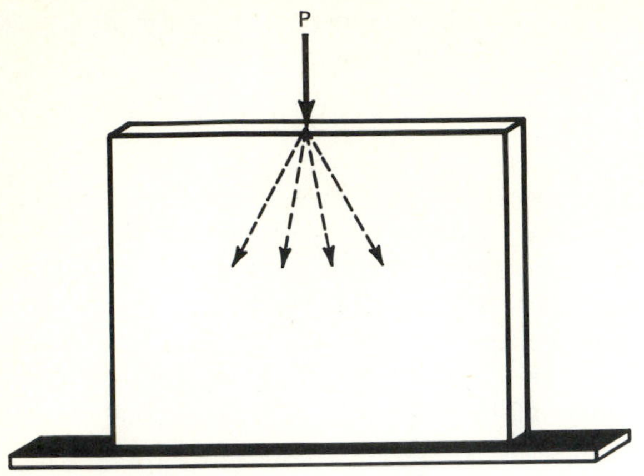

Figure 5–69. Concentrated Load Behavior

stiffest load path. The angle of distribution depends on the material, with common assumptions shown in Fig. 5–70. Flow of loads around openings in walls is complex and often wall "strips" are utilized to deal with the analysis of such problems.

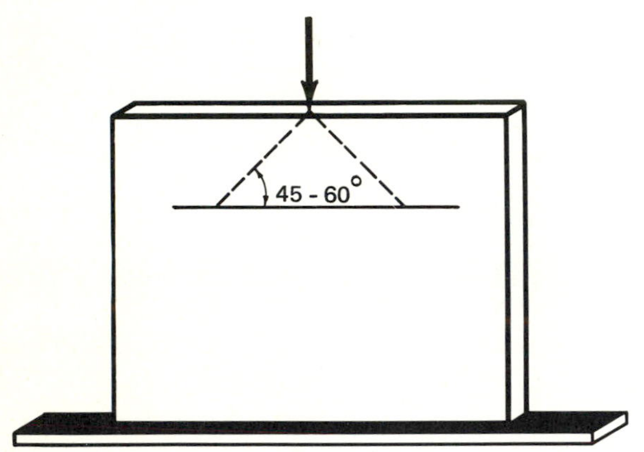

Figure 5–70. Distribution of Concentrated Loads

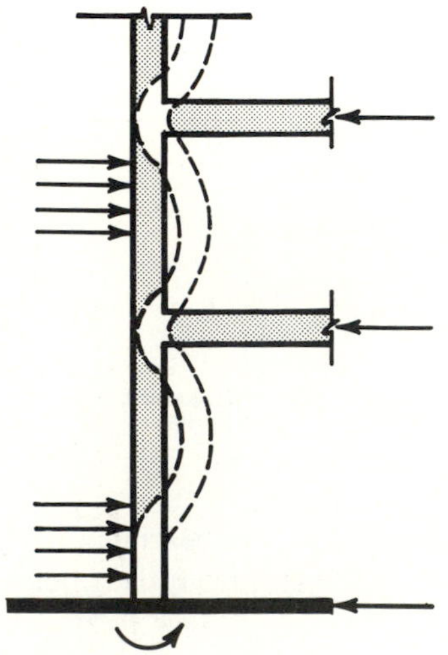

Figure 5–71. Lateral Load Behavior

As *lateral distribution* subsystems, walls transmit wind and earthquake loads (due to their own mass) to the structural frame (Fig. 5–71). These loads, together with gravity loads from the wall's own weight or superimposed floor loads, turn the wall into a series of beam-columns. Continuity at floor and roof connections may be provided, which achieves some structural efficiency in carrying the bending forces. These exterior walls may be subjected to either inward or outward lateral loads and adequate ties to the structural frame must be provided.

The *shear wall* acts as a vertical cantilever beam to provide *lateral support* and experiences both bending and shear deflection (Fig. 5–72). The lateral loads carried by the shear wall (if transmitted by a diaphragm system) depend on the relative rigidities of the wall and the diaphragm. Loads applied to the wall produce flexural and shear forces for which the wall materials must be designed. In addition, overall stability of the wall against overturning must be checked.

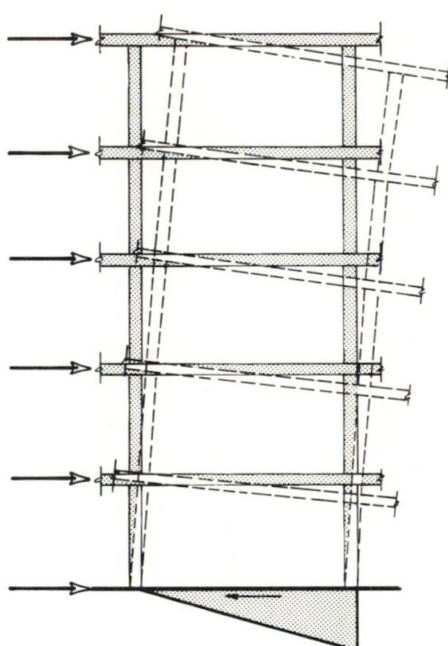

Figure 5–72. Shear Wall Behavior

Shear walls may consist of a single wall surface or may include transverse "flange elements" such as shown in Fig. 5–73. Walls on four sides of a building may function as an equivalent "tube" and behave as shown in Fig. 5–74. Tube subsystems involve some amount of "shear lag" around the corners that causes stresses to differ from the pure bending theory.

Approximate Analysis

The determination of moments, shears, and axial forces in walls involves few new concepts. Axial loads are normally calculated on a tributary width basis, including

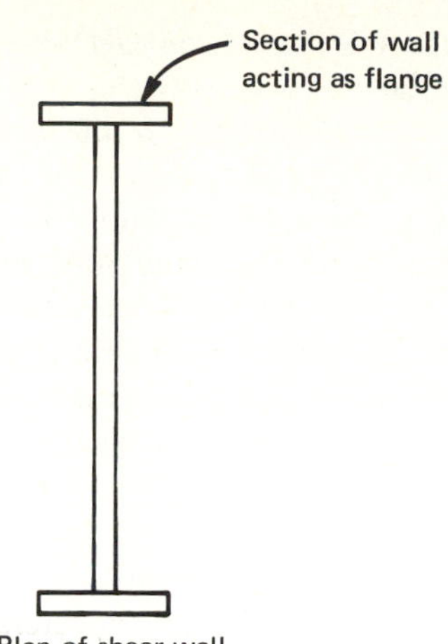

Figure 5–73. Types of Shear Walls

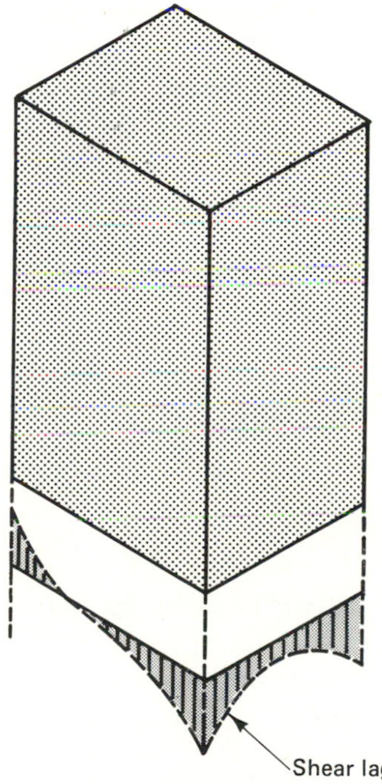

Figure 5–74. Tubular Shear Walls

appropriate live load reductions. Gravity load moments are calculated directly based on eccentric loads, or on the basis of frame action if the floor (roof) wall connections are rigid. Transverse wind load moments are calculated either on the basis of one-way beam action or two-way slab action. The slab coefficients of Table 5–6 may be used for the two-way case.

Analysis of shear walls deserves some discussion. Lateral loads applied to the shear wall depend on the arrangement of the building and the structural concept. Figure 5–75 shows several possible shear wall arrangements. Some of these involve twisting forces which add additional lateral loads to the shear wall. This is especially true for earthquake loads when the center of rigidity is different from the center of mass.

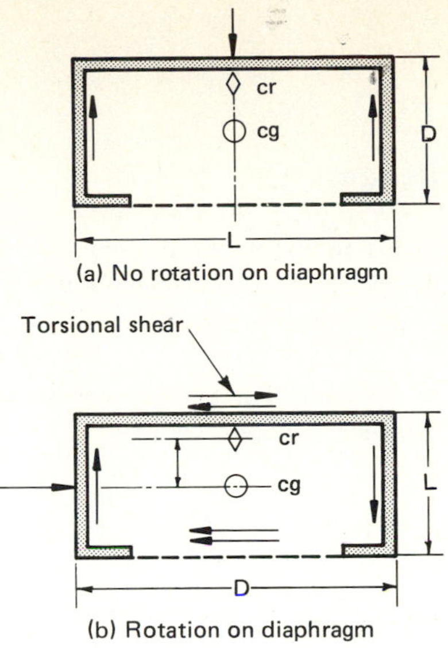

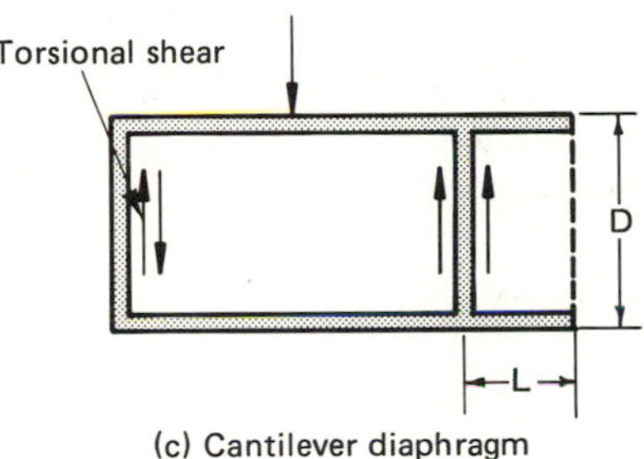

Figure 5–75. Shear Wall Arrangements

Once loads are determined, maximum shear can be calculated at a section near the base (Fig. 5–76) and flexural stresses calculated based on a MC/I distribution (or on a cracked section basis). Offsetting axial stresses should be included in the flexural analysis. Stability against overturning can be checked by summing the overturning and resisting moments about one edge of the wall.

Figure 5–76. Shear Wall Analysis

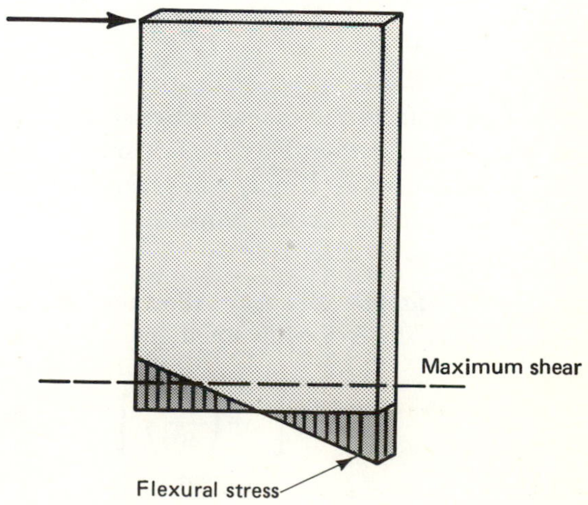

Approximate Design

The design of walls includes several semiempirical approaches that are specified for each type of wall material. Several of these are included here.

Reinforced Concrete Walls. The design axial load strength for reinforced concrete walls is

$$\phi P_{nw} = 0.55\phi f_c' A_g \left[1 - \left(\frac{kl_c}{32h}\right)^2\right] \qquad (5\text{–}33)$$

where $\phi = 0.70$
f_c' = concrete strength, psi (MPa)
l_c = vertical distance between supports, in. (mm)
A_g = gross area of wall, in.^{2} (mm^2)
h = overall wall thickness (not less than $\frac{1}{25}$ of unsupported height or width), in. (mm)
k = effective length factor

Concentrated loads or reactions can be considered distributed over a width of wall not to exceed the center-to-center distance between loads, nor the width of bearing plus four times the wall thickness.

Shear and moment forces perpendicular to the wall are treated similar to slab design. Shear strength in the plane of the wall is

$$V_n = \phi V_c + V_s \qquad (5\text{–}34)$$

where V_n = total shear strength (not to exceed $10\sqrt{f_c'}\, hd$)
$V_c = 2\sqrt{f_c'}\, hd$
$d = 0.8 \times$ wall length

$$V_s = \frac{A_v f_y d}{S_2}$$

A_v = area of horizontal shear reinforcement within distance S_2
$\phi = 0.85$

In-plane moment forces may be resisted by just the uncracked concrete sections if axial loads are high enough to offset flexural tension. If not, reinforcing steel can be sized based on cracked section properties.

Concrete Masonry Walls. Allowable *axial load* stress is given by the following equation:

$$F_A = 0.225 f_m' \left[1 - \left(\frac{h}{40t}\right)^3\right] \qquad (5\text{–}35)$$

where F_A = allowable stress on net cross section, psi (MPa)
f_m' = compressive strength of masonry (875 to 2400 psi)
h = effective height, in. (mm)
t = effective nominal thickness, in. (mm)

Allowable *compressive flexural* stresses for loads perpendicular to the wall are governed by the following equation:

$$\frac{f_a}{F_a} + \frac{f_m}{F_m} \leq 1.0 \quad (1.33 \text{ for } W \text{ or } E) \qquad (5\text{–}36)$$

where f_a = computed axial stress
F_a = allowable axial stress given by Eq. 5–35
f_m = computed flexural compressive stress
F_m = allowable flexural compressive stress $= 0.33 f_m'$

For nonreinforced masonry, the computed flexural compressive stress, f_m, is based on the section modulus for the uncracked section. For reinforced masonry, the computed flexural stress is based on the transformed area of the net cracked section, but an equivalent section modulus can be derived as follows.

Given the basic equation for reinforced masonry compressive stress,

$$f_m = \frac{M}{bd^2}\frac{2}{kj}$$

where M = bending moment
b = width of section
d = distance from compressive face to reinforcing steel
k = coefficient defining the depth, kd, to the neutral axis
$j = 1 - k/3$

$$S = \frac{bd^2}{2}(kj) \qquad (5\text{–}37)$$

Appendix 5B contains section modulus values for both nonreinforced and reinforced masonry assemblies, with ranges for reinforced sections based on minimum and maximum reinforcement percentages.

Allowable *flexural tensile* stresses for nonreinforced masonry are given by the following equation:

$$f_t - f_a \leq F_t \quad (1.33F_t \text{ for } W \text{ or } E) \qquad (5\text{–}38)$$

Flexural tensile stresses for reinforced masonry are obviously taken by the steel, and it usually is not necessary

Table 5–7 Allowable Stress in Concrete Masonry[a,c]

Description		Allowable stresses (psi)	
		Related to f'_m	*Maximum*
Compressive			
Axial	F_a	Eq. 5–35	1000
Flexural	F_m	$0.33f'_m$	1200
Bearing			
On full area	F_a	$0.25f'_m$	900
On one-third area or less	F_a	$0.375f'_m$	1200
Shear			
No shear reinforcement			
Flexural members	v_m	$1.1\sqrt{f'_m}$	50
Shear walls			
$M/Vd_v \geq 1$	v_m	$0.9\sqrt{f'_m}$	34
$M/Vd_v < 1$	v_m	$2.0\sqrt{f'_m}$	40 $(1.85 - M/Vd_v)$
Reinforcement taking entire shear			
Flexural members	v	$3.0\sqrt{f'_m}$	150
Shear walls			
$M/Vd_v \geq 1$	v	$1.5\sqrt{f'_m}$	75
$M/Vd_v < 1$	v	$2.0\sqrt{f'_m}$	45 $(2.67 - M/Vd_v)$
Tension: no tension reinforcement			
Tension normal to bed joints			
Hollow units	F_t	$0.5\sqrt{m_o}$	25
Solid and/or grouted units	F_t	$1.0\sqrt{m_o}$	40
Tension parallel to bed joints in running bond			
Hollow units	F_t	$1.0\sqrt{m_o}$	50
Solid and/or grouted units	F_t	$1.5\sqrt{m_o}$	80
Modulus of elasticity	E_m	$1000f'_m$	2,500,000
Modulus of rigidity	E_v	$400f'_m$	1,000,000

Compressive Strength of Masonry[b]

Compressive strength of the units (psi)	Assumed compressive of masonry, f'_m (psi)	
	Type M and S mortar	*Type N mortar*
Over 1500–2500	1151–1550	875–1100
Over 2500–4000	1551–2000	1101–1250
Over 4000–6000	2001–2400	1251–1350
Over 6000	2400	1350

[a] ACI Committee 531, *Building Code Requirements for Concrete Masonry,* ACI 531–79 (Detroit, Mich.: American Concrete Institute, 1979).

[b] Gross area for masonry of solid units: net area for masonry of hollow units.

[c] m_o = specified 28-day minimum required compressive strength of mortar per ASTM C270, psi
M = design moment, in.-lb.
V = total applied design shear force at section, lb.
d_v = actual depth of masonry in direction of shear, in.
f'_m = specified compression strength of masonry, psi

to check the steel stress during preliminary design. Table 5–7 summarizes allowable stress for concrete masonry. The following example illustrates the approximate design method.

EXAMPLE PROBLEM 5–12: Preliminary Concrete Masonry Wall Design

Select a preliminary concrete masonry wall for the exterior wall of a manufacturing building. The wall will be a bearing wall, supporting a total roof load of 1200 lb/ft (17,512 N/m). Wall height is 25 ft (7.6 m), and wind load is 20 psf (958 N/m²).

Solution

1. *Determine forces.*

$$P = 1200 \text{ lb } (5338 \text{ N}) + \text{wall weight}$$

$$M = \frac{wl^2}{8} = \frac{20(25)^2}{8} = 1562 \text{ ft-lb } (2118 \text{ Nm})$$

$$V = \frac{wl}{2} = \frac{20(25)}{2} = 250 \text{ lb } (1112 \text{ N})$$

2. *Establish masonry strength and allowable stress.* Assume that $f'_m = 1350$ psi (9.3 MPa).

$$F_a = 0.225 f'_m \left[1 - \left(\frac{h}{40t}\right)^3\right]$$

For an 8-in. (203-mm) wall:

$$F_a = 0.225(1350)\left[1 - \left(\frac{25 \times 12}{40 \times 8}\right)^3\right]$$
$$= 53.5 \text{ psi } (0.37 \text{ MPa})$$

For a 12-in. (305-mm) wall:

$$F_a = 0.225(1350)\left[1 - \left(\frac{25 \times 12}{40 \times 12}\right)^3\right]$$
$$= 229 \text{ psi } (1.58 \text{ MPa})$$

$$F_m = 0.33 f'_m = 0.33(1350) = 445 \text{ psi } (3.07 \text{ MPa})$$

3. *Try an 8-in. (203-mm) hollow block.* From Appendix 5B, A = 45.0 in.², S = 80.0 in.³, weight = 50.0 psf.

$$f_m = f_t = \frac{M}{S} = \frac{1562 \times 12}{80}$$
$$= 234 \text{ psi } (1.6 \text{ MPa})$$

$$P \text{ at midheight} = 1200 \text{ lb} + (25/2)50$$
$$= 1825 \text{ lb } (8118 \text{ N})$$

$$f_a = \frac{1825}{45} = 40.6 \text{ psi } (0.28 \text{ MPa})$$

By Eq. 5–32,

$$\frac{40.6}{53.5} + \frac{234}{445} = 1.28 < 1.33 \quad \text{O.K.}$$

By Eq. 5–34,

$$234 \text{ psi} - 40.6 \text{ psi} = 193.4 \text{ psi } (1.33 \text{ MPa})$$
$$> 1.33(25) \quad \text{N.G.}$$

4. *Try a 12-in. (305-mm) hollow block.* From Appendix 5B, A = 70.0 in.², S = 190.0 in.³, weight = 69.0 psf.

$$f_m = f_t = \frac{M}{S} = \frac{1562 \times 12}{190}$$
$$= 98.6 \text{ psi } (0.68 \text{ MPa})$$

$$P \text{ at midheight} = 1200 \text{ lb} + \left(\frac{25}{2}\right) 69$$
$$= 2062 \text{ lb } (9172 \text{ N})$$

$$f_a = \frac{2062}{70.0} = 29.5 \text{ psi } (0.20 \text{ MPa})$$

By Eq. 5–32,

$$\frac{29.5}{229} + \frac{98.6}{445} = 0.35 < 1.33 \quad \text{O.K.}$$

By Eq. 5–34,

$$98.6 \text{ psi} - 29.5 \text{ psi} = 69.1 \text{ psi } (0.48 \text{ MPa})$$
$$> (1.33 \times 25 = 33) \quad \text{N.G.}$$

5. *Try a 12-in. (305-mm) reinforced block.*

$$\text{Max. } f_m = \left(1.33 - \frac{29.5}{229}\right) 445 = 387$$

$$S_{\text{req'd}} = \frac{1562 \times 12}{387} = 48.4$$

By Appendix 5B, 12-in. (305-mm) reinforced block has an available range of 24.4 to 95.4 for S.

Using a squared interpolation, we get

$$A_s = \frac{(48.4 - 24.4)^2}{(95.4 - 24.4)^2} \times (1.50 - 0.033) + 0.033$$
$$= 0.20 \text{ in.}^2/\text{ft}$$

Brick Masonry Walls. The allowable axial load for brick masonry walls for $e/t < \frac{1}{3}$ is given by

$$P = C_e C_s F_a A_g \tag{5–39}$$

where C_e = eccentricity coefficient
C_s = slenderness coefficient
F_a = allowable axial compressive stress ($0.20 f'_m$ for nonreinforced and $0.25 f'_m$ for reinforced)
A_g = gross cross-sectional area

Allowable stresses for brick masonry are given in Table 5–8 and values for C_e and C_s in Table 5–9.

When the ratio e/t exceeds $\frac{1}{3}$, the maximum tensile stress in the masonry is limited to the values given in

Table 5–8 Allowable Stresses for Brick Masonry[a,b]

Description		Allowable stresses (psi) Nonreinforced	Reinforced
Compressive, axial			
Walls	f_m	$0.20f'_m$	$0.25f'_m$
Columns	f_m	$0.16f'_m$	$0.20f'_m$
Compressive, flexural			
Walls and beams	f_m	$0.32f'_m$	$0.40f'_m$
Columns	f_m	$0.26f'_m$	$0.32f'_m$
Tensile, flexural			
Normal to bed joints			
M or S mortar	f_t	36	
N mortar	f_t	28	
Parallel to bed joints			
M or S mortar	f_t	72	
N mortar	f_t	56	
Shear			
No shear reinforcement			
Flexural members	v_m	$0.5\sqrt{f'_m}$ but not to exceed 80	$0.7\sqrt{f'_m}$ but not to exceed 50
Shear walls	v_m	For M or S mortar and not to exceed 28 for N mortar	$0.5\sqrt{f'_m}$ but not to exceed 100
With shear reinforcement taking entire shear			
Flexural members	v		$2.0\sqrt{f'_m}$ but not to exceed 120
Shear walls	v		$1.5\sqrt{f'_m}$ but not to exceed 150
Bond			
Plain bars	u		80
Deformed bars	u		160
Bearing			
On full area	f_m		$0.25f'_m$
On one-third area or less	f_m		$0.376f'_m$
Modulus of elasticity	E_m		$1000f'_m$ but not to exceed 3,000,000 psi
Modulus of rigidity	E_v		$400f'_m$ but not to exceed 1,200,000 psi

Assumed Compressive Strength of Brick Masonry

Compressive strength of units (psi)	Without inspection Type N	Type S	Type M	With inspection Type N	Type S	Type M
14,000 plus	2,140	2,600	3,070	3,200	3,900	4,600
12,000	1,870	2,270	2,670	2,800	3,400	4,000
10,000	1,600	1,930	2,270	2,400	2,900	3,400
8,000	1,340	1,600	1,870	2,000	2,400	2,800
6,000	1,070	1,270	1,470	1,600	1,900	2,200
4,000	800	930	1,070	1,200	1,400	1,600
2,000	530	600	670	800	900	1,000

[a] *Recommended Practice for Engineered Brick Masonry* (McLean, Va.: Brick Institute of America, 1969).

[b] Allowable stresses for inspected brick construction.

Table 5–9 Brick Wall Coefficients[a,b]

Eccentricity Coefficients (C_e)

$\frac{e}{t}$	e_1/e_2: -1	$-\frac{3}{4}$	$-\frac{1}{2}$	$-\frac{1}{4}$	0	$+\frac{1}{4}$	$+\frac{1}{2}$	$+\frac{3}{4}$	$+1$
$0-\frac{1}{20}$ (0.05)	1.00	1.00	1.00	1.00	1.00	1.00	1.00	1.00	1.00
$\frac{1}{12}$ (0.083)	0.90	0.90	0.89	0.89	0.88	0.88	0.87	0.87	0.87
$\frac{1}{8}$ (0.125)	0.82	0.81	0.80	0.79	0.78	0.77	0.76	0.75	0.74
$\frac{1}{6}$ (0.167)	0.77	0.75	0.74	0.72	0.71	0.69	0.68	0.66	0.65
$\frac{5}{24}$ (0.208)	0.73	0.71	0.69	0.67	0.65	0.63	0.61	0.59	0.57
$\frac{1}{4}$ (0.250)	0.69	0.66	0.64	0.61	0.59	0.56	0.54	0.51	0.49
$\frac{7}{24}$ (0.292)	0.65	0.62	0.59	0.56	0.53	0.50	0.47	0.44	0.41
$\frac{1}{3}$ (0.333)	0.61	0.57	0.54	0.50	0.47	0.43	0.40	0.36	0.32

Slenderness Coefficients (C_s)

$\frac{h}{t}$	e_1/e_2: -1	$-\frac{3}{4}$	$-\frac{1}{2}$	$-\frac{1}{4}$	0	$+\frac{1}{4}$	$+\frac{1}{2}$	$+\frac{3}{4}$	$+1$
5.0	1.00	1.00	1.00	1.00	1.00	1.00	1.00	1.00	1.00
7.5	1.00	1.00	1.00	1.00	1.00	0.98	0.96	0.93	0.90
10.0	1.00	0.99	0.98	0.96	0.93	0.91	0.88	0.84	0.80
12.5	0.95	0.94	0.92	0.90	0.87	0.83	0.79	0.75	0.70
15.0	0.90	0.88	0.86	0.83	0.80	0.76	0.71	0.66	0.60
17.5	0.85	0.83	0.81	0.77	0.73	0.69	0.63	0.57	0.50
20.0	0.80	0.78	0.75	0.71	0.67	0.61	0.55	0.48	0.40
22.5	0.75	0.73	0.69	0.65	0.60	0.54	0.47	0.39	
25.0	0.70	0.67	0.64	0.59	0.53	0.47	0.39		
27.5	0.65	0.62	0.58	0.53	0.47	0.39			
30.0	0.60	0.57	0.52	0.47	0.40				
32.5	0.55	0.52	0.47	0.41					
35.0	0.50	0.46	0.41						
37.5	0.45	0.41							
40.0	0.40								

[a] *Recommended Practice For Engineered Brick Masonry* (McLean, Va.: Brick Institute of America, 1969).

[b] e_1 = smaller virtual eccentricity at lateral supports (at either top or bottom of member)
e_2 = larger virtual eccentricity at lateral supports (at either top or bottom of member)

Table 5–8. If these limits are exceeded, the wall must be reinforced and Eqs. 5–36 and 5–37 apply.

Allowable stresses for Eq. 5–36 are:

$$F_a = 0.25 f'_m C_s$$

$$F_m = 0.40 f'_m$$

5–9 BRACING ASSEMBLIES

The final major subsystem is the bracing assembly (Fig. 5–77). Bracing is composed of tension tie and column elements which together provide lateral support, lateral distribution, or stability against element and frame buckling. Bracing can occur in any geometrical plane and usually is incorporated in one of the major subsystem planes of the structural system.

The assumption of braced support points for beam and column elements is fundamental to much of our structural theory. The question, "What constitutes adequate bracing?" is seldom answered and is left largely to our judgment as designers. The discussion here seeks to shed some light on this mysterious subject.

History and Development

The use of individual bracing elements dates back to some of the earliest structures which used tie-backs and struts for support. Complete subsystems of bracing were later developed for bridge structures and towers. The development of framed building structures around the end of

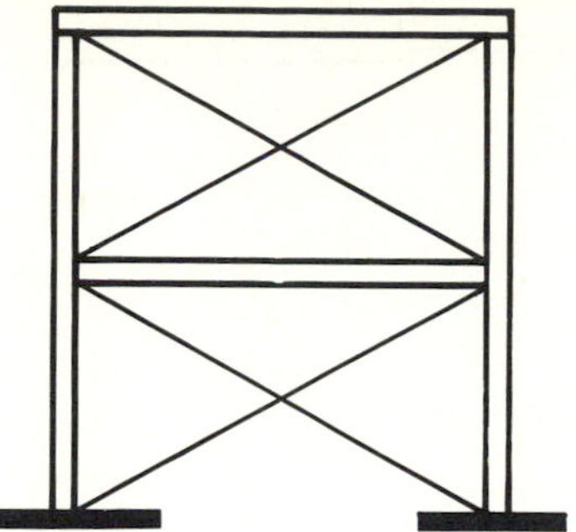

Figure 5–77. Bracing Assembly

the nineteenth century and the gradual decline of the traditional dependence on walls for lateral support led to the introduction of bracing in building structures.

Today the "braced frame" building is a very popular structural system which utilizes the bracing assembly for lateral support. Other uses of bracing for stability become increasingly important as structural elements become stronger and more slender.

Types and Uses

Major types of bracing assemblies are shown in Fig. 5–78. These include vertical wind bracing, bracing for lateral support of trusses, lateral distribution bracing, bracing to distribute wind loads laterally to lateral support points, and lateral bracing of beam elements. The use of bracing to carry lateral wind and earthquake loads is normally the most efficient means of lateral support because only tension and compression forces are involved. In some cases, the use of bracing to resist earthquake loads may provide too much lateral stiffness, and a more ductile frame may be advisable.

Behavior

Diagonal bracing is the most popular bracing assembly because it provides a continuous load path. It is an indeterminate assembly but is normally made determinate by assuming that the compressive diagonal buckles under load, and that only the tension member is then active. Other types of bracing, such as knee bracing and portal bracing, introduce bending forces in connecting beam or column elements. These types rely on the compressive elements for full participation and thus are less efficient than the diagonal form.

Bracing assemblies deform under load as axial compressive or tensile strain takes place. This deformation leads to overall deflections (such as drift) for the braced frame or braced element. The adequacy of bracing, then, becomes not only a question of strength but also of stiffness to limit deflections to acceptable limits. For example, columns in a braced frame are normally assumed immune

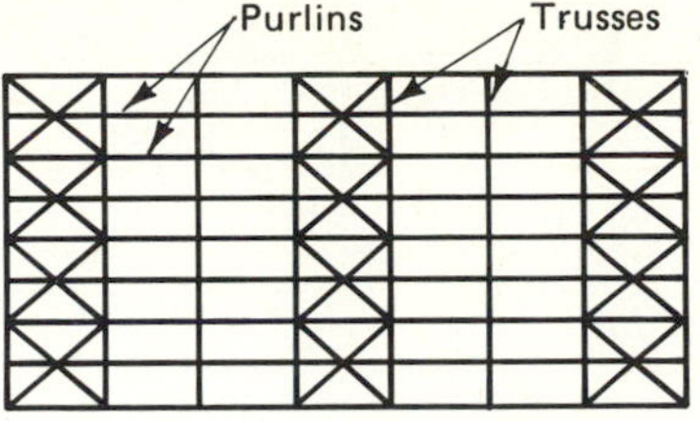

(a) Bracing in plane of top chord

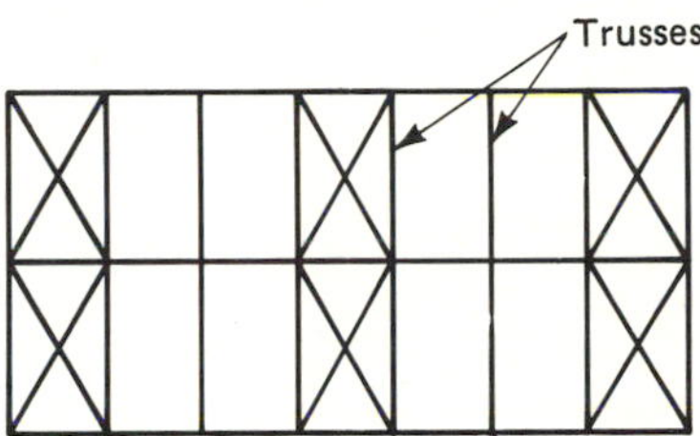

(b) Bracing in plane of bottom chord

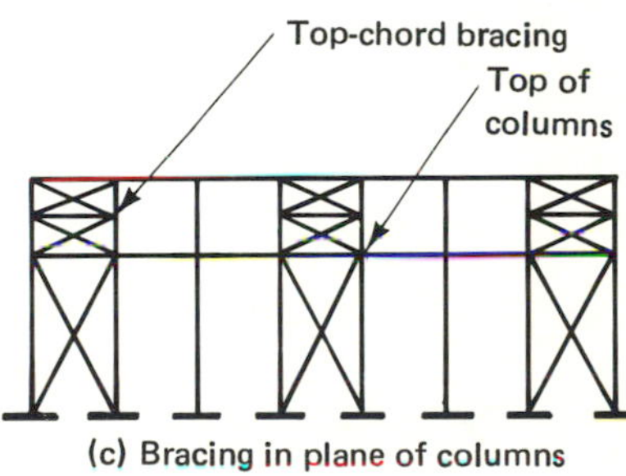

(c) Bracing in plane of columns

Figure 5–78. Types of Bracing

to the $P\Delta$ effect for the sidesway condition. This may not be the case if drift limits are not controlled by adequate bracing stiffness.

Strength and stiffness of bracing are also important in deciding whether adequate lateral support is provided to prevent lateral buckling. Both experimental and analytical tests have demonstrated that *strengths* equal to *$2\frac{1}{2}$%* of the maximum force in the compression flange of the element is sufficient to prevent buckling. *Stiffness* required must be at least *25 times* the lateral bending stiffness of the member to be braced before it can be considered fully effective.

Continuous lateral support for beam elements is commonly provided by floor and roof assemblies as shown in Fig. 5–79. Lightweight steel decks should be checked for adequate shear stiffness. Friction connections can be checked to ensure that the frictional force is adequate to resist the $2\frac{1}{2}$% force.

Approximate Analysis

Analysis of bracing assemblies involves only statics. Knee brace systems and other similar forms are part of indeter-

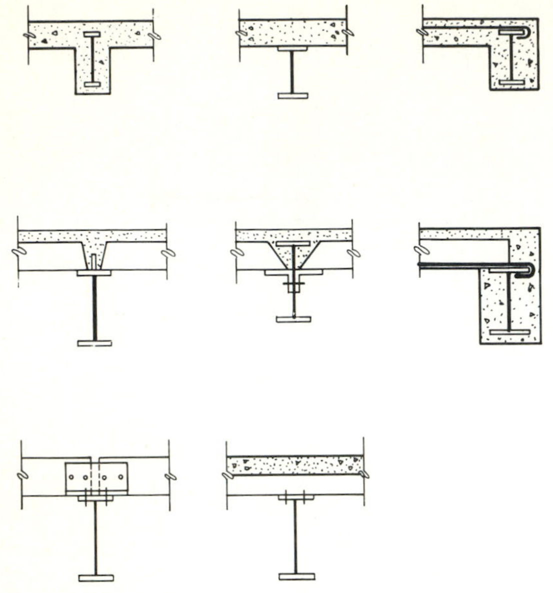

Figure 5–79. Lateral Support Methods

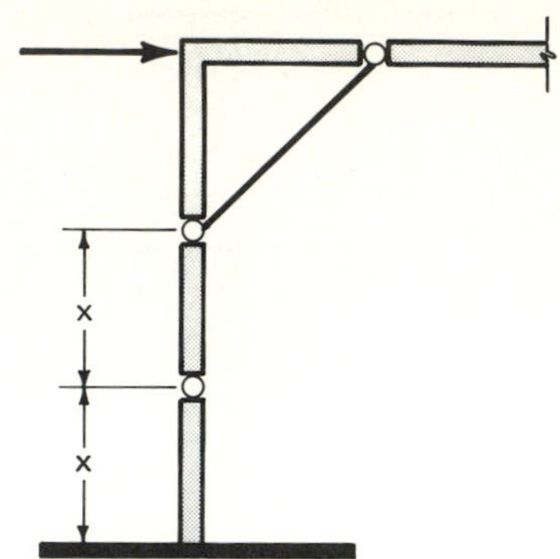

Figure 5–80. Knee Brace Analysis

minate systems, but we can assume inflection points to render these systems determinate. Figure 5–80 illustrates this method for knee-braced frames. Lateral loads to braced bents may in some cases depend on relative rigidities in combination systems such as braced frame/unbraced frame and braced frame/shear wall systems.

Approximate Design

Design of bracing elements follows the procedures for tension ties and column elements discussed earlier. Design choices involve the following:

- Material selection
- Type of bracing assembly
- Number and location of bracing assemblies
- Element proportions

5–10 SUMMARY

In this chapter we covered the second level of structural assembly, the subsystem. The subsystem is a two-dimensional assembly of individual elements, formed in a configuration which equips it to play a major functional role in system stability. Subsystem behavior is an extension of our knowledge of element behavior, while subsystem design reveals an array of choices that add to our decision menu. In the next chapter we will complete the assembly process with the study of structural systems.

PROBLEMS

5–1. There are two options being considered for a post-and-beam subsystem:

1. Columns spaced at 20 ft (6.1 m), with 25 ft (7.6 m) between frames, or
2. Columns spaced at 25 ft (7.6 m), with 20 ft (6.1 m) between frames.

Develop an approximate design for each and compare costs. The dead load is 20 psf (958 N/m²) and the live load is 30 psf (1436 N/m²); column height is 15 ft (4.6 m).

5.2. Develop a preliminary design for members 1, 2, 3, and 4 of the floor assembly shown in Fig. 5–81. Total floor load is 200 psf (9576 N/m²). Use structural steel.

5–3. Develop a preliminary design for the reinforced concrete bridge bent shown in Fig. 5–82. Use 3000-psi (20.7-MPa) concrete and grade 60 steel.

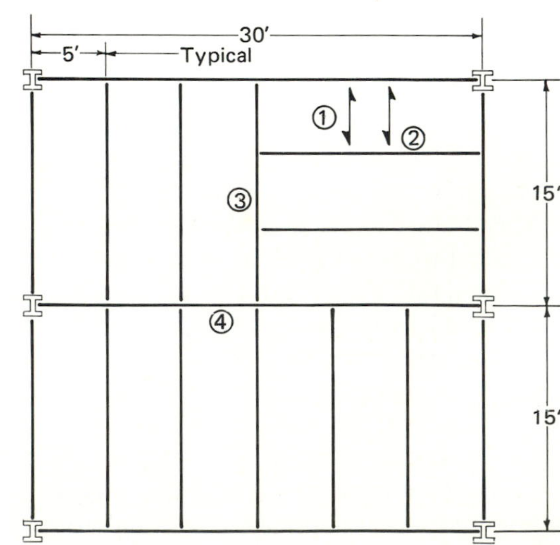

Figure 5–81. Problem 5–2

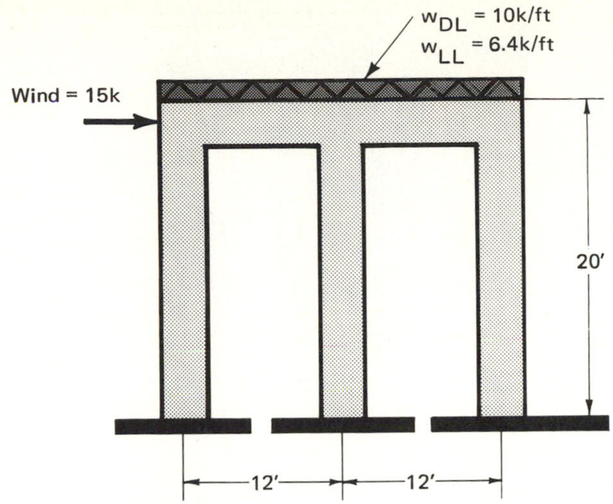

Figure 5-82. Problem 5-3

5-4. The roof plan of a manufacturing building is shown in Fig. 5-83. A 22-gage intermediate rib metal deck has been tentatively selected as the deck assembly to span 6 ft (1.8 m) between bar joist elements. The exterior wall will be metal siding, and the design wind load is 20 psf (958 N/m²). Check to see if the metal deck is capable of acting as a diaphragm for wind in the east-west direction.

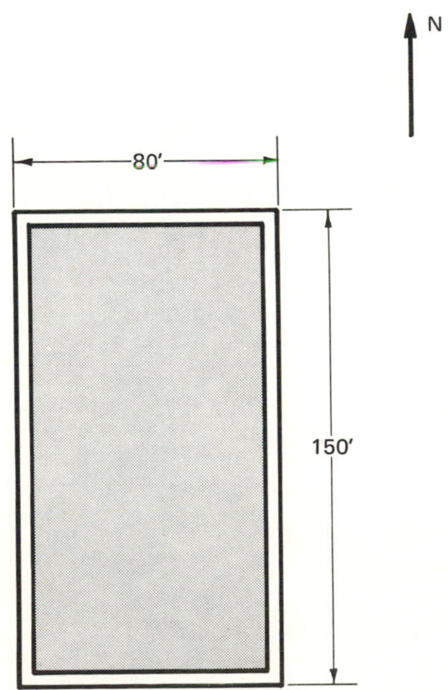

Figure 5-83. Problem 5-4

5-5. Provide a preliminary design for the two-story steel rigid frame shown in Fig. 5-84. Assume A36 steel.

5-6. Provide a preliminary design for the grid system shown in Fig. 5-85. Use structural steel members of A36 steel.

5-7. Design a preliminary space frame to span 200 ft × 200 ft (60.9 m × 60.9 m) between perimeter bearing walls as shown in Fig. 5-86. The roof live load is 30 psf (1436 N/m²).

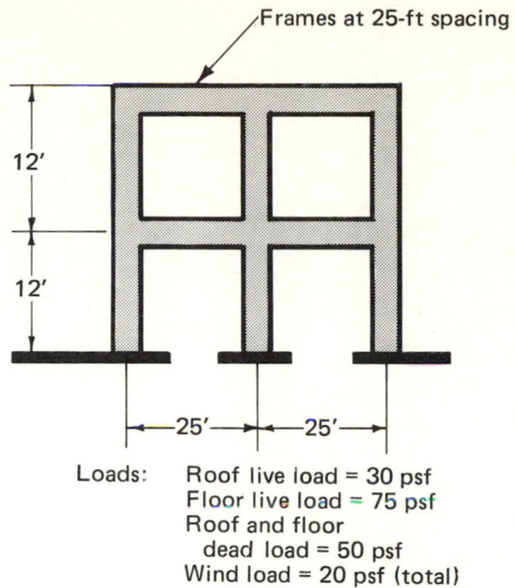

Figure 5-84. Problem 5-5

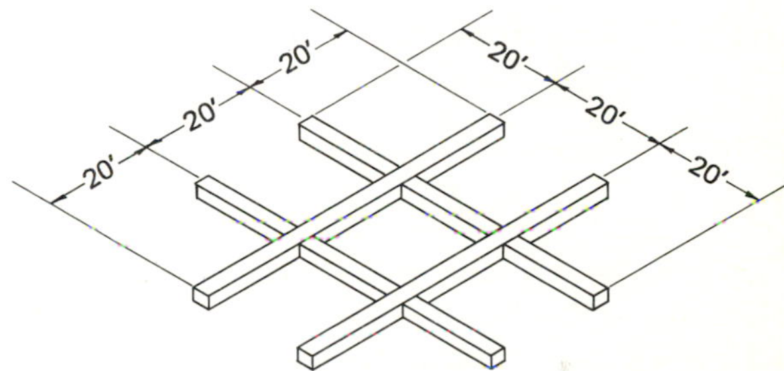

Figure 5-85. Problem 5-6

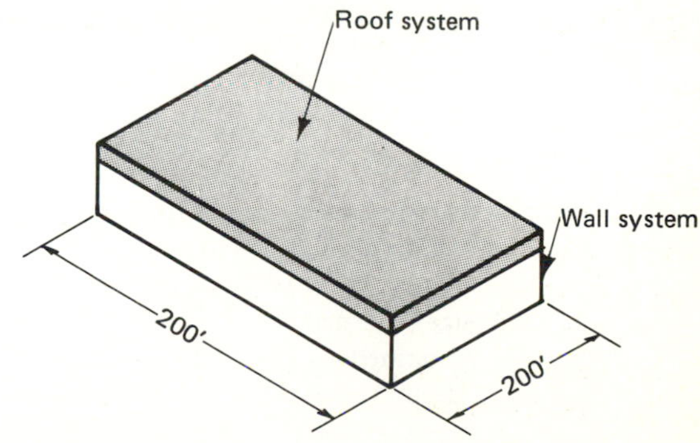

Figure 5-86. Problem 5-7

5-8. Provide a preliminary design for the concrete slab system shown in Fig. 5-87. Use a concrete reinforcement strength of 3000 psi (20.7 MPa) and 60 ksi (413.7 MPa).

5-9. Design a preliminary bearing wall to support an axial load of 10 kips/ft (145,932 N/m). The unbraced wall height is 10 ft (3.05 m), as shown in Fig. 5-88. Use concrete masonry.

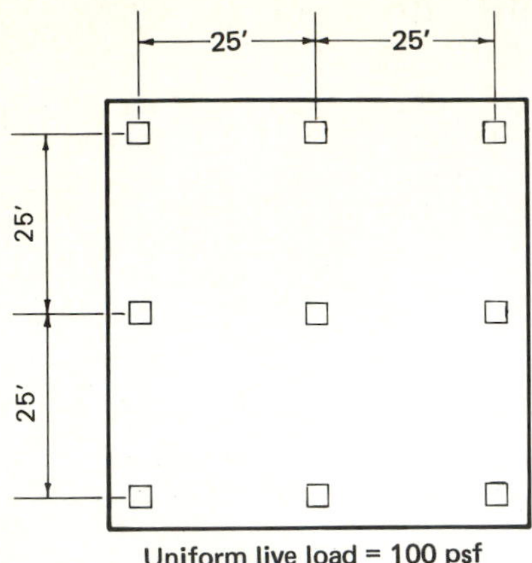

Figure 5–87. Problem 5–8

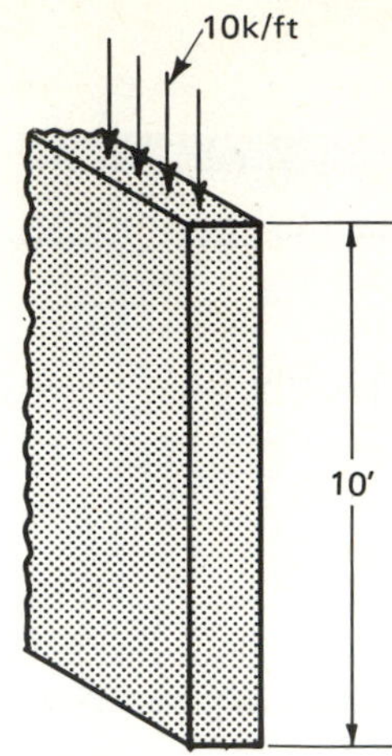

Figure 5–88. Problem 5–9

APPENDIX 5A: Deck Assemblies

Deck	Weight (psf)	Load-span limits[a] (ft) — Total superimposed load (psf) 25	50	75	100	150
Wood[b]						
½-in. plywood, 2 or more spans	1.5	2.67	2.50	2.0	2.0	2.0
		—	1.33	1.33	1.33	1.33
¾-in. plywood, 2 or more spans	2.2	4.0	3.33	2.75	2.5	2.25
		—	2.0	2.0	2.0	2.0
1⅝-in. T&G plank, 2 or more spans	3.5	11	10	8	7	6
		6*	5.5*	5*	4.5*	4*
2⅝-in. T&G plank, red cedar, 2 or more spans	5	—	15.5	13	11.5	9.5
		9.5*	9*	8*	7.5*	6.5*
3⅝-in. T&G plank, red cedar, 2 or more spans	7	—	21	18.5	16	13
		14.5*	12.5*	11.0*	10*	9*
Metal						
22-gage steel deck, narrow rib, 1½ in. deep, 3 or more spans	1.8	7.5	6.0	5.0	—	—
22-gage steel deck, intermediate rib, 1½ in. deep, 3 or more spans	1.8	8.0	6.5	5.5	—	—
18-gage steel deck, wide rib, 1½ in. deep, 3 or more spans	3	10.5	9.0	7.5	7.0	6.0
18-gage steel deck, 4½ in. deep, simple span	4.3	21	18	14.0	13.0	—
20-gage steel deck, 3 in. deep, simple span	2.9	14.0	11.0	9.0	8.0	—
Steel cellular units, 16-gage corrugated + 16-gage plate 3 in. thick, 2½-in. concrete fill, 3 spans	48	—	—	13*	12*	10*

[a] *, limited by deflection.

[b] First row applies to roof construction, second row to floor construction.

APPENDIX 5A: Deck Assemblies (*Continued*)

Deck	Weight (psf)	Load-span limits[a] (ft)				
		Total superimposed load (psf)				
		25	50	75	100	150
18-gage steel deck, wide rib, $1\frac{1}{2}$ in. deep, $2\frac{1}{2}$ in. concrete fill, 3 spans	42	—	11.0	10.0	9.0	8.0
Concrete on metal deck						
Reinforced concrete slab on 24 gage, $1\frac{5}{16}$-in.-deep steel form deck, 4 in. total thickness, negative reinforcement	44	—	6.5	6.5	6.5	6.0
Reinforced concrete on 28 gage, $\frac{9}{16}$-in.-deep steel form deck $2\frac{1}{2}$ in. total thickness	29	—	—	3.0	2.5	2.0
Reinforced concrete with 24 gage, $1\frac{5}{16}$ in. composite metal deck, 5 in. total thickness, 3 spans, negative steel	45	—	14.0	13.0	12.0	10.0
Perlite concrete fill 1:5 on 28 gage, $\frac{9}{16}$-in. deck, $2\frac{1}{2}$ in. total thickness, 3 or more spans	8	4.0*	3.0*	3.0*		
Miscellaneous roof decks						
Wood fiberboard, 2 in. thick, plank form	6	3	3	2.5		
2-in. gypsum concrete slab on 1-in. insulation, bulb tees $32\frac{5}{8}$ in. o.c., tees No. 158 at 1.33 lb/ft	11	6	4.5	3.75	3	
2-in. gypsum concrete slab on 1-in. insulation, bulb tees $32\frac{5}{8}$ in. o.c., tees No. 258 at 4.67 lb/ft	12	12	11	9	8	
Reinforced concrete						
One-way concrete slab, normal weight, simple span f'_c = 3000 psi:						
4 in. deep	50	6.7*	6.7*	6.7*	6.7*	6.7*
5 in. deep	63	8.3*	8.3*	8.3*	8.3*	8.3*
6 in. deep	75	10.0*	10.0*	10.0*	10.0*	10.0*
One-way concrete slab, normal weight, continuous span f'_c = 3000 psi:						
4 in. deep	50	8.0*	8.0*	8.0*	8.0*	8.0*
5 in. deep	63	10.0*	10.0*	10.0*	10.0*	10.0*
6 in. deep	75	12.0*	12.0*	12.0*	12.0*	12.0*

APPENDIX 5A: Deck Assemblies (*Continued*)

Deck	Weight (*psf*)	*Load-span limits*[a] (*ft*) *Total superimposed load* (*psf*) 25	50	75	100	150
One-way pan joist, normal weight, continuous span $f'_c = 4000$ psi:						
(1) 20-in. form + 5-in. rib						
8-in. rib + 3-in. slab	60	19.0*	19.0*	19.0*	19.0*	19.0*
8-in. rib + 4½-in. slab	79	22.0*	22.0*	22.0*	22.0*	22.0*
10-in. rib + 3-in. slab	67	23.0*	23.0*	23.0*	23.0*	23.0*
10-in. rib + 4½-in. slab	85	25.0*	25.0*	25.0*	25.0*	25.0*
12-in. rib + 3-in. slab	74	26.0*	26.0*	26.0*	26.0*	26.0*
12-in. rib + 4½-in. slab	92	29.0*	29.0*	29.0*	29.0*	29.0*
14-in. rib + 3-in. slab	81	30.0*	30.0*	30.0*	30.0*	30.0*
14-in. rib + 4½-in. slab	99	32.0*	32.0*	32.0*	32.0*	32.0*
(2) 20-in. form + 6-in. rib						
16-in. rib + 3-in. slab	94	33.0*	33.0*	33.0*	33.0*	33.0*
16-in. rib + 4½-in. slab	113	36.0*	36.0*	36.0*	36.0*	36.0*
20-in. rib + 3-in. slab	111	40.0*	40.0*	40.0*	40.0*	40.0*
20-in. rib + 4½-in. slab	130	42.0*	42.0*	42.0*	42.0*	42.0*
(3) 30-in. form + 6-in. rib						
8-in. rib + 3-in. slab	56	19.0*	19.0*	19.0*	19.0*	19.0*
8-in. rib + 4½-in. slab	75	22.0*	22.0*	22.0*	22.0*	22.0*
10-in. rib + 3-in. slab	61	22.0*	22.0*	22.0*	22.0*	22.0*
10-in. rib + 4½-in. slab	80	25.0*	25.0*	25.0*	25.0*	25.0*
12-in. rib + 3-in. slab	67	26.0*	26.0*	26.0*	26.0*	26.0*
12-in. rib + 4½-in. slab	85	29.0*	29.0*	29.0*	29.0*	29.0*
14-in. rib + 3-in. slab	72	30.0*	30.0*	30.0*	30.0*	30.0*
14-in. rib + 4½-in. slab	91	32.0*	32.0*	32.0*	32.0*	32.0*
16-in. rib + 3-in. slab	78	33.0*	33.0*	33.0*	33.0*	33.0*
16-in. rib + 4½-in. slab	97	36.0*	36.0*	36.0*	36.0*	33.0*
20-in. rib + 3-in. slab	91	40.0*	40.0*	40.0*	40.0*	39.0*
20-in. rib + 4½-in. slab	109	43.0*	43.0*	43.0*	43.0*	40.0*
Precast concrete						
Prestressed double tee, normal weight, no topping						
8 ft 0 in. × 12 in. deep	37	40.0	34.0	24.0	20.0	16.0
8 ft 0 in. × 14 in. deep	40	46.0	42.0	32.0	28.0	20.0
8 ft 0 in. × 16 in. deep	42	52.0	46.0	40.0	32.0	26.0
8 ft 0 in. × 18 in. deep	45	58.0	52.0	42.0	38.0	28.0
8 ft 0 in. × 20 in. deep	47	64.0	56.0	50.0	40.0	34.0
8 ft 0 in. × 24 in. deep	52	74.0	68.0	58.0	50.0	38.0
8 ft 0 in. × 32 in. deep	74	88.0	88.0	82.0	74.0	62.0
Prestressed double tee, normal weight, 2-in. topping						
8 ft 0 in. × 12 in. deep	62	34.0	32.0	24.0	22.0	20.0
8 ft 0 in. × 14 in. deep	65	36.0	36.0	32.0	30.0	22.0
8 ft 0 in. × 16 in. deep	67	42.0	42.0	34.0	32.0	28.0
8 ft 0 in. × 18 in. deep	70	46.0	44.0	42.0	38.0	28.0
8 ft 0 in. × 20 in. deep	72	52.0	52.0	44.0	40.0	34.0
8 ft 0 in. × 24 in. deep	77	60.0	60.0	52.0	48.0	38.0
8 ft 0 in. × 32 in. deep	99	76.0	76.0	76.0	70.0	62.0

APPENDIX 5A: Deck Assemblies (*Continued*)

Deck	Weight (psf)	Load-span limits[a] (ft) — Total superimposed load (psf) 25	50	75	100	150
Prestressed double tee, lightweight, no topping						
8 ft 0 in. × 12 in. deep	29	40.0	36.0	24.0	22.0	18.0
8 ft 0 in. × 14 in. deep	31	46.0	40.0	34.0	30.0	20.0
8 ft 0 in. × 16 in. deep	33	52.0	48.0	36.0	32.0	28.0
8 ft 0 in. × 18 in. deep	34	60.0	50.0	44.0	34.0	30.0
8 ft 0 in. × 20 in. deep	36	66.0	58.0	46.0	42.0	32.0
8 ft 0 in. × 24 in. deep	40	80.0	68.0	56.0	52.0	40.0
8 ft 0 in. × 32 in. deep	57	100.0	94.0	82.0	74.0	62.0
Prestressed double tee, lightweight, 2-in. topping						
8 ft 0 in. × 12 in. deep	54	36.0	34.0	24.0	22.0	14.0
8 ft 0 in. × 14 in. deep	56	40.0	36.0	34.0	30.0	22.0
8 ft 0 in. × 16 in. deep	58	46.0	40.0	36.0	32.0	28.0
8 ft 0 in. × 18 in. deep	59	50.0	46.0	42.0	34.0	30.0
8 ft 0 in. × 20 in. deep	61	56.0	50.0	46.0	42.0	32.0
8 ft 0 in. × 24 in. deep	65	62.0	60.0	54.0	50.0	38.0
8 ft 0 in. × 32 in. deep	82	82.0	82.0	80.0	72.0	60.0
Prestressed single tee, normal weight, no topping						
8 ft 0 in. × 36 in. deep	74	100.0	96.0	82.0	72.0	60.0
10 ft 0 in. × 48 in. deep	82	104.0	104.0	98.0	84.0	70.0
Prestressed single tee, lightweight, no topping						
8 ft 0 in. × 36 in. deep	57	110.0	100.0	84.0	72.0	0.0
10 ft 0 in. × 48 in. deep		120.0	114.0	100.0	84.0	0.0
Prestressed hollow core slab, 4 ft 0 in. wide × 8 in. deep, normal weight						
Untopped	56	—	—	33.0	30.0	26.0
2-in. topping	81	—	33.0	31.0	30.0	27.0
Prestressed, flat slab, 4 in. thick, normal weight						
Untopped	50	—	—	16.0	14.0	13.0
2-in. topping	75	—	—	17.0	16.0	15.0

APPENDIX 5B: Wall Assemblies

Description	Weight (psf)	Area (in.²/ft)	Section modulus (in.³) Unreinforced	Reinforced[a]	Insulation, R[b]
Single-wythe walls:					
4-in. hollow block	28.0	24.0	24.0	—	
4-in. brick	37.0	43.4	26.2	—	0.47

[a] For block: based on $d = t/2$, $f'_m = 1500$ psi, $A_{s(min)} = 0.33$ in.²/ft, $A_{s(max)} = 1.5$ in.²/ft. For brick: based on $d = t/2$, $f'_m = 1400$ psi, $A_{s(min)} = 0.011$ in.²/ft, $A_{s(max)} = 0.60$ in.²/ft.

[b] Values listed include basic materials and cavity air spaces only. Add 0.68 for inside air surface (still air) and 0.17 for outside air surface (15 mph wind, winter). Also, add resistance of attached materials, such as insulation, gypsum board, etc. Numbers in ( ) are values for cells of block filled with insulation.

APPENDIX 5B: Wall Assemblies (*Continued*)

Description	Weight (psf)	Area (in.^{2}/ft)	Section modulus (in.^{3}) Unreinforced	Section modulus (in.^{3}) Reinforced[a]	Insulation, R[b]
6-in. hollow block	43.0	40.5	50.0	7.8–25.6	
(lightweight)	31.0				1.53(3.72)
6-in. solid block	44.0	67.5	63.0	—	
8-in. hollow block	50.0	45.0	80.0	12.6–44.7	0.98(1.98)
(lightweight)	35.0				1.75(4.85)
8-in. solid block	58.0	91.5	116.0		
10-in. hollow block	47.0	61.0	125.0	18.1–68.0	
(lightweight)	36.0				1.97(5.92)
10-in. solid block	75.0	115.5	185.0		
12-in. hollow block	62.0	70.0	190.0	24.4–95.4	1.16(2.59)
(lightweight)	50.0				2.14(6.80)
12-in. solid block	85.0	139.5	272.0		
Composite walls:					
9-in. with 4-in. brick, both sides	90.0	108.0	128.0	10.8–50.4	
10-in. with 4-in. brick, both sides	100.0	120.0	200.0	13.3–62.2	1.19
12-in. with 4-in. brick, both sides	120.0	144.0	288.0	19.2–89.6	
14-in. with 4-in. brick, both sides	139.0	168.0	392.0	26.1–122.0	
Cavity walls:					
8-in. with 4-in. brick	78.0	87.1	52.4		1.01
10-in. with 4-in. brick, both sides	74.0	87.1	52.4		1.93
10-in. with 4-in. hollow block, both sides	56.0	48.0	48.0		
12-in. with 4-in. brick and 8-in. hollow block	93.0	88.4	106.2		1.42
12-in. with 4-in. brick and 6-in. hollow block	84.0	83.9	76.2		2.28

APPENDIX 5C: Hourly Fire-Resistive Ratings of Typical Construction Assemblies

		Assembly	Ratings and criteria[a,b]
Beams	5c1	Reinforced concrete, no protection, cast in place or precast	4 hr if minimum of $1\frac{1}{2}$-in. cover on reinforcing

[a] Based on Underwriters' Laboratories, Inc., tests.

[b] Grade A concrete is concrete in which coarse aggregate consists of blast-furnace slag, limestone, calcareous gravel, trap rock, burnt clay, or shale, cinders containing not more than 25% of combustible material and not more than 5% of volatile material, and other materials meeting the requirements of the Code and containing not more than 30% quartz, chert, flint, and similar materials.

Grade B concrete is concrete in which the coarse aggregate consists of granite, quartzite, siliceous gravel, sandstone, gneiss, cinders containing more than 25, but not more than 40% of combustible material and not more than 5% of volatile material, and other materials meeting the requirements of the Code and containing more than 30% quartz, chert, flint, and similar materials.

[c] Average thickness of the solid material in the wall.

APPENDIX 5C: Hourly Fire-Resistive Ratings of Typical Construction Assemblies (*Continued*)

	Assembly	*Ratings and criteria*[a,b]
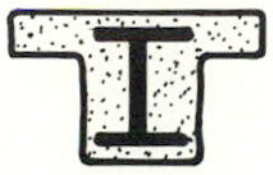 5c2	Steel section, poured concrete cover, stone concrete	1 hr for $1\frac{1}{2}$-in. cover and Grade B concrete to 4 hr for 2-in. cover and Grade A concrete
 5c3	Steel section, lath and plaster cover	2 hr for $\frac{7}{8}$-in. vermiculite-gypsum plaster to 4 hr for 2-in. vermiculite or perlite acoustic plaster
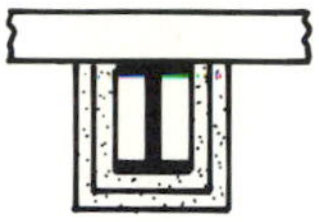 5c4	Steel section, gypsum board cover	2 or 3 hr
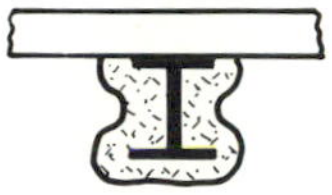 5c5	Steel section, sprayed fiber protection	2 hr for $1\frac{1}{8}$-in. sprayed mineral fibers to 4 hr for $1\frac{7}{8}$-in. sprayed mineral fibers
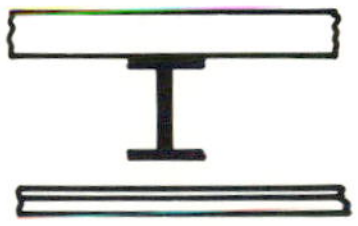 5c6	Steel section, enclosed between floor and ceiling	1 hr for $\frac{5}{8}$-in. gypsum wallboard or rated acoustical tile ceiling to 4 hr for $\frac{3}{4}$-in. perlite-gypsum; special requirements for minimum space between ceiling and beam and for ceiling penetrations
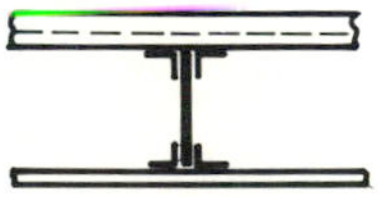 5c7	Bar joist, floor–ceiling assembly	1 hr for 2-in. concrete cover, No. 3 joist chord and $\frac{1}{2}$- to $\frac{3}{4}$-in. acoustical ceiling tile; to 4 hr for $2\frac{1}{2}$-in. concrete cover, No. 5 joist chord, and $\frac{3}{4}$-in. plaster ceiling on metal lath; special requirements for ceiling penetrations
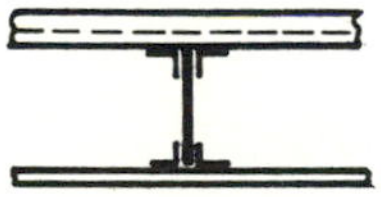 5c7	Bar joist, roof–ceiling assembly	1 hr for steel deck, mineral board insulation, and $\frac{5}{8}$-in. rated ceiling to 3 hr for perlite concrete on steel centering and $\frac{7}{8}$-in. lath and plaster ceiling
Columns		
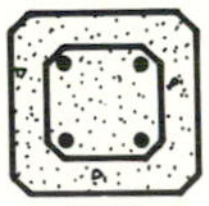 5c8	Reinforced concrete, tied, spiral, or composite	3 to 4 hr depending on type of aggregate, size of column, cover on reinforcing
 5c9	Steel section, poured concrete cover	1 hr for $1\frac{1}{2}$-in. cover of Class B concrete to 4 hr for 2-in. cover of Class A concrete

APPENDIX 5C: Hourly Fire-Resistive Ratings of Typical Construction Assemblies (*Continued*)

	Assembly	*Ratings and criteria*[a,b]
 5c10	Steel section, lath and plaster cover	1 hr for $\frac{1}{2}$-in. sand-gypsum plaster on $\frac{3}{8}$-in. gypsum lath to 4 hr for $1\frac{3}{4}$-in. vermiculite plaster on metal lath
 5c11	Steel section, gypsum wallboard cover	2 hr for 2 layers of $\frac{5}{8}$-in. wallboard to 4 hr for 4 layers
 5c12	Steel section, sprayed protection	$1\frac{1}{2}$ hr for sprayed cementitious fibers to 4 hr for cementitious or mineral fibers
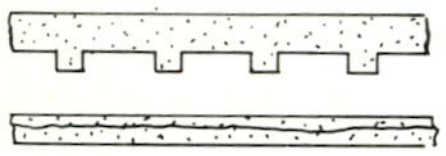 5c13	Steel deck, concrete fill, hung ceiling	$1\frac{1}{2}$ hr for acoustical tile ceiling to 4 hr for plaster ceiling
Floor and Roof Decks		
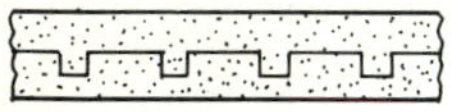 5c14	Steel deck, concrete fill, sprayed fiber on underside of deck	$\frac{3}{4}$ hr for mineral fibers to 4 hr for either mineral fibers or plaster
 5c15	Concrete slab on steel deck, composite action, negative reinforcing in concrete, no protection on underside	$\frac{3}{4}$ to 3 hr depending on thickness of slabs, type of aggregate, etc.
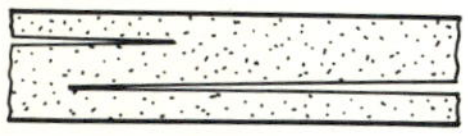 5c16	Reinforced-concrete slab, no protection	1 hr for 3-in.-thick slab to 4 hr for $6\frac{1}{2}$-in. slab
Walls		
	Concrete masonry wall	1 hr for 3.0 in. of equivalent thickness[c] to 4 hr for 6.7 in. of equivalent thickness

REFERENCES

1. ACI Committee 318, *Building Code Requirements for Reinforced Concrete,* ACI 318–77. Detroit, Mich.: American Concrete Institute, 1977.
2. ACI Committee 442, *Response of Multi-story Concrete Structures to Lateral Forces,* SP-36. Detroit, Mich.: American Concrete Institute, 1973.
3. ACI Committee 531, *Building Code Requirements for Concrete Masonry Structures,* ACI 531–79. Detroit, Mich.: American Concrete Institute, 1979.
4. Allison, Horatio, *Practical Steel Design for Buildings.* New York: American Institute of Steel Construction, 1976.
5. Beaufait, Fred W., *Tall Buildings: Planning, Design and Construction.* Nashville, Tenn.: Civil Engineering Program, Vanderbilt University, 1974.
6. Becker, Roy, *Practical Steel Design for Buildings—Seismic Design.* New York: American Institute of Steel Construction, 1976.
7. Beedle, Lynn S., et al., *Structural Steel Design.* New York: The Ronald Press Company, 1964.
8. Borrego, John, *Space Grid Structures.* Cambridge, Mass.: The MIT Press, 1968.
9. Cheong-Siat-Moy, Francis, "Multistory Frame Design Using Story Stiffness Concept," *Journal of the Structural Division, ASCE,* Vol. 102, No. ST-6 (June 1976).
10. Davies, R. M., ed., *Space Structures.* New York: John Wiley & Sons, Inc., 1967.
11. Ferguson, Phil M., *Reinforced Concrete Fundamentals,* 3rd ed. New York: John Wiley & Sons, Inc., 1973.
12. Fintel, Mark, ed., *Handbook of Concrete Engineering.* New York: Van Nostrand Reinhold Company, 1974.
13. Gaylord, Edmund H., Jr., and Charles M. Gaylord, *Structural Engineering Handbook.* New York: McGraw-Hill Book Company, 1968.
14. Kirby, P. A., and D. A. Nethereot, *Design for Structural Stability.* New York: John Wiley & Sons, Inc., 1979.
15. Lothers, John E., *Advanced Design in Structural Steel.* Englewood Cliffs, N.J.: Prentice-Hall, Inc., 1960.
16. Luttrell, Larry D., *Steel Deck Institute Diaphragm Design Manual.* St. Louis, Ill.: Steel Deck Institute, 1981.
17. Mainstone, Roland J., *Developments in Structural Form.* Cambridge, Mass.: The MIT Press, 1975.
18. McCormac, Jack C., *Structural Steel Design.* Scranton, Pa.: International Textbook Co., 1965.
19. Merritt, Frederick S., *Structural Steel Designers Handbook.* New York: McGraw-Hill Book Company, 1972.
20. *Recommended Practice for Engineered Brick Masonry.* McLean, Va.: Brick Institute of America, 1969.
21. Roark, Raymond J., *Formulas for Stress and Strain.* New York: McGraw-Hill Book Company, 1954.
22. Salvadori, Mario, and Matthys Levy, *Structural Design in Architecture.* Englewood Cliffs, N.J.: Prentice-Hall, Inc., 1967.
23. Schueller, Wolfgang, *Highrise Building Structures.* New York: John Wiley & Sons, Inc., 1977.
24. *Seismic Design for Buildings,* TM 5–809–10, NAVFAC P-355, AFM 88–3, Chap. 13. Washington, D.C.: Departments of the Army, Navy, and Air Force, 1973.
25. Wang, Chu-ki, and Charles G. Salmon, *Reinforced Concrete Design,* 3rd ed. New York: Harper & Row, Publishers, 1979.
26. Wiesinger, Frederick P., "Design of Flat Plates with Irregular Column Layout," *Journal of the American Concrete Institute,* February 1973.
27. Winter, George, and Arthur H. Nilson, *Design of Concrete Structures.* New York: McGraw-Hill Book Company, 1972.

6

SYSTEMS

6–1 INTRODUCTION

At this point in our development of the structural system design model, we are ready to complete the study of assembly concepts, as final preparation for creating our own systems in Chapter 7. The third and final level of structural assembly is the *system* (Fig. 6–1).

A structural *system* is a three-dimensional structural form that completely fulfills all project requirements. As an organized assembly of elements and subsystems, systems contain subsystem components that provide each of the five functions (required for strength and stability) that we discussed in Chapter 5. The final assembly of several subsystems into a system configuration is the basis of structural design.

Design at the system level involves a greatly expanded range of choices, both within a particular system configuration (intra-) and between alternative systems (inter-). Our emphasis in this chapter is on the design choices *within* a given system classification, thus allowing us to study individual system types and the behavior of each.

This chapter addresses several classifications of structural systems. Within each classification there are

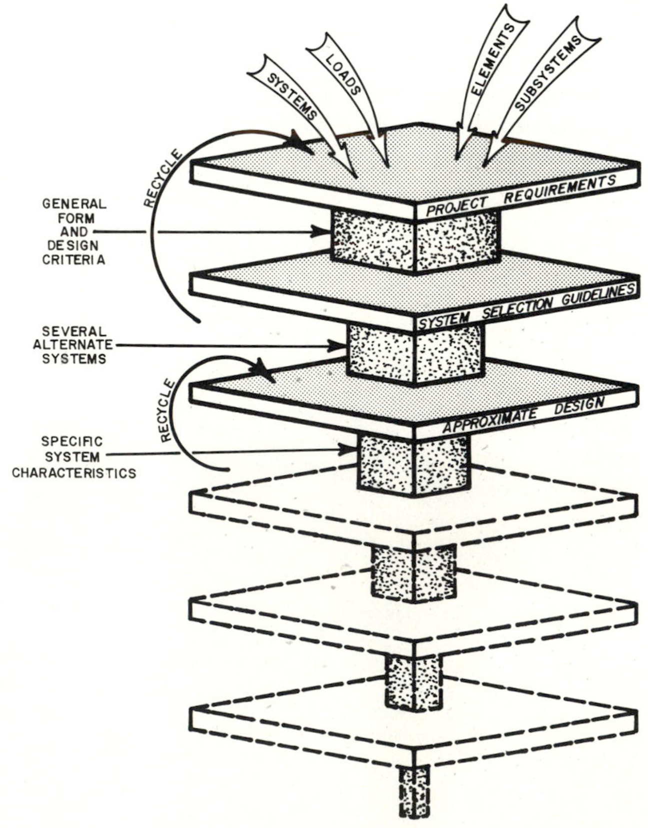

Figure 6–1. System Design Model

an infinite number of possible combinations of subsystems and elements. However, each major classification is assembled based on certain principles that are common to all the possible variations. Our objective is to develop an understanding of these principles and to further expand our knowledge of system characteristics and conceptual design skills. The approach is basically the same: discussion of background, behavior, and development of approximate analysis and design skills. Before beginning our study of system types, a brief description of the design approach is appropriate.

System Design (Intra-)

Design within a system classification involves the following choices:

- Geometrical proportions
- Subsystem choices
- Material selection
- Element proportions

The design approach is an *optimization* process, focusing on achieving the optimum *total* system as opposed to the optimization of any individual subsystem or element. We are thus concerned with cost and function "trade-offs" between various subsystems. Chapter 7 discusses the concept of trade-offs between alternative structural systems, as well as trade-offs with other project systems (e.g., electrical, mechanical, architectural, etc.).

6–2 BRACED FRAMES

Braced frames are structural systems that contain either a rigid frame or post-and-beam frame for vertical support and a bracing assembly or shear wall for lateral support—the presence of the lateral bracing gives the system its name (Fig. 6–2), and any number of different horizontal distribution and lateral distribution subsystems may be used.

The simplicity of the braced frame makes it a very popular system. With lateral stability assured, design focuses only on gravity load design. The braced frame is also generally the most efficient way to carry lateral loads, with distribution of lateral loads to the vertical bracing normally accomplished via diaphragm action of the horizontal distribution subsystem (or by another bracing assembly in the horizontal plane).

Figure 6–2. Braced Frame System

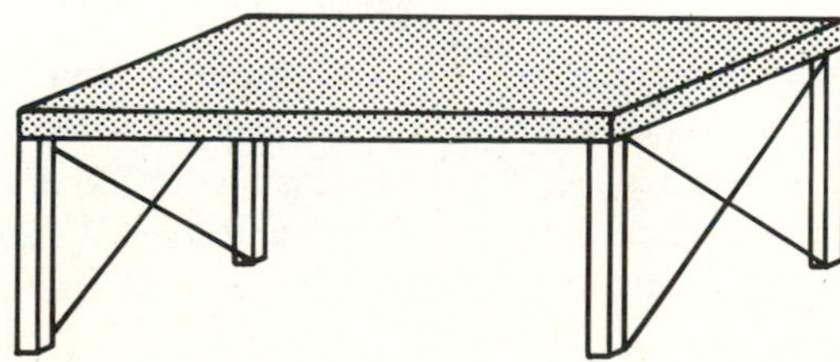

History and Development

The development of braced frame systems for buildings started around 1910–1920. As bearing wall construction gave way to frame construction and as steel became a readily available structural material, various forms of bracing were added to the construction process. The substitution of bracing for the in-filled masonry wall was a natural evolution.

The braced frame became very popular for manufacturing and warehousing facilities. Its use in multistory buildings depended on whether bracing could be effectively worked into the building plan. It certainly became (and still is) the most popular system with structural engineers. Today, the braced frame is one of the major forms of structural assembly and continues to play an important role in all types of structures.

Types and Uses

The major types of braced frames are shown in Fig. 6–3. These include the following:

- Post-and-beam frame with bracing
- Rigid frame with bracing

Figure 6–3. Types of Braced Frame Systems

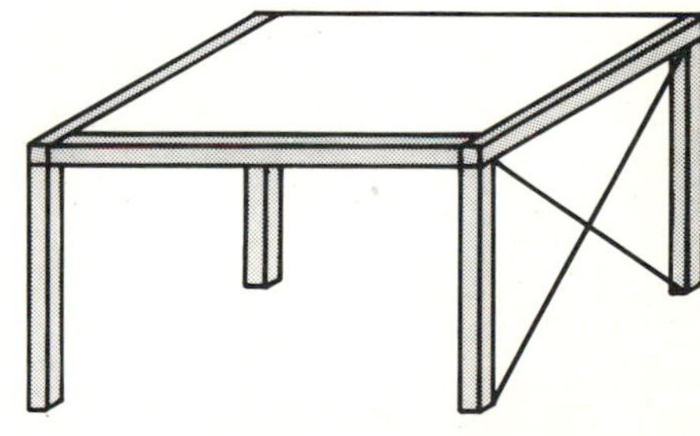

(a) Post and beam with bracing

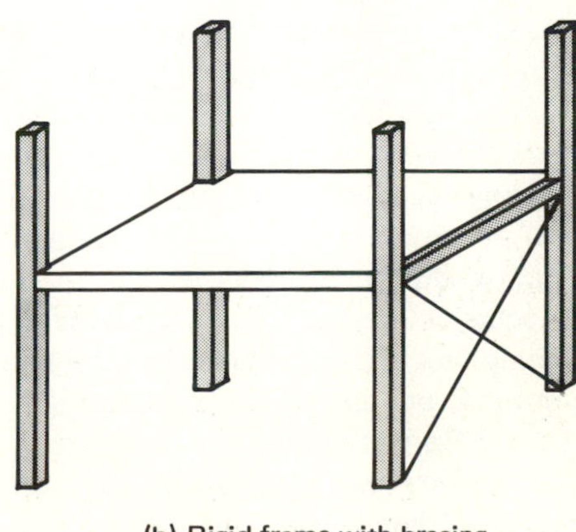

(b) Rigid frame with bracing

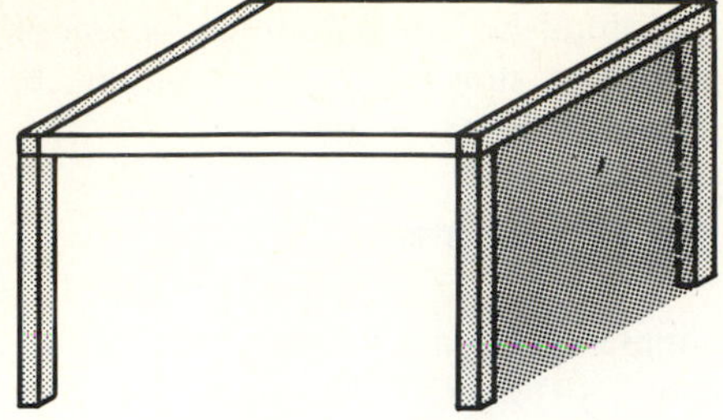

(c) Post and beam with shear wall

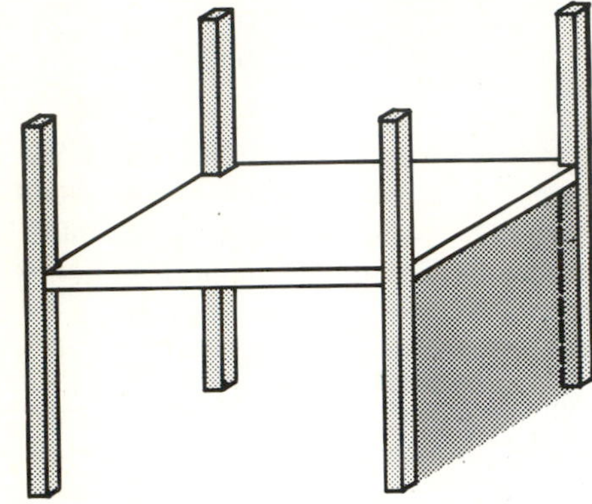

(d) Rigid frame with shear wall

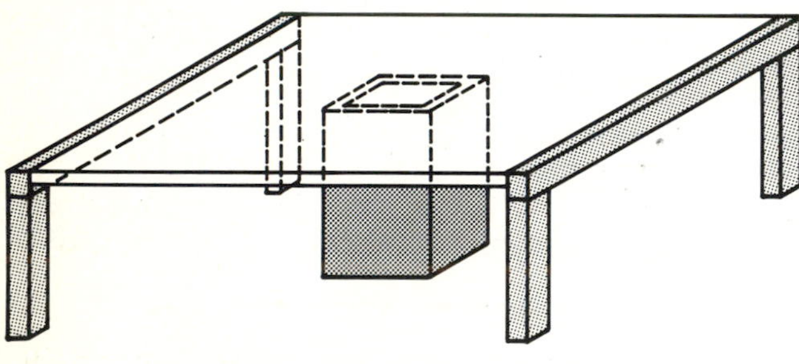

(e) Core shear wall

Figure 6–3. *(Continued)*

- Post-and-beam frame with shear wall
- Rigid frame with shear wall
- Slab subsystem with shear wall
- Core bracing with any of the above
- Exterior wall tubular bracing with any of the above

Various forms of core bracing assemblies and shear walls are common for high-rise office and residential buildings. Core walls often double as shaft walls for elevators and stairs. Exterior wall tubular bracing is used for the taller buildings as an effective way to utilize the facade for structural purposes. (Both core systems and tubular systems are covered in more detail later in this chapter.)

Single bents or tiers of bracing assemblies or shear walls are common for industrial buildings and many types of general-purpose buildings. The post-and-beam frame is the most common vertical support subsystem used for these types of structures. Rigid frames are often used for office buildings, hotels, and similar structures.

Behavior

Gravity Loads. For braced frames in which simple hinged or bearing connections are used, the horizontal distribution subsystem and the vertical support subsystem can simply be uncoupled and only shear forces or bearing forces remain at the intersection boundaries [Fig. 6–4(a)]. The load path is easily visualized as shown. Eccentricity of the boundary forces may induce bending in the vertical support subsystem.

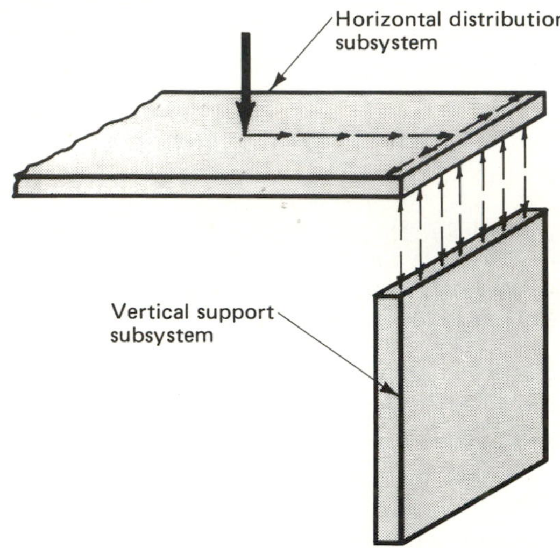

(a) Load path for gravity loads

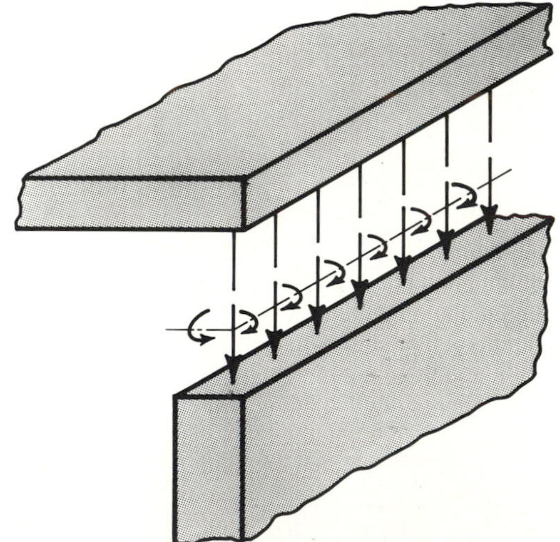

(b) Moment forces at boundary

Figure 6–4. Braced Frame Gravity Forces

Rigidly connected braced frames have both moments and torsional forces at the boundaries in addition to vertical shears and bearing forces. Pattern live loads can thus induce uniaxial and biaxial bending in column elements and torsion in beams and slabs [Fig. 6–4(b)].

Lateral Loads. Lateral loads travel from exterior wall surfaces to floor or roof levels, where they are distributed laterally to the vertical support subsystem [Fig. 6–5(a)]. The uncoupling of the exterior wall yields the model shown in Fig. 6–5(b), with the spring support representing the combined flexibility of the diaphragm and the lateral support subsystem. The uncoupling of the floor and shear wall leaves boundary forces which include shear springs to represent the relative stiffness of the lateral support subsystem [Fig. 6–5(c)].

Figure 6–5. Braced Frame Lateral Forces

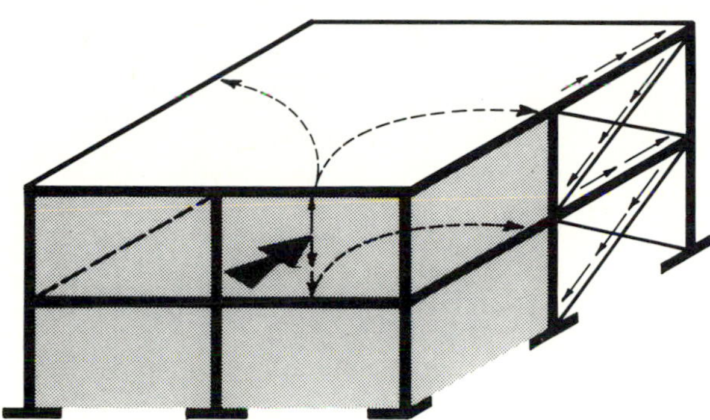

(a) Load path for lateral loads

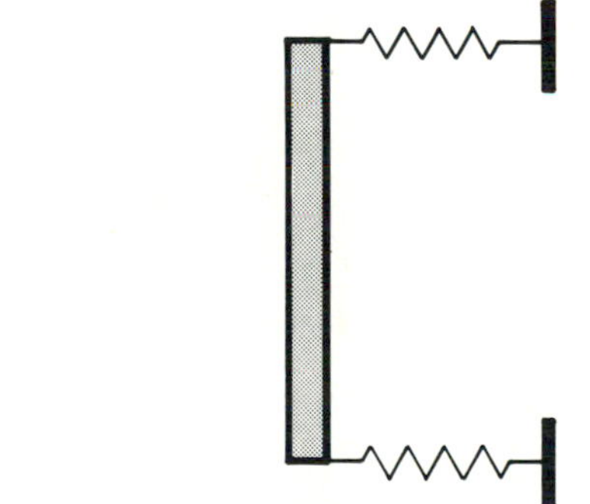

(b) Model for exterior wall

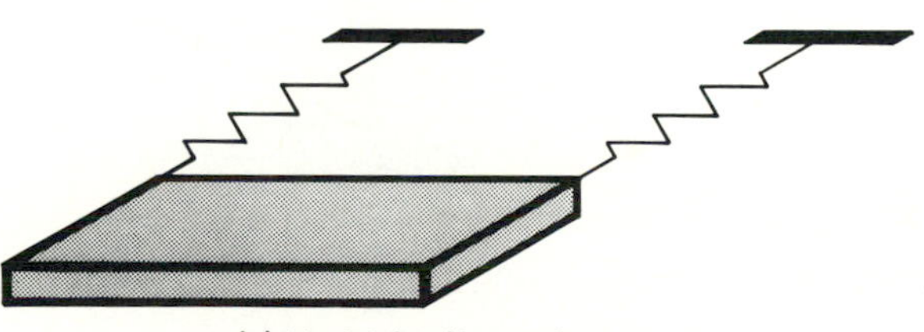

(c) Model for floor subsystem

Approximate Analysis

Braced frames can be analyzed by uncoupling the subsystems and treating each as an independent unit, with appropriate boundary forces acting. A common procedure is to use a "typical bay" for approximate analysis and extrapolate the results to the entire system. Example Problem 6–1 illustrates this process.

Approximate Design

Design of braced frame systems includes the following choices:

- Material selection
- Bay dimensions
- Story height
- Number and location of lateral support assemblies
- Type of horizontal distribution
  - Floor and roof assembly
  - Slab
  - Grid
  - Space frame
- Type of vertical support
  - Post and beam
  - Rigid frame
  - Columns
  - Arch
- Type of lateral distribution
  - Diaphragm
  - Bracing
- Type of lateral support
  - Bracing
  - Shear wall
  - Core
  - Tube
- Type of foundations
  - Spread footings
  - Deep foundations
- Element proportions

Each subsystem, in turn, has another sequence of choices, as discussed in Chapter 5. The following example problem illustrates the preliminary design process.

EXAMPLE PROBLEM 6–1: Preliminary Design of Braced Frame System

Select a preliminary structural system (braced frame) for the warehouse of Example Problem 5–2. The warehouse is 120 ft × 240 ft (36.6 m × 73.2 m) with a clear height of 20 ft (6.1 m). Wind load is 25 psf (1197 N/m²) and live and dead loads are each 20 psf (957 N/m²). Allowable soil bearing pressure is 3000 psf (143,640 N/m²).

Solution

1. *Material selection.* Structural steel is the most efficient flexural material and will generally be the most competitive for this type of structure unless fire protection or availability problems must be considered. Market forces, however, may make other materials equally competitive. Try both structural steel and prestressed concrete systems.
2. *Bay dimensions.* Column spacing of 20 ft (6.1 m) has already been demonstrated by Example Problem 5–2

to be the most economical. Width between bents could be 20, 30, 40, or 60 ft (6.1m, 9.1, 12.2, or 18.3m). Compare 20 ft and 30 ft for steel and 40 ft and 60 ft for prestressed concrete. [*Note:* Minimum span for 12-in. (304.8-mm)-deep double tees is 40 ft (12.2) m), from Appendix 5A.]

3. *Story height.* Given as 20 ft (6.1 m) clear.
4. *Subsystem definition.* Based on Appendix 6A and experience with building types in the geographical area, consider the following schemes:

   - *Scheme 1:* Metal roof deck; bar joists at 5 ft (1.5 m) on center, spanning 20 ft (6.1 m); post-and-beam vertical support with double cantilever continuous beam; columns spaced at 20 ft (6.1 m).
   - *Scheme 2:* Metal roof deck, bar joists at 5 ft (1.5 m) on center spanning 30 ft (9.1 m); post-and-beam vertical support with double cantilever continuous beam; columns spaced at 20 ft (6.1 m).
   - *Scheme 3:* 8 ft (2.4 m) wide by 12 in. (304.8 mm) deep prestressed double-tee roof members spanning 40 ft (12.2 m); post-and-beam vertical support with simple span girder; columns spaced at 20 ft (6.1 m).
   - *Scheme 4:* 8 ft (2.4 m) wide by 20 in. (508 mm) deep prestressed double-tee roof members spanning 60 ft (18.3 m); post-and-beam vertical support with simple span girder; columns spaced at 20 ft (6.1 m).

Lateral support subsystem for all schemes will be bracing assemblies to be sized and located consistent with diaphragm capacity.

5. *Preliminary design of Scheme 1:*
   a. Roof deck: Assume that a 22-gage metal deck (30/4 pattern) with one intermediate side lap connection will meet diaphragm requirements (Example Problem 5–4).
   b. Bar joists:

$$w = 5 \text{ ft (40 psf)} = 200 \text{ lb/ft (6576 N/m)}$$

$$M = \frac{wl^2}{8} = \frac{0.2(20)^2}{8} = 10 \text{ ft-kips}$$

$$= 120 \text{ in.-kips (13,560 Nm)}$$

By Appendix 4F, use 12 H3 at 5.2 lb/ft.

$$5.2 \text{ lb/ft} \div 5 \text{ ft} = 1.04 \text{ psf } (49.8 \text{ N/m}^2)$$

   c. Girder:

$$w = 20 \text{ ft (40 psf)} = 800 \text{ lb/ft (11,674 N/m)}$$

$$M = \frac{wl^2}{10} = \frac{0.8(20)^2}{10} = 32 \text{ ft-kips (43,392 Nm)}$$

$$S_{\text{req'd}} = 32 \times 12/20 \text{ ksi} = 19.2 \text{ in.}^3 \ (314{,}688 \text{ mm}^3)$$

By Appendix 4B, use W12 × 19.

$$19 \text{ lb/ft} \div 20 \text{ ft} = 0.95 \text{ psf } (45.5 \text{ N/m}^2)$$

   d. Columns:
      Axial load:

$$P = 40 \text{ psf}(20 \text{ ft})(20 \text{ ft})$$
$$= 16.0 \text{ kips (71,168 N)}$$

$$kl = 21 \text{ ft (6.4 m)}$$

$$\text{Min. stress at } 57\sqrt{P} = 3.72 \text{ ksi (25.6 MPa)}$$

$$57\sqrt{P} = 57\sqrt{16}$$
$$= 19 \text{ ft; assume that } F_a = 3.2 \text{ ksi}$$

$$A_c = \frac{16.0 \text{ kips}}{3.2 \text{ ksi}}$$
$$= 5.0 \text{ in.}^2 \ (3225 \text{ mm}^2)$$

Use W8 × 17.

$$\frac{17 \text{ lb/ft}(24 \text{ ft})}{20 \text{ ft}(20 \text{ ft})} = 1.02 \text{ psf } (48.8 \text{ N/m}^2)$$

   e. Footings:

$$\text{Column load} = 16 \text{ kips (71,168 N)}$$

$$A_f = \frac{16.0}{3.0 \text{ ksf}} = 5.33 \text{ ft}^2 \ (0.49 \text{ m}^2)$$

Use 2 ft 6 in. × 2 ft 6 in. (0.76 m × 0.76 m).

$$q_{\text{net}} = \frac{P_u}{A_f} = \frac{1.4(8.0) + 1.7(8.0)}{6.25}$$
$$= 3.97 \text{ ksf } (190{,}083 \text{ N/m}^2)$$

$$\frac{\sqrt{f_c'}}{q_{\text{net}}} = \frac{\sqrt{3000}}{3970} = 0.014$$

$$\frac{A_f}{A_c} = \frac{6.25}{1.0} = 6.25, \ r = 12 \text{ in.}, \ a = 9 \text{ in.}$$

$$\frac{d}{r} = 0.4\text{: req'd } d = 0.4(12 \text{ in.}) = 4.8 \text{ in.}$$

$$\frac{a}{d} = 4.4\text{: req'd } d = \frac{4.8 \text{ in.}}{4.42} = 1.08 \text{ in.}$$

Use d = 5 in. (127 mm), t = 5 + 3 = 8 in. (203 mm).

$$\frac{(2.5 \text{ ft})(2.5 \text{ ft})(8 \text{ in.}/12)}{27(400 \text{ sq ft})} = \begin{matrix}0.00039 \text{ cubic yard/sq ft} \\ (0.0032 \text{ m}^3/\text{m}^2)\end{matrix}$$

6. *Preliminary design of Scheme 2:*
   a. Roof deck: Same as Scheme 1.
   b. Bar joists:

$$\text{Uniform load} = 200 \text{ plf (6,576 N/m)}$$

Span = 30 ft (9.1 m)

$$M = \frac{0.20(30)^2}{8} = 22.5 \text{ ft-kips}$$

$$= 270 \text{ in.-kips } (30{,}510 \text{ Nm})$$

By Appendix 4F, use 16 H5 at 7.8 plf.

7.8 lb/ft ÷ 5 ft = 1.56 psf (74.7 N/m²)

c. Girder: By Example 5–2, use W14 × 12.

22 lb/ft ÷ 30 ft = 0.73 psf (34.9 N/m²)

d. Columns: By Example 5–2, use W8 × 20.

$$\frac{20 \text{ lb/ft } (24 \text{ ft height})}{(20 \text{ ft})(30 \text{ ft})} = 0.80 \text{ psf } (38.3 \text{ N/m}^2)$$

e. Footings: By Example 5–2, use 3 ft 0 in. × 3 ft 0 in. × 8 in. (0.9 m × 0.9 m × 203 mm).

$$\frac{(3 \text{ ft})(3 \text{ ft})(8 \text{ in.}/12)}{27(600 \text{ sq ft})} = 0.00055 \text{ cy/sq ft } (0.0045 \text{ m}^3/\text{m}^2)$$

7. *Preliminary design of Scheme 3:*
   a. Deck assembly: 8 ft wide × 12 in. deep (2.4 m × 304 mm) double tee.

Dead load = 37 psf

Total DL = 37 + 8 = 45 psf (2154 N/m²)

b. Girder: Uniform superimposed load = (45 psf + 20 psf) 40 ft = 2.6 kips/ft.

Determine moments: Assume girder weight = 350 plf.

$$M_o = \frac{wl^2}{8} = \frac{0.35(20)^2}{8}$$

$$= 17.5 \text{ ft-kips } (23{,}730 \text{ Nm})$$

$$M_d = \frac{wl^2}{8} = \frac{40(0.45)(20)^2}{8}$$

$$= 90 \text{ ft-kips } (122{,}040 \text{ Nm})$$

$$M_l = \frac{wl^2}{8} = \frac{40(0.20)(20)^2}{8}$$

$$= 40 \text{ ft-kips } (54{,}240 \text{ Nm})$$

Assume that f'_c = 5000 psi (34.5 MPa), f'_{ci} = 3500 psi (24.1 MPa). By Table 4–6:

$$F_T = 2392 \text{ psi}, \qquad F_B = 2104 \text{ psi}$$

$$S_T = \frac{0.20M_u + M_d + M_l}{F_T}$$

$$= \frac{(0.20 \times 17.5 + 90 + 40)12}{2.39}$$

$$= 670 \text{ in.}^3 \ (10{,}988{,}000 \text{ mm}^3)$$

$$S_B = \frac{0.20M_u + M_d + M_l}{F_B}$$

$$= \frac{(0.20 \times 17.5 + 90 + 40)12}{2.1}$$

$$= 763 \text{ in.}^3 \ (12{,}513{,}200 \text{ mm}^3)$$

Select member from Appendix 4C. Use 24 in. wide × 20 in. deep inverted tee beam.

c. Column:

Girder weight = 0.35(20 ft) = 7.0 kips

Roof DL = 0.045 psf(40)(20) = 36.0 kips

$$P_u = 1.4(7.0 + 36.0) + 1.7(0.02)(40)(20)$$

$$= 87.4 \text{ kips } (388{,}755 \text{ N})$$

Assume 3000 psi (20.7 MPa).

Max. allowable stress = 2.77 ksi (19.1 MPa) at $6.13\sqrt{P}$

$6.13\sqrt{P} = 6.13\sqrt{87.4} = 57.3$ in. (4.78 ft)

Min. allowable stress = 1.31 ksi (9.0 MPa) at $18.3\sqrt{P}$

$18.3\sqrt{P} = 18.3\sqrt{87.4} = 171$ in. (14.3 ft)

Since kl = 21 ft, adopt minimum column size to achieve kl/r = 70.

$$h = \frac{kl}{21} = \frac{21 \times 12}{21} = 12 \text{ in.}$$

Use 12 in. × 12 in. column (305 mm × 305 mm) (4% reinforcement).

d. Footing:

Column load = 59 kips

$$A_f = \frac{59}{3.0} = 19.6 \text{ ft}^2$$

Use 4 ft 6 in. × 4 ft 6 in. (1.4 mm × 1.4 mm).

$$q_{\text{net}} = \frac{P_u}{A_f} = \frac{87.4}{20.25} = 4.32 \text{ ksf}$$

$$\frac{\sqrt{f'_c}}{q_{\text{net}}} = \frac{\sqrt{3000}}{4320} = 0.0127$$

$$\frac{A_f}{A_c} = \frac{20.25}{1.0} = 20.25, \ r = 12 \text{ in.}, \ a = 33 \text{ in.}$$

Table 6–1 Example Problem 6–1

Alternative	*Deck (\$/sf)*	*Joists (\$/sf)*	*Girder (\$/sf)*	*Columns (\$/sf)*	*Footing (\$/sf)*	*Total (\$/sf)*
Scheme 1	0.79	0.42	0.48	0.52	0.06	2.27
Scheme 2	0.79	0.62	0.37	0.41	0.08	2.27
Scheme 3	2.06		1.25	0.04	0.11	3.46
Scheme 4	3.50		0.83	0.03	0.15	4.51

$$\frac{d}{r} = 0.6\text{: req'd } d = 0.6(12 \text{ in.}) = 7.2 \text{ in.}$$

$$\frac{a}{d} = 3.93\text{: req'd } d = \frac{33 \text{ in.}}{3.93} = 8.4 \text{ in.}$$

Use d = 9 in., t = 9 + 3 = 12 in. (305 mm)

$$\frac{(4.5)(4.5)(12/12)}{27(40)(20)} = 0.00094 \text{ cy/sq ft}$$

8. *Preliminary design of Scheme 4:*
   a. Deck assembly: 8 ft wide × 20 in. (2.4 m × 508 mm) deep double tee.

   $$\text{DL} = 47 \text{ psf}$$

   $$\text{Total DL} = 47 + 8 = 55 \text{ psf } (2633 \text{ N/m}^2)$$

   b. Girder:

   Uniform superimposed load
   $$= (55 \text{ psf} + 20 \text{ psf})(60 \text{ ft}) = 4.5 \text{ kips/ft}$$

   Using procedure similar to step 7, use 24 in. wide × 24 in. deep inverted tee beam.

   c. Column:

   $$\text{Girder weight} = 0.45(20 \text{ ft}) = 9.0 \text{ kips}$$

   $$\text{Roof DL} = 55 \text{ psf}(60)(20) = 66 \text{ kips}$$

   $$P_u = 1.4(9 + 66) + 1.7(0.02)(20)(60) = 145.8 \text{ kips}$$

   Using procedure similar to step 7, use 12 in. × 12 in. column (4% reinforcement).

   d. Footing:

   $$\text{Column load} = 99 \text{ kips}$$

   $$A_f = \frac{99}{3.0} = 33 \text{ ft}^2$$

   Use 5 ft 9 in. × 5 ft 9 in. Using procedure similar to step 7:

   $$t = 14 \text{ in. } (356 \text{ mm})$$

   $$\frac{5.75(5.75)(14.0/12)}{27(60)(20)} = 0.0011 \text{ cy/sq ft}$$

9. *Preliminary design of lateral support subsystem:*
   Total wind shear at end wall:

   $$V = 275 \text{ lb/ft}(120 \text{ ft}) = 33.0 \text{ kips } (146{,}784 \text{ N})$$

   For single bay of bracing:

   $$\text{Tension tie force} \simeq 33\sqrt{2} = 46.6 \text{ kips}$$

   $$\text{Column load} = \frac{33 \text{ kips } (20 \text{ ft})}{20 \text{ ft}} = 33.0 \text{ kips } (146{,}784 \text{ N})$$

   If exterior columns are made the same size as interior columns, their capacity is:

   $$P = 16 \text{ kips} \times 1.33 = 21.3 \text{ kips}$$

   $$\text{Gravity load} = -8.0 \text{ kips}$$

   13.3 kips available for wind (59,158 N)

   For two bays of bracing, column wind load will equal 33/2 = 16.5 kips. Probably two bays of bracing is reasonable.

10. *Tabulation and selection.* Table 6–1 provides a summary of the unit costs for each alternate, expressed in terms of dollars per building square foot. Appendix 7A cost data was used for this example. Schemes 1 and 2 have the lowest unit cost.

6–3 UNBRACED FRAMES

Unbraced frames are systems that rely on the bending stiffness of column and beam and slab elements to resist lateral loads (Fig. 6–6). The primary lateral support system is the rigid frame. The rigid frame also normally provides the vertical support, although a combination with post-and-beam frames is possible. Horizontal distribution can be provided by any type of floor or roof assembly or slab system, and lateral distribution is accomplished by either bracing or diaphragm action. Single-story un-

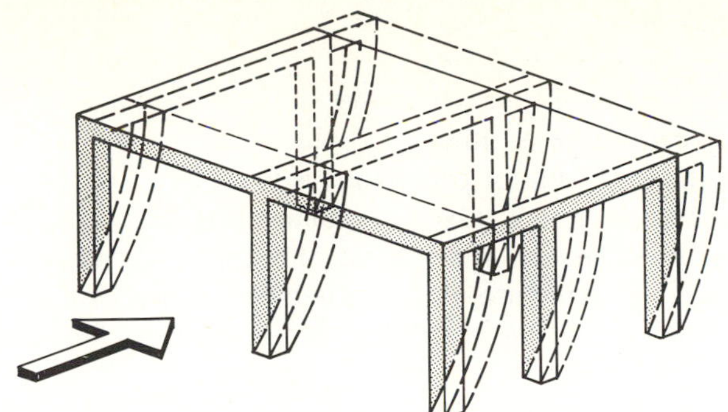

Figure 6–6. Unbraced Frame System

braced frames are called portal frames, and this system is extensively used in the prefabricated building industry. Many types of unbraced frames are typically braced in their weak direction and are, thus, actually a combination of the braced and unbraced frame assembly.

The unbraced frame is indeterminate and requires preliminary assumptions of member sizes before analysis can be performed. Response to lateral forces is a major design concern. An entire area of analysis and design—involving the fields of structural dynamics, wind and earthquake engineering, plastic design, beam-column behavior, and behavior of connections—is devoted to behavior of unbraced frame systems.

History and Development

Unbraced frame construction probably began with reinforced concrete systems since it was very natural (due to the inherent joint fixity at the monolithic connections) to allow the frame action to carry lateral loads. These concrete systems were also normally very stiff, and there was initially little problem with carrying lateral loads for moderate height buildings. Unbraced frame systems for steel buildings were slower to develop, as lack of understanding of connections and indeterminate frame behavior complicated the analysis. The development of welding techniques and high-strength bolted connections greatly enhanced the use of unbraced frames in steel.

The difficulty in working bracing systems into building plans and the popularity of this new framing concept promoted the use of the unbraced framed system for many high-rise buildings during the 1940–1950 period. Close examination of the cost of these taller building frames shows that (in many cases) an unreasonable premium was paid to carry lateral loads. This fact has lead many engineers to reexamine the unbraced frame concept and to limit its use to buildings 25 stories or less. In its place has emerged the tube system and other types of braced frame systems for taller buildings.

Types and Uses

The major types of unbraced frame systems are shown in Fig. 6–7. These include the following:

- Single-story portal frames
- Knee-braced frames
- Rigid frame with floor and roof assemblies
- Rigid frame (strong direction) and post and beam (weak direction)
- Concrete slab systems
- Fixed-base cantilever systems
- Rigid frame/shear wall combination systems
- Framed tubes incorporating aspects of both the shear wall (braced frame) and the rigid frame (unbraced frame)

Unbraced frames are primarily used where lateral loads can be carried by the frame without undue premium cost. They are also very valuable systems for providing sufficient energy absorption to withstand severe earthquake forces. They may be the only choice when braced system concepts cannot be worked into the overall building plan.

Figure 6–7. Types of Unbraced Frame Systems

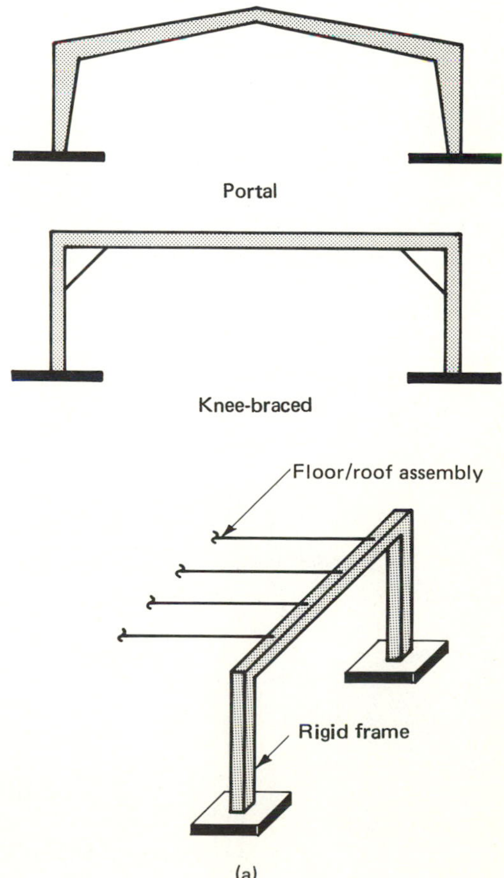

Figure 6–7. *(Continued)*

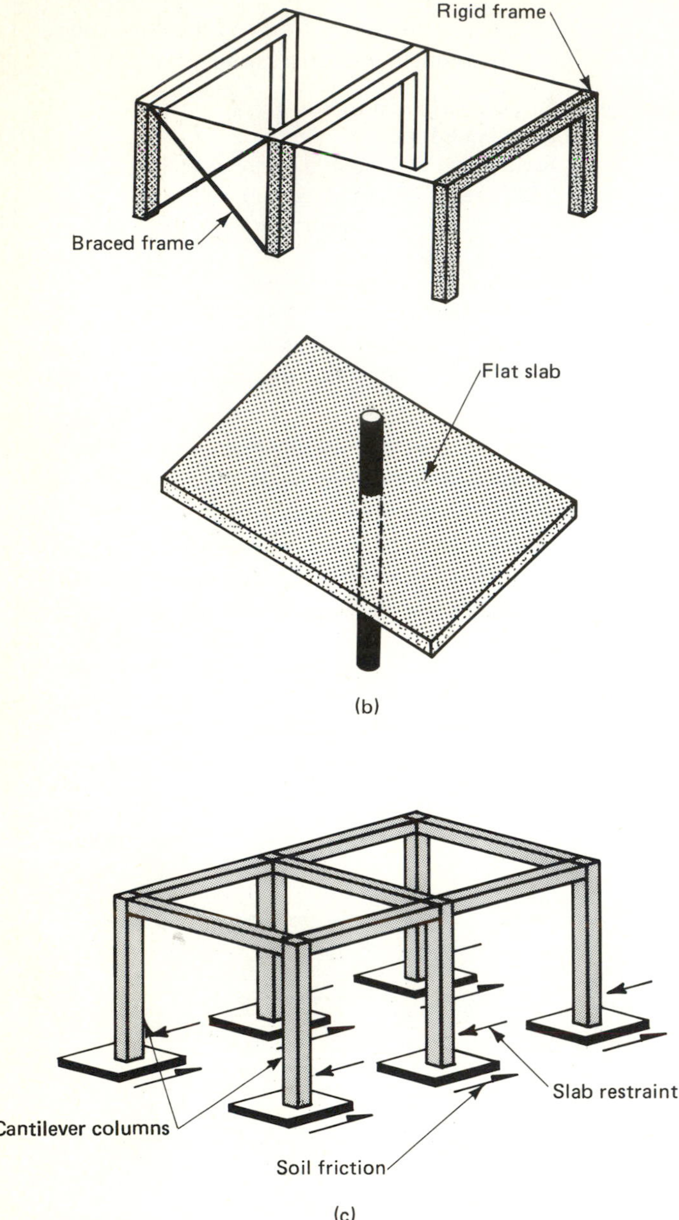

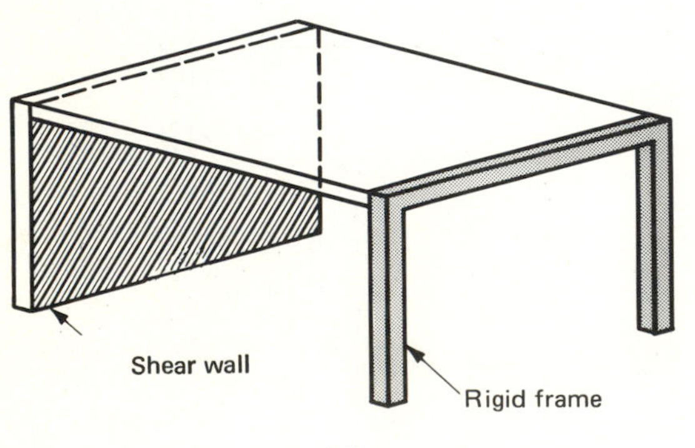

Behavior

Gravity Loads. Gravity load behavior of the unbraced frame is similar to that of the braced frame, and the load path for gravity loads is similar. The nature of the floor or roof subsystem connection to the vertical support subsystem determines the types of boundary forces that come into play. If these connections are rigid or continuous, the possibility for biaxial bending and torsion exists.

Once the gravity loads are on the rigid frame, they are dispersed to the columns via shear in the girders, which creates bending moments in both the girders and the rigidly connected columns (Fig. 6–8). Additional bending moments are created in the columns and girders due to the bending deformation of the columns and the resulting $P\Delta$ effect. Uniform gravity loads produce double curvature bending in the columns and pattern live loads produce single curvature bending.

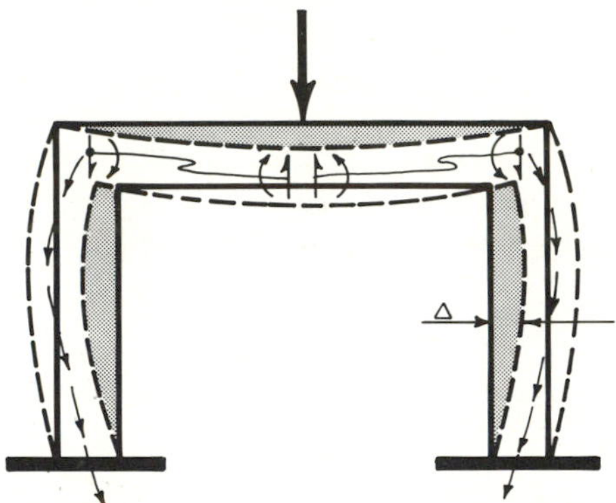

Figure 6–8. Gravity Load Behavior of Unbraced Frame Systems

Lateral Loads. Lateral loads travel from wall surfaces to the lateral distribution subsystem (diaphragm or lateral bracing assembly) and in turn to the rigid frames (Fig. 6–9). The rigid frame transmits the lateral load to the ground via the shear capacity of the columns, which also produces bending moments in both the columns and the girders. Additional bending moments are created by the presence of the gravity load acting at the eccentricities produced by the lateral frame deflection ($P\Delta$ effect). Columns and girders take on the characteristic S-shape curvature under the action of lateral forces.

Loads to each rigid frame are affected by the geometrical arrangement of the structure and the relative stiffness of the diaphragm and lateral support components. The model for this behavior is shown in Fig. 6–10, which simulates the diaphragm as a continuous beam with a moment of inertia (I_D) calculated to produce deflections equivalent to that of the diaphragm. The supports for the model are springs, with assigned stiffness based on the lateral stiffness of the rigid frames. The

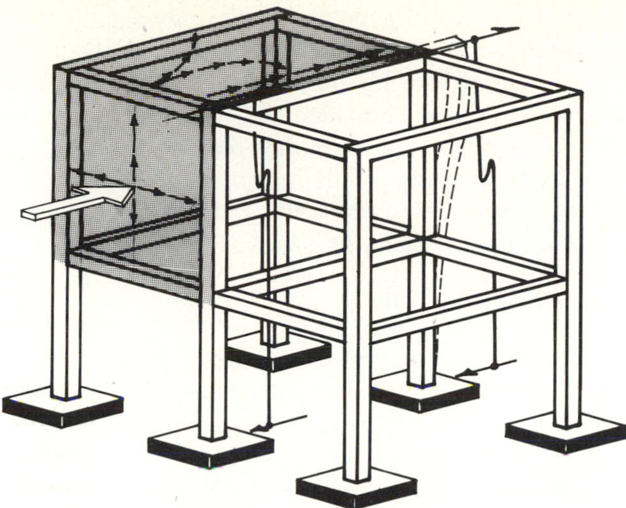

Figure 6-9. Lateral Load Behavior of Unbraced Frame Systems

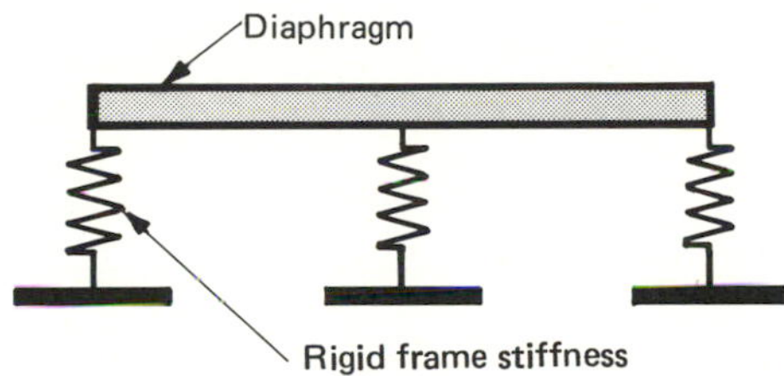

Figure 6-10. Diaphragm/Rigid Frame Model

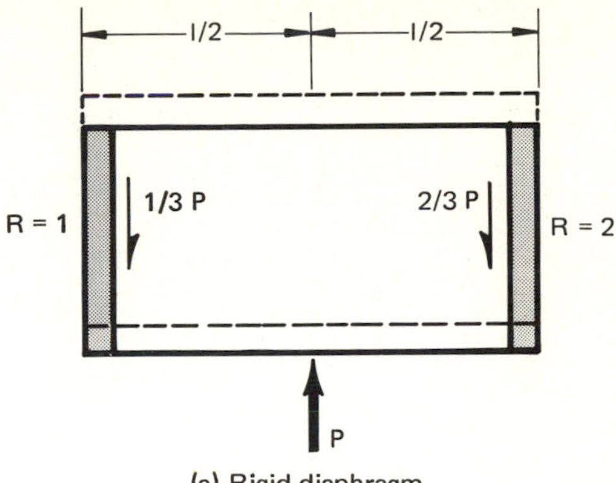

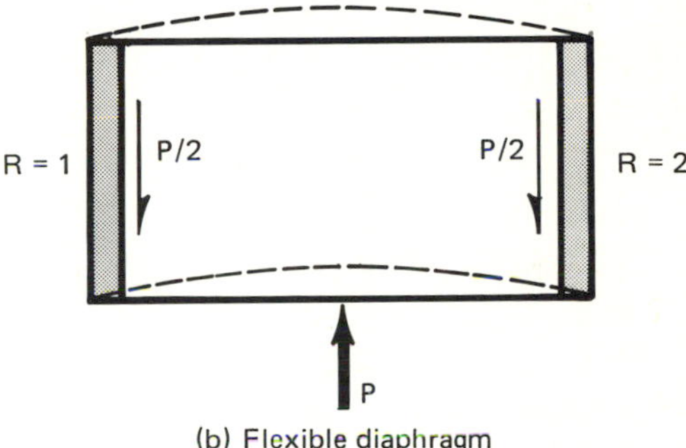

Figure 6-11. Lateral Load Distribution to Rigid Frames

two extremes of lateral load distribution are represented by the rigid-diaphragm case and the very flexible diaphragm case shown in Fig. 6–11. Both these cases represent determinate models: lateral loads are dependent only on spring stiffness for the rigid case; lateral loads are based on tributary width for the very flexible case. Intermediate cases are indeterminate and require solution by indeterminate analysis methods for beams on elastic supports. However, for preliminary design purposes, it is satisfactory to calculate the load distribution for each extreme and make appropriate assumptions based on these limits. The following example illustrates this concept.

EXAMPLE PROBLEM 6–2: Preliminary Lateral Load Distribution to Unbraced Frames

Determine the distribution of lateral load to the three single-story rigid frames shown in Fig. 6–12. Member properties are shown for the girder and column elements and the column bases are hinged. The floor diaphragm is a concrete slab.

Solution

1. *Rigid frame stiffness.* Determine relative stiffness by applying a 1-kip (4448-N) load to each frame and calculating the lateral deflection.

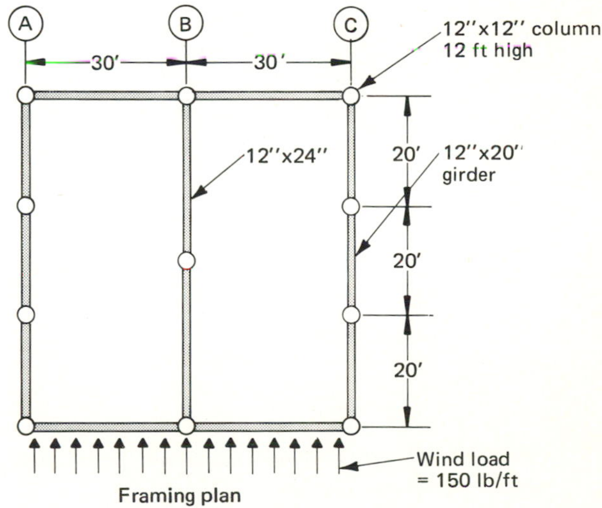

Figure 6-12. Example Problem 6–2

For frames *A* and *C*, assuming top-story condition (by Eqs. 5–13 to 5–15):

$$\Delta = \frac{H}{S_T}; \qquad \frac{1}{S_T} = \frac{1}{S_c} + \frac{1}{S_g}$$

$$S_c = \frac{12E \, \Sigma \, I_c}{h^3}$$

$$= \frac{12(3 \times 10^3)(4 \times 1728)}{(12 \times 12)^3}$$

$$= 83.3 \text{ kips/in. } (14{,}587 \text{ N/mm})$$

$$S_g = \frac{12E\Sigma Ig/l}{h^2}$$

$$= \frac{12(3 \times 10^3)(6) \times 8000/(20 \times 12)}{(12 \times 12)^2}$$

$$= 347.2 \text{ kips/in. } (60{,}801 \text{ N/mm})$$

$$\Delta = \frac{H}{S_T} = \frac{1 \text{ kip}}{83.3} + \frac{1 \text{ kip}}{347.2}$$

$$= 0.015 \text{ in. } (0.38 \text{ mm})$$

For frame *B:*

$$S_c = \frac{12(3 \times 10^3)(3 \times 1728)}{(12 \times 12)^3}$$

$$= 62.5 \text{ kips/in. } (10{,}945 \text{ N/mm})$$

$$S_g = \frac{12(3 \times 10^3)(4) \times 13{,}824/(30 \times 12)}{(12 \times 12)^2}$$

$$= 266.7 \text{ kips/in. } (46{,}704 \text{ N/mm})$$

$$\Delta = \frac{1 \text{ kip}}{62.5} + \frac{1 \text{ kip}}{266.7} = 0.020 \text{ in. } (0.51 \text{ mm})$$

Relative stiffness calculation:

Set frame $B = 1.0$

$$\text{Frames } A \text{ and } C = \frac{0.020}{0.015} = 1.33$$

2. *Rigid Diaphragm Assumption.* Distribution is in proportion to relative stiffness.

$$\text{Load to frames } A \text{ and } C = \frac{1.33}{3.66}(60 \text{ ft} \times 0.15)$$

$$= 3.27 \text{ kips } (14{,}545 \text{ N})$$

$$\text{Load to Frame } B = \frac{1.0}{3.66}(60 \text{ ft} \times 0.15)$$

$$= 2.46 \text{ kips } (10{,}942 \text{ N})$$

3. *Flexible Diaphragm Assumption.* Distribution is in proportion to tributary width.

Load to frames A and $C = (15 \text{ ft} \times 0.15)$
$= 2.25$ kips (10,008 N)

Load to frame $B = (30 \text{ ft} \times 0.15)$
$= 4.5$ kips (20,016 N)

4. *Conclusion.* The limits for a preliminary design assumption have been established. For this example, assume that the concrete slab represents a rigid diaphragm and use the load distribution calculated in step 2.

Lateral loads on unbraced frames produce static and dynamic deflections which are related to the stiffness of the structure, the forcing frequency of the dynamic load, and damping. Both wind and earthquake loads are dynamic in nature, although often reduced to equivalent static loads for design. It is important not to overlook the true dynamic response of unbraced frames when using equivalent static techniques.

Approximate Analysis

The approximate analysis of unbraced frames systems is normally accomplished by uncoupling the horizontal distribution subsystem and the vertical support subsystem in cases where the rigid frame provides vertical support. With concrete slab systems, however, the two subsystems are linked together for analysis, utilizing the ACI Code equivalent frame technique. The approximate analysis methods of Sections 5–4 and 5–7 are applicable to these two cases. The neglect of the out-of-plane torsional forces in the rigid-frame case is considered conservative, and these forces are not normally considered in analysis (Fig. 6–13).

Figure 6–13. Analysis of Unbraced Frame Systems

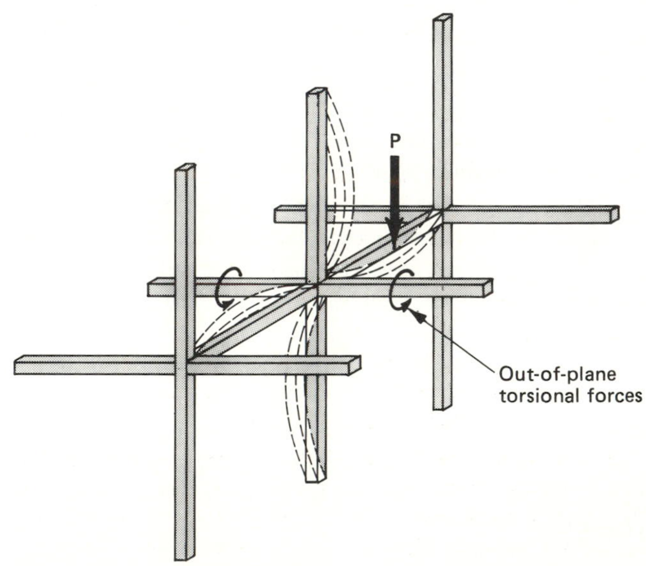

Approximate Design

Design of unbraced framed systems involves the following choices:

- Material selection
- Bay dimensions
- Story heights
- Number and location of lateral support subsystems
- Type of horizontal distribution subsystem
  - Floor and roof assembly
  - Concrete slab subsystem

- Type of lateral support subsystem
  - Rigid column and girder frame
  - Flat slab and column frame
  - Portal frame
  - Knee-braced frame
- Type and stiffness of lateral distribution subsystem
  - Diaphragm
  - Bracing assembly
- Type of vertical support subsystem
  - Rigid frame
  - Columns (beamless slab)

Unbraced frame systems are often selected when no other means of carrying the lateral load is available. The following example problem illustrates the design process.

EXAMPLE PROBLEM 6–3: Preliminary Design of Unbraced Frame System

The owner of the warehouse in Example Problem 6–1 has decided to lengthen the building from 240 ft (73.1 m) to 400 ft (121.9 m). He also now needs at least 40 ft (12.2 m) between columns in order to accommodate storage needs. Select a preliminary structural system.

Solution

1. *Project requirements:*

Function:	Plan area 400 ft × 120 ft (121.9 m × 36.6 m)
	Clear height = 20 ft (6.1 m)
	Clear spacing between columns = 40 ft (12.2 m)
Esthetic:	None
Serviceability:	Deflection: $L/240$
	Slope roof to prevent ponding.
	No fire protection required.
	Assume no special durability problems.
Construction:	Economy is paramount.

2. *Loads:*
   a. *Live load:* 20 psf (957.6 N/m²).
   b. *Dead load:* (Calculate for each system).
   c. *Wind load:* 25 psf (1197 N/m²).
3. *General.* The length of the warehouse now prohibits the use of the roof as a diaphragm to transmit wind loads to the end walls. Lateral loads must now be taken by unbraced interior frames. Three unbraced systems offer potential: (1) knee-braced steel frames with columns at 40 ft (12.2 m); (2) rigid portal steel frames spanning 120 ft (36.6 m); and (3) prestressed concrete fixed-base cantilever system with columns at 40 ft (12.2 m).
4. *Subsystem Definition.* Consider all three potential systems as alternates to compare.
   - *Scheme 1:* 22-gage metal roof deck with bar joists at 5 ft (1.5 m) on center spanning 40 ft (12.2 m) to steel girders; knee-braced steel columns at 40 ft (12.2 m) on center.
   - *Scheme 2:* 22-gage metal roof deck with steel purlins at 5 ft (1.5 m) on center spanning 20 ft (6.1 m) to steel rigid frame. Rigid frame spans 120 ft (36.6 m).
   - *Scheme 3:* Prestressed double-tee roof members spanning 40 ft (12.2 m) to inverted tee girder; fixed-base cantilever columns at 40 ft (12.2 m) on center.
5. *Preliminary design of Scheme 1:*
   a. *Metal deck:* Use 22 gage (Appendix 5A).
   b. *Bar joists:* assume dead load = 20 psf (957.6 N/m²).

$$\text{Uniform load, } w = 5 \text{ ft (40 psf)}$$
$$= 200 \text{ plf (2918 N/m)}$$
$$M = \frac{wl^2}{8} = \frac{200(40)^2}{8}$$
$$= 40{,}000 \text{ ft-lb}$$
$$= 480 \text{ in.-kips (54,240 Nm)}$$

By Appendix 4F, use 22 H7 at 10.7 lb/ft.

$$\frac{10.7}{5} \text{ ft} = 2.14 \text{ psf (31.2 N/m}^2\text{)}$$

c. *Wind analysis:* Assume that, wind load = 25 psf (1197 N/m²).

$$\text{Load to each bent} = (25 \text{ psf})(11 \text{ ft})(40 \text{ ft})$$
$$= 11.0 \text{ kips (48,928 N)}$$

Assume that knee braces connect to columns at 16 ft (4.9 m) above the base and extend to the girders at a 45° angle.

By portal method (Fig. 6–14):

$$\text{Shear at int. col.} = \frac{11.0}{3}$$
$$= 3.67 \text{ kips (16,324 N)}$$

Assuming hinged base:

$$\text{Col. moment} = 3.67 \text{ kips (16 ft)}$$
$$= 58.7 \text{ ft-kips (79,597 Nm)}$$
$$\text{Girder moment} = 1.83 \text{ kips (16 ft)}$$
$$= 29.3 \text{ ft-kips (39,730 Nm)}$$

d. *Girder:*
   (1) Loads:

$$\text{Uniform gravity load} = 40 \text{ ft}(0.04)$$
$$= 1.6 \text{ kips/ft (23,349 N/m)}$$
$$M = \frac{wl^2}{10} = \frac{1.6(40)^2}{10}$$
$$= 256 \text{ ft/kips (347,136 Nm)}$$
$$\text{Wind moment} = 29.3 \text{ ft-kips}$$
$$= 11.4\% \qquad \text{Neglect!}$$

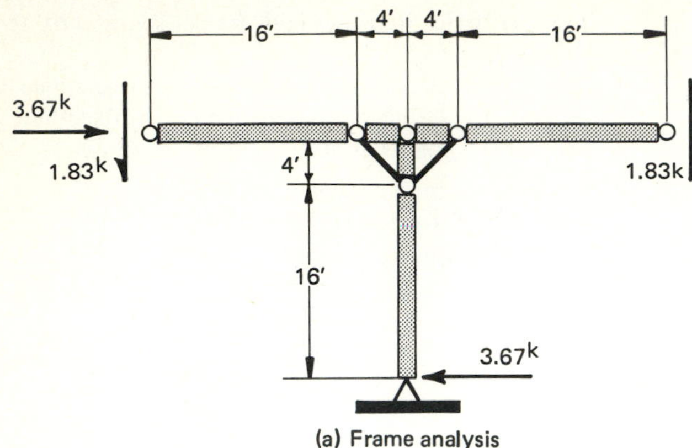

(a) Frame analysis

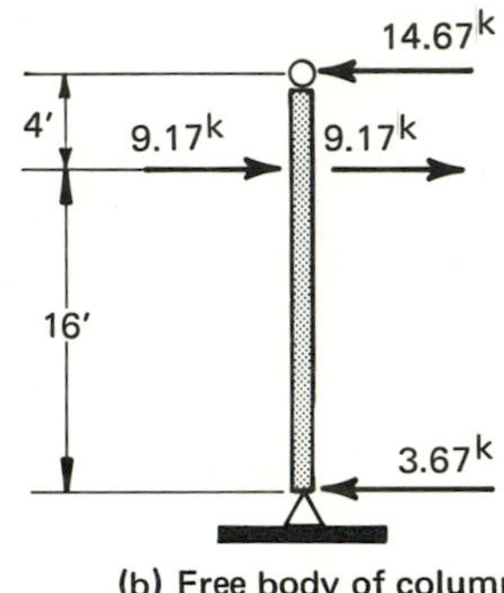

(b) Free body of column

Figure 6–14. Portal Analysis of Scheme 1, Example Problem 6–3

(2) Selection: Assume that bracing provided by joists at 5-ft intervals and F_b = 24.0 ksi (165.5 MPa).

$$S_{\text{req'd}} = \frac{M}{F_b} = \frac{256 \times 12}{24 \text{ ksi}}$$
$$= 128 \text{ in.}^3 \ (2.09 \times 10^6 \text{ mm}^3)$$

Use W24 × 62.

$$\frac{62}{40 \text{ ft}} = 1.55 \text{ psf } (74.2 \text{ N/m}^2)$$

e. *Column:*

(1) Loads:

$$\text{Axial load} = 40 \text{ psf}(40 \text{ ft})(40 \text{ ft})$$
$$= 64.0 \text{ kips } (284{,}672 \text{ N})$$
$$\text{Wind moment} = 58.7 \text{ ft-kips } (79{,}597 \text{ Nm})$$

By Eq. 4–74,

$$A_{\min} = \frac{P}{F_a} + \frac{(2.6)M}{dF_b}$$

(2) Determine F_a

$$F_{a\,(\max)} = 22.0 \text{ ksi } (151.7 \text{ MPa})$$
$$F_a(22.6\sqrt{P}) = 9.37 \text{ ksi } (22.6\sqrt{P} = 15 \text{ ft})$$
$$F_a(57\sqrt{P}) = 3.72 \text{ ksi } (57\sqrt{P} = 38 \text{ ft})$$

For kl = 1.0(20 ft) = 20 ft (6.1 m), select F_a = 8.0 ksi (55.2 MPa).

(3) Determine F_b by Eq. 4–49,

$$F_b = \sqrt{\frac{10{,}320\ M}{ld^2}} \leq 0.60F_y$$
$$= \sqrt{\frac{10{,}320(58.7)}{20(10)^2}}$$
$$= 17.4 \text{ ksi } (120 \text{ MPa})$$

(4) Calculate minimum area. Reduce loads by 0.75 for wind,

$$A_{\min} = \frac{64(0.75)}{8.0} + \frac{2.6(58.7 \times 12)(0.75)}{10(17.4)}$$
$$= 13.9 \text{ in.}^2 \ (8965 \text{ mm}^2)$$

(5) Check moment magnification. Assume that drift (Δ/h_s) is limited to 0.0025.

$$Q = \frac{\Sigma P\Delta}{Hh_s} = \frac{(3 \times 64)(0.0025)}{11.0} = 0.044 > 0.04$$

But assume that no magnification is necessary.

(6) Select section. Appendix 4A shows that a W10 × 45 has an area of 13.2 in.² and L_u = 22.8 ft. Use W10 × 45 (13.2 in.²)

(7) Check deflections.

$$\Delta = \frac{H}{S_T} = \frac{H}{S_c} + \frac{H}{S_g}$$

$$S_c = \frac{12E\ \Sigma\ I_c}{h^3}$$
$$= \frac{12(30 \times 10^3)(4 \times 248)}{(20 \times 12)^3}$$
$$= 25.8 \text{ kips/in. } (4518 \text{ N/mm})$$

$$S_g = \frac{12E\ \Sigma\ I_g/L}{h^2}$$
$$= \frac{12(30 \times 10^3)(6 \times 1550/480)}{(20 \times 12)^2}$$
$$= 121 \text{ kips/in. } (21{,}189 \text{ N/mm})$$

$$\Delta = \frac{11.0}{25.8} + \frac{11.0}{121}$$
$$= 0.52 \text{ in. } (13.2 \text{ mm})$$

$$\Delta/h = \frac{0.52}{20 \times 12}$$

$$= 0.002 < 0.0025 \quad \text{O.K.}$$

$$\frac{45 \text{ lb/ft} \times 20 \text{ ft}}{40(40 \text{ ft})} = 0.56 \text{ psf } (26.8 \text{ N/m}^2)$$

f. *Knee braces:*

$$\text{Axial load} = 9.17\sqrt{2}$$
$$= 12.96 \text{ kips } (57{,}646 \text{ N})$$
$$kl = 4\sqrt{2} = 5.65 \text{ ft } (1.72 \text{ m})$$

Use two angles 2 in. × 2 in. × 3/16 in. at 4.88 lb/ft.

$$\frac{4.88(5.65 \times 2)}{40(40)} = 0.034 \text{ psf } (1.63 \text{ N/m}^2)$$

g. *Footing:*

$$\text{Column load} = 64.0 \text{ kips } (284{,}672 \text{ N})$$

$$A_f = \frac{64}{3} = 21.3 \text{ ft}^2 \ (1.98 \text{ m}^2)$$

Use 4 ft 9 in. × 4 ft 9 in (1.45 m × 1.45 m) (A = 22.6 ft²). Determine footing depth by Table 4–12:

$$q_{\text{net}} = \frac{P_u}{A_f} = \frac{1.4(32) + 1.7(32)}{22.6} = 4.39 \text{ ksf}$$

$$\frac{\sqrt{f'_c}}{q_{\text{net}}} = \frac{\sqrt{3000}}{4390} = 0.0125$$

$$\frac{A_f}{A_c} = \frac{22.6}{1.0} = 22.6, \ r = 12 \text{ in.}, \ a = 22.5 \text{ in.}$$

$$\frac{d}{r} = 0.6\text{: req'd } d = 0.6(12) = 7.2 \text{ in.}$$

$$\frac{a}{d} = 3.93\text{: req'd } d = \frac{22.5}{3.93} = 5.7 \text{ in.}$$

Use d = 7.5 in., t = 7.5 + 3 = 11.5 in. Use t = 12 in. (25.4 mm).

$$\text{Quantity} = \frac{4.75(4.75)(12/12)}{27(40)(40)}$$
$$= 0.0005 \text{ cy/sq ft } (0.0041 \text{ m}^3/\text{m}^2)$$

4. *Preliminary design of Scheme 2*
   a. *Metal deck:* Use 22 gage (Appendix 5A).
   b. *Purlins:* Assume that dead load = 20 psf (957.6 N/m²).

$$\text{Total load} = 40 \text{ psf } (1915 \text{ N/m}^2)$$

$$\text{Uniform load} = 5(40) = 200 \text{ plf } (2{,}918 \text{ N/m})$$

$$M = \frac{wl^2}{8} = \frac{0.2(20)^2}{8}$$
$$= 10 \text{ ft-kips } (13{,}560 \text{ Nm})$$

Assume full lateral support and F_b = 22.0 ksi (151.7 MPa) for channel section.

$$S_{\text{req'd}} = \frac{10 \times 12}{22} = 5.4 \text{ in.}^3 \ (88{,}506 \text{ mm}^3)$$

Use MC10 × 8.4.

$$\frac{8.4 \text{ lb/ft}}{5 \text{ ft}} = 1.68 \text{ psf } (80.4 \text{ N/m}^2)$$

c. *Rigid frame analysis* (Fig. 6–15):
   (1) Uniform gravity load = 20 (0.04) = 0.8 kip/ft (11,674 N/m).
   (2) Moments:
   At midspan:

$$+M = \frac{wl^2}{8} = \frac{0.8(80)^2}{8}$$
$$= 640 \text{ ft-kips } (867{,}840 \text{ Nm})$$

$$\text{Thrust } (T) = 40.0 \text{ kips } (177{,}920 \text{ N})$$

At column:

$$-M = 32 \text{ kips } (20 \text{ ft}) + \frac{0.8(20)^2}{2}$$

Figure 6–15. Analysis of Scheme 2, Example Problem 6–3

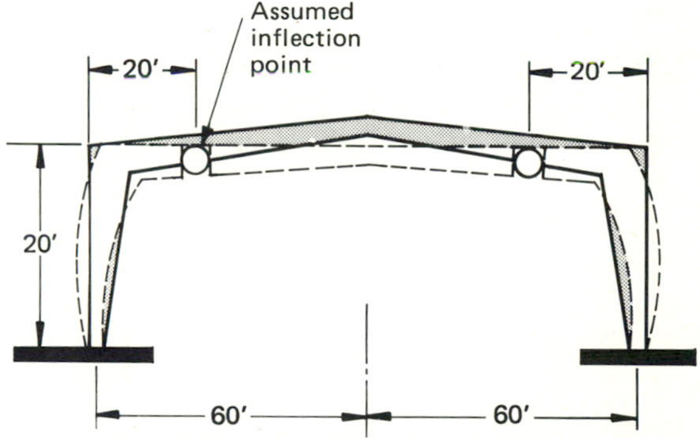

(a) Rigid frame with assumed inflection points

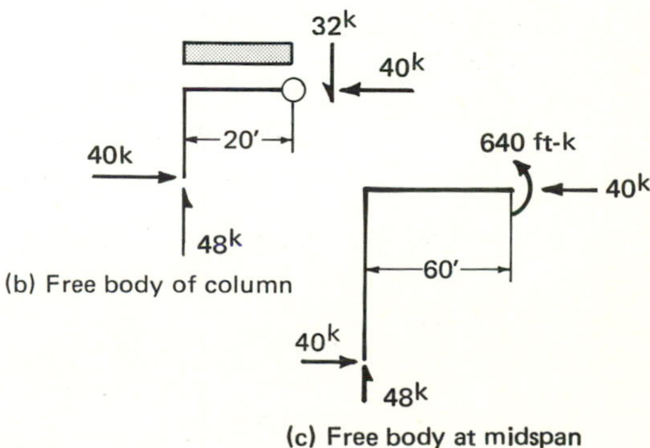

(b) Free body of column

(c) Free body at midspan

$$= 800 \text{ ft-kips } (1{,}084{,}800 \text{ Nm})$$

$$\text{Thrust } (T) = 40.0 \text{ kips } (177{,}920 \text{ N})$$

(3) Wind load:

$$\text{Load} = 25 \text{ psf } (11 \text{ ft})(20 \text{ ft}) = 5.5 \text{ kips } (24{,}464 \text{ N})$$

Assume that one-half goes to each column.

$$\text{Col. } M = 2.75(20 \text{ ft}) = 55 \text{ ft-kips } (74{,}580 \text{ Nm})$$

$$\text{Girder } M = 55 \text{ ft-kips } (74{,}580 \text{ Nm})$$

d. *Girder and column:* Neglect combined stresses, since axial force is small. Assume that $F_b = 22.0$ ksi (151.7 MPa).

At midspan:

$$S_{\text{req'd}} = \frac{640 \times 12}{22} = 349 \text{ in.}^3 \ (5.7 \times 10^6 \text{ mm}^3)$$

Use W33 × 118 (Appendix 4B).

At haunch:

$$S_{\text{req'd}} = \frac{800 \times 12}{22} = 436 \text{ in.}^3 \ (7.1 \times 10^6 \text{ mm}^3)$$

Use W33 × 141 (Appendix 4B).

$$\begin{aligned}\text{Average girder weight} &= 118(105 \text{ ft}) + 141(15 \text{ ft}) \\ &= 14{,}505 \text{ lb} \\ \text{Average column weight} = 118(20 \text{ ft})2 &= \underline{\ \ 4{,}720 \text{ lb}} \\ &\quad\ 19{,}225 \text{ lb}\end{aligned}$$

$$\text{Weight} = \frac{19225 \text{ lb}}{120(20)} = 8.0 \text{ psf } (116.7 \text{ N/m}^2)$$

e. *Footing:*

$$\text{Column load} = 48 \text{ kips } (213{,}504 \text{ N})$$

$$A_f = \frac{48}{3} = 16 \text{ ft}^2 \ (1.49 \text{ m}^2)$$

Use 4 ft 0 in. × 4 ft 0 in. (1.2 m × 1.2 m).

$$q_{\text{net}} = \frac{P_u}{A_f} = \frac{1.4(24) + 1.7(24)}{16} = 4.65 \text{ ksf}$$

$$\frac{\sqrt{f'_c}}{q_{\text{net}}} = \frac{\sqrt{3000}}{4650} = 0.012$$

$$\frac{A_f}{A_c} = \frac{16}{1.0} = 16.0, \ r = 12 \text{ in.}, \ a = 18 \text{ in.}$$

$$\frac{d}{r} = 0.5\text{: req'd } d = 0.5(12) = 6.0 \text{ in.}$$

$$\frac{a}{d} = 3.93\text{: req'd } d = \frac{18}{3.93} = 4.58 \text{ in.}$$

Use d = 6.0 in., t = 6 + 3 = 9 in. (228 mm).

$$\text{Quantity} = \frac{4.0(400)(9/12)}{27(20)(60)} = 0.0003 \text{ cy/sq ft}$$

5. *Preliminary design of Scheme 3*

a. *Deck members:* Superimposed load:

$$\begin{aligned}\text{Live load} &= 20.0 \text{ psf} \\ \text{Dead load} &= \underline{\ \ 8.0} \text{ psf (built-up roof)} \\ &\quad 28.0 \text{ psf total } (1340 \text{ N/m}^2)\end{aligned}$$

By Appendix 5A, use 8 ft 0 in. wide × 12 in. deep prestressed double tee members.

$$\text{Weight} = 37 \text{ psf } (1771 \text{ N/m}^2)$$

b. *Girder:* Assume that total dead load, excluding girder = 45.0 psf (2154 N/m²).

$$\begin{aligned}\text{Uniform load} &= (45 + 20)\text{psf}(40 \text{ ft}) \\ &= 2.6 \text{ kips/ft } (37{,}942 \text{ N/m})\end{aligned}$$

Using procedure similar to Example Problem 6–2, use 24 in. wide × 20 in. deep inverted tee beam.

$$\text{Weight} = 350 \text{ plf}$$

c. *Columns:*

(1) Calculate ultimate loads.

$$\begin{aligned}\text{Dead load} &= 0.045(40)(40) + 0.35(40) \\ &= 86 \text{ kips } (382{,}528 \text{ N}) \\ \text{Live load} &= 0.02(40)(40) \\ &= 32 \text{ kips } (142{,}336 \text{ N}) \\ P_u &= 1.4(86) + 1.7(32) \\ &= 174.8 \text{ kips } (777{,}510 \text{ N})\end{aligned}$$

$$\text{Wind shear} = \frac{11.0}{4} = 2.75 \text{ kips } (12{,}232 \text{ N})$$

$$\begin{aligned}M_u(\text{wind}) &= 1.7(2.75)(20 \text{ ft}) \\ &= 93.5 \text{ ft-kips } (126{,}786 \text{ Nm})\end{aligned}$$

No gravity moment exists; therefore, assume minimum eccentricity according to ACI 10.11.5:

$$\begin{aligned}e &= 0.6 + 0.03h = 0.6 + 0.03(14) \\ &= 1.02 \text{ in.} = 0.09 \text{ ft} \\ M_u &= (\text{gravity}) = 1.4(86)(0.09) + 1.7(32)(0.09) \\ &= 15.7 \text{ ft-kips } (21{,}289 \text{ Nm})\end{aligned}$$

(2) Determine minimum column size before magnification.

Effective length $(k) = 2.0$ for cantilever condition

Limit kl/r to 80 or $h = 2.0(20 \times 12)/(0.3 \times 80) = 20$ in. Use 0.75 reduction for wind. By Eq. 4–79,

$$P'_u = 174.8 - 0.30(20)(20 \times 0.8)2.77$$
$$= -91.1 \qquad \text{Assume} = 0$$

(3) Calculate magnifier for gravity moment, δ_b.

$$\beta_d = \frac{M_D}{M_T} = 1.0$$

for minimum eccentricities

$$\frac{P_u(1+\beta_d)}{A_g} = \frac{174.8(1+1.0)}{20 \times 20} = 0.87 \text{ ksi}$$

From Fig. 4–47,

$$\frac{kl}{h} = \frac{2.0(20 \times 12)}{20} = 24$$

$$\frac{\delta}{c_m} = 3.2$$

(4) Calculate magnifier for wind moment.

$$\Delta_u = \frac{H_u l^3}{3EI} = \frac{0.75(1.7)(2.75)(20 \times 12)^3}{3(3 \times 10^3)(0.8 \times 13333)}$$
$$= 0.50 \text{ in. } (12.8 \text{ mm})$$

$$Q_u = \frac{\Sigma P_u \Delta}{H_u h_s} = \frac{(0.75)(4)(174.8)(0.50)}{(0.75)(1.7)(11.0)(20 \times 12)}$$
$$= 0.08 > 0.04$$

$$\delta_s = \frac{1}{1 - Q_u} = \frac{1}{1 - 0.08} = 1.09$$

$$A = \frac{0(0.75)}{2.74} + \frac{(3.2)(6)(15.7 \times 12)}{3.13(20)} + \frac{1.09(6)(93.5 \times 12)}{3.13(20)}$$

$$A = 175 \text{ in.}^2 \ (112{,}877 \text{ mm}^2)$$
$$< 400 \text{ in.}^2 \text{ provided}$$

Use a 20 in. × 20 in. column with 2% reinforcement.

(5) Check deflection:

$$\Delta = \frac{Hl^3}{3EI}$$

$$\Delta = \frac{2.75(20 \times 12)^3}{3(4 \times 10^3)(2880)} = 1.10 \text{ in.}$$

$$\frac{\Delta}{h} = \frac{1.10}{20 \times 12} = 0.0045 \qquad \text{(Assume O.K.)}$$

d. *Footing:*

Column load = 118 kips

$$A_f = \frac{118}{3} = 39.3 \text{ ft}^2$$

Use 6 ft 6 in. × 6 ft 6 in. Using a procedure similar to that of Example Problem 6–2,

$$t = 11 \text{ in.} + 3 \text{ in.} = 14 \text{ in.}$$

$$\frac{6.5(6.5)(14/12)}{27(40)(40)} = 0.0011 \text{ cy/sq ft}$$

6. *Tabulation and Selection.* Table 6–2 provides a summary of the quantities and unit costs for each alternate. Scheme 1 has the lowest unit cost and is, therefore, the logical choice.

Table 6–2 Summary Costs for Example Problem 6–3

Alternate	*Deck*	*Secondary member*	*Girder*	*Column*	*Footing*	*Total*
Scheme 1						
Quantity	1.0 sf/sf	2.14 psf	1.53 psf	0.56 psf	0.0005 cy/sf	
Cost	$0.61/sf	$0.86/sf	$0.78/sf	$0.29/sf	$0.07/sf	$2.61/sf
Scheme 2						
Quantity	1.0 sf/sf	1.68 psf	8.0 psf		0.0003 cy/sf	
Cost	$0.61/sf	$1.00/sf	$4.1/sf		$0.04/sf	$5.75/sf
Scheme 3						
Quantity	1.0 sf/sf	—	350 plf	0.102 cy/lf	0.0011 cy/sf	
Cost	$2.06/sf		$1.25/sf	$0.03/sf	$0.15/sf	$3.49/sf

6–4 BEARING WALL SYSTEMS

Bearing wall systems are characterized by wall subsystems that provide both vertical support and lateral support (Fig. 6–16). The horizontal distribution subsystem is generally a floor or roof assembly but may also be a concrete slab. Although many different systems may include the bearing wall subsystem, the (bearing wall) system classification is used where bearing walls contribute the major portion of the vertical and lateral support. Bearing walls typically occur at regular intervals or in a modular fashion in the building plan.

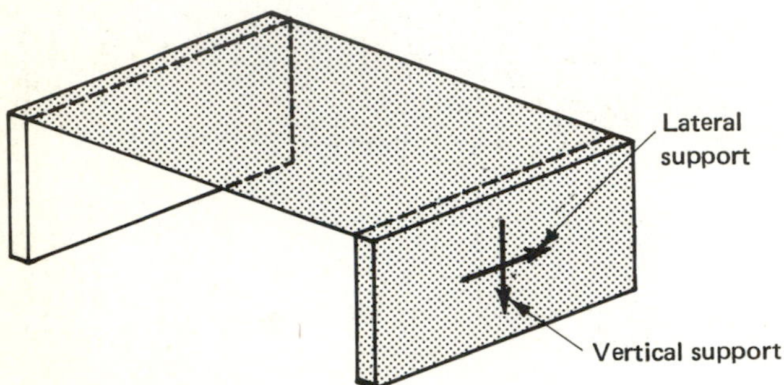

Figure 6–16. Bearing Wall System

Bearing walls are typically located so as to coincide with exterior walls, partition locations, and vertical shaft locations (elevators and stairs), with their size, shape, and location very dependent on architectural considerations and overall building function. Exterior bearing walls also provide insulation, a function that requires attention to energy requirements.

History and Development

Bearing wall systems were probably the first complete structural systems devised by the early builders. The simplicity of constructing a wall of stone or mud and laying timbers across for floor or roof support has always made this system a natural part of the built environment. The extension of bearing walls to greater heights and loads was initially accommodated by proportionately increasing the size of the wall. Only much later did developments in engineering analysis and materials science provide a more complete understanding of wall behavior and initiate a reversal of this process.

Almost oblivious to this early history, builders at one point had reduced bearing walls to mainly single-story applications. Now with wall behavior better understood, the bearing wall system has reemerged as a competitive alternative for many multistory buildings. It continues to be a very popular system for low-rise commercial and institutional buildings.

Types and Uses

Some of the major types of bearing wall systems are shown in Fig. 6–17. These include the following:

- Masonry walls
- Precast concrete walls
- Cast-in-place concrete walls
- Wood stud walls
- Metal stud walls
- Site-cast "tilt-up" walls

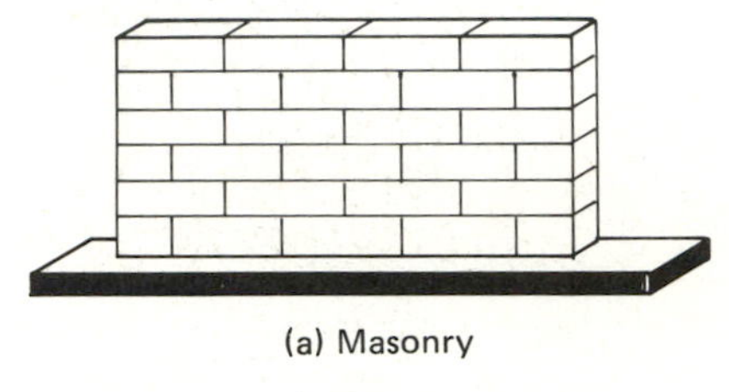

(a) Masonry

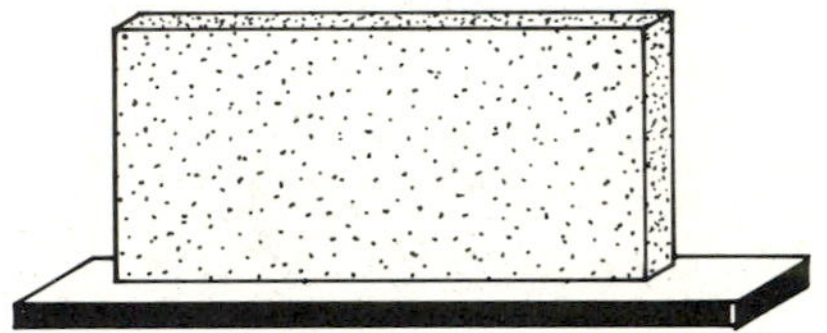

(b) Precast concrete; cast-in-place concrete; tilt-up site-cast concrete

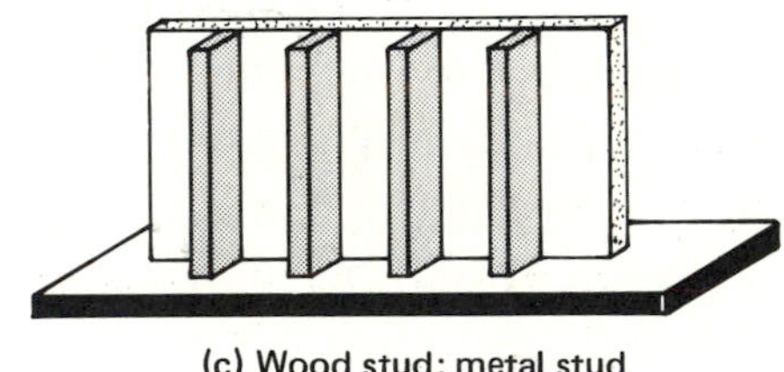

(c) Wood stud; metal stud

Figure 6–17. Types of Bearing Wall Systems

Masonry construction is extensively used for residential and commercial construction, including apartments, dormitories, retail stores, and offices. Precast concrete wall systems are used for manufacturing and warehousing as well as for high-rise office buidings and apartment buildings. Tilt-up wall systems are primarily used for single-story structures in the light manufacturing or commercial category. Popular floor and roof assemblies for bearing wall construction include precast plank, cast-in-place one-way or two-way slabs, bar joists, and composite precast and cast-in-place slab systems. Precast bearing wall systems also typically use precast elements for the floor or roof subsystems, including prestressed double tees, hollow core plank, and similar elements.

Behavior

Gravity Loads. A typical load path for a bearing wall system is shown in Fig. 6–18. Linkage forces normally existing at the boundaries include shear, bearing, and moment. Moment forces are created by pattern live loads on interior bays and uniform loads on exterior bays.

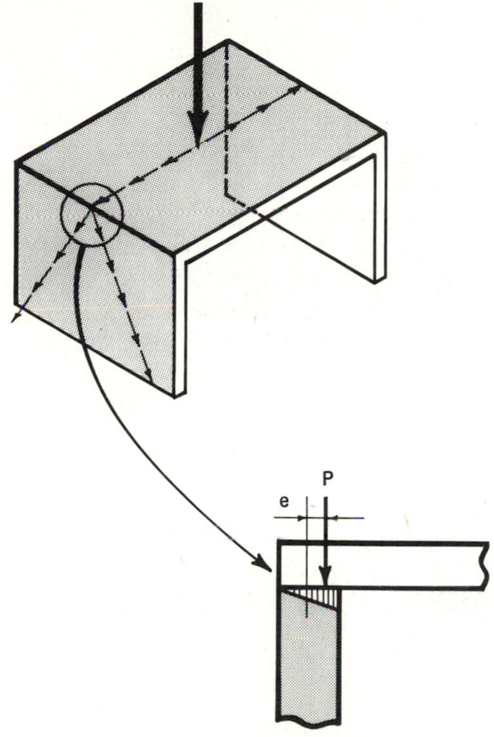

Figure 6–18. Load Path for Gravity Loads: Bearing Wall Systems

The magnitude of the moment forces depends on the degree of fixity at the floor (roof) and wall connection. For pure *hinged* connections, the moment is the product ($P \cdot e$) of the slab load and its eccentricity about the wall centerline. Moments transfer into the wall in proportion to relative stiffness, usually one-half above the joint and one-half below. Figure 6–19 illustrates the overall moment transfer into a multistory bearing wall.

Where wall/floor joints are *fixed,* the system acts as a rigid frame. The rigid frame subassembly model can be used to distribute unbalanced joint moments to the walls above and below the joint (Fig. 6–20). Complete joint fixity is often difficult to achieve without exceeding the flexural tensile strength of the slab subsystem. Partial fixity (semirigid connections) can be assumed, with joint moments set equal to the flexural capacity of the floor members. This approach is analogous to the plastic hinge concept used with limit design methods, which assumes that plastic rotation occurs when the material reaches its yield strength.

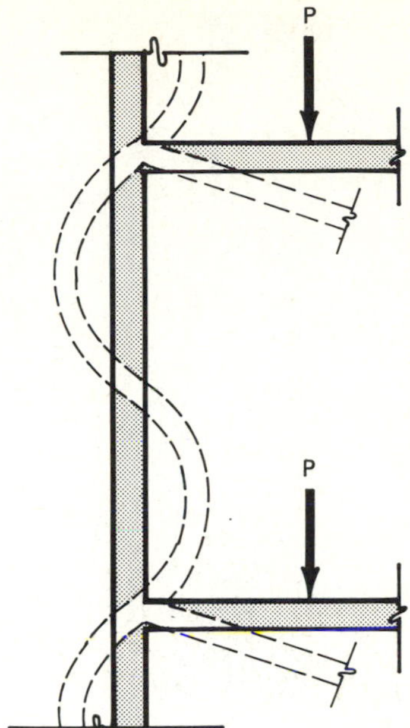

Figure 6–19. Gravity Load Wall Moments for Hinged Connections

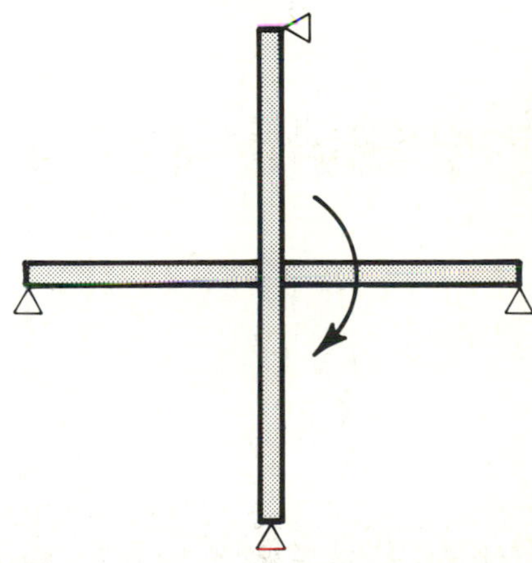

Figure 6–20. Gravity Load Wall Moments for Fixed Connections

Lateral Loads. Lateral wind loads on bearing wall systems travel from exterior wall surfaces to the lateral distribution diaphragm and into the shear walls (Fig. 6–21). In so doing, two sets of wall forces are created: *transverse* shear and bending in walls receiving the wind and *parallel* shear, flexure and overturning of shear walls serving as lateral support. Lateral earthquake loads introduce the same sets of forces, with the exception that all walls (exterior and interior) experience transverse bending and shear (Fig. 6–22).

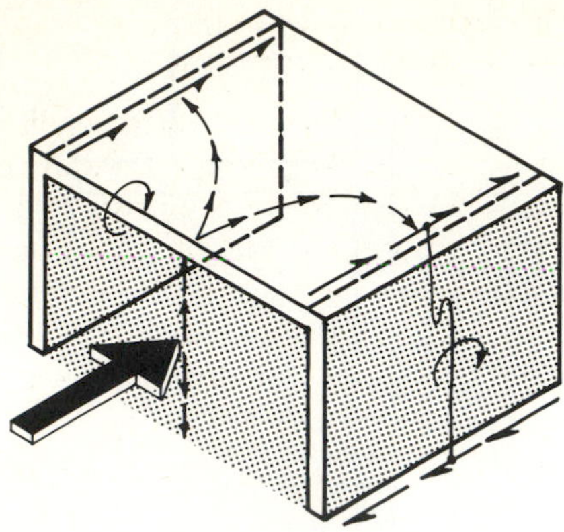

Figure 6–21. Lateral Load Path for Bearing Wall Systems

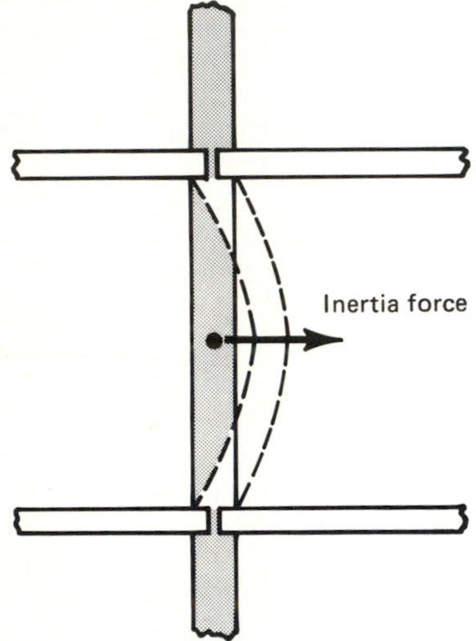

Figure 6–22. Earthquake Loads on Walls

The *transverse* set of forces is resisted by the wall acting as a continuous beam spanning between floors (Fig. 6–23). These forces create stresses in the same plane as the gravity load forces, and loading combinations of the two must be considered.

Figure 6–23. Transverse Forces on Walls

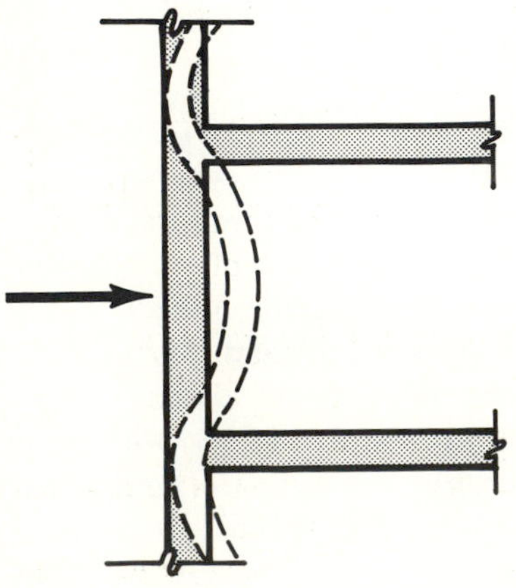

The magnitude of the *parallel* set of forces on the shear walls depends on the relative stiffness of the diaphragm and the various shear walls, and on the relative location of the center of rigidity, center of mass, and center of load. The stiffness of a shear wall can be calculated by the following equation, which assumes that the wall acts as a vertical cantilever beam.

$$\Delta = \Delta_m + \Delta_v = \frac{Ph^3}{3E_mI} + \frac{1.2Ph}{AE_v} \qquad (6\text{–}1)$$

where Δ_m = deflection due to bending, in. (mm)
Δ_v = deflection due to shear, in. (mm)
P = lateral forces, lb (N)
h = height, in. (mm)
A = cross-sectional area, sq in. (mm^2)
I = moment of inertia of wall in direction of bending, $in.^4$ (mm^4)
E_m = modulus of elasticity in compression, psi ($N/m^2 \times 10^{-6}$)
E_v = modulus of elasticity in shear, psi ($N/m^2 \times 10^{-6}$)

Walls penetrated by openings can be treated as a series of individual piers between openings with the wall stiffness equal to the sum of the individual pier stiffness (Fig. 6–24).

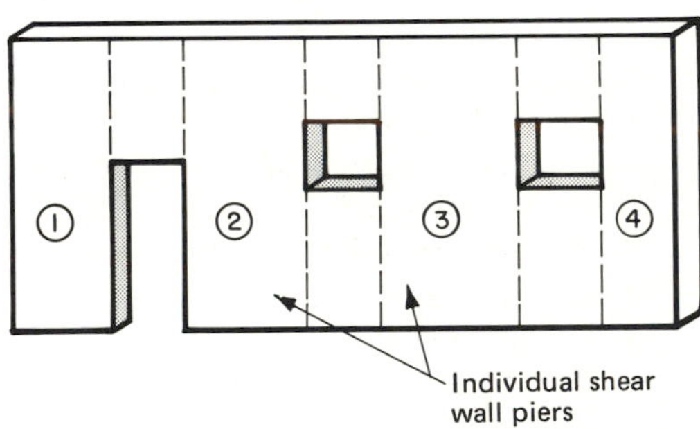

Figure 6–24. Stiffness of Wall with Openings

Parallel shear forces on shear walls depend on diaphragm and wall stiffness, similar to the unbraced frame example (Example Problem 6–2). Unsymmetrical shear wall arrangements, or eccentricity between the applied load and the center of rigidity of the shear wall group, creates a twisting moment on the structural system which must be resisted by additional shears in the walls (Fig. 6–25). Example Problem 6–4 illustrates this behavior.

Progressive collapse. Bearing wall systems are particularly susceptible to progressive collapse failures due to accidental overloading caused by explosions, construction loads, and so on. This subject is currently receiving considerable study, much of it centered around the improve-

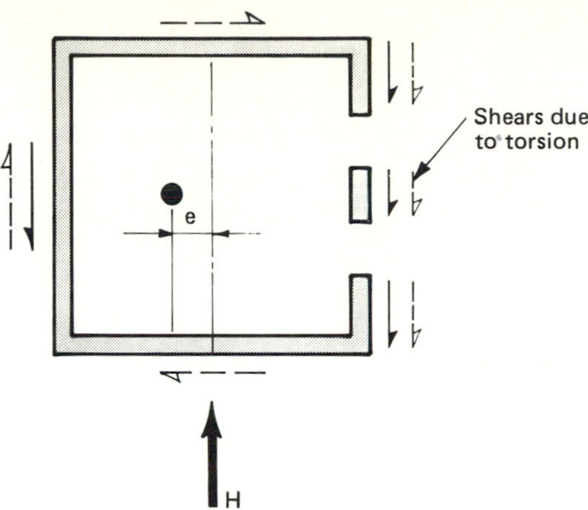

Figure 6–25. Torsion on Shear Wall Group

ment of connections and alternate load paths for these types of systems.

Approximate Analysis

Approximate analysis of bearing wall systems is typically done by selecting a typical bay for study. One of several loading conditions may create maximum wall stresses:

1. Exterior wall (transverse forces)
   a. Gravity axial load
   b. Gravity moments and shears
   c. Wind (earthquake) moments and shears
2. Exterior and interior wall (parallel forces)
   a. Gravity axial load
   b. Wind (earthquake) shear
   c. Wind (earthquake) flexure
   d. Wind (earthquake) overturning
3. Interior wall (transverse forces)
   a. Gravity axial load
   b. Gravity moments and shears
   c. Earthquake moments and shears

A typical exterior wall and a typical interior wall are usually selected for analysis once preliminary loads have been determined. Preliminary lateral loads can usually be determined by assuming a rigid diaphragm and distribution of loads in proportion to the individual wall rigidities.

Approximate Design

Design choices for bearing wall system include the following:

- Material selection
- Type and location of bearing walls
- Story height
- Bay dimensions
- Floor and roof assembly
- Type of wall/floor connection

Walls are normally located corresponding to partition locations in the building. Selected bearing walls must also be designated as shear walls to carry lateral loads. Shear walls are required in each direction, unless some other method of lateral support can be used for stability. Rigid frame action between the walls and floors is occasionally relied on for lateral stability in a direction perpendicular to transverse bearing walls (Fig. 6–26). The following example illustrates the design process.

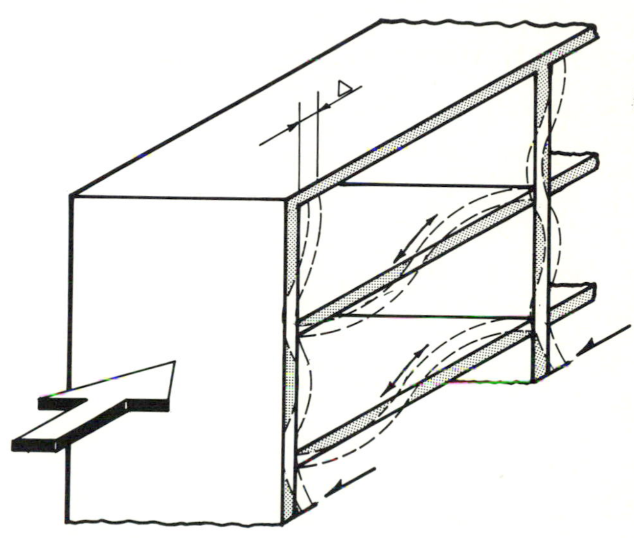

Figure 6–26. Alternate Bearing Wall Stability

EXAMPLE PROBLEM 6–4: Preliminary Bearing Wall System Design

Select a preliminary structural system for the four-story motel shown in Fig. 6–27. Minimum floor to ceiling height is 7 ft 8 in. (2.34 m). The building is located in Washington, D.C. Allowable soil bearing is 3000 psf (143,640 N/m²).

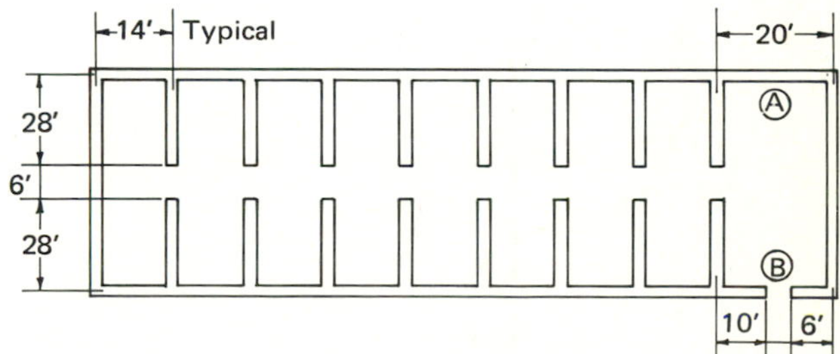

Figure 6–27. Example Problem 6–4

Solution

1. *Project requirements:*
   Function: Plan area 62 ft × 160 ft (18.9 m × 48.8 m)
   Building height = four stories

	Floor to ceiling minimum = 7 ft 8 in. (2.34 m)
	Partition walls at 14-ft (4.27-m) spacing
	Central corridor 6 ft (1.8 m) in width
Esthetics:	Assume that brick walls will be exposed and block walls will be painted.
Serviceability:	Deflection: $L/240$
	No fire protection required.
	Assume no special durability problems.
Construction:	Economy is important.
	Masonry skills are acceptable in area.

2. *Design loads:*
   Uniform live load:

$$\text{Rooms} = 40 \text{ psf } (1915 \text{ N/m}^2)$$

Wind load: 70 mph (Fig. 3–50).

Rather than compute detailed wind loads for preliminary design, assume an overall uniform wind load as follows:

$$w = 10 \text{ psf}(1.3) = 13 \text{ psf}$$

3. *Material selection.* Possible wall materials include brick masonry, concrete block masonry, and precast wall panels. For this example, compare brick and block schemes.
4. *Subsystem definition.* Based on the floor plan, the two alternatives are walls at 14 ft (4.27 m) or walls at 28 ft (8.53 m). These walls can be used as both bearing walls and shear walls (assuming that floor diaphragm capacity exists). Shear walls to carry lateral loads in the longitudinal direction appear possible at one end of the building. Several deck assemblies are possible:

   - One-way continuous concrete slab spanning 14 ft (4.27 m) with slab depth of 8 in. (203 mm).
   - 8-in. (203-mm) hollow-core prestressed, precast concrete slab spanning 28 ft (8.53 m).
   - 4-in. (101-mm) prestressed, precast flat slab spanning 14 ft (4.27 m).

   Since brick has higher compressive strength than block, select a brick alternate with walls at 28 ft (8.53 m) and a block alternate with walls at 14 ft (4.27 m).

   - *Scheme 1:* 8-in. (203-mm) brick bearing walls at 28 ft (8.53 m) on center; 8-in. (203-mm) hollow-core precast, prestressed floor and roof members; longitudinal shear walls at end of building. Brick strength $f'_m = 1400$ psi (9.7 MPa).
   - *Scheme 2:* 8-in. (203-mm) normal weight, hollow concrete block walls at 14 ft (4.27 m) on center; one-way continuous concrete slab, 8 in. (203 mm) depth; longitudinal shear walls at end of building. Block strength $f'_m = 1350$ psi (9.3 MPa).

5. *Preliminary design of Scheme 1:*
   a. *Loads:*

$$\text{Dead load: 8-in. hollow-core plank} = 56 \text{ psf}$$
$$\text{8-in. brick wall} = 79 \text{ psf}$$
$$\text{Built-up roof and insulation} = 8 \text{ psf}$$

   b. *Interior wall (transverse forces):*
   Maximum axial dead load:

$$\text{Assume total wall height} = 33 \text{ ft}$$
$$\text{Walls} = 33 \text{ ft}(79 \text{ psf}) = 2607$$
$$\text{Floors} = 56 \text{ psf}(28 \text{ ft})3 = 4704$$
$$\text{Roof} = 64 \text{ psf}(28 \text{ ft}) = \underline{1192}$$
$$8503 \text{ lb/ft}$$
$$(124{,}085 \text{ N/m})$$

   Maximum axial live load:

$$\text{Floors} = 40 \text{ psf}(28 \text{ ft})3 = 3360$$
$$\text{Roof} = 20 \text{ psf}(28 \text{ ft}) = \underline{560}$$
$$3920 \text{ lb/ft}$$
$$(57{,}205 \text{ N/m})$$

   Moment due to unbalanced loading, assuming hinged connections:

$$M = Pe = 14 \text{ ft}(40 \text{ psf})(2/3 \times 4 \text{ in.})$$
$$= 1493 \text{ in.-lb/ft } (553 \text{ Nm/m})$$

   Distribute one-half to each wall at joint and calculate flexural stress:

$$f_b = \frac{\pm Mc}{I} = \pm\frac{0.5(1493)4}{512}$$
$$= \pm 5.8 \text{ psi } (39.9 \text{ kPa})$$

   Calculate axial stress in lower-story wall:

$$f_a = \frac{8503 + 3920}{8(12)} = 129 \text{ psi}(889 \text{ kPa})$$
$$< 0.20 f'_m \text{ (280 psi)}$$

   By inspection, stresses are O.K.
   (For general case of approximate design, use interaction equation: $(f_b/F_b) + (f_a/F_a) < 1.0$ and allow no tension in wall.)

   c. *Exterior wall (transverse forces):* Lower-story wall is O.K. by inspection of interior wall results. Examine top-story wall for wind.
   Calculate loads:

$$\text{Roof dead load} = 64 \text{ psf}(14 \text{ ft}) = 896 \text{ lb/ft}$$

$$\text{Wall load at } 1/2 \text{ height} = 79 \text{ psf}(7.67/2) = 302 \text{ lb/ft}$$

$$1198 \text{ lb/ft } (17{,}482 \text{ N/m})$$

$$\text{Live load} = 20 \text{ psf}(14 \text{ ft}) = 280 \text{ lb/ft } (4086 \text{ N/m})$$

Calculate stresses:

$$\text{Axial dead load stress } (f_a) = \frac{1198}{8(12)} = 12.5 \text{ psi}$$

Assume wind load on wall = 0.8(10) = 8 psf.

$$\text{Wind moment} = \frac{wl^2}{8} = \frac{8(7.67)^2}{8} = 58.8 \text{ ft-lb}$$

$$\text{Wind stress } (f_b) = \frac{Mc}{I} = \frac{(58.8 \times 12)4}{512} = 5.5 \text{ psi}$$

$$\text{Net stress} = 12.5 - 5.5 = 7.0 \text{ psi } (48.2 \text{ kPa})$$

No tension in wall. Wall checks.

d. *Interior shear wall (parallel forces)*:

(1) Calculate maximum wind shear, assuming rigid diaphragm and distribution to shear walls in proportion to their rigidity:

$$V = 13 \text{ psf}(33 \text{ ft})(28 \text{ ft}) = 12{,}012 \text{ lb } (53{,}429 \text{ N})$$

$$\text{Shear stress} = \frac{12012}{8(56 \times 12)} = 2.23 \text{ psi } (15.4 \text{ kPa})$$

$$\text{Allowable stress} = 0.5\sqrt{f'_m} = 18.7 \text{ psi} > 2.33 \quad \text{O.K.}$$

(2) Check stability:

Overturning moment:

$$M_{OT} = 12{,}012 \text{ lb}(33 \text{ ft}/2) = 198{,}198 \text{ ft-lb } (268{,}756 \text{ Nm})$$

Resisting moment:

$$M_R = 8503 \text{ lb/ft}(56 \text{ ft})(56 \text{ ft}/2) = 13{,}332{,}704 \text{ ft-lb } (18{,}079{,}146 \text{ Nm})$$

$$\text{Stability ratio} = \frac{M_{OT}}{M_R} = \frac{13{,}332{,}704}{198{,}198} = 67.2 > 1.5 \quad \text{O.K.}$$

$$I(\text{wall}) = \frac{8/12(56)^3}{12} = 9756 \text{ ft}^4 \ (84.2 \text{ m}^4)$$

(3) Check stresses:

$$\text{Flexural stress} = \frac{Mc}{I} = \frac{198{,}198(28)}{9756} = 568 \text{ lb/ft}^2 = 3.95 \text{ psi } (27.3 \text{ kPa})$$

$$\text{Dead load stress} = \frac{8503}{8(12)} = 88.5 \text{ psi } (610.2 \text{ kPa})$$

By inspection, flexural stress is O.K.

e. *Longitudinal shear wall (Fig. 6–28):* The two walls are of different rigidities, which means the center of rigidity will not occur at the center of the building. Assume that rigidities are proportional to I values for each wall.

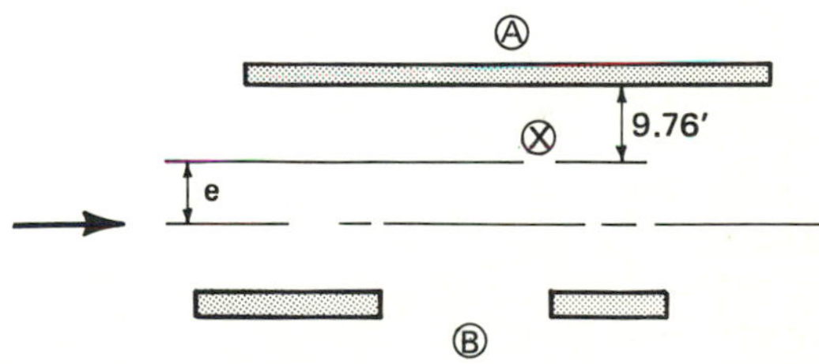

Figure 6–28. Longitudinal Shear Wall: Example Problem 6–4

$$I(\text{wall } A) = \frac{8/12(20)^3}{12} = 444 \text{ ft}^4 \ (3.83 \text{ m}^4)$$

$$I(\text{wall } B) = \frac{8/12(10)^3}{12} + \frac{8/12(8)^3}{12} = 83 \text{ ft}^4 \ (0.72 \text{ m}^4)$$

$$\text{Center of rigidity} = 444(\bar{y}) - 83(62 - \bar{y}) = 0$$

$$\bar{y} = 9.76 \text{ ft}$$

$$\text{Torsional wind moment} = P \cdot e = 13 \text{ psf}(62 \text{ ft})(33 \text{ ft})(31 - 9.76) = 564{,}941 \text{ ft-lb } (77{,}211 \text{ Nm})$$

Forces due to torsional moment are distributed to walls in proportion to their relative rigidities and their distance from the center of rigidity.

Table 6–3 Summary of Costs for Example Problem 6–4

Alternate	*Floor*	*Wall*	*Footing*	*Diff. cost to other systems*[a]	*Total*
Scheme 1[b]					
Quantity	1.0 sf/sf	0.29 sf/sf	0.0056 cy/sf		
Cost	\$3.00/sf	\$1.66/sf	\$0.64/sf	+\$1.36	\$6.66
Scheme 2[c]					
Quantity	1.0 sf/sf	0.59 sf/sf	0.008 cy/sf		
Cost	\$4.93/sf	\$0.93/sf	\$0.93/sf		\$6.79

[a] Cost of intermediate stud walls.

[b] Exposed brick assumed.

[c] Painted block assumed.

$$F_B = F_A\left(\frac{83}{444}\right)\frac{52.24}{9.76} = 1.0(F_A)$$

$$F_A(9.76) + F_A(52.24) = 564{,}941$$

$$F_A = 9111 \text{ lb } (40{,}525 \text{ N}) \quad \text{(torsional shear)}$$

$$\text{Direct shear in wall } B = \frac{83}{444+83}(13 \text{ psf})(33)(62) = 4189 \text{ lb } (18{,}632 \text{ N})$$

$$\text{Total shear in } B = 4189 + 9111 = 13{,}300 \text{ lb } (5915 \text{ N})$$

$$\text{Shear stress} = \frac{13{,}300}{8(16 \times 12)} = 8.65 \text{ psi } (59.3 \text{ kPa}) < 18.7 \quad \text{O.K.}$$

f. *Footing:*

Maximum wall load (without LL reduction):

$$P = 8503 + 3920 = 12{,}423 \text{ lb/ft } (181{,}291 \text{ N/m})$$

$$\text{Width of footing} = \frac{12{,}423}{3000} = 4.14 \text{ ft } (1.26 \text{ m})$$

Use 4 ft 3 in. wide × 12 in. deep (1.30 m × 305 mm).

$$\frac{4.25(62)(12/12)}{27(28)62} = 0.0056 \text{ cy/sq ft}$$

6. *Preliminary design of scheme 2 is accomplished in a similar manner.*
7. *Tabulation and selection.* Table 6–3 lists the quantities and unit costs for both schemes. Scheme 2 is the more economical.

Example Problem 6–4 introduced the concept of differential cost to other systems (other than structural) for the first time. This step is an example of optimization of the total project, not just the structural system. The complete development of this approach is contained in Chapter 7.

6–5 SHELLS

Shells are curved or folded slabs whose thicknesses are small compared with their other dimensions. They may be visualized as containing arch, cable, and ring elements or as a series of finite elements (Fig. 6–29). Shells gain their strength by virtue of the three-dimensional development of their surfaces, with a resulting ability to carry loads primarily through in-plane stresses rather than through bending. Certain types of shells may not constitute complete structural systems by themselves, but we will classify these as systems because of the three-dimensional nature of their geometry. Other shells may provide complete enclosure and structural support.

The variety of curved shells is such that classification by *type of curvature* and *method of generation* helps establish identity. *Synclastic* shells have their curvature

Figure 6–29. Shell Systems

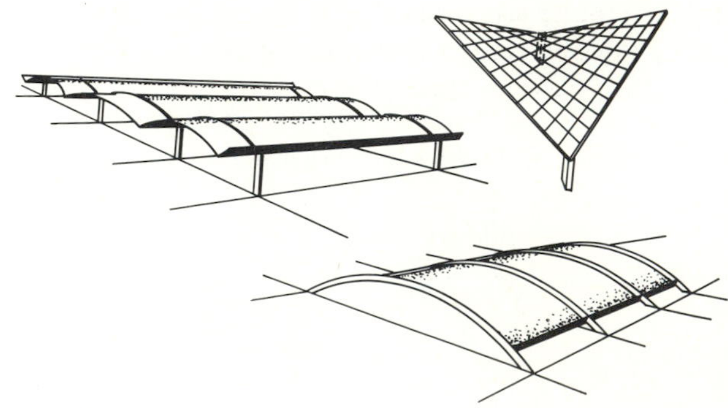

(at a point) of the same sign in all directions (Fig. 6–30), also called positive double curvature. For example, a dome is a synclastic surface. *Developable* shells have their curvature (at a point) of the same sign in all but one direction, in which it is zero, also called singly curved surfaces. A cylinder is a developable surface (Fig. 6–31).

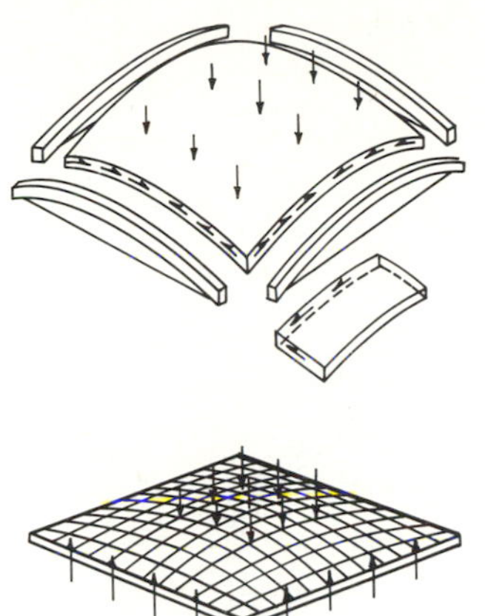

Figure 6–30. Synclastic Shells

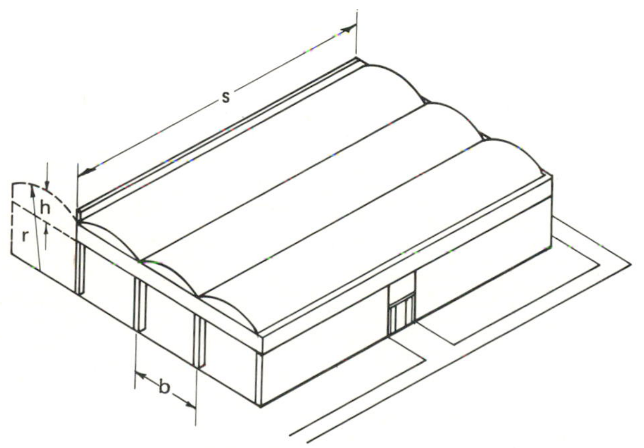

Figure 6–31. Developable Shells

These surfaces can be flattened into a plane without stretching or shrinking. *Anticlastic* shells have their curvature (at a point) positive in certain directions and negative in others, also called negative double curvature. A saddle surface results from this curvature (Fig. 6–32).

Two basic methods for generating shell curvature

Figure 6–32. Anticlastic Shells

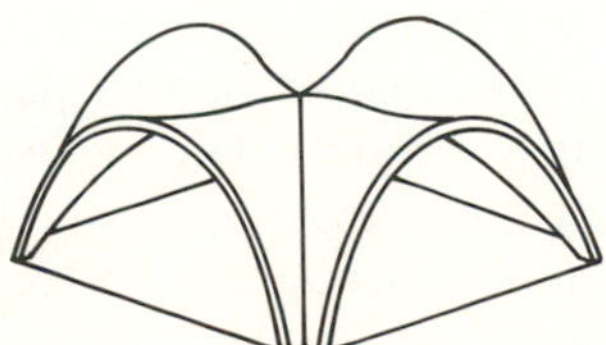

are rotation and translation. Surfaces of *rotation* are formed by revolving a plane curve about an axis in the plane (Fig. 6–33). Surfaces of *translation* are formed by moving a plane curve along some other plane curve (Fig. 6–34).

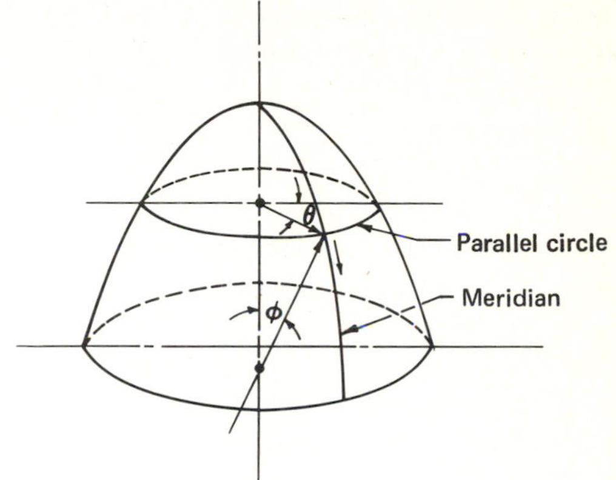

Figure 6–33. Rotational Shells

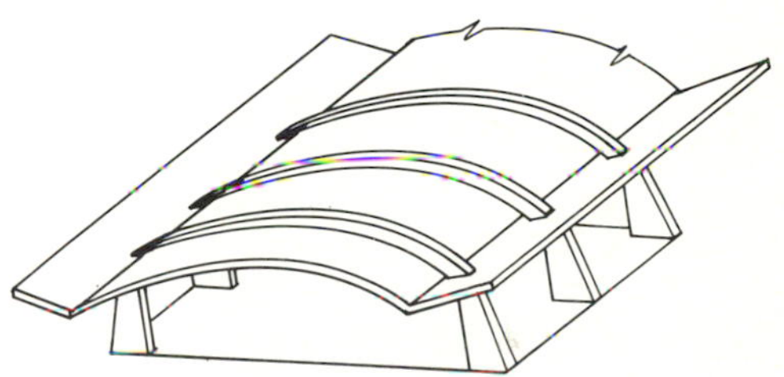

Figure 6–34. Translational Shells

History and Development

Shells occur in nature in a variety of forms, including eggshells, shells of nuts, and plants and leaves of various types. All of these forms represent nature's attempt at efficient use of material. The first shell constructed by people was probably the simple hut built with light timbers and mud. Low circular walls of stone, which may have been covered with domes of mud or mudbrick, have survived since 5000 B.C. and stone was being used for domes by 1500 B.C.

The earliest surviving concrete domes date back to 100 B.C. in Rome, and shell construction (domes) was extensively used in the reconstruction of Rome between A.D. 64 and the completion of the Pantheon in A.D. 128. The thickness of the crown of these early domes was usually between $\frac{1}{10}$ and $\frac{1}{15}$ of the radius, which was not much less than that of the masonry arch of earlier periods. Better understanding of the inherent strength possessed by the double curvature surface might have permitted a reduction of these thicknesses to about $\frac{1}{200}$ of the radius, but this understanding awaited later development.

Shell construction continued during the Byzantine period (A.D. 395–1453), and the first theory of shells was proposed by Aron in 1874. The first shell to be designed and built with the knowledge of structural mechanics was the roof of an experimental planetarium constructed in Jena (Germany) in the early 1920s. Most of the basic possibilities for using reinforced concrete to construct shell forms were explored by the mid-1930s.

Types and Uses

Domes are synclastic surfaces formed by the rotation of a circular, elliptic, or parabolic curve about a central vertical axis (Fig. 6–35). The dome form was the earliest type of shell and continues to be popular for large roof structures. The ribbed dome and Schwedler dome (Fig. 6–36) are used in addition to the continuous concrete type. Half-arches are the principal elements of the ribbed dome, often connected at the crown by a compression ring. The Schwedler dome contains half-arches and a series of horizontal rings with diameter increasing with distance from the crown.

Figure 6–35. Domes

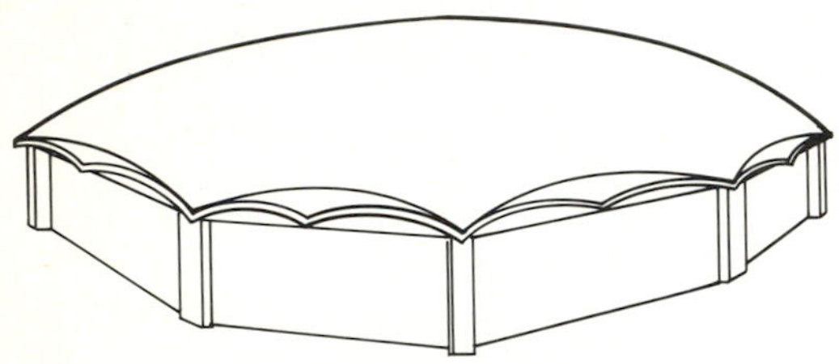

Figure 6–36. Types of Domes

(a) Radial rib-type dome

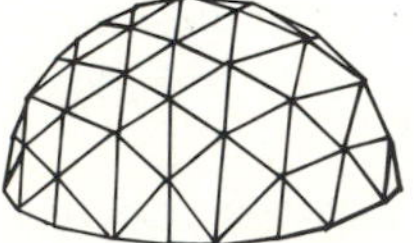

(b) Geodesic dome

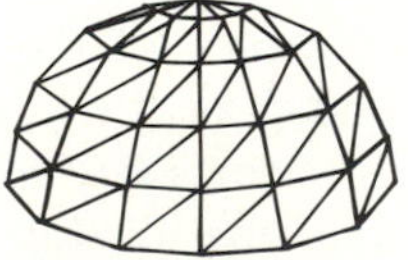

(c) Schwedler dome

(d) Zeiss-Dywidag dome

Cylindrical shells are developable surfaces formed by the translation of a circular curve along a straight line at the vertex (Fig. 6–37). The barrel shell is another name for a commonly used cylindrical shell, often used for industrial and commercial building roofs for spans up to 300 ft (91.4 m).

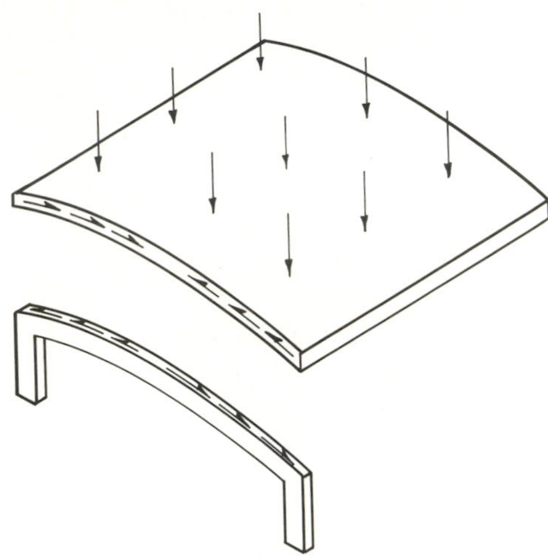

Figure 6–37. Cylindrical Shells

Hyperbolic paraboloids are anticlastic shells (negative double curvature) which are formed by the translation of a parabolic curve with downward curvature on another parabola with upward curvature. The resulting surface is saddle-shaped and is used for roof spans of 60 to 100 ft (18.3 to 30.4 m). Several arrangements of the hyperbolic paraboloid are possible, including the inverted umbrella, the saddle, the umbrella, and the hyperbolic paraboloid dome (Fig. 6–38).

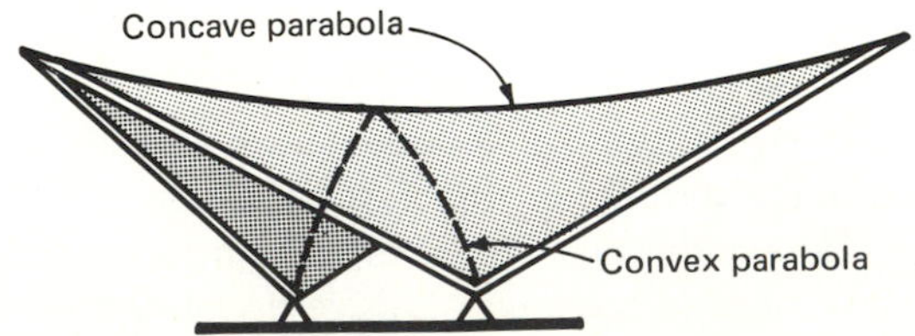

Figure 6–38. Hyperbolic Paraboloid Shells

Folded plates are planar structural slabs inclined to each other and connected along their longitudinal edges. Loads are distributed in two directions: in the transverse direction with the plates acting as slabs spanning between adjacent plates, and in the longitudinal direction with each plate acting as a girder. Typical folded plate arrangements include the V-type, the three-plate arrangement, the butterfly, and the Z-type (Fig. 6–39).

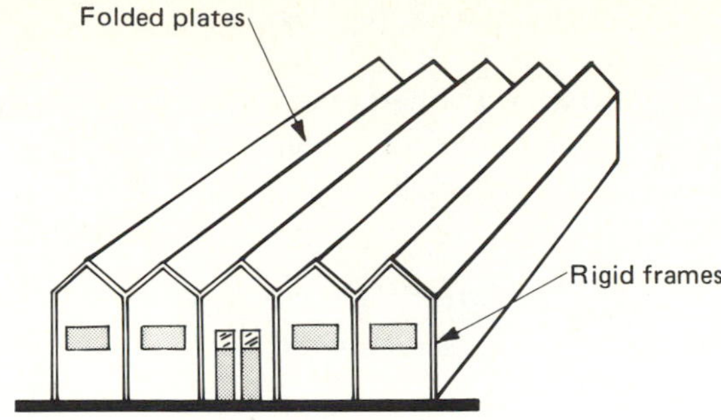

Figure 6–39. Folded Plates

Behavior

Shells achieve their strength from the curvature, which allows them to carry loads primarily in compression, tension, and tangential shear (Fig. 6–40). Termed *membrane* stresses, since they act in the plane of the shell, the tension and compression forces act much like a two-way cable or two-way arch system, while the shears carry load by a mechanism that has no counterpart in either cable or arch behavior. Figure 6–40 shows a rectangular element cut out of a curved membrane, with four sides that are not parallel, but askew in space. The resulting geometrical twist provides an excess of upward force and gives the membrane its load-carrying capacity by shearing action within its own surface. A shell under any smooth load will develop only membrane stresses if support reactions act in the plane tangent to the shell at the boundary, and the displacements at the boundary (due to the strains caused by the membrane stresses) are not restrained.

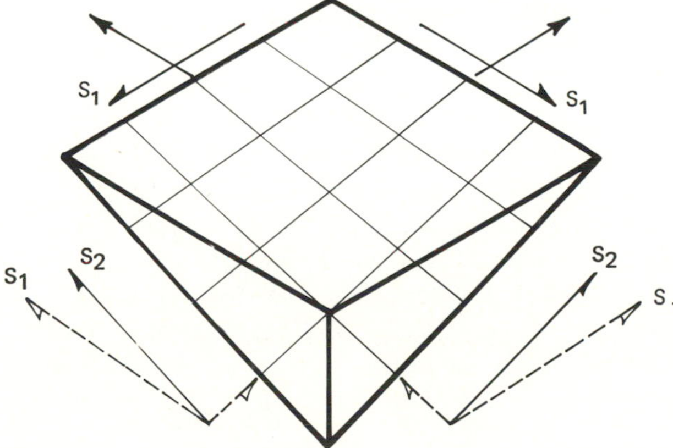

Figure 6–40. Shell Stress

Ideal membrane conditions are seldom realized, and bending forces are consequently introduced in the neighborhood of the shell boundary (Fig. 6–41) when displacements are applied which cancel the free edge displacements of the pure membrane. These bending forces are called *boundary disturbances.* Actual shell stress behavior is a combination of membrane action and bending forces at the boundaries.

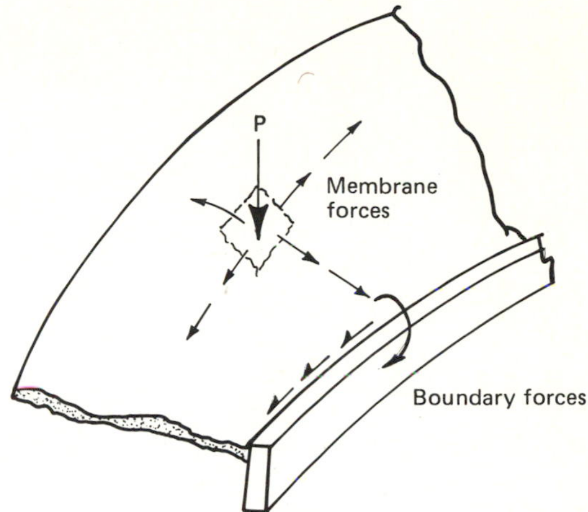

Figure 6–41. Boundary Bending Forces

Since shells are made up of thin compression elements, *buckling instability* can occur in the form of either asymmetrical buckling or snap-through buckling (Fig. 6–42). Generally, buckling is not a problem for most civil engineering types of shells.

Figure 6–42. Shell Buckling

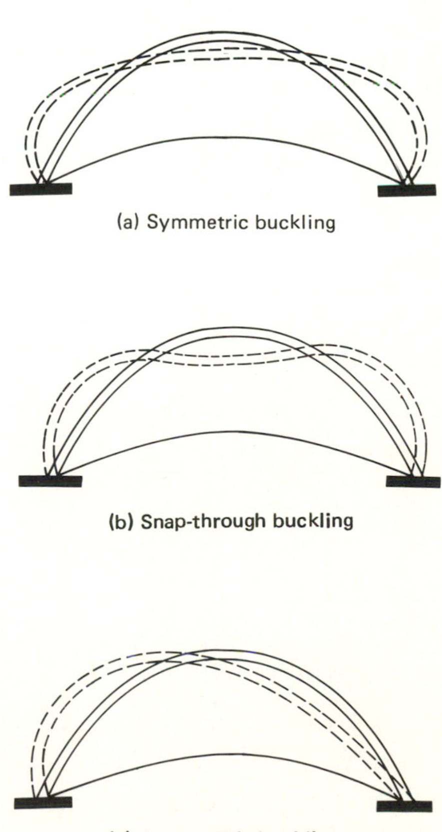

Approximate Analysis

The analysis of shells involves determination of both membrane stresses and bending forces at the boundaries. Approximate formulas for the major types of shells are presented in this section.

Figure 6–43. Dome Shell Forces

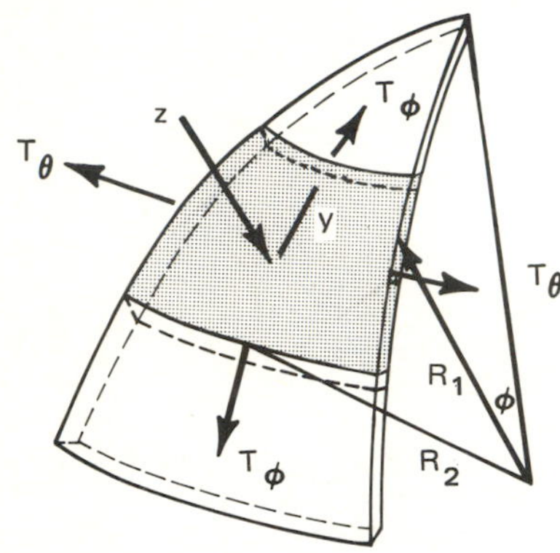

Domes. Membrane Stresses. If a shell of rotation (circular dome) is loaded by its own weight or by a symmetrical snow load, no shear stress will be developed in any meridional section because of symmetry, so that a piece of shell cut by two adjoining meridians and two adjoining parallels will be maintained in equilibrium by only two internal forces, the meridional force T_ϕ and the parallel or hoop stress T_θ. In Fig. 6–43 the radius of curvature of the meridian is R_1, the radius of the parallel is $R_2 \sin \phi$, and the normal component of the vertical load is Z. Equilibrium of the shell in the direction perpendicular to its surface requires that

$$\frac{T_\phi}{R_1} + \frac{T_\theta}{R_2} + Z = 0 \qquad (6\text{–}2)$$

For the case of spherical domes, where $R_1 = R_2 = R$ (radius of the sphere), this equation reduces to

$$T_\phi + T_\theta + RZ = 0 \qquad (6\text{–}3)$$

Also,

$$T_\phi = \frac{-W}{2\pi R_2 \sin^2 \phi} \qquad (6\text{–}4)$$

where W = resultant of all loads from top of shell to the parallel considered

$W = 2\pi R^2 \omega(1 - \cos \phi)$ for uniform load per square foot of dome

$W = 2\pi R^2 \omega'(\sin \phi - \phi \cos \phi)$ for load increasing form 0 at crown to ω'

Once T_ϕ is obtained from this last equation, T_θ can be derived from Eq. 6–2.

To understand the nature of dome stresses, we can examine these equations for a half-sphere whose weight per unit area is ω. From Eq. 6–4,

$$T_\phi = \frac{-2\pi\omega R^2(1 - \cos \phi)}{2\pi R \sin^2 \phi} \qquad (6\text{–}5)$$

$$= -\omega R \text{ at the boundary}$$

$$(\text{where } \phi = 90°, \sin \phi = 1)$$

The meridional stress is *compressive* since it is negative in sign. From Eq. 6–3, since the Z component of the weight ω equals $\omega \cos \phi$,

$$T_\theta = -RZ - T_\phi$$

$$= -R\omega \cos \phi - \left[\frac{-R\omega(1 - \cos \phi)}{\sin^2 \phi}\right] \qquad (6\text{–}6)$$

$$= \omega R \text{ at the boundary}$$

$$(\text{where } \phi = 90°, \cos \phi = 0)$$

The hoop stress T_θ is *tensile* at the boundary, since it is positive, and reinforcement will have to be provided at the boundary.

If the same stress analysis is carried out for a parallel at an angle ϕ, it will be found that the meridional stress T_ϕ is always compressive while the hoop stress is compressive up to an angle equal to 52° from the top, and is tensile below that angle. This indicates that spherical shells will develop only compressive stresses if their angle is less than 52°, while tensile hoop stresses will appear when the angle is over 52°.

Similar results can be established for the case of a uniform snow load on a semispherical shell. The meridional stress is always compressive and equal $-\omega R/2$ at the boundary. The hoop stress is compressive from the top down to 45°, and becomes tensile from then on, reaching a maximum value equal to $+\omega R/2$ at the boundary.

If the shell is terminated so that ϕ is smaller than 90°, the meridional thrust will have an outward component that must be resisted by a ring girder. The ring tension can be calculated by the following equation:

$$P_T = \frac{\omega \cos \phi}{2\pi \sin \phi} \qquad (6\text{–}7)$$

Boundary stresses. The bending moment and shear present at the boundary can be calculated from the following approximate equations for spherical shells with completely fixed boundaries.

For uniform live load:

$$M = 0.145qRh \text{ (in.-lb/in.) (Nm/m)}$$

$$V = -0.38\sqrt{\frac{h}{R}}\, qR \qquad (6\text{–}8)$$

For dead load:

$$M = 0.29\omega Rh \qquad (6\text{–}9)$$

$$V = -0.76\sqrt{\frac{h}{R}}\, \omega R$$

where q = uniform live load, psf hor. proj. (N/m²)
h = thickness of shell, ft (m)
ω = dead load, psf shell surface (N/m²)
R = shell radius, ft (m)

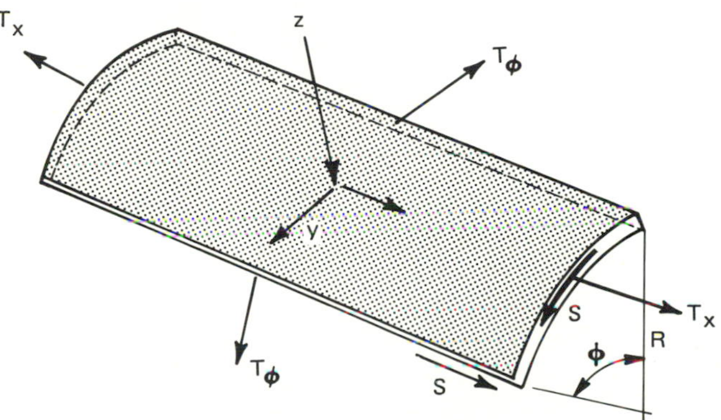

Figure 6–44. Cylindrical Shell Forces

Cylindrical Shells. Membrane stresses. A small element cut out of a cylindrical shell has sides parallel and perpendicular to the cylinder axis, and the load applied to the element can be split into two components: Z perpendicular to the shell surface and Y tangent to the shell cross section (Fig. 6–44). The component Z must be balanced by the stress T_ϕ in the transverse direction, analogous to the equilibrium of a cable under its own weight, and the equation giving the stress T_ϕ per unit length is

$$T_\phi = -ZR \qquad (6\text{–}10)$$

where R is the radius of the cylinder at the point considered. If Z acts into the cylinder, as is the case for dead and live loads, the stress T_ϕ will always be compressive (conventionally given a negative sign), whereas it will be tensile (positive sign) for loads directed outward.

The other two equations of equilibrium that determine T_x and S are not as simple to derive. The equations governing circular cylinders under uniform snow load, q (psf horizontal projection), are given below. These values may be used to approximate dead load stresses as well.

$$T_\phi = -qR \cos^2 \phi \qquad (6\text{–}11)$$

$$T_x = -\frac{3}{2} qR \left(\frac{L}{R}\right)^2 \left[\frac{1}{4} - \left(\frac{x}{L}\right)^2\right] \cos 2\phi$$

$$S = -\frac{3}{2} qR \left(\frac{x}{R}\right) \sin 2\phi$$

$$\text{Max. } T_\phi = -qR$$

$$\text{Max. } T_x = \pm\frac{3}{8} qR \left(\frac{L}{R}\right)^2$$

$$\text{Max. } S = \pm\frac{3}{2} qL$$

As can be seen from the formulas, the maximum value of the longitudinal stress, T_x, occurs at the middle of the shell and is compressive at the top and tensile at the edge. The transverse stress T_ϕ is independent of the location along the axis, is maximum at the top and zero at the edge, and is always compressive. The shear S is maximum at the shell ends at an angle of 45° measured from the top. Figure 6–45 illustrates these patterns.

Figure 6–45. Cylindrical Shell Stress Diagrams

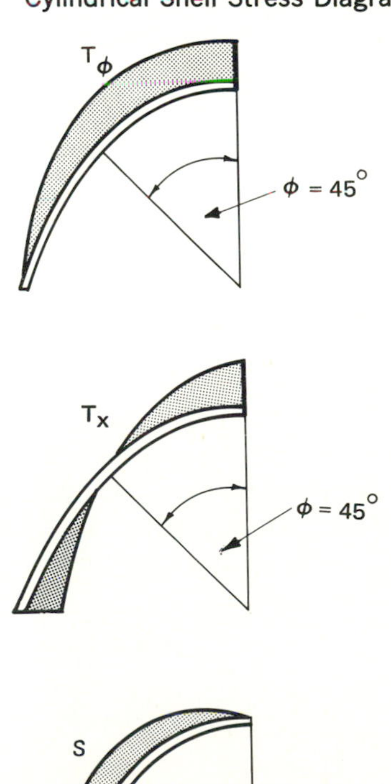

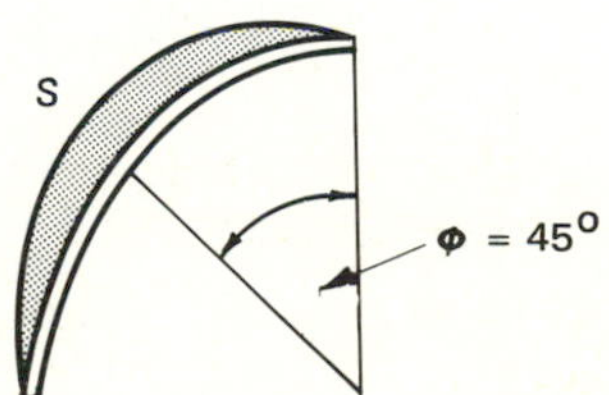

Beam method approximation. By considering the shell as a simple beam of semi-circular cross section, under snow load q, maximum longitudinal stresses and the shear can be approximately determined by simple beam theory:

$$T_x = \frac{My}{I} \quad (6\text{–}12)$$

$$S = \frac{VQ}{I} \quad (6\text{–}13)$$

where M = simple beam bending moment
$= \pi\omega RL^2/8$ for a uniform load per unit area
$= qL^2/8$ for a uniform load on the horizontal projection
I = moment of inertia of curved cross-section
y = distance from neutral axis to extreme fiber
V = shear force
Q = static moment of the section above the fiber where shear force is desired

Values for I, y, and Q may be calculated from the following equations (Fig. 6–44):

$$I = R^3 t\left(\alpha + \sin\alpha\cos\alpha - \frac{2\sin^2\alpha}{\alpha}\right)$$

$$Y_1 = R\left(1 - \frac{\sin\alpha}{\alpha}\right), \qquad Y_2 = R\left(\frac{\sin\alpha}{\alpha} - \cos\alpha\right)$$

$$Q_1 = 2\beta R^2 t\left(\frac{\sin\beta}{\beta} - \cos\beta\right)$$

$$\cos\beta = \frac{R - Y_1}{R}$$

With this method, the stresses T_x and S obtained at the top of the shell and at its edge are identical with the values given by Eq. 6–11. Once these stresses are calculated, the T_ϕ stress may be obtained by Eq. 6–10 and a rough evaluation of the required shell thickness and reinforcement can be obtained very quickly. The beam behavior holds for other loads and support conditions provided that the shell length is more than five times its radius.

Boundary stresses. Boundary bending moment for cylindrical shells is given by Eq. 6–9 for shells fixed into stiffeners between two shells. When the shell is hinged at the stiffener, the following equations apply:

$$M_{\max} = 0.092qRh \text{ (ft-lb/ft)}$$
$$V = 0.38\sqrt{\frac{h}{R}}\, qR \text{ (lb/ft)} \quad (6\text{–}14)$$

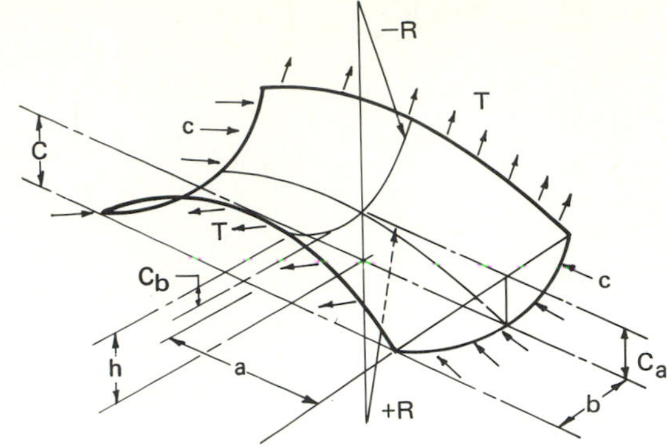

Figure 6–46. Hyperbolic Paraboloid Forces

Hyperbolic Paraboloids. *Membrane stresses.* The membrane stresses (Fig. 6–46) in a shallow rectangular hyperbolic paraboloid, of rises $c_1 = ah/2b$ and $c_2 = -bh/2a$, small in comparison with a and b, are determined by the following equations:

$$\omega = \frac{C}{R} + \frac{-C}{-R} = \frac{2C}{R}, \qquad C = \frac{\omega R}{2} = -T \quad (6\text{–}15)$$

The radius of curvature of the parabolas at the saddle point equals

$$R \approx \frac{1}{d^2 z_a/dx^2}\bigg]_{x=0} = \frac{ab}{h} \quad (6\text{–}16)$$

and the membrane forces in the x and y direction are

$$C = \omega\frac{ab}{2h}; \qquad T = -\omega\frac{ab}{2h} \quad (6\text{–}17)$$

The equations show that the load on a shallow hyperbolic paraboloid is supported by funicular arch action in the x direction and by funicular cable action in the y-direction through membrane stresses which are constant over the entire shell surface. The membrane stresses developed by the shell, which are also constant through its thickness, have a constant value at all points of the shell and the material is ideally utilized.

The equal compression and tension in the x and y directions are equivalent to identical shears at 45° to the x and y axes. The membrane shears (S) are given by

$$S = \pm\omega\frac{ab}{2h} \quad (6\text{–}18)$$

Boundary stresses. Hyperbolic paraboloids are typically supported along straight lines at 45° to x and y, along

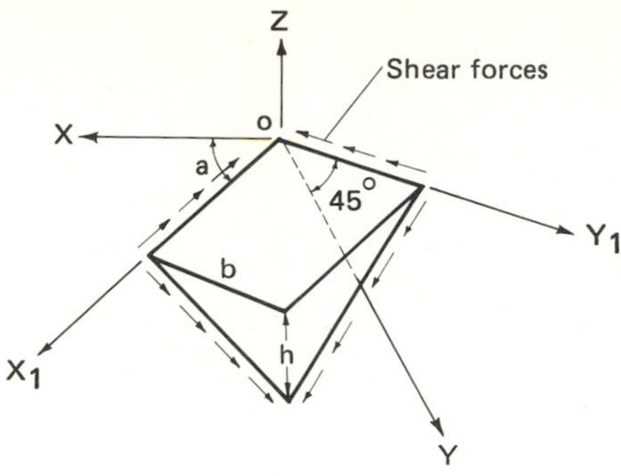

Figure 6–47. Hyperbolic Paraboloid: Boundary Forces

which the unit forces are the pure shears S. The shears on the boundary of an element of paraboloid with sides a and b at 45° to the x and y axes are shown in Fig. 6–47. The shears are provided by boundary beams, which are acted upon by shears equal and opposite to those at the shell boundary. The beams are compressed by the accumulation of shears from O to A and O to B, and from C to A and C to B. The maximum compressive forces in the beams OA, CB, and OB, CA are given, respectively, by

$$F_1 = C\sqrt{a^2 + h^2} = \omega \frac{ab}{2h}\sqrt{a^2 + h^2};$$
$$F_2 = \omega \frac{ab}{2h}\sqrt{b^2 + h^2} \tag{6–19}$$

Bending stresses in a hyperbolic paraboloid cannot be evaluated by cylindrical theory for lack of curvature along the supported boundary. However, the bending stresses in the distributed area near the stiffeners are generally of the same order of magnitude as the membrane shear stresses.

Folded Plates. Folded plate action is a combination of transverse and longitudinal beam action. The load is first transferred to the folds by beam action of the slabs in the transverse direction [Fig. 6–48(a)] and then to the end frames by longitudinal beam action of the slabs [Fig. 6–48(b)]. The analysis of folded plates can be simplified by first considering the plates to act as continuous slabs in the transverse direction, with supports at each fold line [Fig. 6–48(a)]. Approximate *transverse moments* are then

$$M_T = \pm \frac{\omega a h}{10} \tag{6–20}$$

where M_T = transverse bending moment, ft-lb (Nm)
h = transverse depth of slab on incline, ft (m)
ω = uniform load psf of surface area (N/m²)
a = horizontal projection of h

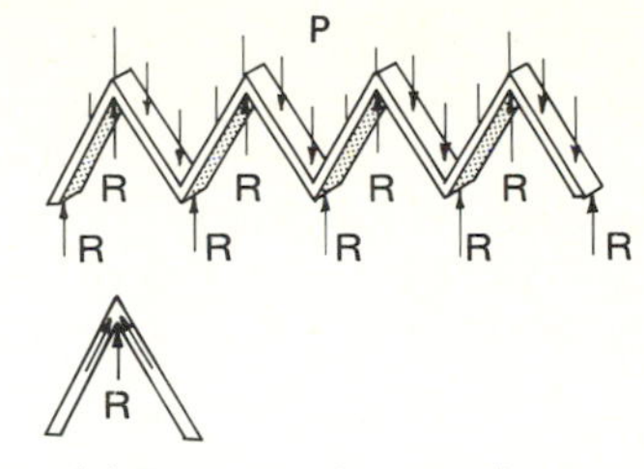

(a) Transverse beam action

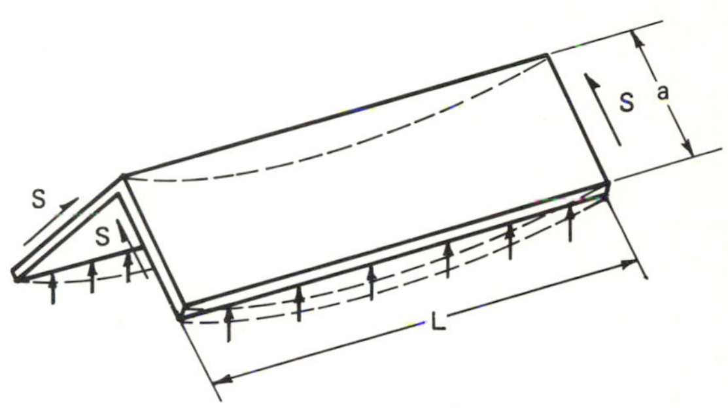

(b) Longitudinal beam action

Figure 6–48. Folded Plate Forces

Near boundaries, this transverse moment is increased by a factor of approximately 5 for very flexible edge members to 1.0 for very stiff edge members.

Longitudinal moments can be initially calculated by uncoupling the plates, resolving the reactions at the folds into components acting in the plane of the plates and calculating their flexural stresses according to traditional beam theory.

For a typical interior plate in a series of equal spacing and slopes,

$$M_L = \frac{\omega h \sin\phi\, L^2}{8} \tag{6–21}$$

where M_L = longitudinal bending moment, ft-lb (Nm)
h = transverse depth of slab, ft (m)
ϕ = slope angle
L = span of slab, ft (m)

For unequal spacing or varying slopes, the vertical load must be resolved into components using other trigonometric relationships.

For typical interior plates, this procedure will approximate the longitudinal bending stress. Near boundaries, however, adjoining plates will have unequal stresses at common joints when this method is used. Unbalanced joint stresses can be balanced by assuming a distribution factor for each plate. If the plates at a joint are of constant section throughout, the unbalanced stress should be distributed in proportion to the reciprocal of the plate areas as follows:

$$\text{Stress stiffness factor} = \frac{1}{ht}$$

where h = transverse depth of slab, ft (m)
t = slab thickness, ft (m)

A carryover factor of $-\frac{1}{2}$ may be used for distribution to the adjoining edge of each plate.

After the bending stresses have been adjusted by distribution, the shears may be computed from

$$T_n = T_{n-1} - \frac{f_{n-1} + f_n}{2} A_n \qquad (6\text{–}22)$$

where T_n = shear at joint n, kips (N)
f_n = bending stress at joint n, ksi (MPa)
A_n = cross-section area of plate n, sq in. (mm^2 $\times 10^{-6}$)

The shear of a boundary edge is usually zero and provides a starting point for the application of Eq. 6–22. For a simply supported, uniformly loaded, folded plate, the shear stress f_v at any point on an edge n is approximately

$$f_v = \frac{T_{\max}}{1.5Lt}\left(\frac{1}{2} - \frac{x}{L}\right) \qquad (6\text{–}23)$$

where f_v = shear stress, ksi (N/mm^2)
x = distance from a support, ft (m)
t = web thickness of plate, in. (mm)
L = longitudinal span of plate, in. (mm)

Approximate Design

Design of shell systems includes the following considerations:

- Material selection
- Type of shell surface
- Shell geometry
- Boundary conditions
- Shell thickness
- Element proportions
- Reinforcement

Approximate design guidelines are readily available for most standard shell types (Appendix 6A). Irregular shells of special curvature should be studied first with simple models (e.g., soap membrane, suspended fabric, etc.) to develop a feeling for the optimum shell surface. This preliminary study can then be followed by more sophisticated model tests and finite-element analysis techniques.

Minimum membrane reinforcement. Bars are required in each direction with a minimum area of reinforcement per unit width equal to specified temperature and distribution steel. Minimum area of reinforcement in one direction is 0.20%, and the sum of areas of reinforcement in two directions should not be less than 0.60%. Two layers of steel are required when the slab thickness exceeds 5.5 in. (140 mm).

Maximum membrane reinforcement. Maximum reinforcement, percentage ρ, can be calculated as follows:

$$\rho = \frac{A_s}{A_c} = 0.6\frac{f_c'}{f_y} \qquad \text{for } f_c' < 4000 \text{ psi (28 MPa)} \qquad (6\text{–}24)$$

$$\rho = \frac{16.8}{f_y} \qquad \text{for } f_c' \geq 4000 \text{ psi (28 MPa)}$$

The following examples illustrate the approximate design of several standard shell types.

EXAMPLE PROBLEM 6–5: Preliminary Design of Hyperbolic Paraboloid

A proposed concrete roof consists of four hyperbolic paraboloid sections (Fig. 6–49) with a total span of 100 ft × 80 ft (30.5 m × 24.4 m). The shell is supported at the four corners and carries a snow load of 40 psf (1915 N/m^2). Determine the shell thickness and the required dimensions of the boundary beams for an allowable concrete compressive stress of 1.0 ksi (6.89 MPa).

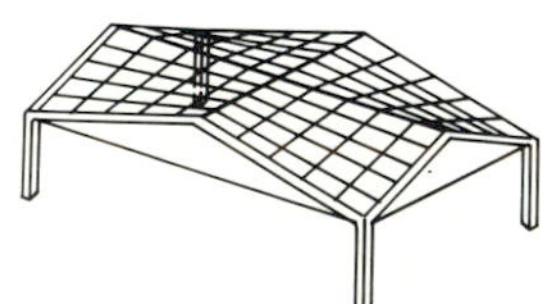

Figure 6–49. Example Problem 6–5

Solution

1. *Assume shell thickness and rise.* From Appendix 6A, try t = 3 in. (76.2 mm).

$$h = \frac{0.003\ ab}{t} = \frac{0.003(50)(40)}{(3/12)} = 24 \text{ ft (7.3 m)}$$

2. *Determine loads.*

Snow load = 40 psf (1915/Nm^2)

Dead load = 150(3/12) = 50 psf (2394 N/m^2)

3. *Calculate membrane forces.* By Eq. 6–15,

$$C = -T = \frac{\omega ab}{2h}, \qquad a = 50 \text{ ft}, \quad b = 50 \text{ ft}.$$

$$C = \frac{90(50)(40)}{2(24)} = 3750 \text{ lb/ft (54,724 N/m)}$$

4. *Check stresses.*
   a. Compression:

$$f_c = \frac{C}{A} = \frac{3750}{3(12)}$$
$$= 104 \text{ psi } (0.72 \text{ MPa}) < 1000 \quad \text{O.K.}$$

   b. Tension

$$A_s = \frac{T}{f_s} = \frac{3{,}750}{20{,}000} = 0.19 \text{ in.}^2/\text{ft } (402 \text{ mm}^2/\text{m})$$

$$\rho = \frac{A_s}{A_c} = \frac{0.19}{3(12)} = 0.005$$

$$\text{Max. } \rho = \frac{0.6(3000)}{60{,}000} = 0.03 \quad \text{O.K.}$$

Use 0.19 in.² in the tension direction. Use minimum in the other direction = 0.002(36) = 0.07 in.²/ft (148 mm²/m).

5. *Determine edge beam size:*

$$F_1 = C\sqrt{a^2 + h^2}$$
$$= 3750\sqrt{(50)^2 + (24)^2} = 208 \text{ kips } (925{,}184 \text{ N})$$
$$F_2 = 3750\sqrt{(40)^2 + (24)^2}$$
$$= 175 \text{ kips } (778{,}400 \text{ N})$$
$$\text{Maximum area req'd} = \frac{208}{1.0}$$
$$= 208 \text{ in.}^2 \ (134{,}160 \text{ mm}^2)$$

Use a 12 in. × 18 in. beam (304 mm × 457 mm).

EXAMPLE PROBLEM 6–6: Preliminary Dome Design

Select a preliminary size for a spherical dome to span 200 ft (60.9 m) and support a live load of 30 psf (1426 N/m²). Assume that the maximum allowable concrete stress is 200 psi (1.4 MPa). See Fig. 6–50.

Figure 6–50. Example Problem 6–6

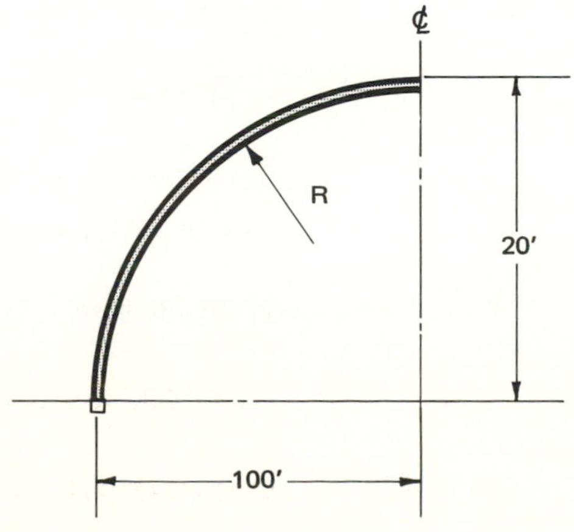

Solution

1. *Assume shell thickness and rise.* From Appendix 6A, try t = 4 in. (102 mm).

$$h = \frac{200 \text{ ft}}{10} = 20 \text{ ft } (6.1 \text{ m})$$

$$R^2 = (100)^2 + (R - 20)^2, \qquad R = 260 \text{ ft } (79.2 \text{ m})$$

2. *Determine loads.*

$$\text{Live load} = 30 \text{ psf } (1436 \text{ N/m}^2)$$
$$\text{Dead load} = (4/12)150 = 50 \text{ psf } 2394 \text{ N/m}^2)$$
$$\omega = 50 + 30 = 80 \text{ psf } (3830 \text{ N/m}^2)$$

3. *Calculate membrane stress.*

$$\sin\phi = \frac{100}{260} = 0.385, \qquad \sin^2\phi = 0.148$$
$$\phi = 22.6°, \qquad \cos\phi = 0.923$$

By Eq. 6–5,

$$T_\phi = -2\pi\omega R^2(1 - \cos\phi)/2\pi R\sin^2\phi$$
$$T_\phi = \frac{-2(3.14)(.080)(260)^2(1 - 0.923)}{2(3.14)(260)(0.148)}$$
$$= -10.8 \text{ kips/ft (compressive)}$$
$$f_c = \frac{10{,}800}{12(4)}$$
$$= 225 \text{ psi } (1.55 \text{ MPa}) > 200 \text{ psi } (1.4 \text{ MPa})$$

By Eq. 6–6,

$$T_\theta = -R\omega\cos\phi - [-R\omega(1 - \cos\phi)/\sin^2\phi]$$
$$T_\theta = -(260)(0.080)(0.923) - (-10.8)$$
$$= -8.4 \text{ kips/ft } (122{,}582 \text{ N/m})\text{(compressive)}$$

4. *Calculate boundary stresses.* By Eq. 6–9,

$$M = 0.29\ \omega\text{Rh}$$
$$M = 0.29(80)(260)(4/12)$$
$$= 1508 \text{ in.-lb/in. } (6.7 \text{ Nm/mm})$$
$$f_b = \frac{Mc}{I} = \frac{1508(2)}{5.33} = 565 \text{ psi } (3.9 \text{ MPa})$$

Reinforcement can easily handle.

5. *Calculate ring tension.* By Eq. 6–7:

$$W = 2\pi R^2\omega(1 - \cos\phi)$$
$$= 2(3.14)(260)^2(0.08)(1 - 0.923)$$
$$= 2615 \text{ kips } (11{,}631{,}520 \text{ N})$$

$$P_T = \frac{2615(0.923)}{2(3.14)(0.385)}$$

$$= 998 \text{ kips } (4{,}439{,}104 \text{ N})$$

$$A_s = \frac{998 \text{ kips}}{20 \text{ ksi}} = 49.9 \text{ in.}^2 \ (32{,}185 \text{ mm}^2)$$

6. *Summary.* Stress is slightly high in step 3. Assume $4\frac{1}{2}$ in. (114 mm) shell thickness for preliminary design.

EXAMPLE PROBLEM 6–7: Preliminary Cylindrical Shell Design

Select a preliminary cylindrical shell (barrel shell) to span 100 ft (30.5 m) under a snow load of 30 psf (1436 N/m²). See Fig. 6–51.

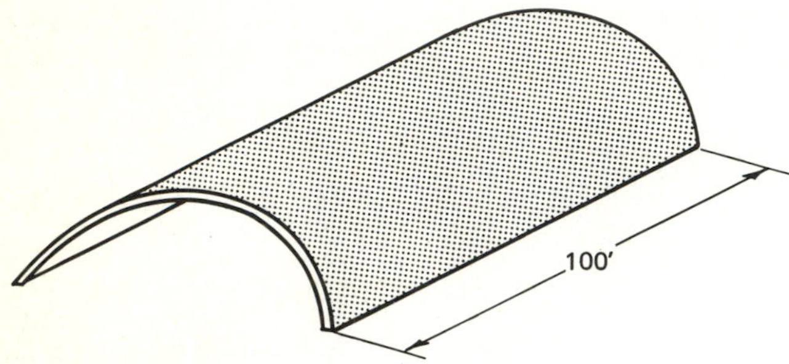

Figure 6–51. Example Problem 6–7

Solution

1. *Assume shell thickness and rise.* From Appendix 6A, try t = 4 in. (102 mm). Assume that W = 30 ft (9.1 m), h = 100 ft/10 = 10 ft (3.05 m).

$$R^2 = (15)^2 + (R - 10)^2,$$

$$R = 16.3 \text{ ft } (4.97 \text{ m})$$

2. *Determine loads.*

$$\text{Live load} = 30 \text{ psf } (1{,}436 \text{ N/m}^2)$$

$$\text{Dead load} = (4/12)(150) = 50 \text{ psf } (2{,}394 \text{ N/m}^2)$$

$$\text{Uniform load} = 2\alpha\omega R = 2(1.17)(80)(16.3)$$

$$= 3051 \text{ lb/ft } (44{,}523 \text{ N/m})$$

3. *Calculate forces.*

$$M_x = \frac{\omega l^2}{8} = \frac{3.05(100)^2}{8} = 3812 \text{ ft-kips } (5{,}169{,}072 \text{ Nm})$$

$$V_x = \frac{\omega l}{2} = \frac{3.05(100)}{2} = 152.5 \text{ kips } (678{,}320 \text{ N})$$

4. *Calculate section properties.*

$$\sin\alpha = \frac{15}{R} = \frac{15}{16.23} = 0.92 \qquad \alpha = 1.17 \text{ rad}$$

$$\cos\alpha = 0.39$$

$$Y_1 = R\left(1 - \frac{\sin\alpha}{\alpha}\right)$$

$$= 16.3\left(1 - \frac{0.92}{1.17}\right) = 3.48 \text{ ft } (1.06 \text{ m})$$

$$Y_2 = R\left(\frac{\sin\alpha}{\alpha} - \cos\alpha\right)$$

$$= 16.3\left(\frac{0.92}{1.17} - 0.39\right) = 6.46 \text{ ft } (2.0 \text{ m})$$

$$I = R^3t\left(\alpha + \sin\alpha\cos\alpha - \frac{2\sin^2\alpha}{\alpha}\right)$$

$$I = (16.3)^3\left(\frac{4}{12}\right)\left(1.17 + (0.92)(0.39) - \frac{2(0.92)^2}{1.17}\right)$$

$$I = 118.3 \text{ ft}^4 \ (1.02 \text{ m}^4)$$

$$\cos\beta = \frac{R - Y_1}{R} = \frac{16.3 - 3.48}{16.3} = 0.79$$

$$\beta = 0.66, \qquad \sin\beta = 0.61$$

$$Q = 2\beta R^2 t\left(\frac{\sin\beta}{\beta} - \cos\beta\right)$$

$$Q = 2(0.66)(16.3)^2(4/12)\left(\frac{0.61}{0.66} - 0.79\right)$$

$$= 15.7 \text{ ft}^3 \ (0.44 \text{ m}^3)$$

5. *Calculate membrane stresses.* By Eq. 6–13,

$$T_x = \frac{My}{I}$$

Compression:

$$T_x = \frac{3812(3.48)}{118.3} = 112 \text{ kips/ft}^2 = 779 \text{ psi } (5.4 \text{ MPa})$$

Tension:

$$T_x = \frac{3812(6.46)}{118.3} = 208 \text{ kips/ft}^2 \ (9{,}948{,}215 \text{ N/m}^2)$$

By Eq. 6–14,

$$S = \frac{VQ}{I} = \frac{152.5(15.7)}{118.3} = 20.2 \text{ kips/ft } (294{,}782 \text{ N/m})$$

By Eq. 6–11,

$$T_\phi = -\omega R\cos^2\alpha$$

$$= 80(16.3)(0.39)^2 = 198 \text{ lb/ft } (2889 \text{ N/m})$$

$$f_c = \frac{198}{12(4)} = 4.1 \text{ psi } (28.2 \text{ KPa})$$

6. *Calculate boundary forces.* Assume that the shell is

hinged at the stiffener. By Eq. 6–15,

$$M = 0.092qRt$$
$$= 0.092(80)(16.3)(\tfrac{1}{3})$$
$$= 39.9 \text{ ft-lb/ft (178 Nm/m)}$$

$$V = 0.38\sqrt{\frac{t}{R}}\,qR$$

$$= 0.38\sqrt{\frac{1/3}{16.3}}(80)(16.3) = 70.9 \text{ lb/ft (1034 N/m)}$$

7. *Determine steel reinforcement.*
   Longitudinal:

$$T_x = 208 \text{ kips/ft}^2 \text{ (9,948,215 N/m}^2\text{)}$$

Assume linear variation of stress from shell edge to neutral axis.

$$\text{Arc distance} = \beta R = 0.66(16.3)$$
$$= 10.8 \text{ ft (3.3 m)}$$

$$\text{Total force} = \tfrac{1}{2}(208)(10.8)(\tfrac{1}{3})$$
$$= 374 \text{ kips (1,663,552 N)}$$

$$A_s = \frac{374}{20.0} = 18.7 \text{ in}^2 \text{ (12,062 mm}^2\text{)}$$

$$\rho = \frac{A_s}{A_c} = \frac{18.7}{(10.8 \times 12 \times 4)} = 0.036 \quad \text{(slightly high)}$$

Transverse: No reinforcement required for axial compression. For moment:

$$A_s = \frac{M}{f_s jd} = \frac{39.9 \times 12}{(20{,}000)(0.9)2} = 0.013 \text{ in.}^2\text{/ft}$$

$$\rho = \frac{A_s}{A_c} = \frac{0.013}{12(4)} = 0.0002$$

Use $A_{s\,(\min)} = 0.002(48) = 0.096$ in.²/ft (203 mm²/m).
Diagonal. Determine shear stress at $x = l/4$ as an indicator of required diagonal steel.

$$V \text{ at } l/4 = 152.5 - 25(3.05) = 76.3 \text{ kips (339,382 N)}$$

$$S = \frac{76.3(15.7)}{118.3} = 10.1 \text{ kips/ft (147,391 N/m)}$$

$$A_s = \frac{10.1}{20.0} = 0.51 \text{ in.}^2\text{/ft (1,079 mm}^2\text{/m)}$$

6–6 TENSILE STRUCTURES

Tensile structures are systems constructed primarily with tensile components, such as cables, nets, and membranes (Fig. 6–52). These structures are flexible and carry loads via in-plane tension and membrane forces. In contrast to shells, which are rigid and do not change their basic shape when loaded, tensile structures change their geometry as required to maintain tension forces, similar to the behavior of the cable element.

(a)

(b)

Figure 6–52. Tensile Structures [(a) Courtesy of Bethlehem Steel; (b) reprinted from *Tensile Structures* by Frei Otto by permission of the MIT Press, Cambridge, Mass.; English translation: Vol. 1, copyright 1967; Vol. 2, copyright 1969, by the Massachusetts Institute of Technology]

The assembly of tensile elements into three-dimensional form creates a unique structural system, indeterminate in its behavior and nonlinear in its deflection patterns. These systems occur in nature as spider webs and were first used by people for fishing nets. Together with shells, tensile systems represent three-dimensional, curved development of space—a fairly recent arrival as a practical structural engineering system, and requiring experience and skill to design. They are extremely efficient in carrying load, usually across horizontal spans, and are popular for long-span roof systems.

History and Development

The oldest examples of tensile structures are tents and suspension bridges. Tents have been used for thousands of years using fabrics made of fibers whose strength has only recently been surpassed. The basic types of tents have changed little over the centuries, with only the familiar circus tent acquiring its shape in the last century.

The first cable suspension bridge was built around 1816. The gravity suspension bridge was perfected by John Roebling and has influenced the design of all large suspension structures. His most important designs were the Ohio Bridge in Cincinnati and the Brooklyn Bridge in New York.

The application of tension-loaded surface structures to major buildings started during the 1930s with the development of cable roof systems. Several major cable structures were developed over the next 50 years, including the Raleigh Arena (1953), the Olympic gymnasiums in Tokyo (1964), Dulles Airport in Washington, D.C. (1962), the French and U.S. pavilions at the Brussels World's Fair (1958), and Madison Square Garden (1967). The development of computer and physical modeling techniques provided structural engineers with the tools to continue the development of this structural form.

Fabric structures (air-supported and tensile) are the most recent tensile structures to be introduced to the building industry. Initially used for radar installations and exhibition structures, these systems have now been used on several large arenas and stadium projects.

Types and Uses

Major classifications of tensile structures include the following:

- Assemblies of single cables
- Cable nets
  - Prestressed
  - Nonprestressed
- Pneumatic structures
  - Air-supported
  - Air-inflated
- Membrane structures

Single cable assemblies such as those shown in Fig. 6–53, have been used either radially in conjunction with ring elements or in parallel form, with anchorage provided

Figure 6–53. Cable Roof Structures

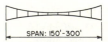

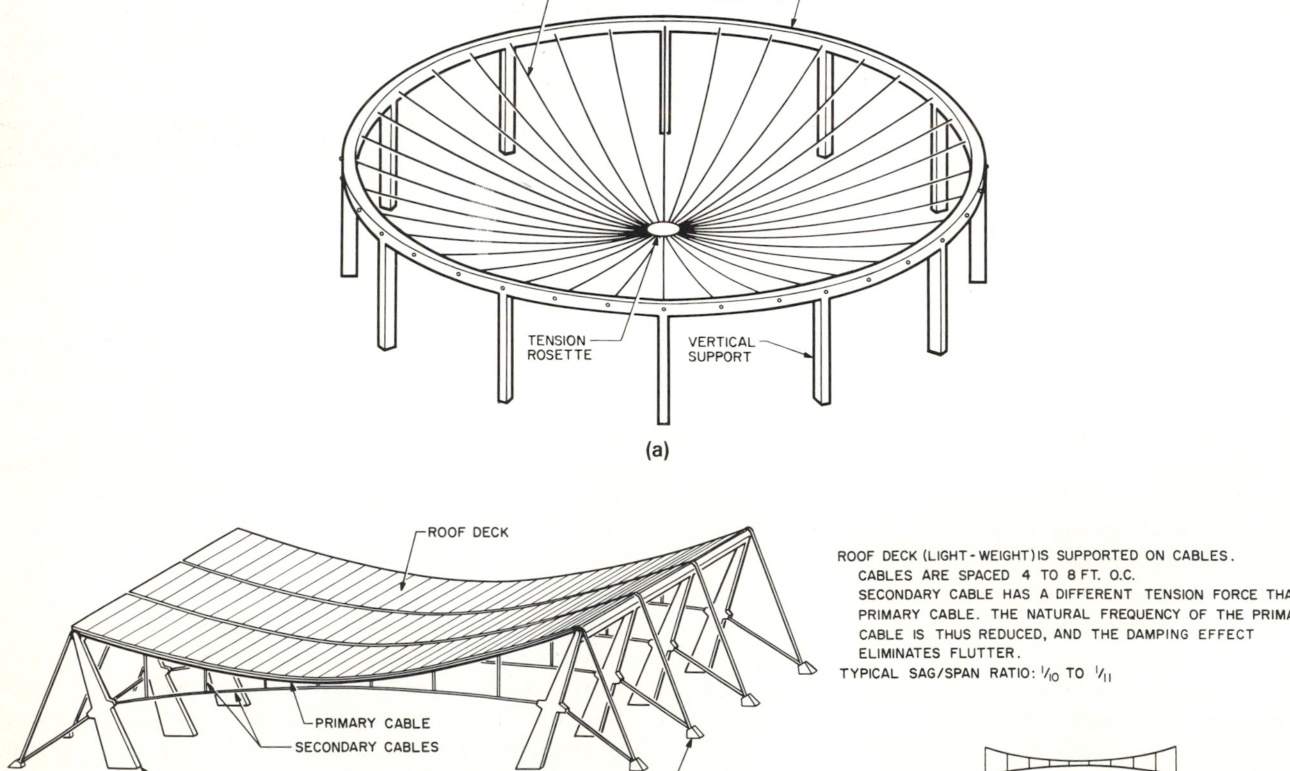
CABLE
CIRCULAR COMPRESSION RING
TENSION ROSETTE
VERTICAL SUPPORT
(a)
ROOF DECK
PRIMARY CABLE
SECONDARY CABLES
PIER
ANCHOR
ROOF DECK (LIGHT-WEIGHT) IS SUPPORTED ON CABLES.
CABLES ARE SPACED 4 TO 8 FT. O.C.
SECONDARY CABLE HAS A DIFFERENT TENSION FORCE THAN PRIMARY CABLE. THE NATURAL FREQUENCY OF THE PRIMARY CABLE IS THUS REDUCED, AND THE DAMPING EFFECT ELIMINATES FLUTTER.
TYPICAL SAG/SPAN RATIO: 1/10 TO 1/11
SPAN: 150'-300'
(b)

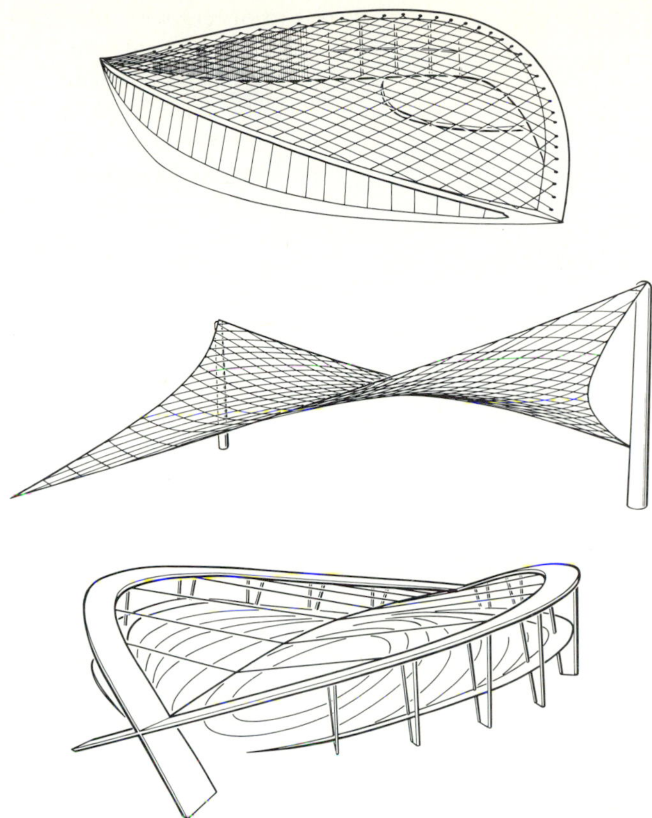

Figure 6–54. Cable Nets (From *Cable Roof Structures,* Bethlehem Steel, 1968)

by tie-downs or abutments. The suspension bridge falls into this classification. These systems are relatively easy to analyze and are popular for arenas and other long-span applications.

Cable nets are interlocking assemblies of cables arranged in a particular geometrical mesh; several examples are shown in Fig. 6–54. *Prestressed* cable nets have negative double curvature surfaces (saddle shaped) and achieve strength and stability (against flutter) via opposing tension forces between the two sets of cables. *Nonprestressed* nets normally have both sets of cables with similar curvature and rely on other methods for achieving stability (e.g., dead weight, etc.).

Membranes are "flexible stretched skins" that are used for a wide variety of forms (Fig. 6–55), often in conjunction with either single cables or cable nets that provide reinforcement. Membranes and cable nets are often grouped together for study since their behavior is similar.

Pneumatic structures are either air-supported or air-inflated. *Air-supported* systems rely on internal air pressure to support a covering membrane (Fig. 6–56). *Air-inflated* systems consist of self-enclosed membranes that form structural members (Fig. 6–57). Pneumatic structures have received wide publicity in recent years.

Figure 6–55. Membrane Structures [Reprinted from *Tensile Structures* by Frei Otto by permission of the MIT Press, Cambridge, Mass.; English translation: Vol. 1, copyright 1967; Vol. 2, copyright 1969, by the Massachusetts Institute of Technology]

Figure 6–56. Air Supported Pneumatic Structure (Courtesy of Owens-Corning Fiberglas Corporation)

All of these tensile systems have been primarily used for roof systems with either long spans or scuptural architectural features. Pneumatic structures are generally economical at spans exceeding 300 ft (91.4 m), and improvement in fabric materials has made the other systems applicable to almost any span as long as cable anchorage can be accomplished economically.

Behavior

Tensile structures have very little compressive or bending strength and consequently can carry loads only by tension. They avoid compressive and bending forces by either changes in geometry (Fig. 6–58) or by pretension sufficient to counteract the compressive forces (Fig. 6–59). Assemblies of single cables and nonprestressed nets avoid compression by changing shape, the behavior that we studied in Chapter 4.

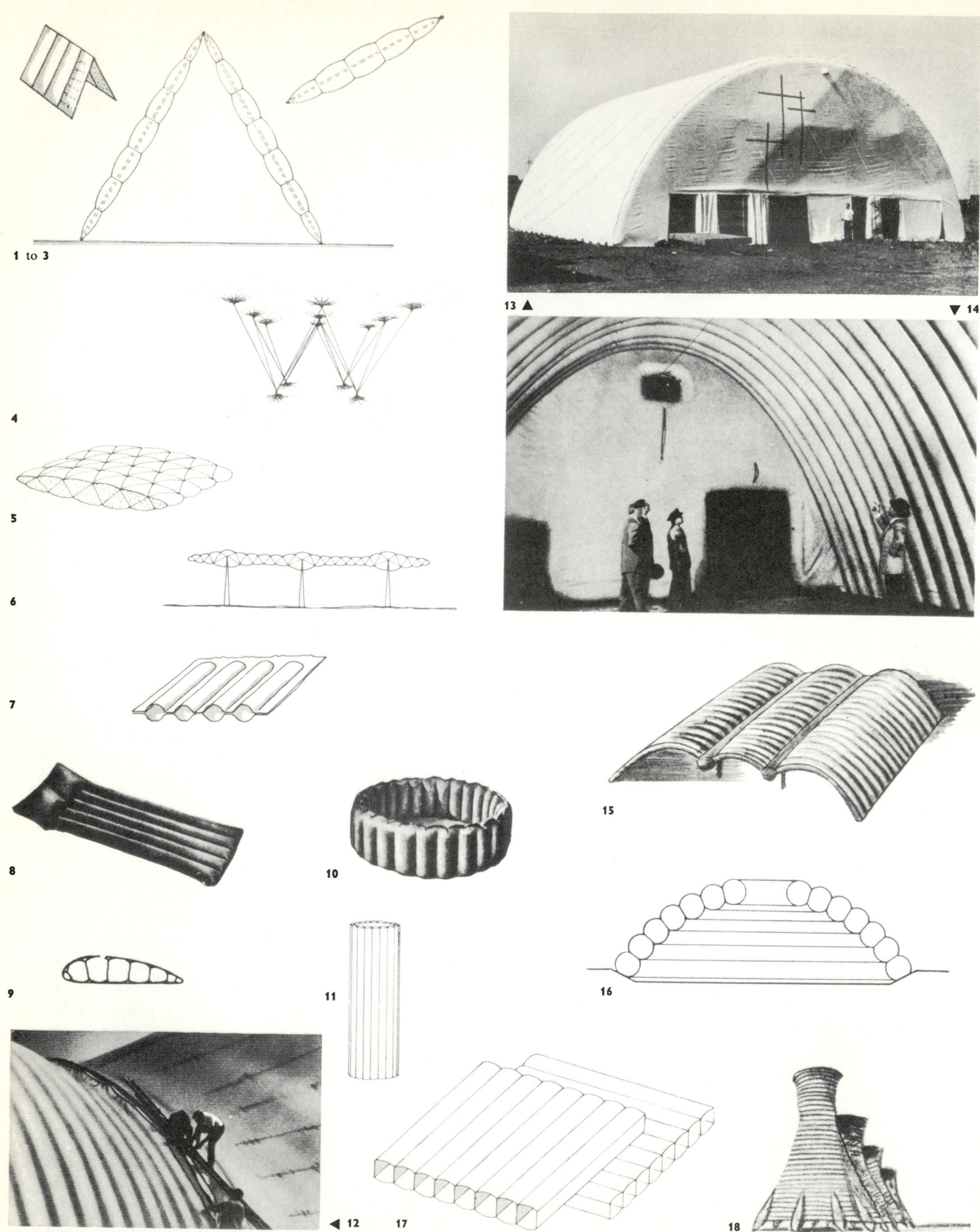

Figure 6–57. Air-Inflated Pneumatic Structure [Reprinted from *Tensile Structures* by Frei Otto by permission of the MIT Press, Cambridge, Mass.; English translation: Vol. 1, copyright 1967; Vol. 2, copyright 1969, by the Massachusetts Institute of Technology]

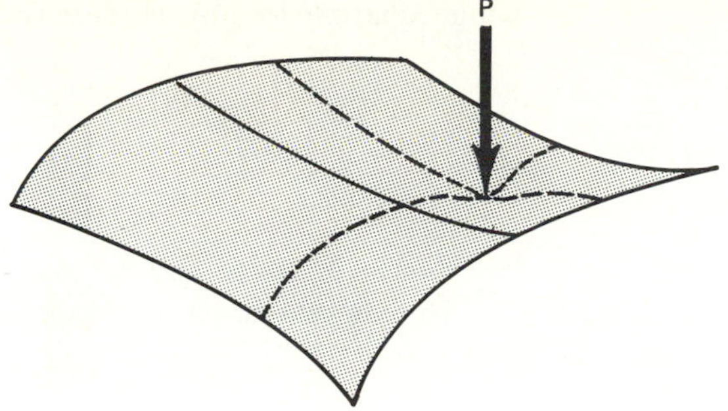

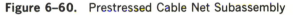

Figure 6–58. Changes in Geometry Under Load

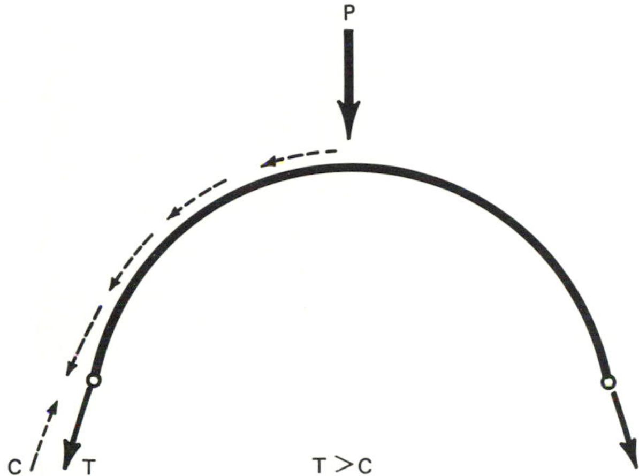

Figure 6–59. Pretension to Counteract Compression

We can study the behavior of *prestressed cable nets* by isolating a two-cable subassembly as shown in Fig. 6–60. Under load *P,* the upper cable elongates and carries tension load, while the lower cable is compressed and becomes shorter. It is obvious that the lower cable must have an initial tension in order that both cables carry load. The initial tension in the lower cable is obtained by pulling down on the upper cable; therefore, both cables are pretensioned. Total cable forces are equal to the sum of initial pretension loads plus loads due to external effects such as dead load, snow, and wind. These forces can be expressed as follows:

$$H_1 = H_{10} + H_{1P} \tag{6–25}$$
$$H_2 = H_{20} + H_{2P}$$

where H_O = initial tension
H_P = loads due to external effects

Equilibrium considerations for the initial state (before the application of external load) require that the vertical force carried by each cable be equal. Therefore, the following expression must be true:

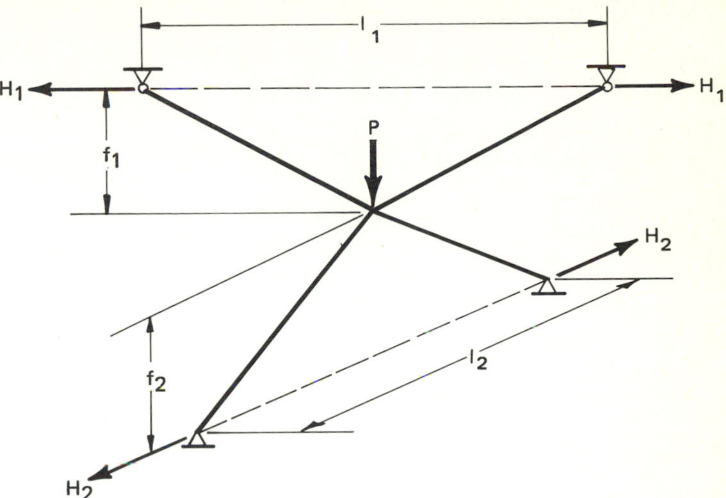

Figure 6–60. Prestressed Cable Net Subassembly

$$\frac{H_{10}}{H_{20}} = k = \frac{f_2\, l_1}{l_2\, f_1} \tag{6–26}$$

where f = cable sag
l = cable span
k = ratio of initial tensions

The ratio of the initial tensions is thus known when the shape of the structure is given. The magnitude of the initial tension is arbitrary, but must be chosen high enough to ensure that every cable remains in tension under any conceivable loading condition. This pretensioning of cable nets is a very effective method of eliminating flutter instability due to wind.

Pneumatic and *membrane* structures achieve their strength from the same membrane behavior that is characteristic of shell systems, but in tension rather than compression. Utilizing the principle of the prestressed cable net, these structures are constructed by tensioning cables against membrane fabric. This pretensioning can be accomplished with air pressure (Fig. 6–61) in the case of air-supported systems or by counteracting geometrical surfaces.

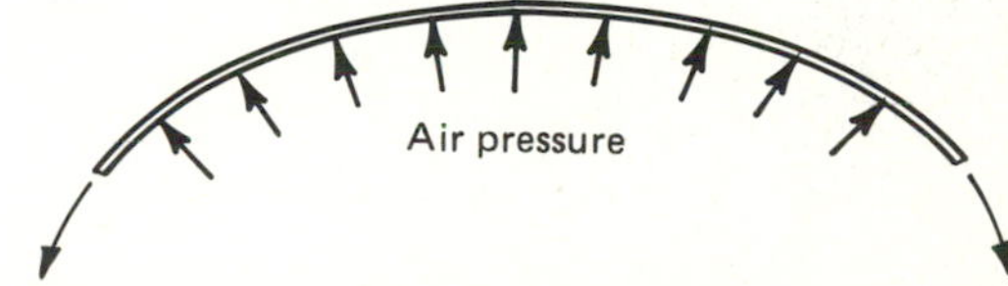

Figure 6–61. Pretensioned Air-Supported Structure

The geometry of these tension systems presents two special problems. First, the overall shape must be such that no "wrinkles" exist in the fabric material. For air-supported systems, this requires the shape to conform

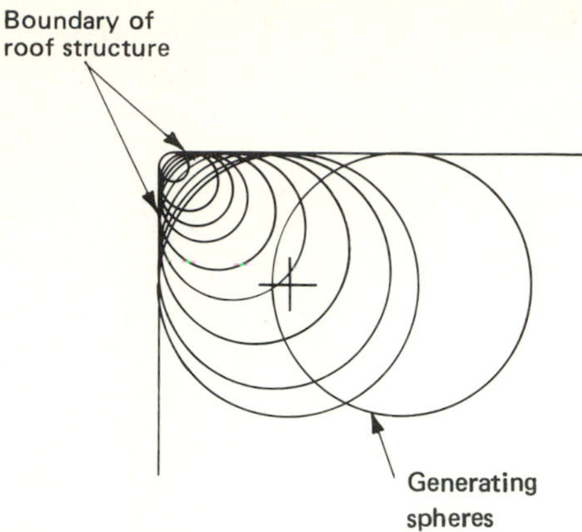

Figure 6–62. Surface Generation for Air-Supported Systems

to surfaces capable of being generated by a series of spheres, whose radii change as necessary to form the boundary and whose centers lie on a common curve (Fig. 6–62).

Second, the stretching of the fabric material during installation, particularly in the fill direction, means that the initial fabricated dimensions of the fabric will not be the same as the final in-service dimensions. Either the fabricated geometry must be set and the final geometry assumed or calculated, or vice versa. Either method is complicated by nonlinear behavior and complex mathematical modeling.

Approximate Analysis

We can use the methods developed in Chapter 4 for the analysis of *single-cable* assemblies.

Approximate formulas[1] for orthogonal *cable nets* start with the following equation for the surface (Fig. 6–63):

Figure 6–63. Surface Geometry of Cable Nets

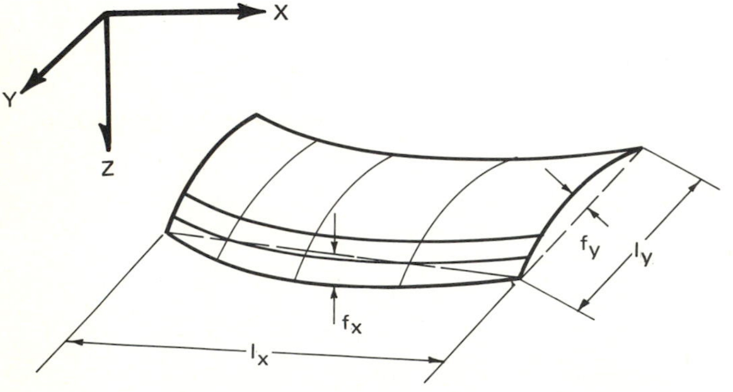

$$Z = -\frac{k_x}{2} X^2 + \frac{k_y}{2} Y^2 \qquad (6\text{–}27)$$

Z, X, and Y are coordinates in space and k_x and k_y are cable curvature coefficients which can be expressed approximately as follows:

$$k_x = \frac{8f_x}{l_x{}^2}, \qquad k_y = \frac{8f_y}{l_y{}^2} \qquad (6\text{–}28)$$

where f = cable sag
l = cable span

The general equation of equilibrium for cable nets is

$$H_{xo}k_x - H_{yo}k_y = 0 \qquad (6\text{–}29)$$

or

$$\frac{H_{xo}}{H_{yo}} = k = \frac{k_y}{k_x} > 0 \qquad (6\text{–}30)$$

where H_o is the initial tension. Equation 6–30 is in the general form of Eq. 6–26, which was derived for a simple two-cable net.

Cable stresses are affected by relative rigidities expressed by the following ratio:

$$\frac{E_xF_x}{E_yF_y} = \mu \qquad (6\text{–}31)$$

where E = modulus of elasticity of cable
F = cross-sectional area of a single cable ÷ distance in plan between cables

The ratio of total vertical surface load carried by each set of cables can be expressed as follows:

$$X_x = \frac{1}{1 + \mu k^2}, \qquad X_y = 1 - X_x \qquad (6\text{–}32)$$

where μ is given by Eq. 6–31 and k by Eq. 6–30.

The horizontal tensions in each direction are then

$$H_{xP} = X_x \frac{P_z}{k_x} \qquad (6\text{–}33)$$

$$H_{yP} = -X_y \frac{P_z}{k_y} = -\mu k H_{xP} \qquad (6\text{–}34)$$

[1] Reprinted from Frei Otto, *Tensile Structures*, by permission of the MIT Press, Cambridge, Mass., 1973, pp. 164–168. English translation: Vol. 1, copyright 1967; Vol. 2, copyright 1969, by the Massachusetts Institute of Technology.

The total cable tensions are found by adding the initial tension to the tension created by external loads as follows:

$$H_x = H_{xo} + H_{xP} \tag{6–35}$$

The deflection in the center of the cable net is approximately

$$w \simeq (1.15 \text{ to } 1.35)\frac{X_x P_z}{k_x^2 E_x F_x} \tag{6–36}$$

These formulas do not account for nonuniform loads such as those caused by wind, snow, etc. The formulas are applicable to nonprestressed nets if k_y is changed to $-k_y$. The following example illustrates the preliminary analysis process.

EXAMPLE PROBLEM 6–8: Preliminary Analysis of Cable Net

Determine approximate cable stresses for the pretensioned cable net shown in Fig. 6–64. The net is supported by a circular compression ring that has been tilted to produce a double curvature surface. Assume that initial tension is 4000 lb/ft (58,373 N/m) and live load is 20 psf (958 N/m²).

Figure 6–64. Example Problem 6–8

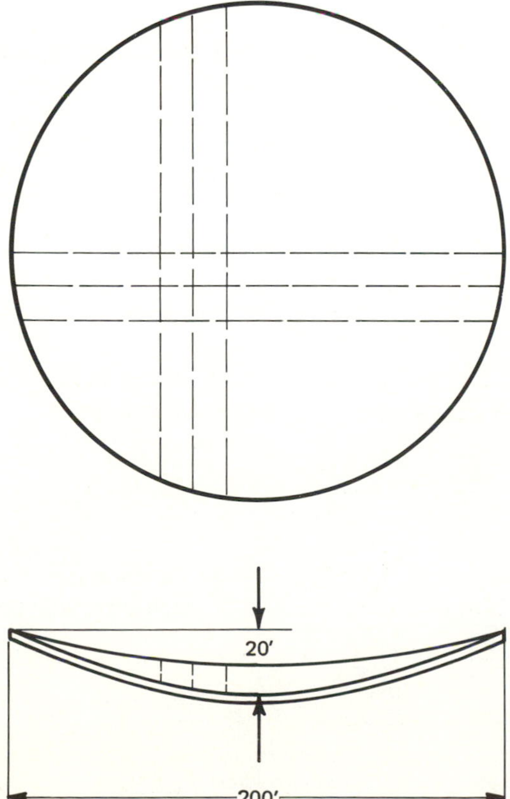

Solution

1. *Calculate the curvature.* Curvature is the same in both directions:

$$k_x = k_y = \frac{8f}{l^2} = \frac{8(20)}{(200)^2} = \frac{1}{250}$$

2. *Calculate the relative rigidities.*

$$\mu = \frac{E_x F_x}{E_y F_y} = 1.0, \qquad \frac{k_y}{k_x} = k = 1.0$$

3. *Calculate the vertical load ratio.*

$$X_x = \frac{1}{1 + \mu k^2} = \frac{1}{1 + (1.0)(1.0)^2} = 0.5$$

4. *Calculate the horizontal tensions due to live load.*

$$H_{xP} = \frac{X_x P_z}{k_x} = \frac{0.5(20)}{1}(250) = 2500 \text{ lb/ft } (36{,}483 \text{ N/m})$$

$$H_{yp} = -\frac{X_y P_z}{k_y} = -\frac{0.5(20)}{1}(250)$$
$$= -2500 \text{ lb/ft (compression)}$$

5. *Calculate the total stress.*

$$H_x = H_{xo} + H_{xP} = 4000 + 2500$$
$$= 6500 \text{ lb/ft } (94{,}856 \text{ N/m})$$

$$H_y = H_{yo} + H_{yP} = 4000 - 2500$$
$$= 1500 \text{ lb/ft } (21{,}890 \text{ N/m})$$

Both sets of cables remain in tension.

Pneumatic and *membrane* structures act much like shells once they are in final position and loaded with service loads. The membrane equations used for shell analysis can be used for approximate analysis.

Approximate Design

Design of tensile structures involves the following choices:

- Material selection
- Type of system
- Surface geometry
- Boundary conditions
- Layout of cables, net
- Pretension forces
- Element properties

Models and soap-bubble techniques are excellent starting points for determining surface geometries and compatible boundary conditions. Almost any conceivable surface can be generated with systems of cable nets and/or membranes. Approximate design is usually accomplished with models and preliminary membrane calculations.

6-7 COMBINATION SYSTEMS

Our earlier classification of structural systems is helpful for study purposes, but actual design may employ concepts of several systems in achieving an economical solution. For example, we may employ a shear wall (braced frame) in combination with a rigid frame (unbraced frame) to jointly resist lateral loads, or we may use closely spaced columns at the exterior wall to act as a type of rigid frame/wall combination called a "tube." Designers of high-rise buildings often combine features of different systems in order to carry lateral load as efficiently as possible, and there are numerous other examples of combination systems which have been employed. The following sections discuss some of the more common ones.

Frame/Shear Wall

Rigid frames and shear walls, or rigid frames and shear trusses, may be used to resist lateral loads when shear walls alone do not provide sufficient strength and/or stiffness. Figure 6–65 illustrates this combination system, which is used for high-rise buildings up to 40 to 50 stories. The load distribution depends on the relative stiffness of the two lateral support subsystems as well as on the diaphragm stiffness.

Core Shear Wall/Belt Truss

Core shear walls can be considerably strengthened by tying them to exterior columns for additional overturning resistance (Fig. 6–66). Belt trusses, placed at the top of the building and in some instances at intermediate heights, restrict the rotation of the core and the system no longer acts as a pure cantilever. Bending moments are reduced due to the greater transformation of lateral forces to axial forces.

Wall Beam/Staggered Truss

Story high beams or trusses may be utilized as "girders" for an unbraced frame system called the *wall beam* or *staggered truss.* The deep girders span from exterior wall to exterior wall and are located at interior partition lines,

Figure 6–65. Frame/Shear Wall Systems

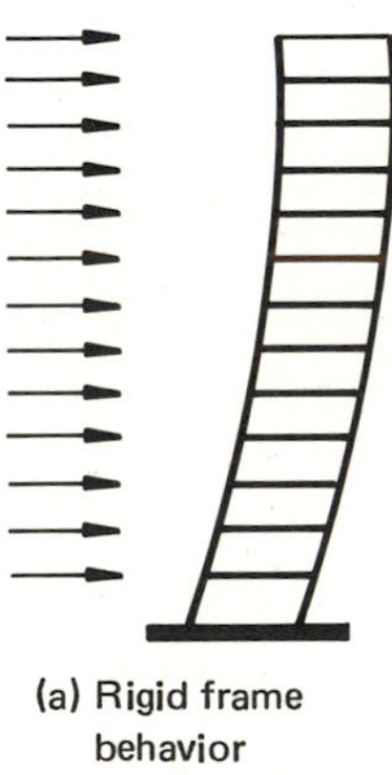
(a) Rigid frame behavior

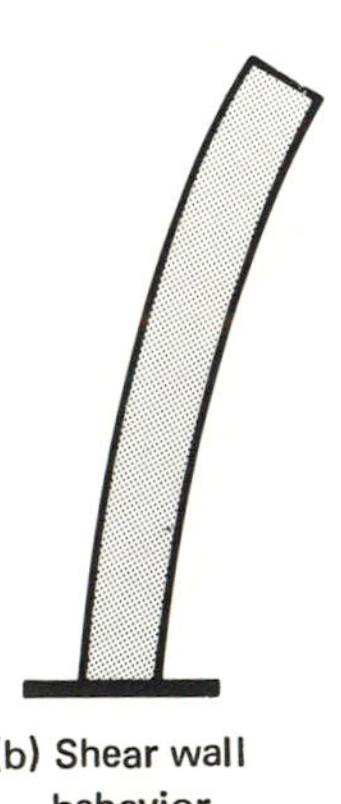
(b) Shear wall behavior

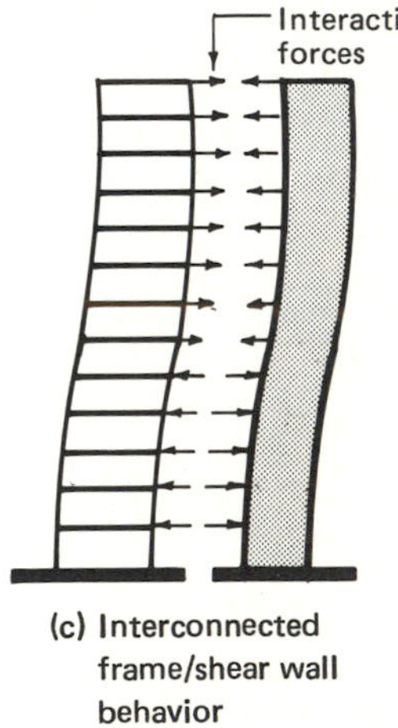

(c) Interconnected frame/shear wall behavior

Figure 6–66. Core Shear Wall/Belt Truss Systems

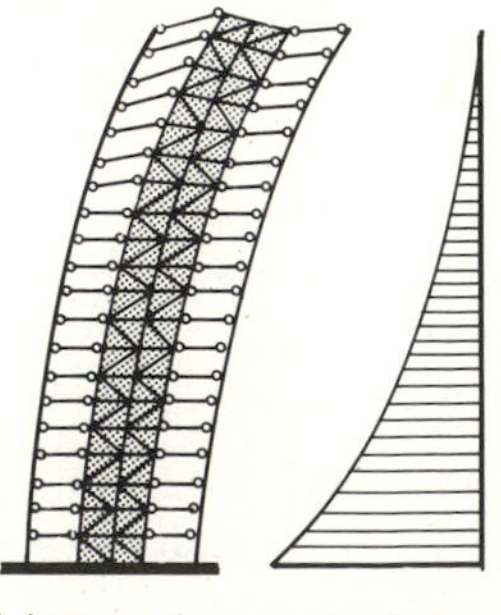
(a) Braced core behavior

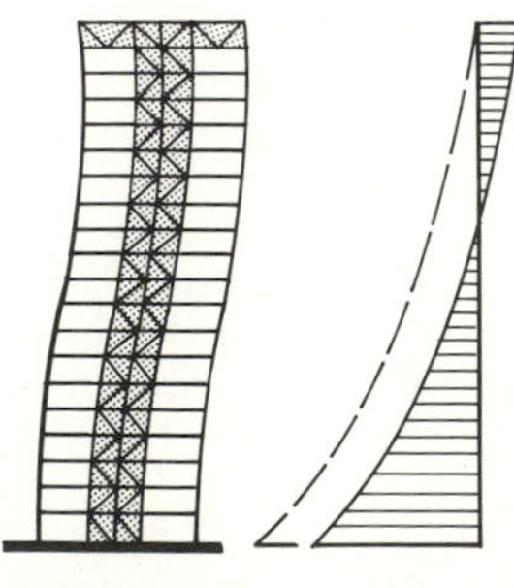
(b) Braced core tied to exterior columns with belt truss

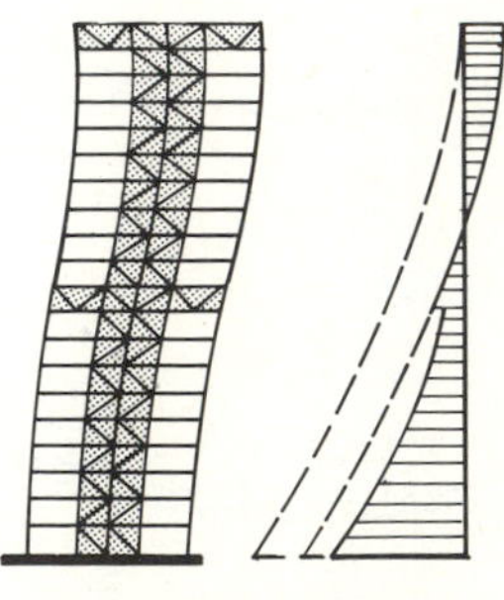
(c) Braced core tied to exterior columns (roof and midheight with belt truss)

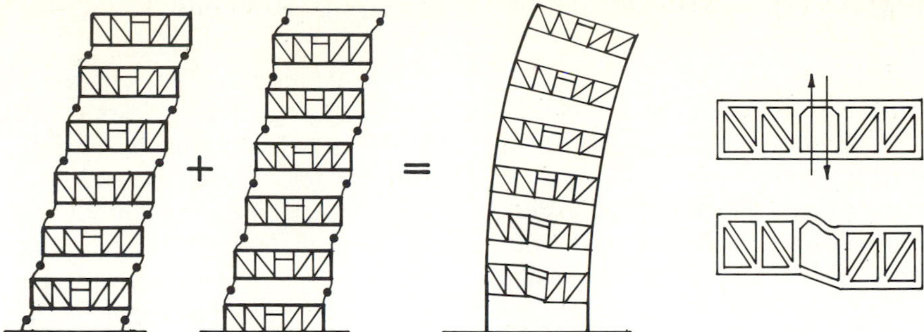

Figure 6–67. Wall Beam/Staggered Truss System

with openings provided for corridor passage. The openings in the girders create a "vierendeel truss" type of behavior which must be considered during design. The columns carry only axial loads, due to the bridging action of the floors acting as rigid diaphragms (Fig. 6–67). This system is suitable for hotel and housing construction from 10 to 30 stories.

Tubular Systems

The concept of utilizing the facade of buildings to resist lateral loads, introduced by Fazlur Khan during the 1960s, assumes that the exterior wall system acts as a closed hollow tube which cantilevers from the ground (Fig. 6–68). A *framed tube* consists of closely spaced columns and deep spandrel beams which act as a rigid frame, while a *diagonal braced tube* uses diagonal bracing on the exterior wall. Interior cores or braced sections may be tied to the exterior tube when needed to provide additional strength, creating a *tube-in-tube* system.

Figure 6–68. Tubular System

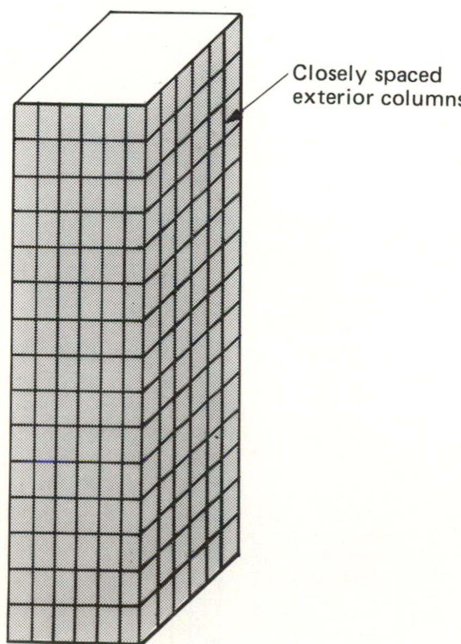

Tubular systems, although acting similar to cantilever beams, experience shear lag at the corners, which produces stress diagrams similar to that of Fig. 6–69. An approximate technique for preliminary analysis is to assume that the tube is replaced by equivalent channels as shown in Fig. 6–70. This technique attempts to correct for the shear lag effect.

Figure 6–69. Shear Lag in Tubular Structures

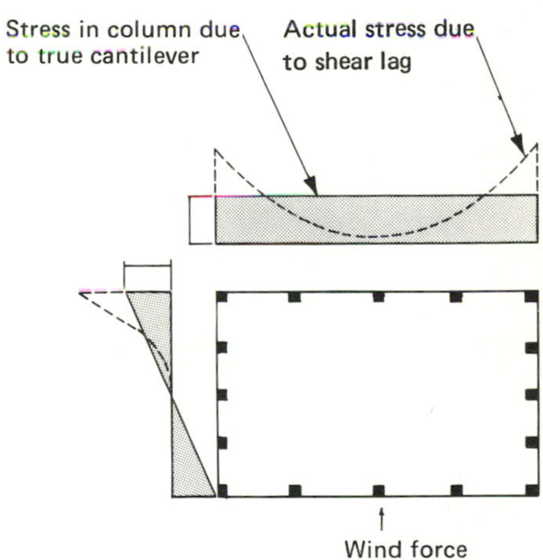

Figure 6–70. Channel Model of Tube System

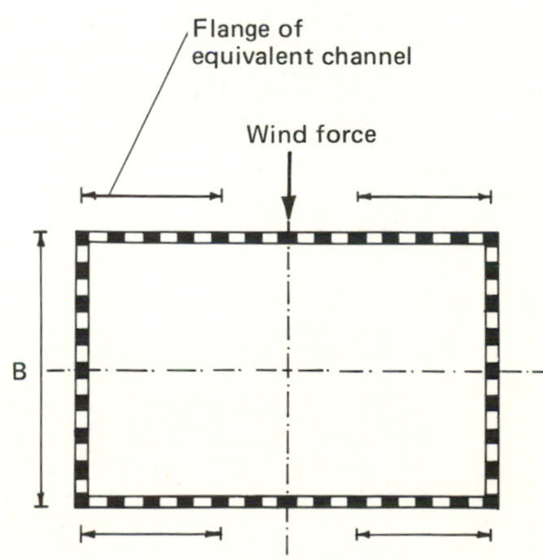

6–8 SUMMARY: SYSTEMS FOR HIGH-RISE BUILDINGS

High-rise building structures provide space for offices, apartments, or hotel guest rooms, functions that focus on people and their living environment, including vertical and horizontal circulation through the structure, lifesafety features, and comfort considerations. Before a high-rise structure can be justified, however, it must represent a sound financial investment.

The financial feasibility of high-rise structures is directly influenced by the ratio of net usable area to gross area, since rental income is normally based on net area. In addition, design requirements for lateral load can have significant cost impact. As buildings become taller, lateral loads due to wind and/or earthquake more directly control the structural design and add a premium to the building cost. Fire safety becomes a critical factor for taller structures, with close coordination required to correlate protection for the structure with sprinkler systems and means of egress.

The key design considerations for high-rise structures can be summarized as follows:

- Vertical circulation
- Fire safety
- Ratio of net area to gross area
- Lateral load
- Economy

System Descriptions

The structural form that typically is selected to respond to these design requirements is square or rectangular in plan and includes a shaft containing vertical circulation elements (e.g., elevators, etc.) and service areas which may be located near the center of the building or adjacent to one of the perimeter walls. The remaining floor area is kept open for tenant use, with spacing between columns generally not a major design consideration. This arrangement of space requires any lateral support elements, such as shear walls or bracing, to be located either in conjunction with the vertical shaft or at the perimeter wall. Height-to-width ratios are usually between 5 and 7 for optimum space utilization and structural efficiency.

Typical systems for high-rise buildings are shown in Fig. 6–71. *Braced frame* systems are usually the most competitive for any height structure as long as the bracing elements can be incorporated naturally in the structure, with no loss of functional space, and can serve other functional purposes, such as shaft walls. If the addition of bracing presents a functional or cost problem, *unbraced frame* systems are a competitive alternative for buildings less then 10 to 12 stories. Above this height, lateral loads begin to dominate, and some type of bracing is required, either in the form of shear wall, X-bracing, or tubular system concepts. Structural steel and reinforced concrete are the primary materials used for high-rise structures

Figure 6–71. Typical High-Rise Building Systems

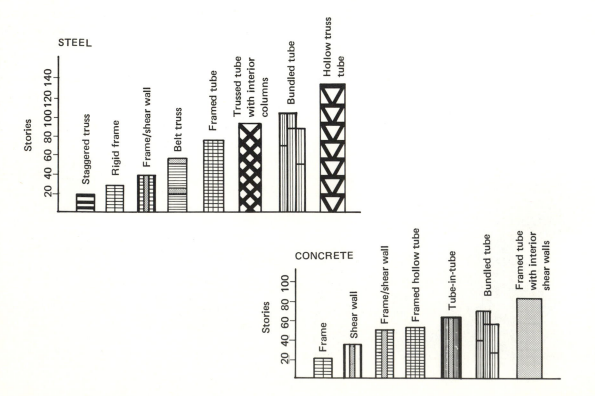

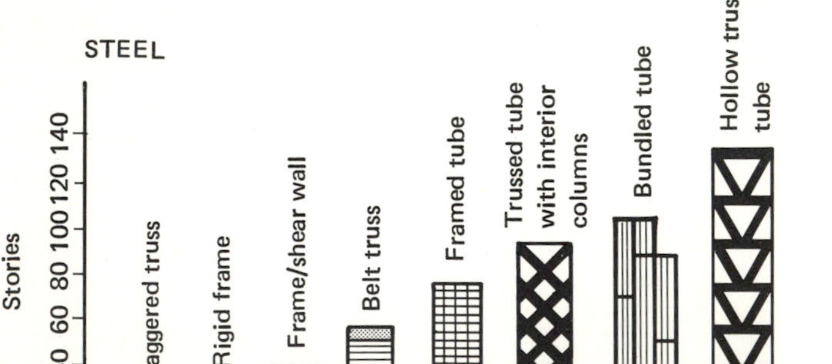

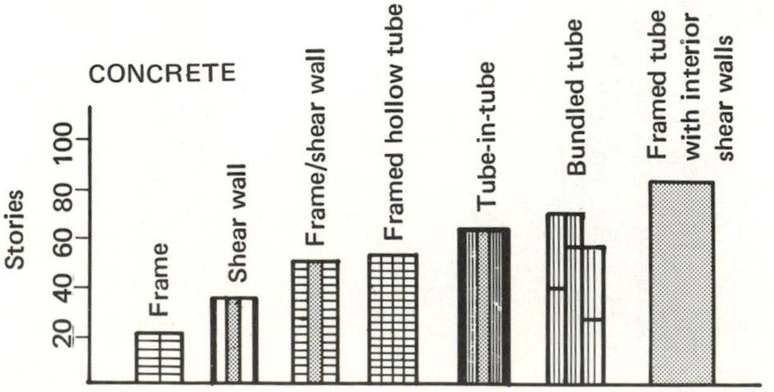

Figure 6–72. Steel High-Rise Building (Courtesy of Bethlehem Steel)

(Fig. 6–72), with precast concrete products also used up to a height of 10 to 12 stories. Bearing wall systems of brick and/or concrete masonry are popular systems for hotel or apartment use up to heights of 10 to 12 stories (Fig. 6–73). Composite systems of concrete and structural steel (particularly for column elements) are finding increasing use for high-rise structures above 35 stories.

Figure 6–73. Bearing Wall System (Courtesy of Portland Cement Association)

Floor systems for high-rise structures usually function as both horizontal distribution and lateral distribution subsystems. Since the depth of this subsystem directly influences the total building height, minimizing the depth usually becomes a major design consideration. Lightweight steel construction of bar joists and metal deck is popular for low-rise buildings, with some type of concrete system generally preferred for the taller structures. Popular concrete systems include one-way pan joists, one-way slab supported on beams, and flat slabs. Flat slabs and precast plank are popular for hotel and apartment use, in which the floor subsystem serves also as a ceiling for the space below.

6–9 SUMMARY: SYSTEMS FOR INDUSTRIAL BUILDINGS

Industrial buildings contain manufacturing, distribution, and warehousing operations. Structures range from single-story warehouse buildings to multistory process buildings. The physical arrangement of the manufacturing equipment, the support of equipment, and the flow of materials from the entrance to the plant to the shipping point are the major functional requirements. Although manufacturing operations may vary from the relatively static environment of semiconductor assembly to the dynamic environment of heavy cranes and forklifts handling bulk materials, the emphasis on manufacturing efficiency is paramount.

Many industrial structures support vibrating machinery, cranes, forklift trucks, and similar dynamic operations. Supporting members must be designed for these special loadings and vibrations must often be isolated from sensitive controls located in other areas of the plant. Heavy-duty operations such as these, often in conjunction with corrosive elements, add importance to durability and maintenance considerations.

Construction time is especially critical for industrial companies, whose products must reach the marketplace as quickly as possible. Structural cost is relatively minor compared to the large sums of capital invested in machinery. Structural flexibility to accommodate future expansions, shifts of equipment, and future loads is very important.

The key design considerations for industrial buildings can be summarized as follows:

- Manufacturing efficiency
- Construction time
- Flexibility
- Dynamic loading
- Maintenance
- Economy

System Descriptions

The utilitarian structures in industrial buildings usually have rectangular forms, with great expanses of horizontal space provided for assembly, warehousing, and shipping operations and vertical towers provided for distillation columns and for vertical "batch feed" of raw materials into the manufacturing process. The single-story portions typically have vertical clearance requirements of 20 to 25 ft (6.1 to 7.6 m), with heights to 60 ft (18.3 m) required for major crane-handling operations. Column spacings typically range from 30 to 40 ft (9.1 to 12.2 m).

The nature of industrial buildings has led to almost exclusive use of structural steel for the tower buildings and the large majority of the single-story buildings, although prestressed concrete is a competitive alternative for single-story warehousing and distribution buildings. *Braced frame* systems of the type illustrated in Fig. 6–74 are widely used for both single-story and multistory industrial buildings. Diaphragm action can often be utilized in lieu of horizontal bracing in the plane of the roof. *Unbraced* systems are used mainly for crane bays (Fig. 6–75), which typically utilize heavy trusses at the roof line for rigid frame action with the crane columns.

Figure 6–74. Braced Frame Industrial Structures

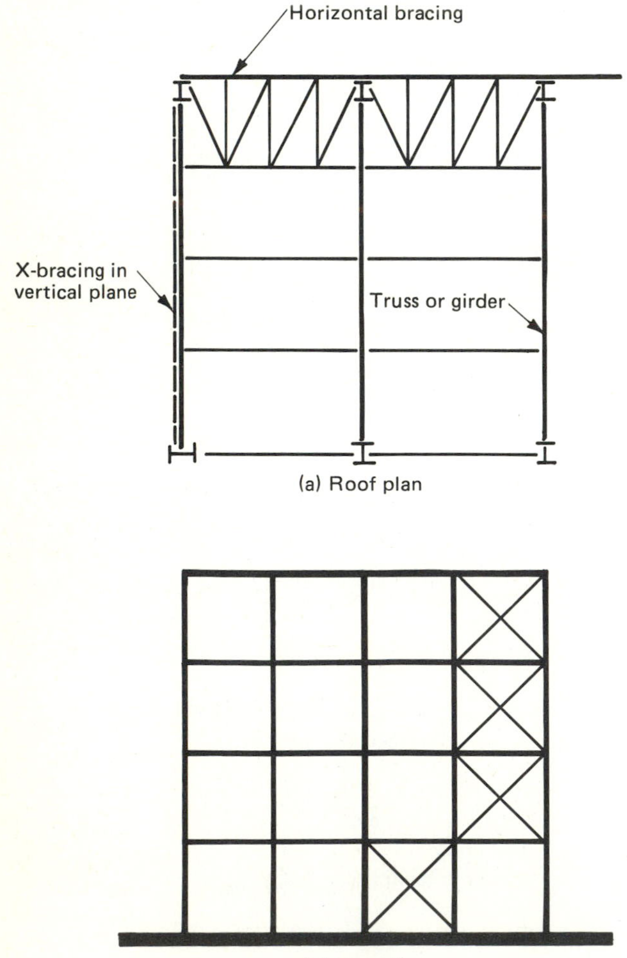

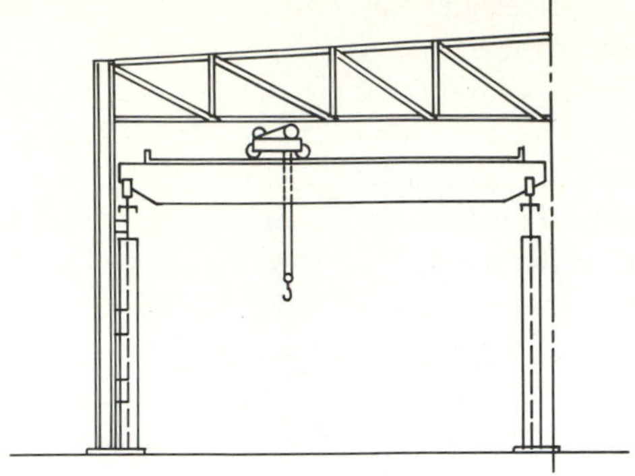

Figure 6–75. Industrial Crane Bay

Floor subsystems, either on grade or elevated are normally concrete, with support for elevated floors provided by steel floor beams at 8 to 10 ft (2.4 to 3.0 m) spacing. Slabs on grade are particularly susceptible to cracking and wear and require special attention, including special mix designs and surface treatments. Metal grating is used for platforms, mezzanines, and similar floors.

Roof subsystems are typically lightweight construction, with bar joists, steel girders, and metal deck heading the selection list. Continuous girder construction, known as "double cantilever construction," is widely favored for its economy and simplicity (Fig. 6–76). Other popular roof decks are precast concrete slabs and double-tee members.

Figure 6–76. Double-Cantilever Roof Construction

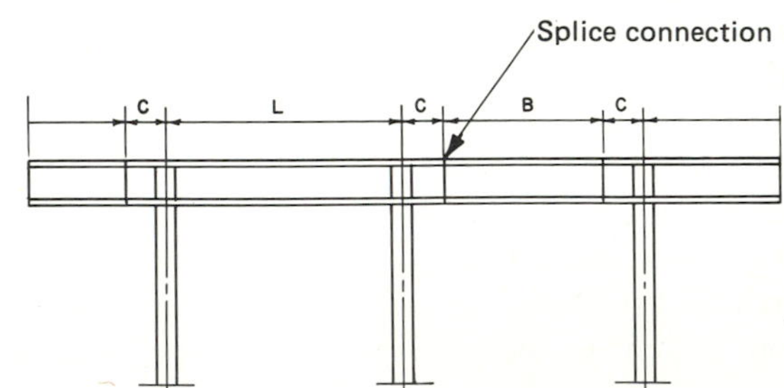

6–10 SUMMARY: SYSTEMS FOR LONG—SPAN ROOFS

Appropriately described by their name, long-span roof systems are used to achieve large expanses of column-free area required by the building function, for buildings such as arenas, exhibition halls, stadiums, and aircraft hangars. In addition to the column-free horizontal space, these structures normally require large vertical clearances as well. The design requirement primarily becomes one of achieving these large spans at minimum cost.

The structural form of long-span roof systems usually establishes the architectural form as well, making this structural system highly visible and thus important esthetically. The desire to be architecturally and structurally innovative has led to many novel designs for long-span structures—several of which have experienced problems due to design—and/or construction deficiencies. Quality control over both the design and construction process is extremely critical for these structures which typically enclose thousands of people at one time.

The key design considerations for long-span roof systems can be summarized as follows:

- Large horizontal spans
- Vertical clearance
- Economy
- Esthetics
- Safety

System Descriptions

It is not practical to span large distances with bending-type elements. Therefore, subsystems that incorporate tension and compression elements are most suitable for long-span structures. These include trusses, space frames, arches, shells, and tensile structures. Arch systems and space frames have been popular systems for the 200 to 250 ft (61 to 76.2 m) spans commonly required for arenas, coliseums, and exhibition halls (Fig. 6–77). As these spans have increased to cover stadiums and similar large spans, cable systems, domes, and pneumatic structures have replaced the space frame and arch as more economical choices (Fig. 6–78).

The surface material used to cover these roof systems is often an integral part of the overall stability system for the structure, providing diaphragm stiffness, lateral bracing for compression elements, and ballast to dampen the effects of wind. Deep-rib metal deck and lightweight concrete slabs are popular roof deck materials. Figure 6–79 shows a hanger structure which incorporates a combination of "stressed skin" metal deck and cable pretensioning to achieve the long cantilever span.

Figure 6–77. Arch Roof (Courtesy of Koppers, Inc.)

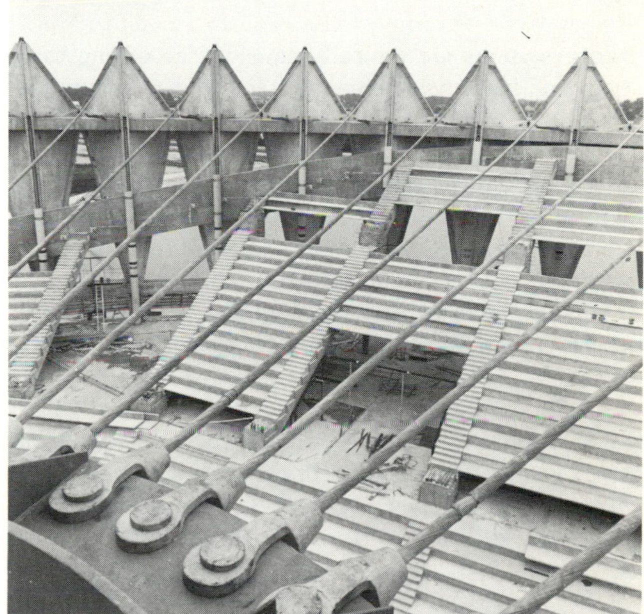

Figure 6–78. Cable Roof (Courtesy of Bethlehem Steel)

Figure 6–79. Hanger Structure (Courtesy of Bethlehem Steel)

6–11 SUMMARY: SYSTEMS FOR BRIDGES

Bridges carry motor and rail vehicles across highways, streams and rivers, and mountainous terrain. In addition to carrying the specified vehicle loading, vertical clearance underneath must be provided for land traffic, river traffic, and floods. Safety is a critical design factor, and weather exposure dictates the use of durable and low-maintenance systems.

Most bridge structures are prominent landmarks with significant visual impact, making appearance an important criterion. Economy is achieved via a careful balancing of substructure and superstructure costs, varying spans and framing schemes as necessary to achieve this balance. Usually, the optimum balance is achieved when the costs for substructure and superstructure are approximately equal.

The key design considerations for bridge systems can be summarized as follows:

- Vertical and horizontal clearance
- Safety
- Esthetics
- Maintenance
- Economy

System Descriptions

Bridge systems are typically classified according to (1) length of span, (2) framing system, and (3) continuity. *Short-span* bridges include spans up to 110 ft (33.5 m); *medium-span* bridges cover spans from 110 to 350 ft (33.5 to 107 m) and *long-span* bridges exceed 350 ft (107 m) in length. Various framing systems include *slab, stringer, girder, box girder, arch, truss,* and *suspension types.* The continuity classification includes *simple, cantilever,* and *continuous* construction. The following sections describe several major bridge systems according to framing type.

Slab Bridges. Slab construction is shown in Fig. 6–80. Applicable to short spans, the actual one-way slab may be constructed with reinforced concrete or precast elements. Continuous construction is more economical for spans exceeding 20 ft (6.1 m) for cast-in-place slabs.

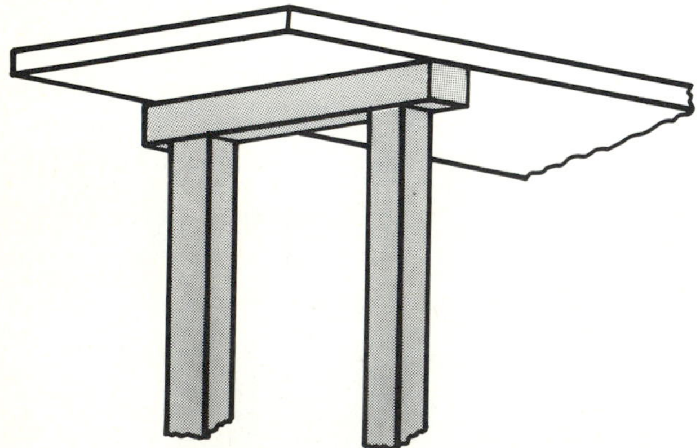

Figure 6–80. Slab Bridges

Stringer Bridges. Stringer construction (Fig. 6–81) consists of parallel beam sections at spacings of 7 to $8\frac{1}{2}$ ft (2.1 to 2.6 m), with one-way slabs spanning the

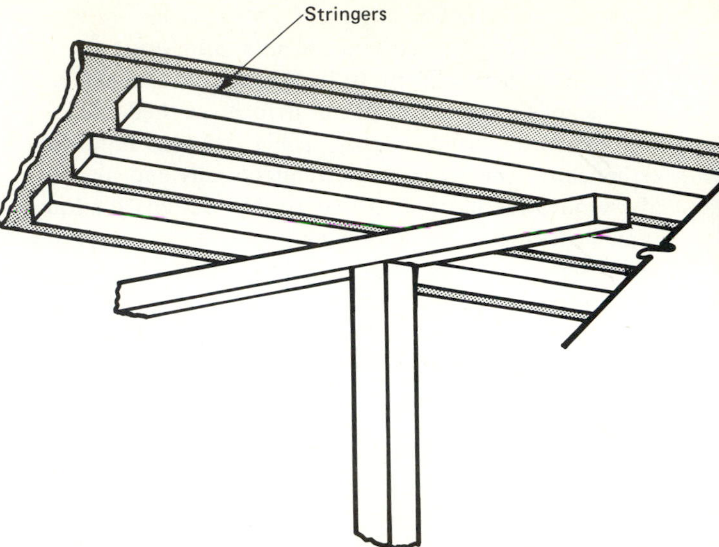

Figure 6–81. Stringer Bridges

transverse direction. Normally four stringers are the optimum number for normal-width highway bridges. Actual stringer material may be rolled steel beams, plate girders, or precast I-girders. Composite construction is normally employed for spans exceeding 60 to 80 ft (18.3 to 24.4 m) and continuous construction beyond 110 ft (33.5 m).

Girder Bridges. Girder construction (Fig. 6–82) includes a three-level system consisting of deck slab, floor beams, and longitudinal girders. (*Note:* The term "girder" is often used to refer to heavy beam sections used in either stringer or girder construction.) True girder construction is normally used for medium-span bridges of 175 to 350 ft (53 to 107 m).

Figure 6–82. Girder Bridges

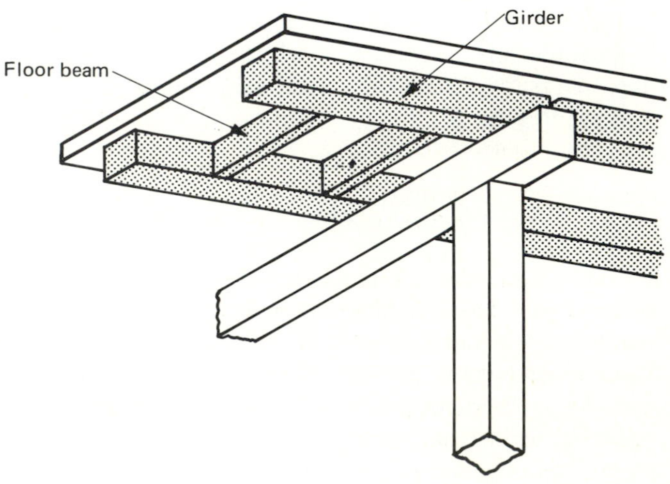

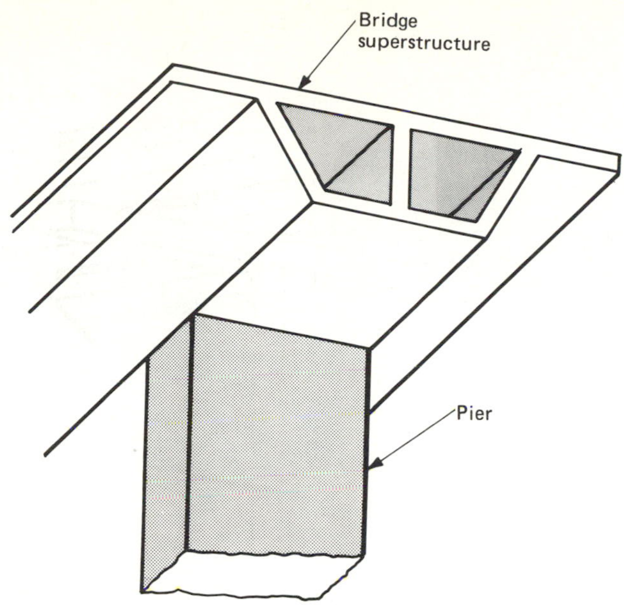

Figure 6–83. Box Girder Bridges

Box Girder Bridges. Box girders (Fig. 6–83) can be constructed of either steel or concrete and are used for medium- and long-span bridges. Segmental construction (precast or cast in place) is a form of box girder construction that employs staged post-tensioning and is very competitive for spans of 500 to 700 ft (152 to 213 m) (Fig. 6–84).

Figure 6–84. Segmental Construction (Courtesy of Oregon Department of Transportation)

Truss Bridges. Trusses are used only for very long spans (Fig. 6–85). The Warren truss is the most popular parallel chord form, with the K-truss normally used for spans greater than 320 ft (97.5 m). *Deck, through,* and *half-through* systems may be used, with the deck system preferred for highway bridges with no vertical clearance requirements underneath.

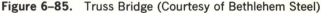

Figure 6–85. Truss Bridge (Courtesy of Bethlehem Steel)

Floor systems for truss bridges normally consist of floor beams spanning transversely between panel points of the two parallel trusses, with stringers running longitudinally between floor beams. The deck system may be one or two-way concrete slab or steel decking material.

Cable-Supported Bridges. Suspension bridges may be either cable-suspended (Fig. 6–86) or cable-stayed (Fig. 6–87). The cable-suspended bridge has the deck and other vertical loading suspended from the main cables at relatively short intervals, with the main cables relatively flexible and possessing a curved funicular shape. Suspension bridges are economical for spans exceeding 1000 ft (305 m).

Cable-stayed bridges utilize taut cables connecting pylons to a span to provide intermediate support for single or multiple box girders. A cable-stayed bridge is stiffer than the suspension type and provides an economical solu-

Figure 6–86. Cable-Supported Suspension Bridge (Courtesy of Triborough Bridge and Tunnel Authority)

Figure 6–87. Cable-Stayed Suspension Bridge (Courtesy of T. Y. Lin International, Consulting Engineers, San Francisco)

tion for spans intermediate between those suited for box girders and suspension systems. The structure acts like a continuous girder over the piers, with intermediate elastic supports provided by the cable stays.

PROBLEMS

6–1. Braced Frame. Prepare a preliminary design for a two-story office building with dimensions of 200 ft × 200 ft (61 m × 36.6 m). Assume braced frame design and compare a structural steel system with a reinforced concrete system. Use design loads from Chapter 3 for location in Atlanta, Georgia.

6–2. Unbraced Frame. Prepare a preliminary design for a single-story warehouse with dimensions of 200 ft × 600 ft (61 m × 183 m) and a clear height of 20 ft (6.1 m). Assume unbraced frame design and compare at least two systems. Use design loads from Chapter 3 for location in Dallas, Texas.

6–3. Bearing Wall. Prepare a preliminary design for a three-story school building located in Charlotte, North Carolina. The building has 30-ft (9.1-m)-wide classrooms on either side of an 8-ft (2.4-m)-wide central corridor. The building length is 150 ft (45.7 m) and the story height is 12 ft (3.7 m). Use design loads from Chapter 3 and compare at least two schemes.

6–4. Shell. Prepare a preliminary design for a reinforced concrete dome shell to span 150 ft (45.7 m) over an auditorium. Assume a uniform snow load of 20 psf (957.6 N/m^2).

6–5. Cable Roof. Prepare a preliminary design for a single-layer cable roof system of the radial type shown in Fig. 6–53. The diameter is 250 ft (76.2 m). Assume a uniform snow load of 30 psf (1436 N/m^2) and a concentrated load at the center ring of 4000 lb (17,792 N).

APPENDIX 6A: STRUCTURAL SYSTEMS DESIGN DATA BASE

System/Factor	*Economic load/span range*	*Depth/span ratio*	*Spacing/height range*	*Considerations*
I. Vertical support subsystems				
A. Load-bearing walls				
Brick	50–400 psi (gross area)	1/18–1/22	Total height 12–20 stories	
Concrete block	100–400 psi (net area)	1/18–1/22	Total height 10–12 stories	Consider higher strength block and/or reinforced masonry above 4–5 stories
Brick-block	100–400 psi (net area)	1/18–1/22	Total height 10–12 stories	Normally only block wythe can be engaged for gravity loads

APPENDIX 6A: *(Continued)*

System/Factor	Economic load/span range	Depth/span ratio	Spacing/height range	Considerations
Reinforced concrete	600–1000 psi	1/25	70–80 stories	Consider standardized forming systems (slip forming, tunnel forms, etc.)
Precast concrete	250 kips/ft (max. ultimate)	1/25	1–15 stories	Finish largely determines cost; speed of erection definite advantage
Wood stud	35–40 psi (gross wall area)		1–3 stories	Residential-type application
Metal stud	10 kips/stud (max.)		1–3 stories	Light commercial and residential buildings
B. Columns				
Steel	Max. 4315 kips (rolled, 36 ksi)	1/40	20, 25, 30 ft o.c.	Fireproofing generally required; normally 2 stories between splices
Reinforced concrete	Max. 3900 kips (ultimate)	1/20	18–25 ft centers	Consider high strengths (5–8 ksi) for heavier loads
Precast concrete	Max. 3500 kips (ultimate)	1/20	20–30 ft	Connection details (haunches, etc.) add considerably to cost
Timber	200–400 psi	$l/d < 35–40$	20–30 ft	Available lumber lengths dictate post-and-lintel-type construction
Steel–concrete composite				Lally-type columns or special design, particularly exterior columns
Laminated wood	400–800 psi	$l/d < 35–40$		
C. Post-and-beam frames	20–60 ft	1/20	20–40 ft	Require lateral support for stability
D. Rigid frame				Consider for multistory use up to 20–25 stories or up to height at which wind forces can no longer be absorbed in $33\frac{1}{3}\%$ allowable increase, or drift dictates design
Steel	40–200 ft 1 story 20–40 ft multistory		16–40 ft	Consider plastic design (single- and multistory) for max. economy
Concrete	60–100 ft 1 story 20–40 ft multistory			Most concrete construction is inherently rigid
Prestressed concrete				Multistory applications generally dictate post-tensioning
Timber	30–80 ft	15–25 ft		Available lumber lengths limit splice lengths, etc.
Laminated wood	40–250 ft			Great visual appeal but probably more expensive than steel
E. Arch			Rise/span	Consider one-dimensional system or three-dimensional (dome) with compression ring at top

APPENDIX 6A: *(Continued)*

System/Factor	*Economic load/span range*	*Depth/span ratio*	*Spacing/height range*	*Considerations*
Steel	Consider for spans exceeding 120 ft		1/5	
Reinforced concrete				
Laminated wood	Three-hinged: 40–250 ft		1/4–1/6	For relatively high-rise applications
	Two-hinged: 50–200 ft		1/4–1/6	For relatively low-rise applications
II. Horizontal distribution: *roof and floor support subsystems—primary*				
A. Beams				Consider double-cantilever system for roofs; pass-through of HVAC possible by web openings or stub-girder system
Steel	20–50 ft rolled 50–200 ft plate girders	1/20	20–40 ft	
Reinforced concrete	16–32 ft	1/16–1/21	20–40 ft	Normally cast monolithic with columns, creating rigid frame
Prestressed concrete	16–50 ft	1/10–1/20	20–60 ft	Depth controlled by connections to secondary members
Timber	6–20 ft floor 6–40 ft roof	1/20–1/30		
Laminated wood	6–40 ft floor 10–100 ft roof	1/20–1/30		Consider uncut lengths (field cut) for light structures or difficult shop details
Steel–concrete composite	25–50 ft	1/24	20–40 ft	20–30% weight reduction versus noncomposite
Post-tensioned concrete	50–100 ft	1/26–1/35	20–60 ft	Consider for long spans or when excessive release strengths are required for precast–prestressed; can also speed erection time versus conventional reinforced; better camber control
B. Trusses				
Steel	60–150 ft 150–250 ft	1/10–1/12	20–25 ft spacing 40–50 ft spacing with secondary trusses	Natural for pitched roofs; use camber for trusses exceeding 80 ft; extra depth facilitates passage of HVAC ducts
Timber	40–150 ft	1/6–1/10	12–20 ft	Natural for pitched roofs
Prestressed concrete	40–100 ft	1/10		Consider for esthetic and/or fire protection reasons
C. Wall beam				Consider for buildings > 10 stories
Steel	40–80 ft	8–10 ft	12–24 ft	Generally requires discipline of apartment, hotel-type layout

APPENDIX 6A: *(Continued)*

System/Factor	Economic load/span range	Depth/span ratio	Spacing/height range	Considerations
Concrete	40–80 ft	8–10 ft	12–24 ft Spacing generally dictated by economics of deck assembly	
Roof and floor support subsystems—secondary				
A. Beams				
Steel	20–40 ft	1/20	6–20 ft o.c.	
Reinforced concrete	20–40 ft	1/16–1/21	6–20 ft centers	Consider standard forming systems when establishing size and spacing (i.e., 8 ft 0 in. spacing, long forms, etc.)
Steel–concrete composite	20–40 ft	1/24	6–20 ft o.c.	Generally more economical than noncomposite for longer spans
Timber	10–20 ft		Joist at 12, 16, 24 in. o.c. Beams: 3–5 ft o.c.	Consider available lengths and stress grades
Laminated wood	10–40 ft			
Prestressed concrete	20–40 ft			
B. Joists				
Steel	12–120 ft Floors: < 40 ft Roofs: 25–120 ft	1/20	2 ft 0 in. o.c. floors 6 ft 0 in. o.c. roofs	Large open areas may give vibration problems; normally very economical in combination with corrugated metal deck; hard to beat for most roof applications
Plywood				Light commercial and residential
Truss–joist				Light commercial and residential
C. Cable	100–600 ft		Sag/span: 1/10	Economical solution to large column free areas; consider vibration and flutter; anchorage details affect economy substantially
D. Space Frame	100–600 ft	1/15–1/25		Economy dictated by connection details; consider limit analysis for max. economy; depth savings versus one-way systems
E. Trusses				
Timber	20–60 ft		24–40 in. o.c.	Light commercial and residential; natural for pitched roofs
Steel	40–60 ft			Consider for longer spans
Roof and floor support subsystems—deck assemblies				
A. One-dimensional				Consider diaphragm properties if critical for lateral support system

APPENDIX 6A: *(Continued)*

System/Factor	Economic load/span range	Depth/span ratio	Spacing/height range	Considerations
Plywood	2–4 ft			
Timber planks	8–20 ft			2-, 3-, 4-in. decking provides good insulation, fire resistance, and appearance
Pan joist	20–40 ft	1/16–1/18.5		Uniform depth possible with slab band system for girders
Corrugated metal decks (composite or noncomposite)				$\frac{9}{16}$ in.–28 gage popular for 2 ft 0 in. joist systems $1\frac{1}{2}$ in.–22 gage popular for joist roof systems 2 and 3 in. popular for longer-span floor systems
Floor plank	10–30 ft light to medium loads	1/30–1/40 floor 1/40–1/50 roof		Consider potential problems with differential camber; cutouts for mechanical and electrical systems
Precast tees	40–80 ft double-tee 60–100 ft single-tee	1/25–1/35 floor 1/35–1/40 roof		Consider potential problems with differential camber; cutouts for mechanical and electrical systems
Reinforced concrete slab	6–15 ft medium to heavy loads	1/40–1/50 post-tensioned		Consider using metal deck as either a form or to act as positive moment steel
Stressed-skin panels	16–24 ft floor 16–36 ft roof			Primarily for roof systems
Concrete I/D system				Light loads—residential application
Concrete spread joist				
Cementitious plank	2–4 ft			Primarily for roof applications; consider for acoustical value
Cementitious plank on bulb tees	2–8 ft			Primarily for roof applications; consider for acoustical value
Poured Gypsum deck on bulb tees	2–4 ft			Primarily for roof applications; consider for acoustical value
B. Two-dimensional				Consider standard forming systems (flying forms, etc.) for max. economy; post-tensioning can increase spans and speed erection; useful for irregular column layout
Flat slab	20–30 ft reinforced	1/32		
Waffle slab	25–40 ft medium to heavy loads	1/32–1/40		Consider for longer spans and heavier loads; visual appeal eliminates ceiling if mechanical and electrical can be accommodated

APPENDIX 6A: *(Continued)*

System/Factor	*Economic load/span range*	*Depth/span ratio*	*Spacing/height range*	*Considerations*
Voided waffle slab		1/32		Consider for integration of mechanical and electrical system
Pipe col-flat plate	12–16 ft light loads	1/32		Light residential
Two-way conc. slab	12–30 ft	1/32		
Flat plate	15–25 ft light to medium loads	1/32		Consider for multistory use where depth can yield savings; natural for residential (apartments, dorms, etc.)
C. Three-dimensional				Consider pneumatically applied concrete for thin shells; also prestressing; formwork major factor in cost
Shells			Rise/span	
Dome	50–400 ft	Thickness 3–4 in.	1/5–1/10	
Hyperbolic Paraboloid	60–100 ft	$ht/ab < 0.003$ $h =$ rise $t =$ thickness $a = \frac{1}{2}$ length $b = \frac{1}{2}$ width	1/3–1/6	
Barrel shell	30–350 ft	Thickness 3–4 in.	1/10	Short barrel acts much like one-way slab supported by arch ribs; long barrel acts much like deep curved beam
Folded plate	50–700 ft		1/8–1/5	
Lamella	30–300 ft		1/5	
Tensile structures			Sag/span	
Single cable Assemblies	200–600 ft		1/10	Use with rings in radial pattern or parallel (with anchorage); wind stability a problem
Cable nets	100–600 ft		1/10	Usually saddle shaped to allow pretensioning
Pneumatic structures (air-supported)	300–600		1/10	Initial air pressure normally 5 psf
Membranes	40–600 ft		1/10	Tentlike structures can achieve any conceivable form
III. Lateral Support			Varies with strength of lateral distribution subsystem	
A. Shear walls			Up to 50 stories	Consider using available walls, thereby eliminating special requirements for lateral support
B. Bracing				Generally most economical solution where possible

APPENDIX 6A: *(Continued)*

System/Factor	*Economic load/span range*	*Depth/span ratio*	*Spacing/height range*	*Considerations*
C. Moment frame			Up to 20 stories	May be required or desirable in seismic areas for ductility
D. Tube			30 to 60 stories	Requires closely spaced exterior columns
E. Core				
F. Shear walls and frame			Up to 70 stories	
G. Tube-in-tube			Up to 100 stories	
H. Wall beam/staggered truss			10 to 30 stories	Requires modular layout
IV. Foundation systems				
A. Spread footings	2000–8000 psf			Most economical except where heavy loads and nearness of rock make caissons more economical
B. Piles			3–4 ft o.c.	Select pile type based on driving resistance expected, anticipated length and splicing considerations, friction versus end-bearing strength
Timber	20–25 tons			
Steel	100 tons			
Cast-in-place concrete	60–80 tons			
Precast concrete	80 tons			
C. Caissons	10–7000 tons			
D. Mats				Use to minimize differential settlement and to "float" building
V. Lateral distribution				
A. Diaphragm		Very flexible diaphragms limited to 1/2; semirigid limited to 1/5		
Metal deck				Welding details are critical to performance
Cast-in-place concrete				Rigid diaphragm
Precast				Connections to transfer shear are critical
Plywood				Nailing pattern critical
B. Bracing				Requires discipline to standard bays and framing methods; consider performance-type spec and prebidding
VI. Preengineered systems				
A. Metal building	40–120 ft rigid frame; > 80 ft post and beam		20–30 ft spacing	Normally one-story structures only
B. Precast concrete	40–60 ft		10–20 stories	Consider for office and residential use

APPENDIX 6A: *(Continued)*

System/Factor	*Economic load/span range*	*Depth/span ratio*	*Spacing/height range*	*Considerations*
VII. Bridge systems				
Slab type				
Solid slab	45 ft	1/15–1/24		
Cored slab	40–65 ft	1/15–1/24		
Precast multi-stemmed units	40–80 ft	1/15–1/24		
Stringer type				
Steel beams (simple)	40–80 ft		7–8.5 ft	
Steel beams (continuous)	40–110 ft		7–8.5 ft	Consider composite construction for spans > 65 ft
Plate girders (simple)	80–110 ft		7–8.5 ft	Consider composite construction for spans > 80 ft
Plate girders (continuous)	80–350 ft		7–8.5 ft	Use higher-strength steels above 175 ft
Precast I-girder	20–100 ft	1/18	7–8.5 ft	
Girder type				
Plate girders (continuous)	175–350 ft			Use higher-strength steels
Box girder type				
Twin steel box Girders	110–175 ft			
Reinforced-concrete box girder	80–200 ft	1/16		
Cast-in-place, post-tensioned	80–1,000 ft	1/22		
Precast segmental	100–700 ft	1/18–1/20		
Arch type			Rise/span	
Barrel	50–200 ft	1/70–1/80	1/5–1/6	
Ribbed	100–1,000 ft	1/70–1/80	1/5–1/6	
Tied	100–400 ft	1/70–1/80	1/5–1/6	
Splayed	250–850 ft	1/70–1/80	1/5–1/6	
Truss type				
Warren	300–600 ft	1/15–1/10	Diagonals should be 50–55° from the horizontal and panels spaced 16–32 ft	
K-truss	300–600 ft	1/5–1/10		
Cable-supported type				
Stayed girder	600–1,600 ft	1/60–1/80	Height of pylon: 1/6–1/8 × span	Ratio of side to main spans ~3:7
Cable-suspension	600–1,800 ft	Cable sag: 1/8–1/12		Ratio of side to main spans ~1:4 Stiffening truss depth: 1/60–1/170 × span

REFERENCES

1. ACI Committee 442, *Response of Multi-story Concrete Structures to Lateral Forces,* SP-36. Detroit, Mich.: American Concrete Institute, 1973.
2. ALLISON, HORATIO, *Practical Steel Design for Buildings.* Chicago: American Institute of Steel Construction, 1976.
3. American Institute of Timber Construction, *Timber Construction Manual.* New York: John Wiley & Sons, Inc., 1974.
4. AMRHEIN, JAMES E., *Reinforced Masonry Engineering Handbook.* Los Angeles, Calif.: Masonry Institute of America, 1972.
5. BEAUFAIT, FRED W., *Tall Buildings, Planning, Design and Construction.* Nashville, Tenn.: Civil Engineering Program, Vanderbilt University, 1974.
6. *Cable Roof Structures.* Bethlehem, Pa.: Bethlehem Steel, 1968.
7. *Concrete Thin Shells,* SP-28. Detroit, Mich.: American Concrete Institute, 1971.
8. Council on Tall Buildings and Urban Habitat, *Structural Design of Tall Concrete and Masonry Buildings.* Vol. CB. New York: American Society of Civil Engineers, 1978.
9. Council on Tall Buildings and Urban Habitat, *Structural Design of Tall Steel Buildings,* Vol. SB. New York: American Society of Civil Engineers, 1979.
10. Council on Tall Buildings and Urban Habitat, *Tall Building Systems and Concepts,* Vol. SC: Monograph on Planning and Design of Tall Buildings. New York: American Society of Civil Engineers, 1980.
11. COWEN, HENRY J., and JOHN DICKSON, *Building Science Laboratory Manual.* London: Applied Science Publishers Ltd., 1978.
12. *Design of Barrel Shell Roofs,* ST-77. Chicago: Portland Cement Association, 1954.
13. *Design of Circular Domes,* ST-55. Chicago: Portland Cement Association, n.d.
14. *Elementary Analysis of Hyperbolic Paraboloid Shells,* ST-85. Chicago: Portland Cement Association, 1960.
15. FINTEL, MARK, ed., *Handbook of Concrete Engineering.* New York: Van Nostrand Reinhold Company, 1974.
16. FISHER, ROBERT E., *New Structures.* New York: McGraw-Hill Book Company, 1964.
17. GAYLORD, EDWIN H., JR., and CHARLES M. GAYLORD, *Structural Engineering Handbook.* New York: McGraw-Hill Book Company, 1968.
18. HOWARD, H. SEYMOUR, JR., *Suspended Structures Concepts.* Pittsburgh, Pa.: United States Steel, 1966.
19. KOMENDANT, D. E., *Contemporary Concrete Structures.* New York: McGraw-Hill Book Company, 1972.
20. LEBA, THEODORE, JR., *Design Manual: The Application of Non-reinforced Concrete Masonry Load Bearing Walls in Multi-storied Structures.* Arlington, Va.: The National Concrete Masonry Association, 1969.
21. MAINSTONE, ROLAND J., *Developments in Structural Form.* Cambridge, Mass.: The MIT Press, 1975.
22. MERRITT, FREDERICK, *Structural Steel Designer's Handbook.* New York: McGraw-Hill Book Company, 1972.
23. OTTO, FREI, ed., *Tensile Structures.* Cambridge, Mass.: The MIT Press, 1973.
24. PARME, A. L., and JOHN A. SBAROUNIS, *Direct Solution of Folded Plate Concrete Roofs.* Chicago: Portland Cement Association, n.d.
25. *PCI Design Handbook.* Chicago: Prestressed Concrete Institute, 1978.
26. RAMASWAMY, G. S., *Design and Construction of Concrete Shell Roofs.* New York: McGraw-Hill Book Company, 1968.
27. RAMSEY, C. G., and H. R. SLEEPER, *Architectural Graphic Standards.* New York: John Wiley & Sons, Inc., 1970.
28. *Recommended Practice for Engineered Brick Masonry.* McLean, Va.: Brick Institute of America, 1969.
29. *Roofs with a New Dimension.* Chicago: Portland Cement Association, 1959.
30. SALVADORI, MARIO, and ROBERT HELLER, *Structure in Architecture.* Englewood Cliffs, N.J.: Prentice-Hall, Inc., 1975.
31. SALVADORI, MARIO, and MATTHYS LEVY, *Structural Design in Architecture.* Englewood Cliffs, N.J.: Prentice-Hall, Inc., 1967.
32. SCHUELLER, WOLFGANG, *High-Rise Building Structures.* New York: John Wiley & Sons, Inc., 1977.
33. *Seismic Design for Buildings,* TM 5–809–10, NAV FAC P-355, AFM 88–3, Chap. 13. Washington, D.C.: Departments of the Army, the Navy, and the Air Force, 1973.
34. Task Committee on Air-Supported Structures, *Air-Supported Structures.* New York: American Society of Civil Engineers, 1979.
35. WILSON, FORREST, *Emerging Form in Architecture: Conversations with Lev Zetlin.* Boston: Cahners Books, 1975.
36. The Working Group on Recommendations, *Recommendations for Reinforced Concrete Shells and Folded Plates.* Madrid: International Association for Shell and Spatial Structures, 1979.

7

CREATION OF SYSTEMS

7-1 INTRODUCTION

The first seven chapters of this book have developed a systematic process for conceptual design of structural systems. Starting with the identification of project requirements and loads, and continuing through a building-block process of structural assembly, our objective has been to develop an understanding of structural system behavior and to acquire some conceptual skills along the way (e.g., approximate analysis and design techniques). In order now to implement fully the systems design model (Fig. 7-1), two final parts of the model must be developed: *generation of ideas and concepts,* and a methodology for *comparison of alternate systems.*

To this point, we have developed essentially a collection of factual information, including information regarding requirements, loads, structural materials, and structural assembly. Along with this collection of information, we have expanded our understanding of behavior and acquired certain approximate design skills. A gradually expanding *set of choices* was identified for each level of structural assembly. It is now appropriate to broaden the range of choices further by using the term *ideas* to name more appropriately the almost infinite number of

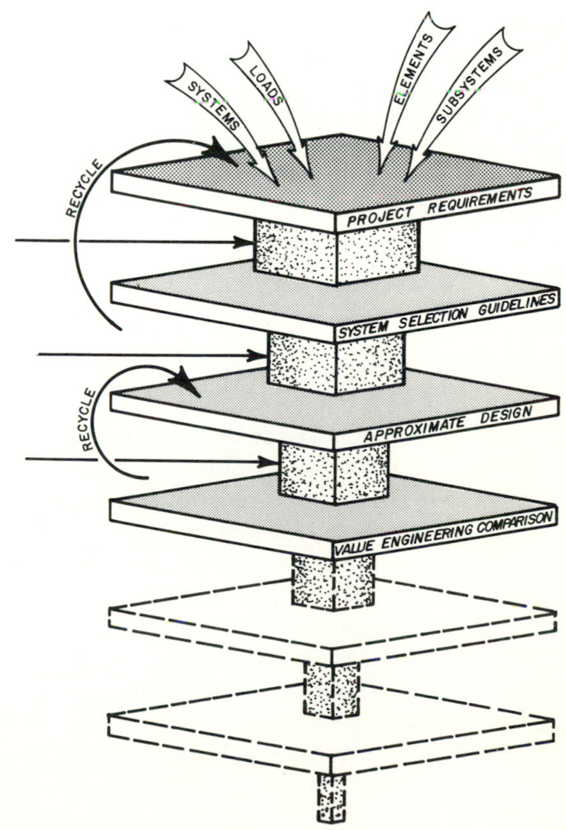

Figure 7-1. Systems Design Model

possibilities that we might consider for structural systems design. *Ideas* are the raw material from which structural systems are conceived and are the sparks that ignite the entire systems design process. With the systems design model, we can systematically screen, evaluate, and refine ideas into a selected system for final analysis and design.

Where do ideas come from? How can more and better ideas be generated? At what point do we evaluate ideas? Ideas are, without question, the most important part of the system design process. The objective of this chapter is to develop some techniques to assist us with idea generation. After the discussion on ideas, the remainder of the chapter concentrates on the methodology for final comparison of alternatives that emerge from the systems design model as potential candidates for selection.

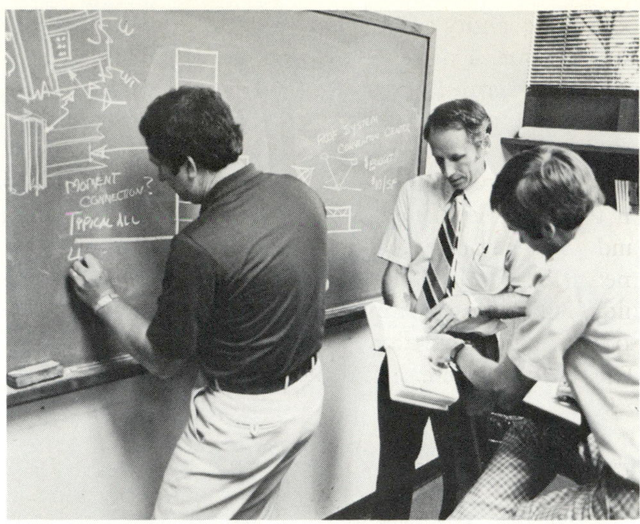

Figure 7–2. Brainstorming Session

7–2 THE CREATIVE PROCESS: GENERATION OF ALTERNATIVES

The subject of creativity has been studied extensively and is the subject of continuing research in both educational and professional circles. It has been successfully demonstrated that creativity can be taught, and there are established techniques for improving our ability to generate both a greater quantity and quality of ideas. It has also been established that very few ideas are truly original. Most ideas are either adaptations, or modifications or a synthesis of other ideas, meaning that we get many of our ideas from our experiences and from the ideas of others. This is one reason that laboratory model experiments can play such a major role in the approach to structural systems design. The sensory experience of observing the model behavior creates mental impressions which later become resources for the generation of ideas.

Sources of Ideas

The objective of a structural system is to enclose space and to carry the load for the least cost. This means that we are continually searching for the most efficient load paths to carry the loads to the foundation; looking for ideas that activate the unique strength of the materials, the structural elements, and subsystems.

One source of ideas is the *data base* for structural systems contained in Appendix 6A. At this point, we can use the data base not only as a screening device but also to indicate what has been done successfully in the past, providing us with ideas for our particular project. Another very effective method for generation of ideas is a group *brainstorming* session (Fig. 7–2), a concept fundamental to the value engineering process. The application of group creativity is based on the premise that, as technology becomes more and more complex and as more projects become truly interdisciplinary in nature, it is unrealistic to believe that one or two individuals can successfully generate the range of ideas that are potentially applicable to a project.

The key to brainstorming activity is to totally exclude evaluation (of ideas) while the creative process is in motion. This allows free thinking, and "piggybacking" of ideas and prohibits our analytical skills from interfering with synthesis and creative thought. If we allow judgment to enter the process during the creation of ideas, it will not only slow down the process but will, in fact, inhibit

Figure 7–3. Graphic Idea Generation

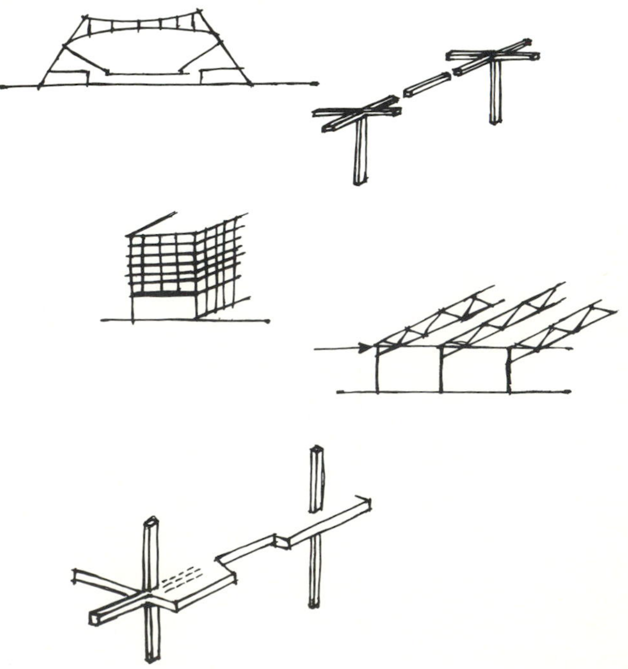

both the quantity and the quality of ideas that the group can generate. This reservation of judgment until later is sometimes termed "green light thinking," with the judgment phase referred to as "red light thinking."

Visual techniques are very effective in increasing the generation of ideas. A person's mind essentially thinks in terms of visual images anyway, and simple sketches and graphic portrayal of ideas can promote idea development (Fig. 7–3). Design needs to be a very visual technique, utilizing crude models, rough sketches—anything to spark the imagination and to allow our minds to work. The following exercise illustrates the group creativity process.

EXAMPLE PROBLEM 7–1: Use of Group Creativity

Assume that preliminary research and fact finding has established the following statement of the design problem to be solved:

An owner has decided to build a 100,000-sq ft (9290-m^2) office building on a downtown city block in Washington, D.C. Zoning requirements limit the maximum height of a building in the area to 15 stories or 150 ft (45.7 m), whichever is less. Preliminary code research indicates that Type II construction will probably be required. The city block is 400 ft × 400 ft (122 m × 122 m). The owner has expressed a desire for a good image for the facility to increase its rentability, which may include a plaza and landscaping at the ground level. He has also expressed a desire for as much on-site parking as possible in order to enhance the convenience to tenants. Based on his investment requirements for the facility, he has decided that a tentative construction cost budget is approximately $5,000,000.

Brainstorm potential solutions to the problem and see if at least 20 ideas can be generated from the group interaction. Use visual techniques (blackboard, sketches, etc.) to record the ideas as you proceed.

7–3 COMPARISON OF ALTERNATIVES

Once we have generated a quantity of ideas, we can put them through the systems design process to isolate two, three, or four systems for approximate analysis and design and more detailed comparisons. In Chapter 6 we compared alternative systems *within* (intra-) a given classification and did not consider either other structural systems or the impact on other project systems, such as architectural or mechanical. Now we are concerned with both: *all* potential structural systems (intersystem comparisons) and the impact on other project systems. Our process for comparing alternatives is the *value engineering* concept, which makes comparisons based on *value* to the owner.

Value Engineering

Value is defined as the benefit gained from the investment in the project. Potential benefits include return on investment, esteem value, functional use, flexibility, level of operation, and maintenance costs. Value can be measured in terms of a cost/worth ratio in which the estimated cost for an alternative is divided by the estimated cost for a standard alternative that would achieve the same function (Fig. 7–4). The ratio is a measure of the increment of additional cost that is being added to the project

Figure 7–4. Cost/Worth Ratio

QTY.	UNIT	COMPONENT	FUNCTION			EXPLANATION	VALUE OF WORTH	ORIGINAL COST
			VERB	NOUN	KIND			
1.5	TON	Steel column	Support	Load	B		$1,000 *	$1,500
250	SF	Paint	Protect	steel	S			$150
5	EA	Openings	Provide	passage	S	steam pipes		$500
20	EA	Ladder rungs	Provide	access	S	Maintenance ladder		$300
						Total	$1,000	$2,450
			Cost/Worth = 2450/1000 = 2.45					

* Alternate concrete column

to derive certain value benefits. With this yardstick, decisions can be made as to whether the value added is worth the additional cost.

Another method of comparing alternatives, and one that we will use, is to develop a weighted list of project criteria and their relative importance to the success of the project. These criteria can be incorporated on a project evaluation form such as shown in Fig. 7–5. Evaluation criteria normally include many of the benefits previously discussed as well as the *initial cost* and *life-cycle cost* of the alternatives, which allows both cost and benefits to be evaluated in one process. Development of system costs is described in the following paragraphs.

Initial Cost. Figure 7–6 illustrates the tabulation of initial costs for one structural system. In accordance with the value engineering methodology, each component is identified by function, dollars per system unit, dollars per building square foot, and differential cost impact on other systems.

The *functional identification* includes a verb and noun description of the purpose of the component and whether this purpose is basic (B) to the structural mission or secondary (S). The basic and secondary description creates a visible means of isolating added costs, allowing later analysis if the total estimated project cost requires reduction.

Figure 7–5. Evaluation Form for Structural Systems

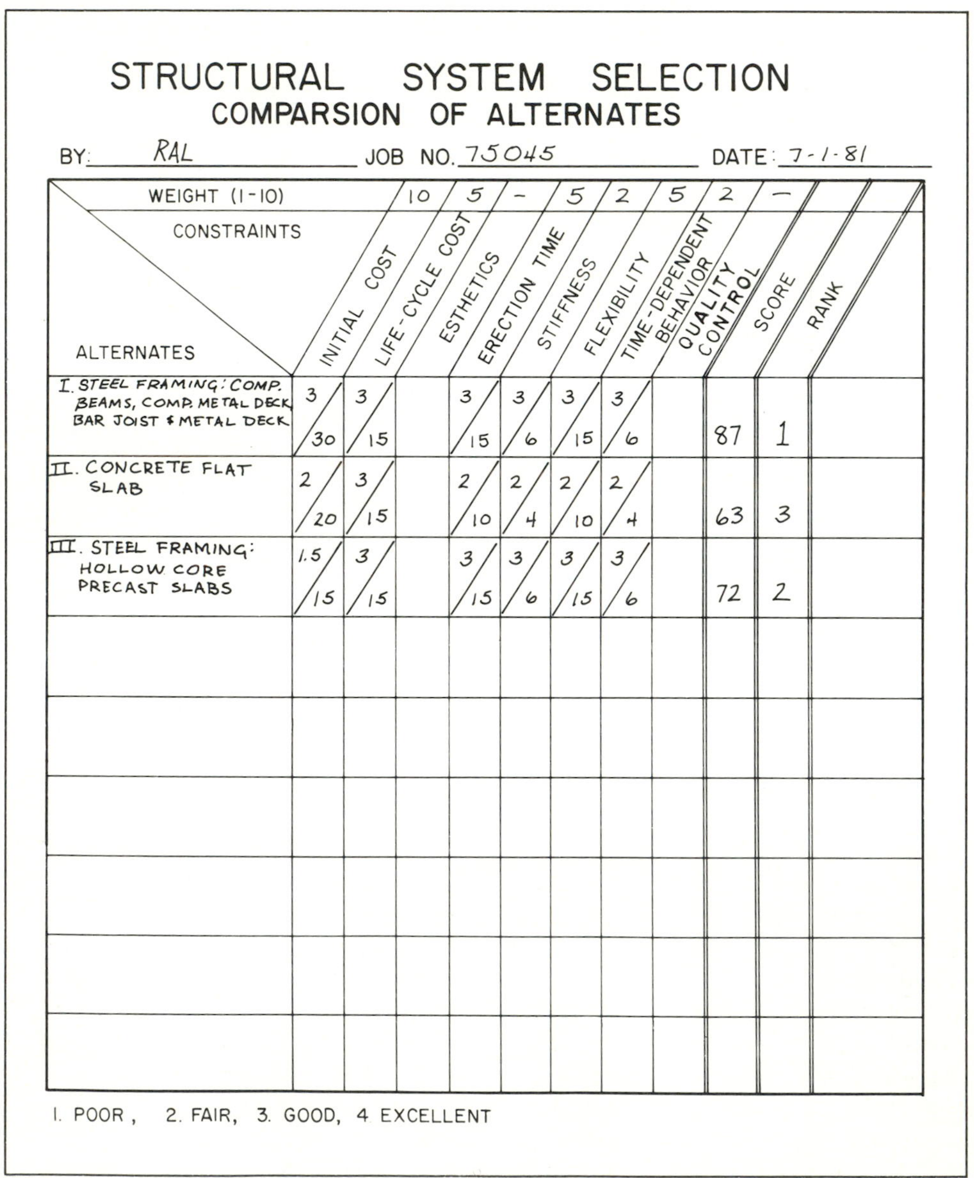

STRUCTURAL SYSTEM SELECTION
COMPARSION OF ALTERNATES

BY: RAL JOB NO. 75045 DATE: 7-1-81

CONSTRAINTS / ALTERNATES	INITIAL COST	LIFE-CYCLE COST	ESTHETICS	ERECTION TIME	STIFFNESS	FLEXIBILITY	TIME-DEPENDENT BEHAVIOR	QUALITY CONTROL	SCORE	RANK
WEIGHT (1-10)	10	5	-	5	2	5	2	—		
I. STEEL FRAMING: COMP. BEAMS, COMP. METAL DECK BAR JOIST & METAL DECK	3/30	3/15		3/15	3/6	3/15	3/6		87	1
II. CONCRETE FLAT SLAB	2/20	3/15		2/10	2/4	2/10	2/4		63	3
III. STEEL FRAMING: HOLLOW CORE PRECAST SLABS	1.5/15	3/15		3/15	3/6	3/15	3/6		72	2

1. POOR, 2. FAIR, 3. GOOD, 4 EXCELLENT

STRUCTURAL SYSTEM SELECTION
INITIAL COST

SYSTEM: COMPLETE FRAMING BY: RAL DATE: 8-1-81 JOB NO. 75045

DESCRIPTION: SCHEME #1: STEEL FRAME, COMP. BEAMS, COMP. DECK. QUANTITY: 60,260 SF

COMPONENT	FUNCTION			REMARKS	\$ SYST. UNIT	\$ BLDG S.F.	±\$ OTHR. SYST.	NET \$/ BLDG. S.F.
	VERB	NOUN	KIND					
BASIC SYSTEM	CARRY	LOADS	B					
	PROVIDE	STAB.	B					
ROOF			B	BAR JOISTS @ 6'O.C. 1½" METAL DECK	.84/SF	.27		
FLOOR			B	4½" CONC. SLAB ON TYPE I MACCOR; COMP. BEAMS	2.65/SF	1.83		
COLUMNS			B	STEEL WF	14.70/LF	.41		
	PROVIDE	FIRE PROT.	S	1. CEILING @ .20 COL. @ .22			+.42	
	PROVIDE	WEATH. PROT.	S	2.				
	PROVIDE	OPNGS.	S					
	PROVIDE	CLEAR.	S					
	PROVIDE	DURAB.	S	3.				
	ENCASE	MECH. EL. SYS.	S	4.				
		HEIGHT	S	5. ADDS 12" DEPTH COMPARED TO FLAT SLAB			+.15	
		"U"	S					
		SOUND	S					
		VIBRA.	S					
		POND'G.	S					
		FDNS. WEIGHT	S					
	CARRY	EQUIP. LOADS	S					
				TOTALS		2.51	.57	3.08

1. INCLUDE COST OF DAMPERS AND HOODS FOR LIGHT FIXTURES FOR FIREPROTECTION BY RATED CEILINGS.
2. INCLUDE COSTS OF FLASHING, CAULKING, WATERPROOFING, WHEN SYSTEM DOUBLES AS AN EXTERIOR WALL.
3. INCLUDE PAINT, PRESERVATIVE TREATMENT, AIR-ENTRAINING, ETC.
4. CELLULAR FLOOR SYSTEMS, EXTRA TOPPING THICKNESS, ETC.
5. INCLUDE HVAC LOADS, WALL COSTS, STAIR COSTS, VERTICAL DUCT, PIPE, & CONDUIT RUN.

Figure 7–6. Initial Cost Tabulation

We obtain dollars per system unit by preliminary design and use of the cost data in Appendix 7A. The conversion to dollars per building square foot (or bridge square foot, etc.) provides a convenient method for tabulating total costs for the system.

Structural systems normally have a cost impact on other project systems, including architectural elements such as walls, ceilings, and story height; mechanical elements such as clearance for ducts and openings and support for equipment; electrical items such as floor penetrations; and so on. A true evaluation of structural alternatives must consider the cost impact on these other systems. This cost impact is included as a differential cost (plus or minus) to the other systems. It is based on establishing one of the structural alternative systems as the *base system,* allowing a relative comparison of the impact of the other structural systems by a + or − index to the base system. For example, if a flat slab system is established as the base system, Fig. 7–6 shows that +\$0.15 must be added to the steel system to account for 12 in. (305 mm) of added depth.

Life-Cycle Cost. Life-cycle cost is defined as the total cost of building and operating a facility for a specified period of time (usually 20 to 30 years). For the structural system, elements of life-cycle cost include:

- Initial investment
- Repair and maintenance
- Replacement
- Operating costs (energy)
- Insurance

Figure 7–7 illustrates the tabulation of life-cycle costs. The energy costs are obviously not a direct structural function, but including them accounts for life-cycle impact on the mechanical and electrical systems.

System Selection

Optimization of all the systems for a project (initial cost) is automatically achieved with the techniques illustrated in Fig. 7–6. After we complete the initial cost estimates and life-cycle cost estimates (Fig. 7–7), we compare the various structural system alternates using the weighted criteria form of Fig. 7–5. Before proceeding with an example, let us review a summary of the total menu of design choices now available to us:

- Material selection
- Geometry (height, width, length)
- Lateral support subsystem
- Horizontal distribution subsystem
- Vertical support subsystem
- Lateral distribution subsystem
- Foundation subsystem

Our approach now is to first *define the problem* by establishing project requirements and loads and second, to begin *solving the problem* by generating ideas

Figure 7–7. Life-Cycle Cost Tabulation

STRUCTURAL SYSTEM SELECTION
LIFE-CYCLE COST

SYSTEM: COMPLETE FRAME BY: RAL DATE: 7-1-81 JOB NO. 75045
DESCRIPTION: SCHEME #1 QUANTITY: 60, 260 SF

CATEGORY	ITEM	REMARKS	LIFE-CYCLE COST
MAINTENANCE	PAINT		
	REPAIR		
	REPLACEMENT		
OPERATING	FUEL		
INSURANCE		PRO RATED FIRE INSURANCE PREMIUM FOR 30 YEARS	
CUSTODIAL			
INVEST. PVT.		$250,000 @ 12% FOR 30 years	
		TOTAL	

for potential systems with the methods we discussed earlier in the chapter. Both the project requirements and the system selection guidelines of Appendix 6A can then be used to screen the potential systems down to a reasonable number for comparison. The following example illustrates the total *system selection process.*

EXAMPLE PROBLEM 7–2: Structural System Selection for a Student Center Building

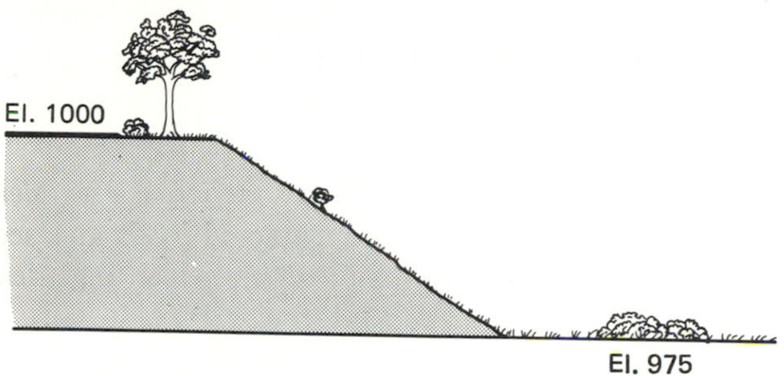

Figure 7–8. Example Problem 7–2

A 50,000-sq ft student center will be built on a college campus in Greenville, South Carolina. The site slopes steeply to an elevation 25 ft (7.6 m) below street level (Fig. 7–8). A complete program of space requirements has established the following blocks of net space that need to be on the same level:

Student activities	8,000 sq ft
Administrative offices	10,000 sq ft
Auditorium	8,000 sq ft
Meeting rooms	5,000 sq ft
Miscellaneous	1,500 sq ft
Net area =	32,500 sq ft (3019 m²)

Assumed net/gross ratio = 0.65

Total gross area = 32,500/0.65 = 50,000 sq ft (4645 m²)

Preliminary soil information indicates varying soil conditions at the lower elevation, with mixtures of sand and clay. Although adjacent buildings are supported on deep foundations, there is a possibility that selective excavation of unsuitable soil material will allow use of shallow spread footings at an allowable bearing pressure of 2500 psf (119,700 N/m²), provided that some amount of differential settlement can be tolerated. If not, firm bearing of 10 tons/sq ft (957,600 N/m²) is available at a depth of approximately 40 ft (12.2 m).

Select a structural system for final design.

Solution

1. *Project requirements:*

Function: 50,000-sq ft (4645-m²) student center. Program defines space affinities. Floor to ceiling height = 9 ft (2.7 m). (*Note:* No bracing between columns permitted as a result of step 3 below.)

Esthetics: Structural system will not be exposed. Architectural form needs to be innovative to establish image for college.

Serviceability: Normal deflection limits (Table 2–1). Sloped roof for drainage. Check floor vibration in open areas. Sound transmission (Table 2–3). Flexible structure if supported on shallow footings. Fire protection (Table 2–6).

Type IV construction permitted for two stories at 18,000 sq ft (1672 m²) each (no sprinklers) and 36,000 sq ft (3344 m²) each (with sprinklers). One story permitted at 18,000 sq ft (1672 m²) (no sprinklers) and 54,000 sq ft (5017 m²) (with sprinklers).

Type II construction permitted for single-story with no limit on area and multistory use up to 80 ft (24.4 m) with no limit on area.

(After step 3, add the following specific requirements:)

Columns:	3 hr
Beam:	2 hr
Floor:	2 hr
Roof:	1 hr

Construction: Construction budget = $2,500,000. Construction schedule = 18 months. Quality of work needs to be excellent.

2. *Loads:*

a. *Live load* (Appendix 3C):

Offices:	50 psf (2394 N/m²)
Campus store:	100 psf (4788 N/m²)
Corridors:	100 psf (4788 N/m²)
Lobby and reception:	100 psf (4788 N/m²)
Assembly:	100 psf (4788 N/m²)
Recreation:	75 psf (3591 N/m²)

Reduce live load by Eq. 3–2.

b. *Minimum roof live load:*

By local code 20 psf (958 N/m²)

c. *Snow load:* By Eq. 3–3,

$$p_f = 0.7 C_e C_t I p_g$$

$$p_g = 10.0 \text{ psf (Fig. 3–14)}$$

$$C_e = 1.0 \text{ (Table 3–2)}$$

$C_t = 1.0$ (Table 3–3)

$I = 1.0$ (Table 3–4)

$p_f = 0.7(1.0)(1.0)(1.0)10.0$

$= 7.0$ psf (287 N/m²)

d. *Wind load:* (*Note:* Complete wind load calculations after the building geometry is determined in step 3 below.)

$V = 75$ mi/hr (120.7 km/hr) (Fig. 3–50)

$I = 1.00$ (Table 3–7)

$(IV)^2 = 1.00(75)^2 = 5{,}625$

$G_h = 1.46$ for building height $\simeq 40$ ft (Table 3–8, Exposure B)

C_p (windward wall) $= 0.8$ (Table 3–9)

C_p (leeward wall at $L/B = 100$ ft/167 ft) $= -0.5$ (Table 3–9)

Pressure Tabulation (k_z values from Table 3–6):

$$P_z = q_z G_h C_p = 0.00256 k_z (IV^2) G_h C_p$$

$$= 0.00256 k_z (5625)(1.46)(0.8)$$

$$= 16.8 k_z$$

$$P_h = 0.00256 k_h (5625)(1.46)(-0.5)$$

$$= -10.5 k_h$$

$$= -10.5(0.57) + 5.9 \text{ psf at } h = 40 \text{ ft}$$

Height	k_z	p_z	p_h
0–15	0.37	+6.2	−5.9
20	0.42	+7.1	−5.9
25	0.46	+7.7	−5.9
30	0.50	+8.4	−5.9
40	0.57	+9.6	−5.9

Assume total $p = 15.0$ psf (718 N/m²) as a uniform load for preliminary design.

3. *Building form and geometry.* With the given site, practical options for the building are two stores at 25,000 sq ft (2323 m²) each or three stories at 16,700 sq ft (1551 N/m²) each. Type II construction will be required in either case, unless the building is sprinklered.

The three-story scheme provides a good match to the program requirements and provides excellent architectural design opportunities. It has a relative disadvantage of concentrating more foundation load over a smaller area, a potential problem in light of the preliminary soils report. Wind loads will be higher but are not a significant factor at this height. An elevator will probably be required.

The two-story scheme will have lower wind loads, less concentration of foundation load, and probably will not require an elevator. However, this scheme will require grouping of functional spaces in less than an ideal arrangement. Also, two stories does not provide enough vertical height to work architecturally with the site.

Note: The evaluation described above is representative of the deliberation by the design team during the early conceptual stage. In this case, the architect selected the three-story scheme and produced a schematic floor plan as shown in Fig. 7–9. The structural bay dimensions were coordinated with the structural engineer and incorporated in the overall dimensional layout. The vertical floor-to-floor dimension was established at 9 ft 0 in. (2.7 m) + 2 ft 6 in. (0.76 m) for duct (Table 2–10) + structural depth. Cantilever floor areas for balconies and column placement do not permit bracing between columns.

4. *System alternatives:*

a. *General:* For Type II construction, the 2-hour rating for the floor subsystem can be obtained with a 4½-in. (114-mm) concrete slab with membrane protection (if unprotected steel beams are used) or with a 5¼-in. (133-mm) slab with contact fireproofing of steel members (Appendix 5C).

Steel framing is known to be competitive in the area and will produce a lightweight system capable of tolerating some degree of differential settlement. Type 3 construction can be used to simplify framing and provide a more flexible structure. Concrete framing will eliminate the fire protection requirements, provide a natural rigid frame, and will save on building height. Because of its rigidity and height, a concrete system will require deep foundations.

After reviewing all the project requirements

Figure 7–9. Example Problem 7–2, Floor Plan

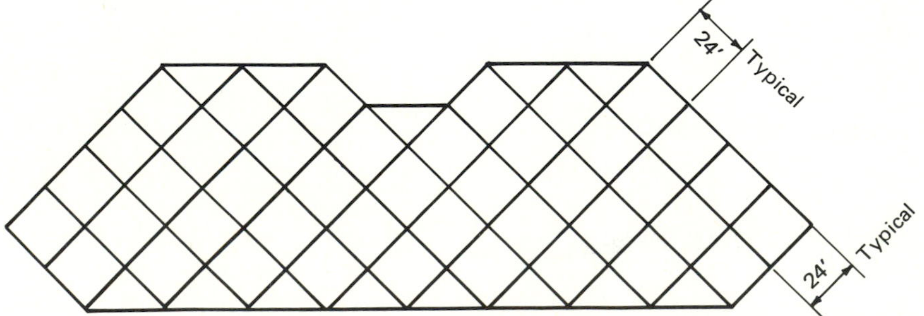

and Appendix 6A, select the following systems for preliminary design and comparisons.

b. *Scheme I:*
   (1) *Horizontal support:* Steel framing, with composite steel beams at 8 ft 0 in. (2.4 m) spacing and supported by composite steel girders; $4\frac{1}{2}$-in. (114-mm) concrete slab on composite metal deck; bar joists and metal deck for roof subsystem.
   (2) *Vertical support:* Post and beam, with steel columns and composite steel girders (simple framing for gravity loads). Columns fire protected.
   (3) *Lateral support:* Type 3 semirigid frame to carry wind loads.
   (4) *Lateral distribution:* Slab should act effectively as diaphragm.
   (5) *Foundation support:* Spread footings.

c. *Scheme II*
   (1) *Horizontal support:* Concrete flat slab, 24 ft × 24 ft (7.3 m × 7.3 m).
   (2) *Vertical support:* Rigid frame formed by flat slab and concrete columns.
   (3) *Lateral support:* Rigid frame.
   (4) *Lateral distribution:* Slab should act effectively as diaphragm.
   (5) *Foundation support:* Caissons drilled to firm bearing strata probably at an average depth of 40 ft (12.2 m).

5. *Preliminary design of Scheme I:*
   a. *Loads:*
      (1) Live Load:

$$\text{Floors} = 100 \text{ psf } (4788 \text{ N/m}^2)$$
$$\text{Roof} = 20 \text{ psf } (958 \text{ N/m}^2)$$

      (2) Dead load:
         Floors:

$4\frac{1}{2}$-in. conc. slab	=	48 psf
Ceiling	=	2 psf
Beam framing	=	5 psf
Misc.	=	3 psf
		58 psf (2777 N/m²)

         Roof:

Built-up roof	=	6 psf
Insulation	=	2 psf
Metal deck	=	3 psf
Ceiling	=	3 psf
Misc.	=	3 psf
Joist framing	=	2 psf
		18 psf (862 N/m²)

   b. *Materials:*
      (1) Concrete: 3000 psi (20.7 MPa).
      (2) Steel: A36, 36,000 psi (248 MPa).
   c. *Floor subsystem:*
      (1) Deck: By Appendix 5A, use 24-gage, $1\frac{5}{16}$-in. (33-mm) composite metal deck.
      (2) Slab:

$$w_u = 1.4(53) + 1.7(100) = 244 \text{ psf } (11{,}682 \text{ N/m}^2)$$

(58 − 5 = 53 for beam framing.)

$$S = \frac{bh^2}{6} = \frac{12(4.5)^2}{6}$$
$$= 40.5 \text{ in.}^3 \ (663{,}795 \text{ mm}^3)$$

$$-M = \frac{wl^2}{11} = \frac{-244(8)^2}{11}$$
$$= -1419 \text{ ft-lb } (-1930 \text{ Nm})$$

$$f_b = \frac{1419 \times 12}{40.5} = 420 \text{ psi } (2.9 \text{ MPa})$$
$$< 1260 \text{ psi } (8.7 \text{ MPa})$$

Since allowable 1260 psi (8.7 MPa) is based on 0.85% reinforcement, assume $A_{s\,(\text{req'd})} = 0.35\%$ (minimum):

$$-A_s = 0.0035(12)(3.75)$$
$$= 0.16 \text{ in.}^2/\text{ft } (339 \text{ mm}^2/\text{m})$$

Extend to one-third point of slab.

      (3) Beam: w = 158 psf (7565 N/m²); full lateral support.

$$M = \frac{wl^2}{8} = \frac{8(0.158)(24)^2}{8}$$
$$= 91.0 \text{ ft-kips } (123{,}396 \text{ Nm})$$

$$S_{\text{req'd}} = \frac{91.0 \times 12}{24} = 45.5 \text{ in.}^3 \ (745{,}745 \text{ mm}^3)$$

For composite beam, assume that minimum d = 0.80 × noncomposite and S = 0.70 × noncomposite:

$$d = 0.80(\tfrac{1}{2} \times 24) = 9.6 \text{ in. } (244 \text{ mm})$$
$$S = 0.70(45.5) = 31.9 \text{ in.}^3 \ (522{,}841 \text{ mm}^3)$$

By Appendix 4B, use W12 × 26.

$$V_h = \frac{A_s f_y}{2} = \frac{7.65(36)}{2} = 137.7 \text{ kips } (612{,}490 \text{ N})$$

According to Table 4–7,

$$\text{No. shear connectors} = \frac{137.7}{11.5}$$
$$= 12 \text{ each side (24 total)}$$

d. *Roof subsystem:*
   (1) Deck: By Appendix 5A, use 22-gage, $1\frac{1}{2}$-in. (38-mm) metal deck.
   (2) Joist: Assume 6 ft 0 in. (1.8 m) spacing.

$$w = 38 \text{ psf}(6 \text{ ft}) = 228 \text{ lb/ft } (3327 \text{ N/m})$$

$$M = \frac{wl^2}{8} = \frac{228(24)^2}{8} = 16{,}416 \text{ ft-lb } (22{,}260 \text{ Nm})$$
$$= 196{,}992 \text{ in.-lb}$$

By Appendix 4F, use 14 H 4 at 6.5 lb/ft (94.8 N/m).

e. *Rigid frame analysis for gravity load:*
   (1) Floor girder:

$$\text{Beam end reaction} = 8(0.158)(24) = 30 \text{ kips } (133{,}440 \text{ N})$$

$$\text{Equiv. } w = \frac{2(30)}{24} = 2.50 \text{ kips/ft } (36{,}483 \text{ N/m})$$

For two concentrated loads (using correction factor):

$$M_g = \frac{1.33\ wl^2}{8} = \frac{1.33(2.5)(24)^2}{8} = 239.4 \text{ ft-kips } (324{,}626 \text{ Nm})$$

   (2) Roof girder:

$$\text{Joist end reaction} = 6(0.038)(24) = 5.47 \text{ kips } (24{,}331 \text{ N})$$

$$\text{Equiv. } w = \frac{3(5.47)}{24} = 0.68 \text{ kip/ft } (9923 \text{ N/m})$$

For three concentrated loads (using correction factor):

$$M_g = \frac{1.33\ wl^2}{8} = \frac{1.33(0.68)(24)^2}{8} = 65.1 \text{ ft-kips } (88{,}276 \text{ Nm})$$

   (3) Column (maximum load at lower story):
   (a) Dead load:

$$\text{Roof} = 24(24)(0.018) = 10.4 \text{ kips}$$
$$\text{Floors} = 2(24)(24)(0.058) = \underline{66.8} \text{ kips}$$
$$P_{DL} = 77.2 \text{ kips } (343{,}386 \text{ N})$$

   (b) Live load:

$$\text{Roof} = 24(24)(0.02) = 11.5 \text{ kips } (51{,}152 \text{ N})$$

$$\text{Floor area} = 2(24)(24) = 1152 \text{ sq ft } (107 \text{ m}^2)$$
$$A_I = 4 \times 1152 = 4608$$

By Eq. 3–1,

$$L = 100\left(0.25 + \frac{15}{\sqrt{4608}}\right) = 47 \text{ psf } (2250 \text{ N/m}^2)$$

$$\text{Minimum} = 0.040(100) = 40 \text{ psf}$$

$$\text{Floors} = 2(24)(24)(47 \text{ psf}) = 54.1 \text{ kips } (240{,}637 \text{ N})$$

$$P_{LL} = 11.5 + 54.1 = 65.6 \text{ kips } (291{,}789 \text{ N})$$

f. *Rigid frame analysis for wind load:* Select frame along column line 2 for wind load in east–west direction. Assume that story height = 9 + 2.5 + 4.5/12 + 16/12 = 13.2 ft; say 13 ft 6 in. (4.1 m).
   Total wind shear at first level:

$$H_T = (2 \times 13.5 + 13.5/2)(300)(0.015 \text{ psf}) = 152 \text{ kips } (676{,}096 \text{ N})$$

Assume rigid diaphragm and distribution to frames in proportion to relative stiffness. Assume stiffness is proportional to number of columns in each frame:

$$\Sigma \text{ stiffness all frames} = 40$$

$$H_2 = \frac{6}{40}(152 \text{ kips}) = 22.8 \text{ kips } (101{,}414 \text{ N})$$

By portal method:
   Wind shear, H, at bottom story column:

$$H = \frac{22.8}{5} = 4.56 \text{ kips } (20{,}283 \text{ N})$$

$$M_c = 4.56(13.5) + 61.6 \text{ ft-kips } (83{,}530 \text{ Nm})$$

Wind shear (V) in girder:

$$V = \frac{Hh}{l} = \frac{4.56(20)}{24} = 3.8 \text{ kips } (16{,}902 \text{ N})$$

$$M_g = \frac{Vl}{2} = \frac{3.8(24)}{2} = 45.6 \text{ ft-kips } (61{,}834 \text{ Nm})$$

g. *Rigid frame design:*
   (1) Floor girder:

$$M_g = +239.4 \text{ ft-kips } (324{,}626 \text{ Nm}) \text{ at midspan}$$

$$M_w = \pm 47.6 \text{ ft-kips } (64{,}546 \text{ Nm}) \text{ at column}$$

$$\text{Min. } d = \frac{l}{2} = 24 \times \frac{12}{2} = 12 \text{ in. (305 mm)}$$

Assume full lateral support, and for composite design:

$$S = \frac{0.70M}{F_b} = \frac{0.70(239.4 \times 12)}{24}$$
$$= 83.8 \text{ in.}^3 \text{ (1,373,482 mm}^3\text{)}$$

By Appendix 4B, use W18 × 50.

$$V_h = \frac{A_s f_y}{2} = \frac{14.7(36)}{2}$$
$$= 264.6 \text{ kips (1,176,941 N)}$$

$$\text{No shear connectors} = \frac{264.6}{11.5}$$
$$= 23.0 \text{ each side (46 total)}$$

(2) Roof girder:

$$M_g = 65.1 \text{ ft-kips (88,276 Nm)},$$
$$M_\omega = \text{negligible}$$

Assume lateral support at 6 ft (1.8 m) and 14 in. (356 mm) depth.

$$F_B = \sqrt{\frac{10{,}320(65.1)}{6(14)^2}}$$
$$= 23.9 \text{ ksi} > 22.0 \text{ ksi max. (151.7 MPa)}$$

$$S = \frac{M}{F_B} = \frac{65.1 \times 12}{22} = 35.5 \text{ in.}^3 \text{ (581,845 mm}^3\text{)}$$

By Appendix 4B, use W14 × 26.

(3) Beam-column at lower story:

$$P_{DL} = 77.2 \text{ kips (343,386 N)}$$
$$P_{LL} = 65.6 \text{ kips (291,789 N)}$$
$$M_{LL} = 0$$
$$M_\omega = 45.6 \text{ ft-kips (64,546 Nm)}$$

(a) Allowable compressive stress:

$$F_{a\,(\max)} = 22.0 \text{ ksi (151.7 MPa)}$$
$$F_a = 9.37 \text{ ksi (64.6 MPa) at } kl$$
$$= 22.6\sqrt{P}$$
$$= 22.6\sqrt{178.6} = 302 \text{ in.}$$
$$= 25.2 \text{ ft (7.7 m)}$$

For kl = 13.5 ft (4.1 m), select F_a = 15.0 ksi (103.4 MPa).

(b) Allowable bending stress:

$$F_b = \sqrt{\frac{10{,}320(45.6)}{13.5(10)^2}} = 18.7 \text{ ksi (128.9 MPa)}$$

(c) Moment magnifier for lateral load: Assume drift limited to 0.0025. By Eq. 4–72,

$$Q = \frac{\Sigma P\Delta}{Hh_s}$$

$$Q = \frac{5(142.8)(0.0025)}{22.8} = 0.070 > 0.04$$

$$\delta_s = \frac{1}{1 - Q} = \frac{1}{1 - 0.078} = 1.08$$

(d) Member size:

$$A = \frac{142.8}{15(1.33)} + \frac{1.08(2.6)(45.6 \times 12)}{18.7(10)1.33}$$
$$= 7.2 + 6.2 = 13.4 \text{ in.}^2 \text{ (8643 mm}^2\text{)}$$

By Appendix 4A, use W10 × 49.

(e) Check deflection:

$$\Delta = \frac{H}{S_T} = \frac{H}{S_c} + \frac{H}{S_g}$$

$$S_c = \frac{12E\,\Sigma\,I_c}{h^3} = \frac{12(30 \times 10^3)(6 \times 272)}{(13.5 \times 12)^3}$$
$$= 138.2 \text{ kips/in. (24,201 N/mm)}$$

$$S_g = \frac{12E\,\Sigma\,I_g/L}{h^2}$$
$$= \frac{12(30 \times 10^3)(10 \times 800/288)}{(13.5 \times 12)^2}$$
$$= 381 \text{ kips/in. (66,720 N/mm)}$$

$$\Delta = \frac{22.8}{138.2} + \frac{22.8}{381} = 0.22 \text{ in. (5.6 mm)}$$

$$\frac{\Delta}{h} = \frac{0.22}{13.5 \times 12} = 0.0014 < 0.0025$$

h. *Foundations:*

$$\text{Max. column load} = 77.2 + 65.6$$
$$= 142.8 \text{ kips (635,174 N)}$$

$$A_f = \frac{142.8}{2.5} = 57.1 \text{ ft}^2 \text{ (5.3 m}^2\text{)}$$

Use 7 ft 6 in. × 7 ft 6 in. footing (2.3 m × 2.3 m).

$$q_{net} = \frac{P_u}{A_f} = \frac{1.4(77.2) + 1.7(65.6)}{56.25} = 3.9 \text{ ksf}$$

$$\frac{\sqrt{f'c}}{q_{net}} = \frac{\sqrt{3000}}{3900} = 0.014$$

$$\frac{A_f}{A_c} = \frac{56.25}{1.0} = 56.25,\ r = 12 \text{ in.},\ a = 39 \text{ in.}$$

By Table 4–12,

$$\frac{d}{r} = 1.0\text{: req'd } d = 1.0(12 \text{ in.}) = 12.0 \text{ in.}$$

$$\frac{a}{d} = 4.42\text{: req'd } d = \frac{39 \text{ in.}}{4.42} = 8.8 \text{ in.}$$

Use d = 12.0 in., t = 12 + 3 = 15 in. (381 mm).

i. *Tabulation of costs:* Use "base costs" for Atlanta area.

(1) Floor:	
(a) 24-gage composite metal deck at 1.0 sf/sf × 0.80	$0.80
(b) Concrete slab = (4.5/12)(1)(1)/27 = 0.014 cy/sf:	
×$42.00 (material)	0.59
×$12.00 (place)	0.17
(c) Finish slab = 0.25/sf	0.25
(d) Reinforcing steel = (5.33 × 12)(0.16)(0.28)/8 = 0.36 psf at 0.37	0.13
(e) Steel beam = 26/8 = 3.25 psf at 0.51	1.66
24 shear connectors at 0.85/(24)(8)	0.10
Total floor	$3.70/sf
(2) Roof:	
(a) 22-gage metal deck = 1.0 sf/sf at 0.66	0.66
(b) Joists = 6.5/6 = 1.08 psf at 0.40	0.43
Total roof:	$1.09/sf
(3) Rigid frame:	
(a) Floor girder = 50/24 = 2.08 psf at 0.51	$ 1.06
46 shear connectors at 0.85/(24)(24)	0.06
(b) Roof girder = 26/24 = 1.08 psf at 0.51	0.55
(c) Column = 49 × 0.51	$24.90/lf
(4) Foundations:	
(a) Spread footing = 7.5(7.5)(15/12)(1/27) = 2.6 cy at 160	$416.00/ea.
(5) Fire protection:	
(a) Membrane ceiling protection:	
Ceiling premium	$0.15
Insulation over lights, etc.	0.15
	$0.30/sf
(b) Column protection: Gypsum board at $10/lf = $10(13.5)/(24)(24)	$0.23/sf

(6) Added depth differential cost:
   (a) Steel system depth = 4½ in. + 18 = 22.5 in.
   (b) Concrete system depth = −9.0 in.
   13.5 in.
   (c) Added wall area
   = (13.5/12)(3 floors)(740 ft)
   = 2497 sf
   (d) Wall cost = 2497 sf at 7.00 = $17,479
   (e) Unit cost = 17,479/50,000 = $0.35/sf

j. System selection form: Figure 7–10 summarizes initial costs.

6. *Preliminary design of Scheme II:*

a. *Slab depth:* Assume 18 in. × 18 in. (457 mm × 457 mm) column.

$$l_n = 24 - \frac{18}{12} = 22.5 \text{ ft } (6.86 \text{ m})$$

By Eq. 5–30,

$$\text{Max. } h = \frac{l_n}{32} = \frac{22.5 \times 12}{32} = 8.44 \text{ in. } (214 \text{ mm})$$

Use 9-in. (228-mm) slab thickness.

b. *Loads:*

(1) Live load:

Floors = 100 psf (4788 N/m²)
Roof = 20 psf (958 N/m²)

(2) Dead load:

Floors:

9-in. concrete slab	= 112.5 psf
Ceiling	= 2.0 psf
Miscellaneous	= 3.0 psf
	117.5 psf (5626 N/m²)

Roof:

Built-up roof	= 6.0 psf
Insulation	= 2.0 psf
9-in. slab	= 112.5 psf
Ceiling	= 2.0 psf
Miscellaneous	= 3.0 psf
	125.5 psf (6009 N/m²)

c. *Materials:*

(1) Concrete: 3000 psi (20.7 MPa)

STRUCTURAL SYSTEM SELECTION
INITIAL COST

SYSTEM: COMPLETE FRAMING BY: RAL DATE: 8-1-81 JOB NO. 75045

DESCRIPTION: SCHEME I QUANTITY: 50,000 SF

COMPONENT	FUNCTION			REMARKS	$ SYST. UNIT	$ BLDG S.F.	±$ OTHR. SYST.	NET $/ BLDG. S.F.
	VERB	NOUN	KIND					
BASIC SYSTEM	CARRY	LOADS	B					
FLOOR	″ ~~PROVIDE~~	″ ~~STAB.~~	B	4½ in. CONC. SLAB ON COMP. METAL DECK + STEEL BEAMS	3.70/SF	2.96		
ROOF	″	″	B	METAL DECK + BAR JOISTS	1.09/SF	.36		
FLOOR GIRDER	″	″	B	COMPOSITE STEEL BEAM	1.12/SF	.75		
ROOF GIRDER	″	″	B	STEEL BEAM	.55/SF	.18		
COLUMN	″	″	B	STEEL	24.90/LF	.58		
	PROVIDE	FIRE PROT.	S	1.				
	PROVIDE	WEATH. PROT.	S	2.				
	PROVIDE	OPNGS.	S					
	PROVIDE	CLEAR.	S					
	PROVIDE	DURAB.	S	3.				
	ENCASE	MECH. EL. SYS.	S	4.				
		HEIGHT	S	5.				
		"U"	S					
		SOUND	S					
		VIBRA.	S					
		POND'G.	S					
FOUNDATION		FDNS. WEIGHT	B ~~S~~	7-ft.-6in. Sq. SPREAD FOOTING	416/EA	.24		
	CARRY	EQUIP. LOADS	S					
				TOTALS		4.57	0.88	5.45

1. INCLUDE COST OF DAMPERS AND HOODS FOR LIGHT FIXTURES FOR FIREPROTECTION BY RATED CEILINGS.
2. INCLUDE COSTS OF FLASHING, CAULKING, WATERPROOFING, WHEN SYSTEM DOUBLES AS AN EXTERIOR WALL.
3. INCLUDE PAINT, PRESERVATIVE TREATMENT, AIR-ENTRAINING, ETC.
4. CELLULAR FLOOR SYSTEMS, EXTRA TOPPING THICKNESS, ETC.
5. INCLUDE HVAC LOADS, WALL COSTS, STAIR COSTS, VERTICAL DUCT, PIPE, & CONDUIT RUN.

Figure 7–10. Example Problem 7–2, Initial Costs for Scheme I

(2) Reinforcing steel: 60,000 psi yield (413.7 MPa)

d. *Slab analysis for gravity load:*

(1) Moment:

$$w_u = 1.4(117.5) + 1.7(100)$$
$$= 334 \text{ psf } (15{,}992 \text{ N/m}^2)$$

By Eq. 5–27,

$$M_o = \frac{w_u l_2 l_n^2}{8} = \frac{0.334(24)(22.5)^2}{8}$$
$$= 507 \text{ ft-kips } (687{,}492 \text{ Nm})$$
$$-M = 0.65(507) = -329 \text{ ft-kips } (446{,}124 \text{ Nm})$$
$$+M = 0.35(507) = +177 \text{ ft-kips } (240{,}012 \text{ Nm})$$

(2) Shear: Assume that $d = 8$ in. (203 mm).

$$b_o = 4(18 + 8) = 104 \text{ in. } (2642 \text{ mm})$$
$$V_u = 0.334\left[24 \times 24 - \left(\frac{18+8}{12}\right)^2\right]$$

$$= 190.8 \text{ kips } (848{,}678 \text{ N})$$

$$\phi V_c = \phi 4\sqrt{f'_c}\ b_o d$$
$$= (0.85)(4\sqrt{3000})(104)8 = 154{,}939 \text{ lb}$$
$$= 154.9 \text{ kips } (688{,}955 \text{ N})$$

Since $\phi V_c < V_u$, use column drop panels. Assume that width = 0.15(24) = 3.6 ft; use 3 ft 6 in. (1.1 m). Use depth = 12 in. (305 mm).

(3) Check shear at $d/2$ from drop panel:

$$b_o = 4(3.5 \times 12 + 8) = 200 \text{ in. } (5080 \text{ mm})$$

$$V_u = 0.334\left[24 \times 24 - \left(3.5 + \frac{8}{12}\right)^2\right]$$
$$= 186.6 \text{ kips } (829{,}997 \text{ N})$$

$$\phi V_c = (0.85)(4\sqrt{3000})(200)8$$
$$= 297{,}961 \text{ lb}$$
$$= 298 \text{ kips } (1{,}325{,}504 \text{ N})$$
$$\phi V_c > V_u \quad \text{O.K.}$$

e. *Column analysis for gravity load:*

(1) Calculate relative stiffness for column (k_c) and slab (k_s).

$$k_c = \frac{4I}{l_c - 2h}, \qquad I = \frac{18(18)^3}{12} = 8748 \text{ in.}^4$$
$$= \frac{4(8748)}{(13.5 \times 12) - 2(9)}$$
$$= 243 \text{ in.}^3 \ (3{,}982{,}770 \text{ mm}^3)$$
$$k_s = \frac{4I}{l}, \qquad I = \frac{(24 \times 12)(9)^3}{12} = 17{,}496 \text{ in.}^4$$
$$= \frac{4(17496)}{24 \times 12} = 243 \text{ in.}^3 \ (7{,}965{,}540 \text{ mm}^3)$$

(2) Calculate unbalanced moment to column:

$$w_d = 1.4(0.117 \text{ ksf}) = 0.164 \text{ kip/ft } (2393 \text{ N/m})$$
$$w_l = 1.7(0.100 \text{ ksf}) = 0.17 \text{ kip/ft } (2480 \text{ N/m})$$
$$M_c = 0.07[(w_d + 0.5w_l)l_2 l_N^2 - w'_d l'_2 (l'_N)^2] \times \frac{\Sigma k_c}{\Sigma k_c + \Sigma k_s}$$
$$M_c = 0.07(0.5 \times 0.17)(24)(22.5)^2 \times \frac{486}{972}$$
$$M_c = 36 \text{ ft-kips } (46{,}104 \text{ Nm})$$

Distribute one-half above and one-half below slab.

(3) Calculate axial load:

(a) Dead load:

Roof = 24(24)(0.125) = 72 kips

Floors = 24(24)(0.118) = 68 kips

P_{DL} = 140 kips (622,720 N)

(b) Live load: Using live load reduction from scheme I:

$$P_{LL} = 65.6 \text{ kips } (291{,}789 \text{ N})$$

f. *Slab/column analysis for wind load:* Reference portal analysis results from scheme I:

$$M_c = 61.6 \text{ ft-kips } (83{,}530 \text{ Nm})$$
$$M_g = 45.6 \text{ ft-kips } (61{,}834 \text{ Nm})$$

g. *Slab design for negative moment:*

$$M_g = 329 \text{ ft-kips } (446{,}124 \text{ Nm})$$
$$M_\omega = 45.6 \text{ ft-kips } (61{,}834 \text{ Nm})$$

Since $0.75(329 + 45.6) < 329$, design for gravity load only.

(1) Column strip: Distribute 75% to column strip:

$$M_{cs} = 0.75(329) = 246.7 \text{ ft-kips } (334{,}525 \text{ Nm})$$

Column strip width = 12 ft (3.7 m)

$$S \text{ provided} = \frac{144(9)^2}{6}$$
$$= 1944 \text{ in.}^3 \ (31{,}862{,}160 \text{ mm}^3)$$

$$f_b = \frac{M}{S} = \frac{246{,}700 \times 12}{1944}$$
$$= 1522 \text{ psi } (10.5 \text{ MPa})$$

Since A_s = 0.85% for 1260 psi (8.7 MPa), assume that A_s = 1.0%.

$$A_s = 0.01(12)(8) = 0.96 \text{ in.}^2\text{/ft } (2031 \text{ mm}^2\text{/m})$$

Extend to one-quarter point of span as an average.

(2) Middle strip: Distribute 25% to middle strip:

$$M_{ms} = 0.25(329)$$
$$= 82.3 \text{ ft-kips } (111{,}599 \text{ Nm})$$

Strip width = 12 ft (3.7 m)

$$S \text{ provided} = \frac{144(9)^2}{6}$$
$$= 1944 \text{ in.}^3 \ (31{,}862{,}160 \text{ mm}^3)$$
$$f_b = \frac{M}{S} = \frac{82{,}300 \times 12}{1944}$$
$$= 508 \text{ psi } (3.5 \text{ MPa})$$

Use $A_{s(\min)}$ = 0.0018(12)(9) = 0.19 in.²/ft (402 mm²/m).

h. *Slab design for positive moment:*

$$M_u = 177 \text{ ft-kips } (240{,}012 \text{ Nm})$$

(1) Column strip: Distribute 75% to column strip:

$$M_{cs} = 0.75(177) = 132 \text{ ft-kips } (178{,}992 \text{ Nm})$$

$$f_b = \frac{M}{S} = \frac{132{,}000 \times 12}{1944} = 814 \text{ psi } (5.6 \text{ MPa})$$

Assume that $A_s = 0.004(12)(8) = 0.38$ in.²/ft (804 mm²/m).

(2) Middle strip: By inspection,

$$A_s = 0.19 \text{ in.}^2/\text{ft } (402 \text{ mm}^2/\text{m}).$$

i. *Column design:*

(1) Loads:

$$P_u = 1.4(140) + 1.7(65.6) = 307 \text{ kips } (1{,}365{,}536 \text{ N})$$

$$M_u(\text{gravity}) = 18 \text{ ft-kips } (2{,}305{,}210 \text{ m})$$

$$M_u(\text{wind}) = 1.7(61.6) = 105 \text{ ft-kips } (142{,}380 \text{ Nm})$$

(2) Trial selection: Assume a 16 in. × 16 in. column (406 mm × 406 mm).

$$A = 156 \text{ in.}^2 (165{,}120 \text{ mm}^2)$$

(3) Check gravity load magnification:

$$\frac{kl}{r} = \frac{1.0(13.5 \times 12)}{0.3(16)} = 33.8 < 34$$

No magnification required for gravity load.

(4) Calculate magnifier for wind moment, with drift assumed limited to 0.0025:

$$Q_u = \frac{\Sigma P_u \Delta_u}{H_u h_s} = \frac{(0.75)(307)(5)(0.0025)}{(0.75)(22.8)(1.7)} = 0.099 > 0.04$$

$$\delta_s = \frac{1}{1 - Q_u} = \frac{1}{1 - 0.099} = 1.11$$

(5) For 3000-psi (20.7-MPa) concrete, $F'_a = 2.74$ ksi (18.9 MPa); $F'_b = 3.13$ ksi (21.6 MPa).

$$P'_u = P_u - 0.30bdf'c = 307 - 0.30(16)(16 \times 0.8)(3.0) = 122.7 \text{ kips } (545{,}770 \text{ N})$$

$$A = \frac{P'_u}{F'_a} + \frac{\delta 6 M_u}{F'_b h} = \frac{122.7(0.75)}{2.74} + \frac{(1.0)(6)(18 \times 12)(0.75)}{3.13(16)} + \frac{1.11(6)(105 \times 12)(0.75)}{3.13(16)} = 178.6 \text{ in.}^2 (115{,}238 \text{ mm}^2)$$

< 256 in.² (165,120 mm²) provided O.K.

Use a 16 in. × 16 in. (406 mm × 406 mm) column with 4% reinforcement.

j. *Foundations:*
Max. working column load = 140 + 65.6 = 205.6 kips (914,509 N)
Since concrete system is too rigid to tolerate expected differential settlement, design caisson system.

$$\text{Caisson area req'd} = \frac{205.6 \text{ kips}}{20 \text{ ksf}} = 10.3 \text{ ft}^2 (0.96 \text{ m}^2)$$

Use 44-in. (1118-mm)-diameter caisson at assumed average length of 40 ft (12.2 m)

k. *Tabulation of costs:*

(1) Floor:

(a) Concrete slab $= \dfrac{(9/12)(1)(1)}{1(27)}$ = 0.028 cy/sf:

× \$42.00 (material)	\$1.18
× \$12.00 (place)	0.34
(b) Finish slab = 0.25/sf	0.25
(c) Form slab at \$2.05/sf	2.05

(d) Reinforcing steel:
Column strip:

$$-A_s = \frac{2(12 \times 12)(0.96)(12)(0.28)}{24(24)} = 1.63 \text{ psf at } 0.37 \qquad 0.60$$

Middle strip:

$$-A_s = \frac{2(12 \times 12)(0.19)(12)(0.28)}{24(24)} = 0.32 \text{ psf at } 0.37 \qquad 0.12$$

Column strip:

$$+A_s = \frac{2(24 \times 12)(0.38)(12)(0.28)}{24(24)} = 1.29 \text{ psf at } 0.37 \qquad 0.48$$

Middle strip $+A_s$ = 0.32 psf at 0.37	0.12
Total floor	\$5.14

(2) Column:

(a) Reinforcing steel:

First-story steel = 0.04(16 × 16)(13.5 × 12)	= 1659 in.³
Top-story steel = 0.02(16 × 16)(13.5 × 12)	= 829 in.³

Assume average $= \frac{1659 + 829}{2}$
= 1244 in.³
Increase 1/3 for splice
= 1244 × 1.3 = 1617 in.³
Add 5% for ties = 80
1697 in.³
Total steel = 1697 in.³ × 0.284 = 482 lb
Cost = 482(0.37)/13.5 $13.21/lf

(b) Concrete = 16(16)(1)/(144)(27)
= 0.066 cy/lf
× $42.00 (material) 2.77/lf
× $14.00 (place) 0.92/lf

(c) Form = 4(16)(1)/(12)
= 5.33 sf/lf at 2.90 15.47/lf
Total column $32.37/lf

(3) Foundations:

(a) Caisson = 44 in. at 40 ft × $110 $110.00/lf

1. *System selection form:* Figure 7–11 summarizes initial costs.

Figure 7–11. Example Problem 7–2, Initial Costs for Scheme II

STRUCTURAL SYSTEM SELECTION
INITIAL COST

SYSTEM: COMPLETE FRAMING BY: RAL. DATE: 8-1-81 JOB NO. 75045
DESCRIPTION: SCHEME II QUANTITY: 50,000 SF

COMPONENT	FUNCTION			REMARKS	$ SYST. UNIT	$ BLDG S.F.	±$ OTHR. SYST.	NET $/ BLDG. S.F.
	VERB	NOUN	KIND					
BASIC SYSTEM	CARRY	LOADS	B					
	PROVIDE	STAB.	B					
FLOOR	CARRY	LOAD	B	9" FLAT SLAB WITH DROP PANELS	5.14/SF	3.44		
ROOF	"	"	B	" "	5.14/SF	1.69		
COLUMN	"	"	B	16 in. × 16. In.	32.37/LF	0.76		
FOUNDATION	"	"	B	44 in. Dia. CAISSON ×40 Ft.	110/LF	2.54		
	PROVIDE	FIRE PROT.	S	1.				
	PROVIDE	WEATH. PROT.	S	2.				
	PROVIDE	OPNGS.	S					
	PROVIDE	CLEAR.	S					
	PROVIDE	DURAB.	S	3.				
	ENCASE	MECH. EL. SYS.	S	4.				
		HEIGHT	S	5.				
		"U"	S					
		SOUND	S					
		VIBRA.	S					
		POND'G.	S					
		FDNS. WEIGHT	S					
	CARRY	EQUIP. LOADS	S					
				TOTALS		8.43		8.43

1. INCLUDE COST OF DAMPERS AND HOODS FOR LIGHT FIXTURES FOR FIREPROTECTION BY RATED CEILINGS.
2. INCLUDE COSTS OF FLASHING, CAULKING, WATERPROOFING, WHEN SYSTEM DOUBLES AS AN EXTERIOR WALL.
3. INCLUDE PAINT, PRESERVATIVE TREATMENT, AIR-ENTRAINING, ETC.
4. CELLULAR FLOOR SYSTEMS, EXTRA TOPPING THICKNESS, ETC.
5. INCLUDE HVAC LOADS, WALL COSTS, STAIR COSTS, VERTICAL DUCT, PIPE, & CONDUIT RUN.

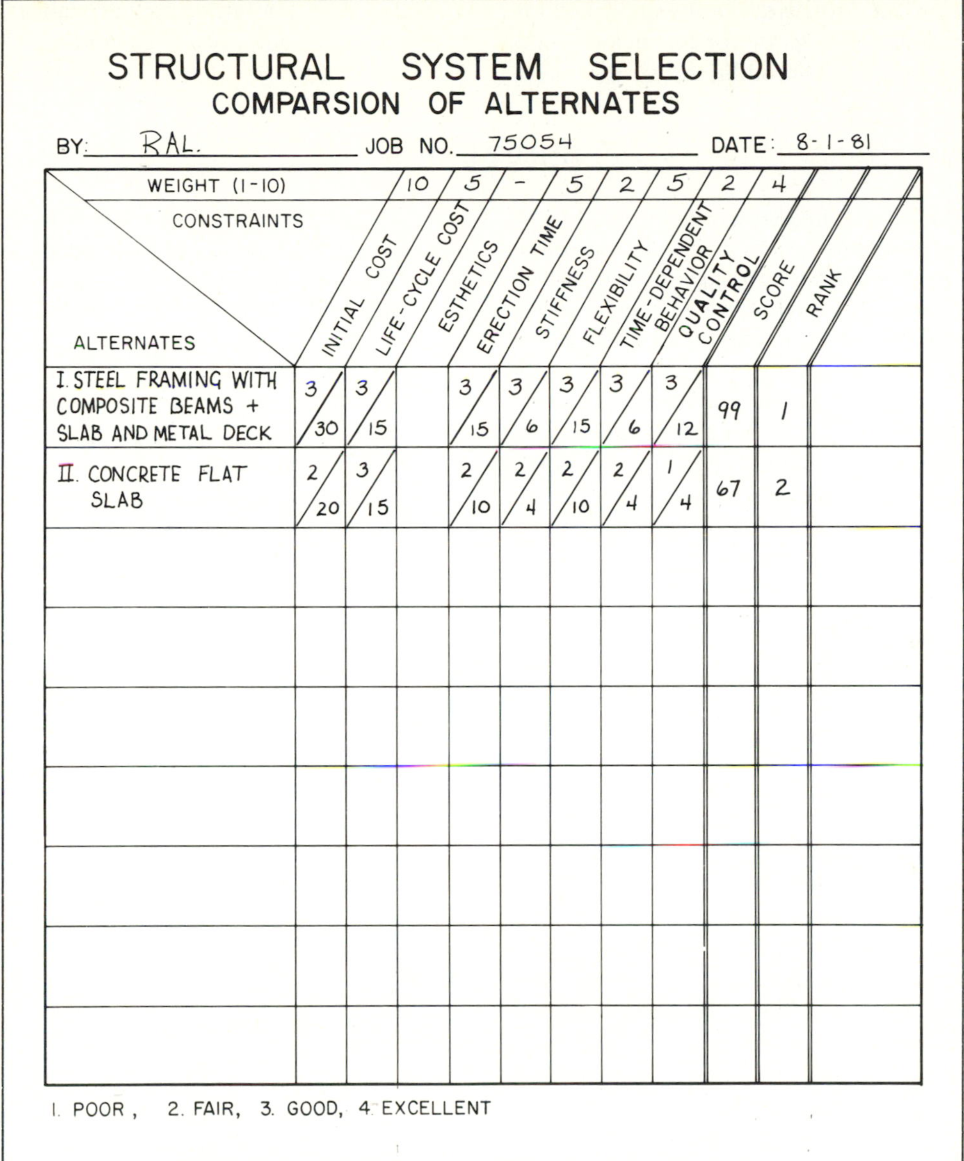

STRUCTURAL SYSTEM SELECTION
COMPARSION OF ALTERNATES

BY: RAL. JOB NO. 75054 DATE: 8-1-81

CONSTRAINTS / ALTERNATES	INITIAL COST	LIFE-CYCLE COST	ESTHETICS	ERECTION TIME	STIFFNESS	FLEXIBILITY	TIME-DEPENDENT BEHAVIOR	QUALITY CONTROL	SCORE	RANK
WEIGHT (1-10)	10	5	-	5	2	5	2	4		
I. STEEL FRAMING WITH COMPOSITE BEAMS + SLAB AND METAL DECK	3/30	3/15		3/15	3/6	3/15	3/6	3/12	99	1
II. CONCRETE FLAT SLAB	2/20	3/15		2/10	2/4	2/10	2/4	1/4	67	2

1. POOR, 2. FAIR, 3. GOOD, 4. EXCELLENT

Figure 7–12. Example Problem 7–2, Comparison of Alternatives

7. *Comparison of alternatives.* With initial costs for both systems tabulated, the next step is to compare their relative value using the system selection form shown in Fig. 7–12. The assigned weights for each evaluation category should be consistent with the project requirements and owner's needs.
8. *Selection of system for final design.* Select Scheme I for final design.

7–4 SUMMARY

In this chapter we have completed the development of the systems design model with the addition of sections covering the generation of alternatives and comparison of alternatives. Each section represents the final phase of two related, but independent, tracks of development.

Generation of alternatives capped the development of a gradually expanding set of choices and design options, represented by the various assemblies of materials, elements, subsystems, and systems. Since these options constitute the "raw material" fed in at the top of the model, the *first* phase of the creative process is now complete.

Comparison of alternatives completed the development of screening and evaluation techniques, represented by requirements, approximate analysis/design methods, and system selection guidelines. This *final* phase of the creative process completes the model by producing one or more systems as candidates for final analysis and design.

We should view the framework now established for structural systems design as a *guide* for creative thought and systematic evaluation. If, in addition, we have now developed an "inituitive feeling" for structural behavior, we are well equipped to produce innovative structural designs.

PROBLEMS

7–1. Create a structural system to carry a trolley with a 10,000-lb (44,480-N) capacity through the yard area of an industrial plant. The trolley will ride on a single-beam rail which will be attached to the structural system. Minimum spacing between any vertical support occurs at road crossings and is 15 ft (4.6 m).

7–2. A 12-in. (305-mm) water line must be supported across a stream that is 70 ft (21.3 mm) in width. Create a structural system to support the pipe.

7–3. A platform must be built to support a piece of mechanical equipment 30 ft (9.1 m) above the ground level. The plan dimensions of the platform are to be 20 ft × 20 ft (6.1 m × 6.1 m) and the equipment weighs 30,000 lb (133,440 N). Create a structural system for the platform.

7–4. A small, single-story office will be 60 ft × 80 ft (18.3 m × 24.4 m) in plan. Create a structural system for the building.

7–5. A local golf course needs golf cart bridges constructed across several streams that run through the golf course. The width of the crossings varies from 15 to 50 ft (4.6 to 15.2 m). Create a structural system for the bridges, using a similar design for all spans.

7–6. A highway sign must be designed to cantilever 20 ft (6.1 m) over the roadway and carry a highway sign that weighs 1000 lb (4448 N). Create a structural system for the sign support.

DESIGN PROJECTS

1. INTRODUCTION

A model has been established for structural systems design which is essentially a screening technique for processing ideas and information regarding potential materials, elements, subsystems, and systems. Approximate analysis and design techniques have been developed as well as a data base of system characteristics to aid us in systematically narrowing the range of choices to specific alternatives for comparison. Once the choices have been narrowed, the approximate design techniques can be utilized to compare alternative systems prior to making a selection of a system for final analysis and design. The book has discussed concepts of structural assembly, starting with the very basic element form and progressing through total system assembly. We are now ready to put these concepts into actual practice by creating a structural system for a specific project. The design project becomes a synthesis of knowledge and ideas as various concepts of load and assembly are brought together into a total design.

2. APPROACH

Perform the design project in teams, with each team responsible for a different project. Review the brief statement of the project given to each team and develop the design based on the techniques developed in the book and interviews with the "client." Each team must develop and present a project report which contains the following major divisions:

Written Report:

a. Detailed description of project requirements and criteria.
b. Design loads.
c. General alternatives considered.
d. Comparison of two alternatives (using value engineering methods).
e. Description of selected system and reasons for selection.

Model Presentation. A presentation model to demonstrate the design solution.

3. JURY PRESENTATION

Each team must make a verbal presentation, including the model, to a jury of not less than three professional engineers. Presentations consist of a 20-minute verbal description of the design project: describing the approach used, comparisons made, and conclusions. The verbal presentation will be followed by a question-and-answer session with members of the professional jury.

The jury will use the following criteria in evaluating the design projects:

a. *Definition of the problem:* How well did the team take the initial project assignment and develop a specific set of criteria and objectives for the design solution?

b. *Thoroughness of design process:* How well did the team search out ideas and systematically evaluate them for possible application to the project?

c. *Detailed comparison of alternatives:* How well did the team compare the selected alternatives for their conformance with the project requirements and criteria?

d. *Quality of design solution:* How well does the selected design meet the project requirements? Is it structurally sound and cost effective?

e. *Quality of presentation:* How well did the design team sell their solution as the best design for the project?

4. DESIGN PROJECTS

A. Arena

The coliseum board in Greenville, South Carolina, wishes to construct a 15,000-seat multipurpose arena that will provide space for the following activities: basketball, ice hockey, circus, road shows, exhibition space for conventions, trade shows, and other similar activities.

The proposed project site slopes from front to rear approximately 20 ft (6.1 m). Preliminary soil borings have indicated that rock occurs across the site at a depth of about 10 ft (3.05 m) below the surface. The rock is considered by the soil consultant to be "rippable" (removable by machine) down to a depth of about 20 ft (6.1 m). Foundations bearing on the upper layers of the weathered rock can be designed for uniform bearing pressures of 6000 psf (277,280 N/m^2). If deeper foundations are used [extending to depth of 20 ft (6.1 m) or more], uniform bearing pressures of 20,000 psf (957,600 N/m^2) are recommended.

The arena must contain a main events floor, seating for 15,000 spectators, concession facilities, concourses serving the spectator area, dressing rooms, storage, and office space. The following space requirements have been established:

Arena floor:	30,000 sq ft (2,790 m^2) (150 ft × 200 ft)
Dressing rooms:	8,000 sq ft (744 m^2)
Storage areas:	3,000 sq ft (279 m^2)
Offices:	8,000 sq ft (744 m^2)
Concessions:	3,000 sq ft (279 m^2)
Public toilets:	3,000 sq ft (279 m^2)

B. High-Rise Office Building

An owner wishes to build a high-rise office building in Charleston, South Carolina. Zoning and parking requirements have already dictated that the building must be 10 stories in height and contain 10,000 sq ft (930 m^2) of *net* usable square footage per floor. Additional square footage per floor will be required for elevator and stair access, mechanical equipment room, toilet areas, and so on. Preliminary soil borings indicate that the flat site is underlain by soft clay that is unsuitable for spread footings and that some type of deep foundation system must be used. The soil consultant recommends some type of timber or prestressed concrete pile system. The soil report contains approximate cohesion factors for use in estimating required pile lengths. The exterior of the building must be designed in accordance with the latest energy code criteria.

C. Industrial Building

A manufacturer of synthetic fibers will locate a new plant facility in Birmingham, Alabama. The facility will contain the following functional areas:

- Single-story office building of 50,000 sq ft (4650 m^2).
- Four-story process building containing 10,000 sq ft (930 m^2) per floor.
- 80,000 sq ft, 400 ft × 200 ft (7440 m^2, 121.9 m × 61 m) for linear portions of the process (stretching, drawing, twisting, etc., and for shipping, receiving, and warehousing).

The single-story portion of the area must have a clear height of 22 ft (6.7 m) and one bay must contain a 50-ton (444,800-N) bridge crane running the length of the facility.

Preliminary soil information indicates that the clay soil can support shallow spread footings with an allowable soil bearing pressure of 3000 psf (143,640 N/m^2).

D. Highway Bridge

A 2000-ft (610-m)-long bridge will be constructed across the Congaree River in Columbia, South Carolina. The distance across the river from embankment to embankment is 800 ft (244 m), with the approach sections making up the remainder of the bridge span. The bridge must be designed to HS 20 loading and a lane width of 40 ft (12.2 m). The height of the roadway above the river bottom is 50 ft (15.2 m). Preliminary soil borings have been performed on both the river banks and the bed of the river and indicate that some type of deep foundation system must be used because of the silt and clay characteristics of the soil. The soils report contains approximate recommended cohesion values for pile foundations.

APPENDIX 7A: Construction Costs

INTRODUCTION

Structural systems design normally includes cost estimates along the way to verify that the design meets project budget requirements. Various cost estimating systems are available to assist with the compilation of estimates. For classroom and professional reference use during conceptual work, it is desirable to have a method that is simple and can be readily updated from year to year (or quarter to quarter) without purchasing a new book. The estimating system described here is a simple manual system providing a data base of unit costs which are indexed to *Engineering News-Records'* quarterly cost report.

Construction cost estimating is a process of estimating *quantities* of material and labor required to construct various structural systems, applying *unit costs,* applying *judgment,* and adding *markups.* The following paragraphs describe this process.

QUANTITIES

Quantities are physical items of construction to which unit costs will be applied to arrive at total construction costs. Units of measurement may be dimensional, such as linear feet (lf), square feet (sf), or cubic feet (cf), or they may be nondimensional, such as each (ea.), lump sum (ls), or allowance (allow). Quantities are relatively easy to obtain once complete working drawings and specifications have been prepared. Prior to this point, however, the engineer or estimator must use "conceptual estimating" to determine approximate quantities. At the conceptual level, this is often done by working with composite items of construction, a process termed *systems estimating.* An example of a composite item of construction is an item described in the estimate as *concrete slab on grade.* This item may actually include subgrade preparation, stone, vapor barrier, formwork, concrete material, concrete placement, reinforcing steel, steel placement, finishing, and curing. Obviously, this degree of detail is not consistent with preliminary design estimates, and the composite item is more appropriate. Both detailed and composite items are included in the data base.

UNIT COSTS

Unit costs are "bare costs" without any markup for overhead and profit. The costs in the data base are indexed to *Engineering News-Record*'s cost report, which appears quarterly in the magazine. Figure 7–13 shows the cost report printed in the June 18, 1981 issue. The index is keyed to 20 major cities in the United States and reflects *current* material prices and union labor rates. Current costs for any project location may be established by selecting an index that represents conditions in the geographical area of the project and dividing it by the *base index,* which was used in establishing each cost item in the data base. Cost for any item of construction is then determined as follows:

$$\text{Item cost} = \text{data base cost} \times \frac{\text{current index}}{\text{data base index}} \times \text{judgment factor}$$

For example, cost for concrete in Kansas City, assuming a current index of 288, would be calculated as follows:

$$\text{Concrete} = \$42.00 \times \frac{288}{292} = \$41.42/\text{cy}$$

The unit cost data base is shown in Fig. 7–14.

JUDGMENT

Both detailed and conceptual estimates require considerable *judgment* in addition to quantity estimates. Judgment corrects so-called "average" unit or system costs to reflect the following:

- *Difficulty of the work:* primarily relates to additional hours of labor required due to accessibility, complexity of construction operations, and expected construction tolerances.
- *Quality of the specified product:* relates to higher-cost materials, which may be selected for reasons such as esthetics, durability, and insulation value.
- *Productivity:* efficiency in the application of labor to construction relates to many factors, such as weather conditions, skill of the available labor force, and motivation.

Judgment factors should be applied to labor costs for difficulty and productivity and applied to material costs for quality.

MARKUPS

The "bare costs" represented by the unit costs must be marked up for overhead and profit to arrive at total construction costs. For work performed by subcontractors, subcontractor markup will normally average around 25%. General contractors will markup subcontract work by another 10 to 15%. Work performed directly by the general contractor will normally be marked up 35 to 40%.

Figure 7–13. *Engineering News-Record* Quarterly Cost Report (Reprinted from the June 18, 1981 issue of *Engineering News-Record* by special permission; copyright 1981 by McGraw-Hill, Inc., New York, NY 10020; all rights reserved)

Construction wage indexes*

	Common labor			*Skilled labor*			*Electrician*			*Equipment operator*			*Mechanical trades*		
	80	*81*	*% chg.*	*80*	*81*	*% chg.*	*80*	*81*	*% chg.*	*80*	*81*	*% chg.*	*80*	*81*	*% chg.*
Atlanta	296	330	+11.5	256	282	+10.2	297	302	+1.7	270	298	+10.4	257	274	+6.6
Baltimore	342	371	+8.5	272	297	+9.2	275	275	0	256	289	+12.9	275	279	+1.5
Birmingham	298	317	+6.4	269	289	+7.4	298	307	+3.0	280	299	+6.8	284	297	+4.6
Boston	295	295	0	254	265	+4.3	255	289	+13.3	276	282	+2.2	262	300	+14.5
Chicago	273	298	+9.2	270	289	+7.0	281	310	+10.3	295	295	0	266	298	+12.0
Cincinnati	310	378	+21.9	296	314	+6.0	263	298	+13.3	298	366	+22.0	310	331	+6.8
Cleveland	297	349	+17.5	294	315	+7.1	281	314	+11.7	271	323	+19.2	288	310	+7.6
Dallas	317	365	+15.1	277	331	+19.5	294	306	+4.1	397	461	+16.1	274	318	+16.1
Denver	310	311	0	280	281	0	332	344	+3.6	312	343	+9.9	324	324	0
Detroit	283	313	+10.6	285	314	+10.2	280	307	+9.6	288	318	+10.4	291	323	+11.0
Kansas City	351	382	+8.8	307	338	+10.0	276	324	+17.4	365	392	+7.4	302	324	+7.3
Los Angeles	311	356	+14.5	300	336	+12.0	305	316	+3.6	287	323	+12.5	275	285	+3.6
Minneapolis	283	283	0	276	276	0	279	314	+12.5	283	283	0	266	312	+17.3
New Orleans	328	348	+6.1	264	297	+12.5	266	276	+3.8	255	275	+7.8	264	282	+6.8
New York	244	257	+5.3	238	250	+5.0	288	309	+10.4	244	256	+4.9	283	301	+6.4
Philadelphia	310	384	+23.9	267	298	+11.6	278	278	0	262	295	+12.6	257	282	+9.7
Pittsburgh	296	328	+10.8	259	282	+8.9	237	260	+9.7	285	309	+8.4	267	289	+8.2
St. Louis	294	294	0	249	276	+10.8	256	278	+8.6	262	292	+11.5	249	255	+2.4
San Francisco	298	344	+15.4	295	314	+6.4	316	347	+9.8	288	291	+1.0	264	278	+5.3
Seattle	308	351	+14.0	291	327	+12.4	351	422	+20.2	276	313	+13.4	344	381	+10.8

* official ENR June indexes including base wages and fringes: base: 1967 = 100

Material price indexes

	Cement bulk			*Ready-mix concrete*			*Crushed stone*		
CEMENT	*Mar. '80*	*Mar. '81*	*% chg.*	*Mar. '80*	*Mar. '81*	*% chg.*	*Mar. '80*	*Mar. '81*	*% chg.*
Atlanta	262	265	+1.1	273	292	+7.0	142	142	0
Baltimore	262	262	0	298	355	+19.1	150	165	+10.0
Birmingham	288	261	−9.4	248	264	+6.5	151	267	+76.8
Boston	288	272	−5.6	291	317	+8.9	270	297	+10.0
Chicago	296	254	−14.2	360	345	−4.2	89	95	+6.7
Cincinnati	254	272	+7.1	270	253	−6.3	159	209	+31.4
Cleveland	252	225	−10.7	207	208	+0.5	275	355	+29.1
Dallas	320	322	+0.6	313	346	+10.5	361	373	+3.3
Denver	268	272	+1.5	226	248	+9.7	—	—	—
Detroit	279	292	+4.7	289	319	+10.4	141	159	+12.8
Kansas City	283	309	+9.2	270	288	+6.7	281	317	+12.8
Los Angeles	336	337	+0.3	316	326	+3.2	—	305	—
Minneapolis	290	314	+8.3	299	317	+6.0	271	288	+6.3
New Orleans	271	278	+2.6	255	294	+15.3	—	—	—
New York	293	293	0	300	367	+22.3	218	279	+28.0
Philadelphia	219	252	+15.1	222	268	+20.7	324	471	+45.4
Pittsburgh	217	236	+8.8	251	251	0	605	605	0
St. Louis	282	282	0	278	259	−6.8	480	414	−13.8
San Francisco	315	329	+4.4	321	340	+5.9	275	288	+4.7
Seattle	325	374	+15.1	272	290	+6.6	201	153	−23.9

LUMBER	Softwood lumber 2" × 4" Jun. '80	Jun. '81	% chg.	Plywood Standard $\frac{5}{8}$" May '80	May '81	% chg.	Plyform $\frac{1}{2}$" May '80	May '81	% chg.
Atlanta	253	277	+9.5	227	257	+13.2	223	241	+8.1
Baltimore	348	365	+4.9	248	350	+41.1	244	233	−4.5
Birmingham	224	237	+5.8	205	247	+20.5	226	257	+13.7
Boston	287	229	−20.2	396	410	+3.5	384	420	+9.4
Chicago	177	206	+16.4	252	307	+21.8	238	276	+16.0
Cincinnati	364	364	0	313	401	+28.1	344	364	+5.8
Cleveland	216	348	+61.1	493	362	−26.6	342	458	+33.9
Dallas	314	314	0	199	263	+32.2	208	241	+15.9
Denver	226	226	0	286	294	+2.8	—	—	—
Detroit	196	225	+14.8	213	259	+21.6	261	291	+11.5
Kansas City	273	302	+10.6	286	308	+7.7	273	297	+8.7
Los Angeles	274	237	−13.5	308	308	0	325	325	0
Minneapolis	211	215	+1.9	332	358	+7.8	333	338	+1.5
New Orleans	245	259	+5.7	198	287	+44.9	188	203	+8.0
New York	331	298	−10.0	384	384	0	354	346	−2.3
Philadelphia	328	305	−7.0	276	331	+19.9	287	314	+9.4
Pittsburgh	240	256	+6.7	267	334	+25.1	292	299	+2.4
St. Louis	341	—	—	341	—	—	298	351	+17.8
San Francisco	355	346	−2.5	307	443	+44.3	307	358	+16.6
Seattle	184	250	+35.9	234	290	+23.9	275	320	+16.4

STEEL	Structural steel shapes Feb. '80	Feb. '81	% chg.	Reinforcing bars Feb. '80	Feb. '81	% chg.
Atlanta	280	290	+3.6	225	225	0
Baltimore	301	301	0	246	246	0
Birmingham	315	341	+8.3	261	258	−1.1
Boston	430	440	+2.3	276	265	−4.0
Chicago	298	347	+16.4	299	293	−2.0
Cincinnati	301	310	+3.0	229	268	+17.0
Cleveland	297	315	+6.1	338	411	+21.6
Dallas	349	388	+11.2	261	275	+5.4
Denver	246	254	+3.3	267	241	−9.7
Detroit	315	336	+6.7	—	—	—
Kansas City	265	275	+3.8	250	250	0
Los Angeles	383	376	−1.8	186	172	−7.5
Minneapolis	280	281	+0.4	212	188	−11.3
New Orleans	301	278	−7.6	227	208	−8.4
New York	304	310	+2.0	251	251	0
Philadelphia	371	419	+12.9	243	225	−7.4
Pittsburgh	333	353	+6.0	238	231	−2.9
St. Louis	—	408	—	204	197	−3.4
San Francisco	340	355	+4.4	211	219	+3.8
Seattle	275	288	+4.7	—	—	—

Source: ENR Construction Economics Dept.; Base 1967 = 100

R—Revised.

Figure 7–14. Unit Cost Data Base

Construction item	Unit	Labor		Material		Total
		Unitcost	*Index*[a]	*Unit cost*	*Index*[b]	
Sitework						
Earthwork						
Hand excavation	cy	27.00	330			
Machine excavation	cy	0.30	274	1.50[c]	290	
Compacted fill by hand	cy	10.0	330			
Compacted fill by machine	cy	0.40	274	1.00[c]	290	
Caisson foundations						
24-in.-diam. caisson (no casing)	lf	5.90	274	11.95	292	
to						
84-in.-diam. caisson (no casing)	lf	14.95	274	150.00	292	
24-in.-diam. caisson (pulled casing)	lf	10.55	274	11.95	292	
to						
84-in.-diam. caisson (pulled casing)	lf	86.00	274	150.00	292	
Cast-in-place concrete piles						
8-in.-diam. 16-gage shell	lf	2.57	274	3.15	292	
to						
16-in.-diam. 16-gage shell	lf	3.59	274	7.70	292	
Precast concrete piles						
10-in. square	lf	2.57	274	5.00	292	
to						
24-in. square	lf	4.08	274	24.00	292	
Steel piles						
8-in. × 8-in. H section	lf	2.81	274	8.65	290	
to						
14-in. × 14-in. H section	lf	3.52	274	24.50	290	
Wood piles						
12-in. butts, 8-in. points, treated	lf	2.87	274	4.00	277	
Reinforced Concrete						
Forms, elevated slab	sf	1.50	282	0.55	257	
Forms, beam	sf	2.42	282	0.65	257	
Forms, columns	sf	2.30	282	0.60	257	
Forms, walls	sf	1.50	282	0.75	257	
Concrete	cy	—	—	42.00	292	
Place concrete, beams	cy	13.00	282	8.30[c]	290	
Place concrete, slabs	cy	7.00	282	5.00[c]	290	
Place concrete, walls	cy	9.00	282	6.00[c]	290	
Place concrete, columns	cy	14.00	282	7.00[c]	290	
Finish concrete	sf	0.25	282	—	—	
Reinforcement	lb	0.15	282	0.22	225	
Composites[d]						
Footings (80)	cy				292	140
Beams (200)	cy				292	350
Columns (700)	cy				292	500
Flat slab (100)	cy				292	195
Flat plate (120)	cy				292	200
Waffle slab (120)	cy				292	230
Pan joist (150)	cy				292	300

[a] 274: equipment operator index; 282: skilled labor index in ENR quarterly cost report; 330: common labor cost index.

[b] 225: reinforcing bar index; 257: standard plywood index; 277: softwood lumber index; 290: structural steel index; 292: ready-mix concrete index.

[c] Equipment cost.

[d] Numbers in parentheses are assumed lb/cy of steel.

Figure 7–14. *(Continued)*

Construction item	Unit	Labor Unit cost	Labor Index[a]	Material Unit cost	Material Index[b]	Total
One-way beam and slab (150)	cy				292	270
Two-way beam and slab (170)	cy				292	260
Walls	cy				292	250
Precast Concrete						
Double tees						
12 in. deep × 8 ft wide	sf	0.41	282	1.65	292	
to						
36 in. deep × 8 ft wide	sf	0.26	282	5.20	292	
Single tees						
28 in. deep × 8 ft wide	sf	0.66	282	3.20	292	
to						
48 in. deep × 8 ft wide	sf	0.44	282	4.10	292	
Floor planks						
4-in.-thick hollow core	sf	0.75	282	1.65	292	
to						
10-in.-thick hollow core	sf	0.59	282	2.40	292	
Columns						
12 in. × 12 in.	lf	17.60	282	50.00	292	
to						
28 in. × 28 in.	lf	22.00	282	70.00	292	
Beams						
12 in. wide × 18 in. deep	lf	6.60	282	36.00	292	
to						
18 in. wide × 36 in. deep	lf	3.30	282	61.00	292	
Wall panels						
4 ft × 8 ft × 4 in. thick (smooth, gray	sf	7.35	282	5.45	292	
to						
30 ft × 10 ft × 6 in. thick	sf	1.01	282	6.60	292	
Masonry						
4-in. face brick	sf	2.78	282	1.38	292	
8-in. brick wall	sf	4.69	282	2.97	292	
4-in. block	sf	1.39	282	0.67	292	
8-in. block	sf	1.63	282	0.95	292	
12-in. block	sf	2.09	282	1.38	292	
Metals						
Structural steel						
Rolled shapes	lb	0.11	282	0.40	290	
Misc. light shapes	lb	0.15	282	0.45	290	
Built-up section	lb	0.12	282	0.45	290	
Trusses	lb	0.12	282	0.45	290	
Pipe	lb	0.13	282	0.57	290	
Tubing	lb	0.13	282	0.59	290	
Bar joists	lb	0.10	282	0.30	290	
Metal deck						
$1\frac{1}{2}$ in. deep						
22 gage	sf	0.13	282	0.53	290	
20 gage	sf	0.14	282	0.65	290	
18 gage	sf	0.16	282	0.95	290	
3 in. deep						
22 gage	sf	0.17	282	0.76	290	

Figure 7–14. *(Continued)*

Construction item	Unit	Labor Unit cost	Labor Index[a]	Material Unit cost	Material Index[b]	Total
20 gage	sf	0.17	282	0.90	290	
18 gage	sf	0.19	282	1.12	290	
1½ in. deep, composite gal., 22 gage to	sf	0.15	282	0.65	290	
1½ in. deep, composite gal., 16 gage	sf	0.19	282	1.12	290	
$\frac{9}{16}$ in. deep, slab form, 28 gage	sf	0.11	282	0.35	290	
Shear connectors						
¾ in. × 3 in.	ea.	0.50	282	0.35	290	
to						
¾ in. × 8 in.	ea.	0.51	282	0.71	290	
Wood						
Light framing						
2 in. and 3 in. wide	bf	0.75	282	0.38	277	
Heavy framing						
3 in. to 8 in. wide	bf	0.25	282	0.54	277	
Plywood						
$\frac{5}{16}$ in. thick	sf	0.16	282	0.25	257	
to						
¾ in. thick	sf	0.21	282	0.44	257	
Planks						
3 in. thick	sf	0.80	282	1.98	277	
4 in. thick	sf	1.03	282	2.64	277	
Trusses						
2 × 4 top chord	lf	0.19	282	0.91	277	
2 × 6 top chord	lf	0.21	282	1.56	277	
Laminated timber						
Straight beams	bf	0.24	282	1.10	277	
Curved members	bf	0.33	282	1.28	277	
Purlins or columns	bf	0.42	282	1.26	277	
Roof deck						
3 in.	sf	0.60	282	2.00	277	
4 in.	sf	0.79	282	2.35	277	

8

CONCLUSION

8–1 SUMMARY

The purpose of structural design is to create structures that respond to the requirements of function, esthetics, serviceability, and constructability. In order to create, we must deal conceptually not only with a vast array of structural system alternatives but also with the behavior of these systems under varying loading conditions. We have seen that conceptual design requires both an understanding of behavior and preliminary methods for sizing of structural members. By integrating these two aspects of conceptual design into an overall systems design model, we have established a method for structural design.

Our objective has been to establish in one volume both a design approach and a reference book for active use during design sessions. The description of approximate analysis and design techniques represents not only a compilation of well-established methods (such as the portal method) but also translates some of our newer knowledge on member and frame behavior into new approximate techniques. The Chapter 3 information on loads is a state of the art description of the most current standards for live loads, wind loads, and earthquake loads. The Chapter 2 description of structural requirements contains current deflection criteria, insulation requirements based on energy use, fire protection requirements, and volume change information. Finally, the systems data base provides guidelines for the initial screening of optional structural systems.

If we now can visualize structural assemblies actively moving under load, trace the path of the load as it flows through the structure, and form systems to resist the load, we have acquired an intuitive "feeling" for structures. With this understanding we can deal with broad design concepts and innovative thought, leading us to better and more economical structural design solutions.

8–2 FUTURE DIRECTIONS

Technology is changing so rapidly that only a momentary "snapshot" can, perhaps, record the trends at any one point in time or provide a basis for reference. At this moment, we should be aware of the emerging technology for space structures (Fig. 8–1), deep-sea structures (Fig. 8–2), and large enclosed building environments

(Fig. 8–3). Behind each of these developments are new breakthroughs in materials science and manufacturing and new concepts of engineering.

As these new frontiers challenge our imagination, by-products of these developments gradually find their way into everyday application, including:

- Computer graphics
- Sandwich panel construction
- Composite materials
- Higher-strength materials
- Limit states design

Somewhere there is an idea, yet to be uncovered, for building ordinary structures in a revolutionary way. Some designers have suggested that nature may provide the clue; others that manufactured systems provide the answer. Whatever the solution, it will most certainly require innovative thought. Perhaps, this book will help.

Figure 8–1. Space Structure (Courtesy of Michael J. Bartus, Jr., "Building Star Scrapers in Orbit," *Civil Engineering Magazine*)

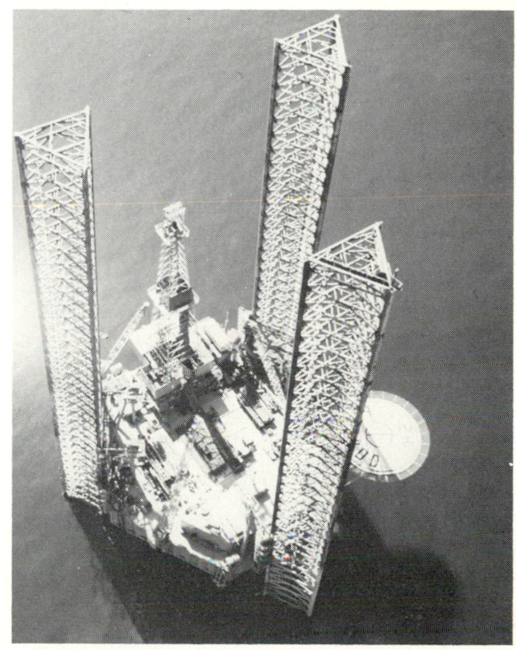

Figure 8–2. Deep-Sea Structure (Courtesy of Tenneco, Inc.)

Figure 8–3. Air-Supported Building Enclosure (Courtesy of United States of America General Services Administration)

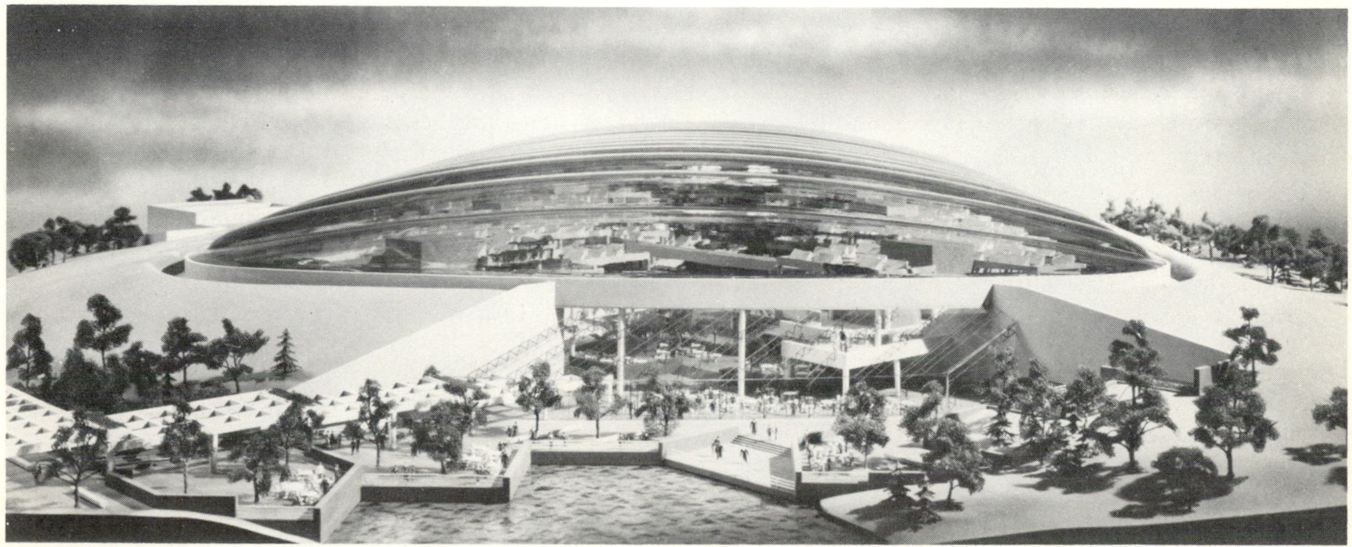

INDEX

C

D

E

F

G

H

I

J

K

L

M

T

U

V

W

Y